Progress in Mathematics

Volume 85

Bruce C. Berndt Harold G. Diamond
Heini Halberstam Adolf Hildebrand
Editors

Analytic Number Theory

Proceedings of a Conference in Honor of Paul T. Bateman

1990 Birkhäuser
Boston · Basel · Berlin

Bruce C. Berndt
Harold G. Diamond
Heini Halberstam
Adolf Hildebrand
Department of Mathematics
University of Illinois at Urbana-Champaign
1409 West Green Street
Urbana, Illinois 61801

Library of Congress Cataloging-in-Publication Data
Analytic number theory : papers from a conference held in honor of
Paul T. Bateman / Bruce C. Berndt . . . [et al.], editors.
p. cm. — (Progress in mathematics ; v. 85)
ISBN-13:978-1-4612-8034-7 e-ISBN-13:978-1-4612-3464-7
DOI: 10.1007/978-1-4612-3464-7
1. Number theory—Congresses. I. Bateman, P. T. II. Berndt, Bruce C., 1939- . III. Series: Progress in mathematics; vol. 85.
QA241.A489 1990
512'.73—dc20

90-905

Printed on acid-free paper.

Softcover reprint of the hardcover 1st edition 1990

Camera-ready copy prepared by the editors using AMSTEX and troff.

9 8 7 6 5 4 3 2 1

PREFACE

On April 25-27, 1989, over a hundred mathematicians, including eleven from abroad, gathered at the University of Illinois Conference Center at Allerton Park for a major conference on analytic number theory. The occasion marked the seventieth birthday and impending (official) retirement of Paul T. Bateman, a prominent number theorist and member of the mathematics faculty at the University of Illinois for almost forty years. For fifteen of these years, he served as head of the mathematics department.

The conference featured a total of fifty-four talks, including ten invited lectures by H. Delange, P. Erdös, H. Iwaniec, M. Knopp, M. Mendès France, H. L. Montgomery, C. Pomerance, W. Schmidt, H. Stark, and R. C. Vaughan. This volume represents the contents of thirty of these talks as well as two further contributions. The papers span a wide range of topics in number theory, with a majority in analytic number theory.

The conference was made possible by grants from the National Science Foundation, the National Security Agency, the Institute for Mathematics and Applications, the College of Liberal Arts and Sciences, and the Department of Mathematics at the University of Illinois. We are grateful to these agencies and institutions for their support. We also extend our thanks to the directors of the Allerton Park Conference Center, Jennifer and Gary Eickman, and their staff for their help in preparing the conference and for ensuring a pleasant stay for the participants. Lastly, we thank Pat Coombs, administrative clerk of the mathematics department at the University of Illinois, for organizational guidance; graduate students Gennady Bachman, Richard Blaylock, and Liang-Cheng Zhang for their help during the conference; and Hilda Britt and Lou Wei for expertly typing most of the papers that appear in this volume.

Urbana, Illinois
February 1990

B. C. B.
H. G. D.
H. H.
A. H.

Paul T. Bateman

TABLE OF CONTENTS

q-Trinomial Coefficients and Rogers-Ramanujan Type Identities

GEORGE E. ANDREWS

To my friend, Paul Bateman, on his seventieth birthday

1. Introduction

There are many proofs [4], [2, Ch. 7] of the celebrated Rogers-Ramanujan identities:

$$1+\sum_{n=1}^{\infty}\frac{q^{n^2}}{(1-q)(1-q^2)\cdots(1-q^n)}=\prod_{n=0}^{\infty}\frac{1}{(1-q^{5n+1})(1-q^{5n+4})}, \quad (1.1)$$

$$1+\sum_{n=1}^{\infty}\frac{q^{n^2+n}}{(1-q)(1-q^2)\cdots(1-q^n)}=\prod_{n=0}^{\infty}\frac{1}{(1-q^{5n+2})(1-q^{5n+3})}. \quad (1.2)$$

However one of the most useful proofs is due to Schur [16, §4]:

If $D_0(q)=D_1(q)=1$ and $D_n(q)=D_{n-1}(q)+q^{n-1}D_{n-2}(q)$ for $n>1$, then

$$D_n(q)=\sum_{\lambda=-\infty}^{\infty}(-1)^{\lambda}q^{\lambda(5\lambda+1)/2}\begin{bmatrix} n \\ \lfloor\frac{n-5\lambda}{2}\rfloor\end{bmatrix}_q, \quad (1.3)$$

where $\lfloor x\rfloor$ is the greatest integer $\leq x$ and

$$\begin{bmatrix} A \\ B\end{bmatrix}_q=\begin{cases}\dfrac{(1-q^A)(1-q^{A-1})\cdots(1-q^{A-B+1})}{(1-q^B)(1-q^{B-1})\cdots(1-q)}, & 0\leq B\leq A,\\ 0, & \text{otherwise.}\end{cases} \quad (1.4)$$

Partially supported by National Science Foundation Grant DMS-870269

Similarly if $D_1^*(q) = D_2^*(q) = 1$ and $D_n^*(q) = D_{n-1}^*(q) + q^{n-1}D_{n-2}^*(q)$ for $n > 2$, then

$$D_n^*(q) = \sum_{\lambda=-\infty}^{\infty} (-1)^\lambda q^{\lambda(5\lambda-3)/2} \begin{bmatrix} n \\ \lfloor \frac{n-5\lambda}{2} \rfloor + 1 \end{bmatrix}_q . \tag{1.5}$$

A brief combinatorial argument reveals that (1.1) follows from (1.3) and (1.2) from (1.5) by letting $n \to \infty$.

There are several points of importance about Schur's proof. They have led to completely new generalizations of the Rogers-Ramanujan identities [**1**], [**3**], [**8**], [**10**], [**11**]. Also the fact that (1.3) and (1.5) are polynomial identities allows for treatment of duality in the Hard Hexagon model [**3**]; such polynomial representations are, at times, essential in these statistical mechanics models [**9**, §2.6].

In [**7**], Baxter and I required comparable polynomial representations related to the modulus 7 instance of B. Gordon's [**14**] generalization of the Rogers-Ramanujan identities. However instead of the Gaussian polynomials playing an essential role, now q-analogs of trinomial coefficients took their place. For example, suppose $Y_m(b;q)$ is the (polynomial) generating function for partitions $a_1 + a_2 + \cdots + a_j$ with $a_i \geq a_{i+1}$, $a_i - a_{i+2} \geq 2$, $a_1 \leq m$, and m appears as a part exactly $3 - b$ times ($b = 1, 2,$ or 3).

Then B. Gordon [**33**, Thm. 1, d=t=3] has proved that

$$Y_\infty(3;q) = \prod_{\substack{n=1 \\ n \not\equiv 0, \pm 3 \bmod 7}}^{\infty} (1-q^n)^{-1}. \tag{1.6}$$

Baxter and I found that [**7**, p.319, eq. (4.7), j=k=1]

$$\begin{aligned} Y_m(3;q) = &\sum_{\mu=-\infty}^{\infty} q^{14\mu^2+\mu} \binom{m; 7\mu; q}{7\mu}_2 \\ &- \sum_{\mu=-\infty}^{\infty} q^{14\mu^2+13\mu+3} \binom{m; 7\mu+3; q}{7\mu+3}_2, \end{aligned} \tag{1.7}$$

where

$$\binom{n; B; q}{A}_2 = \sum_{j \geq 0} \frac{q^{j(j+B)}(q)_n}{(q)_j (q)_{j+A} (q)_{n-2j-A}}, \tag{1.8}$$

and

$$(A)_j = (A;q)_j = (1-A)(1-Aq)\cdots(1-Aq^{j-1}). \tag{1.9}$$

Now a double application of the binomial theorem shows that

$$(1+x+x^{-1})^n = \sum_{k=-n}^{n} \binom{n}{k}_2 x^k$$

$$= \sum_{k=-n}^{n} \left(\sum_{j\geq 0} \frac{n!}{j!(j+k)!(n-2j-k)!} \right) x^k. \quad (1.10)$$

Thus the polynomials in (1.8) are q-analogs of trinomial coefficients. (Note: these are *not* to be confused with the coefficients in $(x+y+z)^n$ which are also often called trinomial coefficients.)

Our object here is to explore other applications of q-trinomial coefficients to polynomials familiar in additive number theory. Perhaps most surprising and intriguing is the fact that the Schur polynomials $D_n(q)$ andd $D_n^*(q)$ can be related to the q-trinomial coefficients also. Namely

$$D_n(q) = \sum_{\mu=-\infty}^{\infty} q^{10\mu^2+\mu} \binom{n; 5\mu; q}{5\mu}_2 - \sum_{\mu=-\infty}^{\infty} q^{10\mu^2-9\mu+2} \binom{n; 5\mu-2; q}{5\mu-2}_2, \quad (1.11)$$

and

$$D_n^*(q) = \sum_{\mu=-\infty}^{\infty} q^{10\mu^2+3\mu} \binom{n; 5\mu+1; q}{5\mu+1}_2 - \sum_{\mu=-\infty}^{\infty} q^{10\mu^2-7\mu+1} \binom{n; 5\mu-2; q}{5\mu-2}_2. \quad (1.12)$$

In Section 2 we shall collect the necessary background on q-trinomial coefficients. In Section 3 we shall prove the above formulas. In Section 4 we consider related partition polynomials due to B. Gordon [**15**] arising from the Göllnitz–Gordon identities [**2**, §7.4]. We conclude with some discussion of the possible application of such identities.

2. Background

In [**7**], q-trinomial coefficients are extensively developed. In this section we shall present without proof those formulae necessary for our subsequent work. All these results are proved fully in [**7**].

We shall need three q-analogs of the trinomial coefficients. Recalling (1.8):

$$\binom{m; B; q}{A}_2 = \sum_{j\geq 0} \frac{q^{j(j+B)}(q)_m}{(q)_j(q)_{j+A}(q)_{m-2j-A}}; \quad (2.1)$$

in addition [**7**, eqs (2.8) and (2.9)]

$$T_0(m,A,q)=\sum_{j=0}^{m}(-1)^j\begin{bmatrix}m\\ j\end{bmatrix}_{q^2}\begin{bmatrix}2m-2j\\ m-A-j\end{bmatrix}_q, \tag{2.2}$$

and

$$T_1(m,A,q)=\sum_{j=0}^{m}(-q)^j\begin{bmatrix}m\\ j\end{bmatrix}_{q^2}\begin{bmatrix}2m-2j\\ m-A-j\end{bmatrix}_q. \tag{2.3}$$

These polynomials possess the following symmetry relations [**7**, eq. (2.15)]:

$$\binom{m;B;q}{-A}_2=q^{A(A+B)}\binom{m;B+2A;q}{A}_2, \tag{2.4}$$

$$T_i(m,-A,q)=T_i(m,A,q)\qquad (i=0,1). \tag{2.5}$$

Next there are four Pascal triangle type formulas that are necessary [**7**, eqs. (2.16), (2.25), (2.28), (2.29)]:

$$\begin{aligned}T_1(m,A,q)&=T_1(m-1,A,q)\\&\quad+q^{m+A}T_0(m-1,A+1,q)+q^{m-A}T_0(m-1,A-1,q),\end{aligned} \tag{2.6}$$

$$\begin{aligned}\binom{m;A-1;q}{A}_2&=q^{m-1}\binom{m-1;A-1;q}{A}_2\\&\quad+q^A\binom{m-1;A+1;q}{A+1}_2+\binom{m-1;A-1;q}{A-1}_2,\end{aligned} \tag{2.7}$$

$$\begin{aligned}\binom{m;B;q}{A}_2&=\binom{m-1;B;q}{A}_2\\&\quad+q^{m-A-1+B}\binom{m-1;B;q}{A+1}_2+q^{m-A}\binom{m-1;B-1;q}{A-1}_2,\end{aligned} \tag{2.8}$$

$$\begin{aligned}\binom{m;B;q}{A}_2&=\binom{m-1;B;q}{A}_2\\&\quad+q^{m-A}\binom{m-1;B-2;q}{A-1}_2+q^{m+B}\binom{m-1;B+1;q}{A+1}_2.\end{aligned} \tag{2.9}$$

Additionally there are two further identities needed [**7**, eq. (2.20]:

$$\begin{aligned}&T_1(m,A,q)-q^{m-A}T_0(m,A,q)\\&\qquad-T_1(m,A+1,q)+q^{m+A+1}T_0(m,A+1,q)=0\end{aligned} \tag{2.10}$$

and [**7**, eq. (2.27) corrected]

$$\begin{aligned}&\binom{m;A;q}{A}_2+q^m\binom{m;A;q}{A+1}_2\\&\qquad-\binom{m;A+1;q}{A+1}_2-q^{m-A}\binom{m;A-1;q}{A}_2=0.\end{aligned} \tag{2.11}$$

In closing this section, we note that [7, paragraph following (2.14)]

$$\binom{m;B;1}{A}_2 = T_0(m,A,1) = T_1(m,A,1) = \binom{m}{A}_2. \qquad (2.12)$$

Thus these polynomials are indeed q-analogs of the trinomial coefficients $\binom{m}{A}_2$.

3. Schur's polynomials

Our object here is the proof of (1.11) and (1.12). In the remarks just prior to (1.3) we find the defining recurrence for $D_n(q)$. Hence we only need to show that the right-hand side of (1.11), which we shall denote $d_n(q)$, satisfies the same conditions.

Clearly

$$d_0(q) = d_1(q) = 1. \qquad (3.1)$$

Now

$$\begin{aligned}
d_n(q) - d_{n-1}(q) & \qquad (3.2)\\
= \sum_{\mu=-\infty}^{\infty} q^{10\mu^2+\mu} & \left(\binom{n;5\mu;q}{5\mu}_2 - \binom{n-1;5\mu;q}{5\mu}_2\right)\\
- \sum_{\mu=-\infty}^{\infty} q^{10\mu^2-9\mu+2} & \left(\binom{n;5\mu-2;q}{5\mu-2}_2 - \binom{n-1;5\mu-2;q}{5\mu-2}_2\right)\\
= \sum_{\mu=-\infty}^{\infty} q^{10\mu^2+\mu} & \left(q^{n-5\mu}\binom{n-1;5\mu-1;q}{5\mu-1}_2 + q^{n-1}\binom{n-1;5\mu;q}{5\mu+1}_2\right)\\
- \sum_{\mu=-\infty}^{\infty} q^{10\mu^2-9\mu+2} & \left(q^{n+5\mu-2}\binom{n-1;5\mu-1;q}{5\mu-1}_2\right.\\
& \left. + q^{n-5\mu+2}\binom{n-1;5\mu-4;q}{5\mu-3}_2\right)
\end{aligned}$$

(by (2.8) applied to the first sum and (2.9) applied to the second)

$$\begin{aligned}
= q^{n-1} \Bigg(& \sum_{\mu=-\infty}^{\infty} q^{10\mu^2+\mu}\binom{n-1;5\mu;q}{5\mu+1}_2\\
& - \sum_{\mu=-\infty}^{\infty} q^{10\mu^2-14\mu+5}\binom{n-1;5\mu-4;q}{5\mu-3}_2\Bigg)
\end{aligned}$$

$$= q^{n-1}\Bigg(\sum_{\mu=-\infty}^{\infty} q^{10\mu^2+\mu}\Bigg(\begin{pmatrix} n-2;5\mu;q \\ 5\mu \end{pmatrix}_2$$

$$+q^{n-2}\begin{pmatrix} n-2;5\mu;q \\ 5\mu+1 \end{pmatrix}_2 + q^{5\mu+1}\begin{pmatrix} n-2;5\mu+2;q \\ 5\mu+2 \end{pmatrix}_2\Bigg)$$

$$-\sum_{\mu=-\infty}^{\infty} q^{10\mu^2-14\mu+5}\Bigg(q^{5\mu-3}\begin{pmatrix} n-2;5\mu-2;q \\ 5\mu-2 \end{pmatrix}_2$$

$$+q^{n-2}\begin{pmatrix} n-2;5\mu-4;q \\ 5\mu-3 \end{pmatrix}_2 + \begin{pmatrix} n-2;5\mu-4;q \\ 5\mu-4 \end{pmatrix}_2\Bigg)\Bigg)$$

(by (2.7) applied to both sums)

$$= q^{n-1}d_{n-1}(q)$$

$$+\sum_{\mu=-\infty}^{\infty} q^{10\mu^2+6\mu+1}\Bigg(\begin{pmatrix} n-2;5\mu+2;q \\ 5\mu+2 \end{pmatrix}_2 - \begin{pmatrix} n-2;5\mu+1;q \\ 5\mu+1 \end{pmatrix}_2$$

$$+ q^{n-5\mu-3}\begin{pmatrix} n-2;5\mu;q \\ 5\mu+1 \end{pmatrix}_2 - q^{n-2}\begin{pmatrix} n-2;5\mu+1;q \\ 5\mu+2 \end{pmatrix}_2\Bigg)$$

$$= q^{n-1}d_{n-2}(q) \qquad \text{(by (2.11))}.$$

Hence $d_n(q)$ fulfills the defining conditions for $D_n(q)$; consequently $d_n(q) = D_n(q)$ and equation (1.11) is established.

Equation (1.12) is proved in almost exactly the same way. The only change is in the first step where (2.9) is applied to the first sum and (2.8) to the second. We therefore omit the details.

4. Göllnitz–Gordon polynomials

In [**15**, p. 741], B. Gordon proves the following two partition theorems.

First Göllnitz–Gordon Identity. *The number of partitions of any positive integer n into parts $\equiv$ 1, 4, or 7, (mod 8) is equal to the number of partitions of the form $n = n_1 + n_2 + \cdots + n_k$, where $n_i \geq n_{i+1} + 2$ and $n_i \geq n_{i+1} + 3$ if n_i is even $(1 \leq i \leq k-1)$.*

Second Göllnitz–Gordon Identity. *The number of partitions of n into parts $\equiv$ 3, 4, or 5 (mod 8) is equal to the number of partitions $n = n_1 + n_2 + \cdots + n_k$ satisfying $n_k \geq 3$ in addition to the inequalities listed in the first Göllnitz–Gordon identity.*

H. Göllnitz [**12**] had found these theorems previously; however he first published them in [**13**].

Both Göllnitz and Gordon base their proof on the following two identities due to Lucy Slater [**17**, eqs. (36) and (34)]:

$$\sum_{n=0}^{\infty} \frac{q^{n^2}(-q;q^2)_n}{(q^2;q^2)_n} = \prod_{n=0}^{\infty} \frac{1}{(1-q^{8n+1})(1-q^{8n+4})(1-q^{8n+7})}, \tag{4.1}$$

and

$$\sum_{n=0}^{\infty} \frac{q^{n^2+2n}(-q;q^2)_n}{(q^2;q^2)_n} = \prod_{n=0}^{\infty} \frac{1}{(1-q^{8n+3})(1-q^{8n+4})(1-q^{8n+5})}. \tag{4.2}$$

In his proof, Gordon [**15**, p. 744] examines the polynomials $S_0(q) = 1+q$, $S_1(q) = 1+q+q^2+q^3+q^4$, and for $n > 1$,

$$S_n(q) = (1+q^{2n+1})S_{n-1}(q) + q^{2n}S_{n-2}(q). \tag{4.3}$$

Gordon [**15**, pp. 744–745] shows that $S_n(q)$ is the generating function for the second type of partitions considered in the first Göllnitz–Gordon identity with the added condition that all parts are $\leq 2n+1$.

To prove the first Göllnitz–Gordon identity one need only show that $S_\infty(q)$ is equal to the right-hand side of (4.1). However no representations of $S_n(q)$ are known which yield the infinite product on the right-hand side of (4.1) directly in the limit. It turns out, however, that

$$S_{n-1}(q) = \sum_{\mu=-\infty}^{\infty} (-1)^\mu q^{4\mu^2+\mu}(T_0(n,4\mu,q) + T_0(n,4\mu+1,q)). \tag{4.4}$$

To prove (4.4), it is convenient to define

$$U(m,A,q) = T_0(m,A,q) + T_0(m,A+1,q). \tag{4.5}$$

Next we note that Gordon's polynomials are equally well-defined by taking $S_{-1}(q) = 1$, $S_0(q) = 1+q$, and then requiring (4.3) to hold for all $n \geq 1$. Let us denote the right-hand side of (4.4) by $s_{n-1}(q)$. Then clearly $s_{-1}(q) = 1$ and $s_0(q) = T_0(1,0,q) + T_0(1,1,q) = 1+q$.

In order to prove the recurrence (4.3), we must use

$$\begin{aligned} U(m,A,q) &- (1+q^{2m-1})U(m-1,A,q) \\ &= q^{m-A}T_1(m-1,A-1,q) + q^{m+A+1}T_1(m-1,A+2,q), \end{aligned} \tag{4.6}$$

a result easily derived from the formulae given in Section 2 and proved in full in [6, Lemma 1].

Hence by (4.6) with $n \geq 2$:

$$\begin{aligned}
&s_{n-1}(q) - (1+q^{2n-1})s_{n-2}(q) - q^{2n-2}s_{n-3}(q) \qquad (4.7)\\
&= \sum_{\mu=-\infty}^{\infty} (-1)^\mu q^{4\mu^2+\mu}\Big(q^{n-4\mu}T_1(n-1,4\mu-1,q)\\
&\qquad + q^{n+4\mu+1}T_1(n-1,4\mu+2,q)\Big)\\
&\quad - \sum_{\mu=-\infty}^{\infty} (-1)^\mu q^{4\mu^2+\mu+2n-2}\Big(T_0(n-2,4\mu,q) + T_0(n-2,4\mu+1,q)\Big)\\
&= \sum_{\mu=-\infty}^{\infty} (-1)^\mu q^{4\mu^2-3\mu+n}\Big(T_1(n-2,4\mu-1,q)\\
&\qquad + q^{n+4\mu-2}T_0(n-2,4\mu,q) + q^{n-4\mu}T_0(n-2,4\mu-2,q)\Big)\\
&\quad + \sum_{\mu=-\infty}^{\infty} (-1)^\mu q^{4\mu^2+5\mu+n+1}\Big(T_1(n-2,4\mu+2,q)\\
&\qquad + q^{n+4\mu+1}T_0(n-2,4\mu+3,q) + q^{n-4\mu-3}T_0(n-2,4\mu+1,q)\Big)\\
&\quad - \sum_{\mu=-\infty}^{\infty} (-1)^\mu q^{4\mu^2+\mu+2n-2}\Big(T_0(n-2,4\mu,q) + T_0(n-2,4\mu+1,q)\Big)\\
&= \sum_{\mu=-\infty}^{\infty} (-1)^\mu q^{4\mu^2+\mu}\Big(q^{n-4\mu}T_1(n-2,4\mu-1,q)\\
&\qquad + q^{2n-8\mu}T_0(n-2,4\mu-2,q) + q^{n+4\mu+1}T_1(n-2,4\mu+2,q)\\
&\qquad + q^{2n+8\mu+2}T_0(n-2,4\mu+3,q)\Big).
\end{aligned}$$

Treat this last expression as four separate sums and shift μ to $\mu-1$ in the third and fourth sums. Hence

$$\begin{aligned}
&s_{n-1}(q) - (1+q^{2n-1})s_{n-2}(q) - q^{2n-2}s_{n-3}(q)\\
&= q^n \sum_{\mu=-\infty}^{\infty} (-1)^\mu q^{4\mu^2-3\mu}\Big(T_1(n-2,4\mu-1,q) + q^{n-4\mu}T_0(n-2,4\mu-2,q)\\
&\qquad - T_1(n-2,4\mu-2,q) - q^{n+4\mu-3}T_0(n-2,4\mu-1,q)\Big)\\
&= 0 \quad \text{(by (2.10))}. \qquad (4.8)
\end{aligned}$$

Thus (4.4) is established.

An exactly similar result holds for the second Göllnitz–Gordon identity. In this case [**15**, p. 745], Gordon defines a sequence of polynomials $T_n(q)$, given by $T_{-1}(q) = 0$, $T_0(q) = 1$, and for $n > 0$

$$T_n(q) = (1 + q^{2n+1})T_{n-1}(q) + q^{2n}T_{n-2}(q). \tag{4.9}$$

In the same manner as above, it is possible to show that

$$T_{n-1}(q) = \sum_{\mu=-\infty}^{\infty} (-1)^\mu q^{4\mu^2+3\mu}\Big(T_0(n, 4\mu+1, q) + T_0(n, 4\mu+2, 1)\Big). \tag{4.10}$$

To close this section we note that the limiting cases of these polynomials are easily seen to be the respective infinite products in the identities (4.1) and (4.2).

In [**6**, eq. (4.16)], we note the limit

$$\lim_{m\to\infty} U(m, A, q) = \prod_{n=1}^{\infty} \frac{(1 + q^{2n-1})}{(1 - q^{2n})}. \tag{4.11}$$

Hence by (4.4),

$$\begin{aligned}\lim_{n\to\infty} S_{n-1}(q) &= \prod_{n=1}^{\infty} \frac{(1 + q^{2n-1})}{(1 - q^{2n})} \sum_{\mu=-\infty}^{\infty} (-1)^\mu q^{4\mu^2+\mu} \\ &= \prod_{n=1}^{\infty} \frac{(1 - q^{8n-2})(1 - q^{8n-6})(1 - q^{8n})(1 - q^{8n-5})(1 - q^{8n-3})}{(1 - q^n)} \\ &\quad \text{(by Jacobi's triple product [2, p. 22, Cor. 2.9])} \\ &= \prod_{n=1}^{\infty} \frac{1}{(1 - q^{8n-1})(1 - q^{8n-4})(1 - q^{8n-7})},\end{aligned} \tag{4.12}$$

and by (4.10),

$$\begin{aligned}\lim_{n\to\infty} T_{n-1}(q) &= \prod_{n=1}^{\infty} \frac{(1 + q^{2n-1})}{(1 - q^{2n})} \sum_{\mu=-\infty}^{\infty} (-1)^\mu q^{4\mu^2+3\mu} \\ &= \prod_{n=1}^{\infty} \frac{(1 - q^{8n-2})(1 - q^{8n-6})(1 - q^{8n})(1 - q^{8n-7})(1 - q^{8n-1})}{(1 - q^n)} \\ &\quad \text{(by Jacobi's triple product [2, p. 22, Cor. 2.9])} \\ &= \prod_{n=1}^{\infty} \frac{1}{(1 - q^{8n-3})(1 - q^{8n-4})(1 - q^{8n-5})}.\end{aligned}$$

5. Conclusion

In a way, this paper is, I hope, the mere beginning of further study of q-trinomial coefficients. The most intriguing question to me is this:

Is there a nice combinatorial explanation of (1.11), (1.12), (4.4) and (4.10)? As was noted in the introduction, the combinatorial extensions of (1.3) have been quite fruitful. The beginning of the explanation of (1.3) requires that we know that $\begin{bmatrix} A \\ B \end{bmatrix}_q$ is the generating function for partitions into at most B parts, each $\leq A - B$. Then a sieve is introduced concerning successive ranks [**1**]. Thus the simplest question is: what are natural partition-theoretic interpretations of any and all the q-trinomial coefficients that would support a sieve-theoretic interpretation of (1.11), (1.12), (4.4) and (4.10)?

REFERENCES

[1] G. E. Andrews, Sieves in the theory of partitions, Amer. J. Math. **94** (1972), 1214–1230.

[2] G. E. Andrews, *The Theory of Partitions*, Vol. 2, Encyclopedia of Mathematics and Its Applications (G.-C. Rota, ed.), Addison–Wesley, Reading, 1976; reissued: Cambridge University Press, London and New York, 1984.

[3] G. E. Andrews, The hard-hexagon model and the Rogers-Ramanujan type identities, Proc. Nat. Acad. Sci. U. S. A. **78** (1981), 5290–5292.

[4] G. E. Andrews, Uses and extensions of Frobenius' representations of partitions, in *Enumeration and Design* (D. M. Jackson and S. A. Vanstone, eds.), Academic Press 1984, pp. 51–65.

[5] G. E. Andrews, On the proofs of the Rogers-Ramanujan identities, in *q-Series and Partitions* (Dennis Stanton, ed.), IMA Volumes in Mathematics and its Applications, Springer–Verlag, New York, 1989, pp. 1–14.

[6] G. E. Andrews, Euler's "Exemplum Memorabile Inductionis Fallacis" and q-trinomial coefficients (to appear).

[7] G. E. Andrews and R. J. Baxter, Lattice gas generalization of the hard hexagon model. III. q-trinomial coefficients, J. Stat. Phys. **47** (1987), 297–330.

[8] G. E. Andrews, R. J. Baxter, D. M. Bressoud, W. H. Burge, P. J. Forrester, and G. Viennot, Partitions with prescribed hook differences, Europ. J. Combinatorics **8** (1987), 341–350.

[9] G. E. Andrews, R. J. Baxter, and P. J. Forrester, Eight-vertex SOS model and generalized Rogers-Ramanujan-type identities, J. Stat. Phys. **35** (1984), 193–266.

[10] D. M. Bresssoud, Extension of the partition sieve, J. Number Th. **12** (1980), 87–100.
[11] W. H. Burge, A correspondence between partitions related to generalizations of the Rogers-Ramanujan identities, Discrete Math. **34** (1981), 9–15.
[12] H. Göllnitz, *Einfache Partitionen* (unpublished), Diplomarbeit W. S. 1960, Göttingen, 65 pp..
[13] H. Göllnitz, Partitionen mit Differenzenbedingungen, J. Reine Angew. Math. **225** (1967), 154–190.
[14] B. Gordon, A combinatorial generalization of the Rogers-Ramanujan identities, Amer. J. Math. **83** (1961), 393–399.
[15] B. Gordon, Some continued fractions of the Rogers-Ramanujan type, Duke Math. J. **31** (1965), 741–748.
[16] I. Schur, Ein Beitrag zur additiven Zahlentheorie und zur Theorie der Kettenbrüche, S.-B. Preuss. Akad. Wiss. Phys.-Math. Kl., 302–321; reprinted in: *I. Schur, Gesammelte Abhandlungen, Vol. 2,* Springer–Verlag, Berlin 1973.
[17] L. J. Slater, Further identities of the Rogers–Ramanujan type, Proc. London Math. Soc. (2) **54** (1952), 147–167.

George E. Andrews
Dept. of Mathematics
The Pennsylvania State University
University Park, PA 16802

Evaluations of Selberg Character Sums

JULIE AUTUORE AND RONALD EVANS

Dedicated to Paul T. Bateman on his 70th birthday

Abstract

The n-dimensional Selberg character sums $L_n(A, B, C)$ are evaluated for all $n \geq 0$ when the character C is trivial or quadratic. Additional character sum evaluations related to integral formulas of Selberg are conjectured.

1. Introduction

Let $GF(q)$ denote the finite field of q elements, where q is a power of an odd prime p. Let 1 and ϕ denote the trivial and quadratic characters on $GF(q)$, respectively. Throughout, A, B, and C will denote complex multiplicative characters on $GF(q)$. By convention, $A(0) = 0$ for all A (even $A = 1$).

Define the Gauss sum $G(A)$ over $GF(q)$ by

$$G(A) = \sum_m A(m)\zeta^{T(m)}, \tag{1.1}$$

where the sum is over all $m \in GF(q)$, $\zeta = \exp(2\pi i/p)$, and T denotes the trace map from $GF(q)$ to $GF(p)$. Define the Jacobi sum $J(A, B)$ over $GF(q)$ by

$$J(A, B) = \sum_m A(m)\, B(1 - m)\,. \tag{1.2}$$

(See [**4**, Chapter 8] for elementary properties of Gauss and Jacobi sums.)

For nonnegative integers n, define the n-dimensional Selberg character sums $L_n(A, B, C\phi)$, $L_n(A, C\phi)$, and $L_n(C\phi)$ by

$$L_n(A,B,C\phi) = \sum_{\substack{F \\ \deg F = n}} A((-1)^n F(0))\, B(F(1))\, C\phi(D_F), \tag{1.3}$$

$$L_n(A, C\phi) = \sum_{\substack{F \\ \deg F = n}} A(F(0))\, C\phi(D_F)\zeta^{T(t_1)}, \tag{1.4}$$

and

$$L_n(C\phi) = \sum_{\substack{F \\ \deg F = n}} C\phi(D_F)\zeta^{T(t_1^2/2 - t_2)}; \tag{1.5}$$

here, each sum is over the monic polynomials F over $GF(q)$ of degree n, D_F denotes the discriminant of F, and the $t_i = t_i(F)$ are defined by

$$F(x) = x^n + t_1 x^{n-1} + t_2 x^{n-2} + \cdots + t_n. \tag{1.6}$$

We use the convention that $D_F = 1$ when F has degree ≤ 1.

The following formulas for Selberg character sums were conjectured in [**2**, (29), (29a), (29b)]; they are analogues of Selberg's integral formulas [**2**, (1), (1a), (1b)]. For all $n \geq 0$,

$$L_n(A, B, C\phi) = \prod_{j=0}^{n-1} \frac{G(C^{j+1})\, G(AC^j)\, G(BC^j)}{G(C)\, G(ABC^{n-1+j})} \tag{1.7}$$

provided that

$$ABC^{n-1+j} \neq 1 \text{ holds for all } j, \ 0 \leq j \leq n-1; \tag{1.8}$$

$$L_n(A, C\phi) = \prod_{j=0}^{n-1} \frac{G(C^{j+1})\, G(AC^j)}{G(C)}, \tag{1.9}$$

and

$$L_n(C\phi) = \prod_{j=0}^{n-1} \frac{\phi(2)\, G(\phi)\, G(C^{j+1})}{G(C)}. \tag{1.10}$$

Conjectured evaluations of $L_n(A, B, C\phi)$ in cases not covered by (1.8) are given in [**3**].

The primary purpose of this paper is to prove the result announced in [**3**, Theorem 1.1], which evaluates $L_n(A, B, C\phi)$ for all $n \geq 0$ in the special case $C^2 = 1$. This is done in §3 (Theorem 3.5). Straightforward modifications of the proof of Theorem 3.5 can be used to evaluate $L_n(A, C\phi)$ and $L_n(C\phi)$ for $C^2 = 1$, thus proving the validity of (1.9) and (1.10) for $C^2 = 1$. Less elementary methods appear to be needed for handling general characters C.

In §2, we present further conjectures related to the conjectures (1.9) and (1.10), inspired by a recent integral formula of Selberg [**1**]. Theorems 2.2 and 2.5 show that some of these conjectures follow from others.

2. Conjectures

Conjecture 2.1. *For all A, B, C and all $n \geq 0$,*

$$\sum_{\substack{F \\ \deg F = n}} A(F(0))B(1+t_1)C\phi(D_F)$$

$$= \begin{cases} \dfrac{G(\overline{B}\,\overline{A}^n\overline{C}^{n(n-1)})}{G(\overline{B})} \displaystyle\prod_{j=0}^{n-1} \frac{G(C^{j+1})G(AC^j)}{G(C)}, \\ \qquad \text{if } B \neq 1, \\ \dfrac{A^n(-1)G(B)}{G(BA^nC^{n(n-1)})} \displaystyle\prod_{j=0}^{n-1} \frac{G(C^{j+1})G(AC^j)}{G(C)}, \\ \qquad \text{if } BA^nC^{n(n-1)} \neq 1. \end{cases} \tag{2.1}$$

The formula above is a character sum analogue of the following recent n-dimensional integral formula of Selberg (see **[1**, (3.5)]):

$$\int \Delta^c \left(1 - \sum_{i=1}^n u_i\right)^{b-1} \prod_{i=1}^n u_i^{a-1} du_1 \cdots du_n$$

$$= \frac{\Gamma(b)\, n!}{\Gamma(b + na + n(n-1)c)} \prod_{j=0}^{n-1} \frac{\Gamma(a+jc)\,\Gamma(c+jc)}{\Gamma(c)}, \tag{2.2}$$

where $a, b, c > 0$ and

$$\Delta = \prod_{1 \leq i < j \leq n} (u_i - u_j)^2 \tag{2.3}$$

and where the integral is over the set of nonnegative u_i with $u_1 + \cdots + u_n \leq 1$.

Theorem 2.2. *If (1.9) is true, then Conjecture 2.1 is true.*

Proof. The sum

$$\sum_{\substack{F \\ \deg F = n}} A(F(0))\, B(-1 - t_1)\, C\phi(D_F) \tag{2.4}$$

is unchanged when B is replaced by $\overline{B}\,\overline{A}^n\overline{C}^{n(n-1)}$. This may be seen by replacing $F(x)$ by $(-1-t_1)^{-n} F(x(-1-t_1))$ when $1 + t_1 \neq 0$. Thus it suffices to prove just the case $B \neq 1$ of (2.1).

For each $w \in GF(q)^*$, replace $F(x)$ by $w^n F(x/w)$ in (1.9) to obtain

$$\sum_{\substack{F \\ \deg F = n}} A(F(0))\, C\phi(D_F)\, \zeta^{T(wt_1)}$$

$$= \overline{A}^n\, \overline{C}^{n(n-1)}(w) \prod_{j=0}^{n-1} \frac{G(C^{j+1})\, G(AC^j)}{G(C)}. \tag{2.5}$$

Multiply both sides of (2.5) by $\overline{B}(w)\zeta^{T(w)}$ and sum on w to obtain

$$\sum_{\substack{F \\ \deg F=n}} \sum_{w} A(F(0))\, C\phi(D_F)\, \overline{B}(w)\, \zeta^{T(w(1+t_1))}$$

$$= \sum_{w} \overline{B}\,\overline{A}^{n}\overline{C}^{n(n-1)}(w)\,\zeta^{T(w)} \prod_{j=0}^{n-1} \frac{G(C^{j+1})G(AC^{j})}{G(C)}. \tag{2.6}$$

Observe that

$$\sum_{w} \overline{B}(w)\,\zeta^{T(w(1+t_1))} = B(1+t_1)\,G(\overline{B}) \tag{2.7}$$

when either $1+t_1 \neq 0$ or $1+t_1 = 0, B \neq 1$. Thus Theorem 2.2 follows from (2.6) in the case $B \neq 1$.

Conjecture 2.3. *For all C and all $w \in GF(q)^*$,*

$$\sum_{\substack{F \\ \deg F=n}} C\phi(D_F)\zeta^{T(w(t_1^2/2-t_2))}$$

$$= \phi^{n}\overline{C}^{n(n-1)/2}(w) \prod_{j=0}^{n-1} \frac{\phi(2)G(\phi)G(C^{j+1})}{G(C)}. \tag{2.8}$$

Note that (2.8) reduces to the conjecture (1.10) when $w = 1$. If $w \in GF(p)^*$, then (2.8) can in fact be deduced from (1.10) by application of an automorphism mapping ζ to ζ^w. If $w = 0$, then (2.8) reduces to

$$\sum_{\substack{F \\ \deg F=n}} C\phi(D_F) = 0\,, \tag{2.9}$$

which is true whenever $C^{n(n-1)} \neq 1$; to see this, replace $F(x)$ by $u^n F(x/u)$ for any $u \in GF(q)^*$ with $C^{n(n-1)}(u) \neq 1$.

Conjecture 2.4. *For $v \in GF(q)^*$ and all B, C with $B \neq 1$,*

$$\sum_{\substack{F \\ \deg F=n}} C\phi(D_F)B(v + t_1^2/2 - t_2)$$

$$= \frac{\phi^n BC^m(v)\, G(\phi^n \overline{B}\,\overline{C}^m)}{G(\overline{B})} \prod_{j=0}^{n-1} \frac{\phi(2)G(\phi)G(C^{j+1})}{G(C)}, \tag{2.10}$$

where $m = \dfrac{n(n-1)}{2}$.

Theorem 2.5. *If Conjecture 2.3 is true, then Conjecture 2.4 is true.*

Proof. Multiply both sides of (2.8) by $\overline{B}(w)\zeta^{T(wv)}$ and sum on w to obtain

$$\sum_{\substack{F \\ \deg F=n}} C\phi(D_F)\sum_{w}\overline{B}(w)\zeta^{T(w(t_1^2/2-t_2+v))}$$

$$=\sum_{w}\phi^n\overline{B}\,\overline{C}^m(w)\zeta^{T(wv)}\prod_{j=0}^{n-1}\frac{\phi(2)G(\phi)G(C^{j+1})}{G(C)}, \qquad (2.11)$$

where $m = n(n-1)/2$. Dividing both sides of (2.11) by $G(\overline{B})$, we obtain (2.10), since $B \neq 1$.

3. Evaluations of $L_n(A,B,1)$ and $L_n(A,B,\phi)$

We will use the following additional notation in this section. Let N_r denote the norm map from $GF(q^r)$ to $GF(q)$. For multiplicative characters χ, ψ on $GF(q^r)$, the Gauss and Jacobi sums $G_r(\chi)$ and $J_r(\chi,\psi)$ on $GF(q^r)$ are

$$G_r(\chi)=\sum_m \chi(m)\zeta^{T(m)},\quad J_r(\chi,\psi)=\sum_m \chi(m)\psi(1-m), \qquad (3.1)$$

where the summations are over $m \in GF(q^r)$ and where T is the trace map from $GF(q^r)$ to $GF(p)$. Let $\lambda(F)$ denote the summand in (1.3), i.e.,

$$\lambda(F) = A((-1)^n F(0))\,B(F(1))\,C\phi(D_F), \qquad (3.2)$$

where $n = \deg F$. Then the generating function $L(z)$ for the Selberg sums $L_n(A,B,C\phi)$ can be written

$$L(z)=\sum_{n=0}^{\infty} L_n(A,B,C\phi)z^n=\sum_F \lambda(F)z^{\deg F}. \qquad (3.3)$$

We will evaluate $L_n(A,B,C)$ for $C^2 = 1$ in Theorem 3.5, by expressing $L(z)$ as an explicit rational function. We conjecture that $L(z)$ is a rational function for *all* A,B,C.

Lemma 3.1. *Let f denote a monic irreducible polynomial over $GF(q)$. Then*

$$C\phi(D_f)=\begin{cases}1, & if \quad C=\phi \\ -(-1)^{\deg f}, & if \quad C=1.\end{cases} \qquad (3.4)$$

Proof. This is trivial for $C=\phi$, so let $C=1$. Let $\alpha_i=\gamma^{q^i}$ $(1\le i\le \deg f)$ be the zeros of f. Since $\alpha := \prod_{1\le i<j\le \deg f}(\alpha_i-\alpha_j) \in GF(q)$ if and only if $\deg f$ is odd, we have $C\phi(D_f)=\phi(D_f)=\phi(\alpha^2)=-(-1)^{\deg f}$.

Lemma 3.2. *Let $C^2 = 1$. If $F = GH$ for monic, relatively prime polynomials G, H over $GF(q)$, then $\lambda(F) = \lambda(G)\,\lambda(H)$.*

Proof. Let $\{\alpha_i\}$, $\{\beta_j\}$ denote the zeros of G, H, respectively. Then $D_F = D_G\, D_H\, u^2$, where

$$u = \prod_{i,j} (\alpha_i - \beta_j). \tag{3.5}$$

Since $u \in GF(q)^*$ and $C^2 = 1$, $C\phi(u^2) = 1$. Thus $C\phi(D_F) = C\phi(D_G)\,C\phi(D_H)$, so $\lambda(F) = \lambda(G)\,\lambda(H)$.

Lemma 3.3. *Let $C^2 = 1$. Then*

$$L(z) = \prod_{f} (1 + \lambda(f) z^{\deg f}), \tag{3.6}$$

where the product is over all monic irreducible f over $GF(q)$.

Proof. By (3.3) and Lemma 3.2,

$$L(z) = \prod_{f} \sum_{i=0}^{\infty} \lambda(f^i) z^{i \deg f}.$$

Since $\lambda(f^i) = 0$ for $i \geq 2$, (3.6) follows.

Lemma 3.4. *The generating function*

$$M(z) = M(A, B; z) = \sum_{k=1}^{\infty} \frac{z^k}{k} J_k(A \circ N_k,\, B \circ N_k) \tag{3.7}$$

has the evaluation

$$M(z) = \begin{cases} \log\left((1 - z)^2 (1 - zq)^{-1}\right) & , \text{ if } A = B = 1 \\ \log(1 + zJ(A, B)) & , \text{ otherwise}. \end{cases} \tag{3.8}$$

Note that $\exp M(z)$ is an L-function [**5**, pp. 338-339].

Proof. The result follows since, by the Hasse-Davenport Theorem [**4**, p. 162], [**5**, p. 197], the Jacobi sum J_k in (3.7) equals $q^k - 2$ if $A = B = 1$, and it equals $-(-J(A, B))^k$ otherwise.

Theorem 3.5. *Let $C^2 = 1$. If $C = 1$, then*

$$L(z) = \begin{cases} (1 + qz)(1 + z)^{-2}, & \text{if } A = B = 1 \\ (1 - zJ(A, B))^{-1}, & \text{otherwise}, \end{cases} \tag{3.9}$$

so that for each $n \geq 0$,

$$L_n(A,B,\phi) = \begin{cases} (-1)^n S(n,q), & \text{if } A = B = 1 \\ J(A,B)^n \quad , & \text{otherwise}, \end{cases} \tag{3.10}$$

where

$$S(n,q) = 1 + n - nq. \tag{3.11}$$

If $C = \phi$, then

$$L(z) = \begin{cases} (1-qz^2)(1-qz)^{-1}(1+z)^{-2} \; , \\ \qquad \text{if } A = B = 1 \\ (1-qz^2)(1-z)^{-1}(1+z)^{-2} \; , \\ \qquad \text{if } A = \phi, B = 1 \\ \qquad \text{or } B = \phi, A = 1 \\ (1-qz^2)(1-z^2)^{-1}(1+(-1)^{(q-1)/2}z)^{-1} \; , \\ \qquad \text{if } A = B = \phi \\ (1+zJ(A,B))(1+z^2J(A^2,B^2))^{-1} \; , \\ \qquad \text{otherwise}, \end{cases} \tag{3.12}$$

so that for each $n \geq 0$,

$$L_n(A,B,1) = \begin{cases} (-1)^n T(n,q) \; , \\ \qquad \text{if } A = B = 1 \\ (-1)^n S([n/2],q) \; , \\ \qquad \text{if } A = \phi, B = 1 \\ \qquad \text{or } B = \phi, A = 1 \\ (-1)^{n(q+1)/2} S([n/2],q) \; , \\ \qquad \text{if } A = B = \phi \\ (-J(A^2,B^2))^{[n/2]} J(A,B)^{n-2[n/2]} \; , \\ \qquad \text{otherwise}, \end{cases} \tag{3.13}$$

where

$$T(n,q) = -n + \sum_{k=0}^{n} (2k+1)(-q)^{n-k}. \tag{3.14}$$

Proof. It suffices to prove (3.9) and (3.12).

By (3.6),

$$\begin{aligned} -\log L(z) &= \sum_{f} \sum_{m=1}^{\infty} z^{m \deg f} (-\lambda(f))^m / m \\ &= \sum_{k=1}^{\infty} \frac{z^k}{k} \sum_{r \mid k} \sum_{\substack{f \\ \deg f = r}} r(-\lambda(f))^{k/r}, \end{aligned} \tag{3.15}$$

where f is always monic and irreducible over $GF(q)$. Thus, by (3.2) and (3.4),

$$-\log L(z) = \begin{cases} \sum_{k=1}^{\infty} \frac{(-z)^k}{k} \sum_{r|k} S_{r,k} \quad , & \text{if } C = 1 \\ \sum_{k=1}^{\infty} \frac{z^k}{k} \sum_{r|k} (-1)^{k/r} S_{r,k}, & \text{if } C = \phi, \end{cases} \tag{3.16}$$

where

$$S_{r,k} = \sum_{\substack{f \\ \deg f = r}} rA^{k/r}\left((-1)^r f(0)\right) B^{k/r}\left(f(1)\right). \tag{3.17}$$

The second formula in (3.16) can be rewritten as

$$-\log L(z) = \sum_{k=1}^{\infty} \frac{z^k}{k} \left\{ 2 \sum_{2r|k} S_{r,k} - \sum_{r|k} S_{r,k} \right\}, \quad \text{if } C = \phi \tag{3.18}$$

By associating elements $\alpha \in GF(q^r)$ of degree r over $GF(q)$ with their minimal polynomials f, we see that

$$S_{r,k} = \sum_{\substack{\alpha \in GF(q^r) \\ \deg \alpha = r}} A^{k/r} \circ N_r(\alpha) B^{k/r} \circ N_r(1-\alpha), \tag{3.19}$$

so

$$S_{r,k} = \sum_{\substack{\alpha \in GF(q^k) \\ \deg \alpha = r}} A_k(\alpha) \quad B_k(1-\alpha), \tag{3.20}$$

where

$$A_k = A \circ N_k \quad , \quad B_k = B \circ N_k . \tag{3.21}$$

It follows from (3.20) and (3.1) that

$$\sum_{r|k} S_{r,k} = J_k(A_k, B_k). \tag{3.22}$$

Similarly, if $k = 2m$,

$$\begin{aligned} \sum_{2r|k} S_{r,k} &= \sum_{r|m} S_{r,2m} \\ &= \sum_{r|m} \sum_{\substack{\alpha \in GF(q^m) \\ \deg \alpha = r}} A_m^2(\alpha) B_m^2(1-\alpha) \\ &= J_m\left(A_m^2, B_m^2\right). \end{aligned} \tag{3.23}$$

If $C = 1$, then by (3.16), (3.22), and (3.7),

$$-\log L(z) = \sum_{k=1}^{\infty} \frac{(-z)^k}{k} J_k(A_k, B_k) = M(-z). \tag{3.24}$$

Thus (3.9) follows from (3.8).

Finally, let $C = \phi$. Then by (3.18), (3.22), (3.23), and (3.7),

$$\begin{aligned} -\log L(z) &= \sum_{m=1}^{\infty} \frac{z^{2m}}{m} J_m\left(A_m^2, B_m^2\right) - \sum_{k=1}^{\infty} \frac{z^k}{k} J_k(A_k, B_k) \\ &= M(A^2, B^2; z^2) - M(A, B; z). \end{aligned} \tag{3.25}$$

Thus (3.12) follows from (3.8).

REFERENCES

1. R. Askey and D. Richards, Selberg's second beta integral and an integral of Mehta, to appear.
2. R. Evans, Identities for products of Gauss sums over finite fields, Enseignement Math. **27** (1981), 197-209.
3. R. Evans and W. Root, Conjectures for Selberg character sums, J. Ramanujan Math. Soc. **3** (1) 1988, 111-128.
4. K. Ireland and M. Rosen, *A classical introduction to modern number theory*, Springer-Verlag, N. Y., 1982.
5. R. Lidl and H. Niederreiter, *Finite Fields,* Addison-Wesley, Reading, 1983.

Julie Autuore
Department of Mathematics
University of California, San Diego
La Jolla, CA 92093

Ronald Evans
Department of Mathematics
University of California, San Diego
La Jolla, CA 92093

Oscillations of Quadratic L-Functions

R. C. BAKER AND H. L. MONTGOMERY

Dedicated to Paul T. Bateman on his seventieth birthday

1. Introduction

All real non-principal characters are of the form $\chi_D(n) = (\frac{D}{n})$ where D belongs to the set $\mathcal{Q}$ of quadratic discriminants, $\mathcal{Q} = \{D : D \text{ is not a square and } D \equiv 0 \text{ or } 1 \pmod 4\}$. The character $(\frac{D}{n})$ is induced by a primitive charater $(\frac{d}{n})$ where d belongs to the set $\mathcal{D}$ of fundamental discrminants,

$$\mathcal{D} = \{d : d \equiv 1 \pmod 4,\ d \text{ squarefree}\}$$
$$\cup \{d : 4 \mid d,\ d/4 \equiv 2 \text{ or } 3 \pmod 4,\ d/4 \text{ squarefree}\}$$

(see §5 of Davenport [6]). Of special interest are the prime discriminants, which in the present context we take to be $\mathcal{P} = \{d \in \mathcal{D} : |d| \text{ is prime}\}$. If $\mathcal{A}$ is a subset of the integers, we let $N_{\mathcal{A}}(x)$ denote the numbers of members of $\mathcal{A}$ whose absolute value does not exceed x. Clearly $N_{\mathcal{P}}(x) \sim x/\log x$ and $N_{\mathcal{Q}}(x) \sim x$ as $x \longrightarrow \infty$. With a little more effect (say by appealing to Lemma 4 below) it can also be shown that $N_{\mathcal{D}}(x) \sim (6/\pi^2)x$ as $x \longrightarrow \infty$.

For $D \in \mathcal{Q}$ we let $L_D(s)$ be the associated L-function, defined to be

$$L_D(s) = \sum_{n=1}^{\infty} \left(\frac{D}{n}\right) n^{-s}$$

for $\Re e(s) > 0$. We are concerned with the way in which $L_D(s)$ wobbles as s tends to 1/2 from above. In this connection it is necessary to establish some conventions concerning the manner in which sign changes are to be counted. Let $S^-(a_1, a_2, \ldots, a_R)$ denote the number of sign changes in the sequence $a_1, a_2, \ldots, a_R$ with zero terms deleted, and let

Research supported in part by National Science Foundation Grant NSF-DMS-85-02804

$S^+(a_1, a_2, \ldots, a_R)$ denote the maximum number of sign changes with zero terms replaced by number of arbitrary sign. Thus for example, $S^-(1,0,1) = 0$ while $S^+(1,0,1) = 2$. In any case, $S^- \le S^+$. If f is a real-valued function defined on an interval (a,b), then $S^\pm(f;a,b)$ denotes the supremum of $S^\pm(f(a_1), f(a_2), \ldots, f(a_R))$ over all finite sequences for which $a < a_1 < a_2 < \cdots < a_R < b$. These conventions are standard (see Karlin [15]). We show that most $L_D(s)$ are very far from being monotonic in the interval $(1/2, 1)$.

Theorem. *Suppose that* $s > 1/2$. *If* $\left|\frac{L_D'}{L_D}(s)\right| > 1/(s-1/2)$ *then set* $A_D(s) = \frac{L_D'}{L_D}(s)$. *Otherwise put* $A_D(s) = 0$. *For* $r = 1, 2, \ldots$ *let* $s_r = 1/2 + \exp(-4^r)$, *put*

$$\mathcal{B}(R) = \{D \in \mathcal{Q} : S^-(A_D(s_1), A_D(s_2), \cdots, A_D(s_R)) < R/300000\},$$

and set

$$q(R) = \limsup_{x \to \infty} \frac{1}{x} N_{\mathcal{B}(R)}(x).$$

Then

$$\lim_{R \to \infty} q(R) = 0.$$

Similarly, if

$$p(R) = \limsup_{x \to \infty} \frac{\log x}{x} N_{\mathcal{B}(R) \cap \mathcal{P}}(x)$$

then

$$\lim_{R \to \infty} p(R) = 0.$$

With more work we could replace s_r by a sequence tending to $1/2$ more slowly. We note several consequences of our main result.

Corollary 1. *Let* K *be a given number. The set of* $D \in \mathcal{Q}$ *for which* $S^-(L_D'; 1/2, 1) \le K$ *has asymptotic density* 0.

It seems likely that $S^-(L_D'; 1/2, 1) \asymp \log\log |D|$ for most D. For $D \in \mathcal{Q}$ let

$$F_D(z) = \sum_{n=1}^{|D|} \left(\frac{D}{n}\right) z^N$$

be the associated Fekete polynomial. Thus

$$\sum_{n=1}^{\infty} \left(\frac{D}{n}\right) z^n = F_D(z)/(1 - z^{|D|}) \tag{1}$$

for $|z| < 1$. By a familiar inverse Mellin transform it follows that

$$L_D(s)\Gamma(s) = \int_0^\infty F_D(e^{-x})(1 - e^{-|D|x})^{-1} x^{s-1}\, dx$$

for $\Re e(s) > 0$. From this identity it is evident that if $F_D(z) > 0$ for $0 < z < 1$ then $L_D(s) > 0$ for $0 < s < 1$. Fekete proposed this as a means to show that quadratic L-functions have no positive real zeros, but Pólya [17] showed that if $D \in \mathcal{Q}$ and $\left(\frac{D}{2}\right) = \left(\frac{D}{3}\right) = \left(\frac{D}{5}\right) = \left(\frac{D}{7}\right) = \left(\frac{D}{11}\right) = -1$ then $F_D(0.7) < 0$. Hence the Fekete Hypothesis (FH)

$$F_D(z) \ge 0 \qquad \text{for } 0 < z < 1$$

fails for a positive proportion of the $D \in \mathcal{Q}$. Subsequently, Bateman, Purdy and Wagstaff [2] showed that $\left(\frac{D}{2}\right) = \left(\frac{D}{3}\right) = \left(\frac{D}{5}\right) = \left(\frac{D}{7}\right) = -1$ does not imply that FH fails for D . More recently, Chowla [3] proposed conjectures which imply FH, and Heilbronn [9], unaware of the earlier literature, gave a different disproof of FH. From our Theorem we are able to derive a stronger form of these negative results.

Corollary 2. *Let the number K be given. The set of $D \in \mathcal{Q}$ for which $S^-(F_D; 0, 1) \le K$ has asymptotic density 0. Similarly, the relative asymptotic density of those $d \in \mathcal{P}$ for which $S^-(F_d; 0, 1) \le K$ is 0.*

It is not known whether there exist infinitely many $d \in \mathcal{D}$ for which FH holds. Kaczorowski [14] related the zeros of $F_D(z)$ to the nature of certain splitting fields. The papers of Wolke [19], and of Baker and Harman [1] are also related to this question.

For $D \in \mathcal{Q}$ and $x > 0$ let

$$S_D(x) = \sum_{1 \le n \le x} \left(\frac{D}{n}\right). \qquad (3)$$

By partial summation we see that the power series in (1) is

$$(1 - z) \sum_{n=1}^{\infty} S_D(n) z^n$$

for $|z| < 1$. Hence if $S_D(x) \ge 0$ for all $x > 0$ the FH is valid for this particular D. Thus from Corollary 2 we see that for almost all $D \in \mathcal{Q}$ there is an $x < |D|$ for which $S_D(x) < 0$. By arguing more carefully from our Theorem, we establish a stronger form of this.

Corollary 3. *For $N \geq 1$ let*

$$\mathcal{G}(N) = \{D \in \mathcal{Q} : S_D(n) \geq 0 \quad \text{for } n = 1, 2, \cdots, N\}.$$

Then

$$\alpha(N) = \lim_{x \longrightarrow \infty} \frac{1}{x} N_{\mathcal{G}(N)}(x)$$

exists , and $\lim_{N \longrightarrow \infty} \alpha(N) = 0$. *Similarly,*

$$\beta(N) = \lim_{x \longrightarrow \infty} \frac{\log x}{x} N_{\mathcal{G}(N) \cap \mathcal{P}}(x)$$

exists, and $\lim_{N \longrightarrow \infty} \beta(N) = 0$.

Before turning to technical details, we first outline the method we use to derive our Theorem. For most $D \in \mathcal{Q}$, one can approximate to $\frac{L'_D}{L_D}(s)$ by an appropriately weighted partial sum of the (possibly divergent) series

$$-\sum_{n=1}^{\infty} \left(\frac{D}{n}\right) \Lambda(n) n^{-s}.$$

By distinguishing between odd and even powers of primes this may be written

$$-\sum_{p} \left(\frac{D}{p}\right) \frac{\log p}{p^s - p^{-s}} - \sum_{p \nmid D} \frac{\log p}{p^{2s} - 1} = -\sum\nolimits_0 - \sum\nolimits_e,$$

say. If p is fixed and D runs over $\mathcal{Q}$, $|D| \leq X$, the distribution of the numbers $\left(\frac{D}{p}\right)$ is asymptotically that of the random variables Z_p for which

$$P(Z_p = -1) = P(Z_p = 1) = \frac{p-1}{2p}, \qquad P(Z_p = 0) = \frac{1}{p}. \tag{4}$$

Moreover, if $p_1, p_2, \cdots, p_K$ are distinct primes, then the quantities $(\frac{D}{p_k})$ are asymptotically independent. Thus it is to be expected that if s is fixed, $s > 1/2$, then $\sum_0$ has an asymptotic distribution function which is the distribution function of the random variable

$$Z(s) = \sum_{p} Z_p \frac{\log p}{p^s - p^{-s}}$$

where the Z_p are indenpendent. Indeed $\frac{L'_D}{L_D}(s)$ has an asymptotic distribution function for every fixed s with $s > 1/2$, and our analysis thus far is similar to that of Elliott [8], who considered $\log L_D(s)$. As $s \longrightarrow (1/2)^+$, the distribution of $Z(s)$ approaches the normal distribution with mean 0 and standard deviation

$$\left(\sum_p \frac{(p-1)(\log p)^2}{p(p^s - p^{-s})^2}\right)^{1/2}.$$

This is approximately $1/(2s-1)$ when s is just a little larger than $1/2$. Moreover, $\sum_e$ is approximately this same size. Let

$$\Phi(x) = (2\pi)^{-1/2} \int_{-\infty}^{x} e^{-u^2/2}\, du \tag{5}$$

denote the cumulative distribution function of the normal random variable with mean 0 and standard deviation 1. Thus it follows that if s is slightly larger than 1/2 then $L'_D(s) < 0$ for a set of $D \in \mathcal{D}$ with relative density aproximately $\Phi(1) = 0.841\ldots$, and $L'_D(s) > 0$ for a set of $D \in \mathcal{D}$ with relative density aproximately $\Phi(-1) = 0.159\ldots$. Further examination reveals that $Z(s)$ depends most heavily on the primes in the vicinity of $\exp(1/(s-1/2))$. More precisely, if we let $Z_T(s)$ denote the truncated sum

$$Z_T(s) = \sum_{u(s)<p\leq v(s)} Z_p \frac{\log p}{p^s - p^{-s}}$$

where $u(s) = \exp((s-1/2)^{-1/2})$, $v(s) = \exp((s-1/2)^{-2})$, then $Z(s)$ is usually near $Z_T(s)$. Now consider a sequence of s_r tending to 1/2 very rapidly, say $s_r = 1/2 + \exp(-4^r)$. Then the intervals $(u(s_r), v(s_r)]$ are disjoint, and hence the variables $Z_T(s_1), \cdots, Z_T(s_R)$ are independent. Consequently we find that this sequence usually has a large number of changes of sign.

2. Basic Lemmas

We begin with four number-theoretic lemmas.

Lemma 1. *Suppose that $X \geq 2$ and $Y \geq 2$. Then for arbitrary real or complex numbers a_n ,*

$$\sum_{\substack{D\in\mathcal{Q}\\ |D|\leq X}} \left|\sum_{n\leq Y} a_n \left(\frac{D}{n}\right)\right|^2 \leq (X + Y^2 \log Y) \sum_{\substack{mn=\square\\ m,n\leq Y}} |a_m a_n|. \tag{6}$$

Moreover, if $2 \leq Y \leq X^{1/3}$ then

$$\sum_{p\leq X} \left|\sum_{n\leq Y} a_n \left(\frac{n}{p}\right)\right|^2 \ll \frac{X}{\log X} \sum_{\substack{mn=\square\\ m,n\leq Y}} |a_m a_n|. \tag{7}$$

For other estimates similar to (6), see Jutila [11,12,13] and Elliott [7,8].

Proof: The left hand side of (6) is

$$\sum_{\substack{1\le m\le Y\\ 1\le n\le Y}} a_m\overline{a_n} \sum_{\substack{D\in Q\\ |D|\le X}} \left(\frac{D}{mn}\right).$$

If mn is a square then the inner sum is $\ll X$, and the contribution of such terms is accounted for on the right hand side of (6). If mn is not a square then by several applications of the Pólya-Vinogradov inequality we find that the inner sum above is $\ll X^{1/2} + Y\log Y$. From the arithmetic-geometric mean inequality we find that $|a_m a_n| \le (|a_m|^2 + |a_n|^2)/2$, so that the contribution of such term is

$$\ll (x^{1/2}Y + Y^2\log Y)\sum_{n\le Y} |a_n|^2 \ll (x^{1/2}Y + Y^2\log Y) \sum_{\substack{mn=\square\\ m,n\le Y}} |a_m a_n|.$$

Using the arithmetic-geometric mean inequality again, we find that $X^{1/2}Y \ll X + Y^2$, and hence the expression on the right above is majorized by the right hand side of (6).

The estimate (7) is a special case of Lemma 9 of Montgomery and Vaughan [16].

Lemma 2. *Suppose that $\sigma > 1/2$. Put $A = 12/(\sigma - 1/2)$, and suppose that $X > X_o(\sigma)$. There is a (possibly empty) set $\mathcal{E}(\sigma) \subset Q$ of 'exceptional discriminants' such that*

$$N_{\mathcal{E}(\sigma)}(X) < X^{1-(\sigma-1/2)/5} \tag{8}$$

and with the property that if $D \in Q\setminus\mathcal{E}(\sigma)$, then $L_D(s) \ne 0$ and

$$\frac{L'_D}{L_D}(s) = -\sum_{n=1}^{\infty}\left(\frac{D}{n}\right)\Lambda(n)n^{-s}e^{-n/x} + O(1/\log|D|)$$

uniformly for $\sigma \le s \le 5/4$, $(\log|D|)^A \le x \le |D|$.

Proof: Let $\mathcal{E}_1(\sigma)$ be the set of those $D \in Q$ for which $L_D(s)$ has at least one zero $\rho = \beta + i\gamma$ in the rectangle

$$\mathcal{R}_1(\sigma) = \{w : (\sigma + 1/2)/2 \le \Re e(w) \le 1,\ \ |\Im m(w)| \le 2(\log 2|D|)^2\}.$$

Suppose that $D \in \mathcal{E}_1(\sigma)$, $|D| \le X$, and that $(\frac{D}{n})$ is induced by $(\frac{d}{n})$. Then the zero of $L_d(s)$ in the half-plane $\sigma > 0$ are precisely the same as those of

$L_D(s)$. Suppose that $U \le |d| \le 2U$. Such a $d \in \mathcal{D}$ induces $\ll X^{1/2}U^{-1/2}$ discriminants $D \in \mathcal{Q}$ with $|D| \le X$. Jutila [11] proved that

$$\sum_{\substack{d\in\mathcal{D}\\ |d|\le U}} N_d(\alpha, T) \ll U^{1-(\alpha-1/2)/2}T^2(\log TU)^{68}$$

where $U \ge 2$, $T \ge 2$ and $N_d(\alpha, T)$ denotes the number of zeros of $L_d(s)$ in the rectangle $\alpha \le \Re e(s) \le 1$, $|\Im m(s)| \le T$. We take $\alpha = (\sigma + 1/2)/2$ and $T = 2(\log 2X)^2$, and find that there are

$$\ll X^{1/2}U^{1/2-(\sigma-1/2)/4}(\log X)^{72}$$

$D \in \mathcal{E}_1(\sigma)$ for which $U \le |d| \le 2U$. We sum over $U = 2^{-k}X$, $k = 1, 2, \cdots$, and deduce that

$$N_{\mathcal{E}_1(\sigma)}(X) \ll X^{1-(\sigma-1/2)/4}(\log X)^{72}.$$

We now take $\mathcal{E}(\sigma) = \mathcal{E}_1(\sigma) \bigcup \{D \in \mathcal{D} : |D| \le \exp(1/(\sigma - 1/2))\}$. Then the above estimate gives (8) for $X > X_o(\sigma)$.

Suppose that $D \in \mathcal{Q} \setminus \mathcal{E}(\sigma)$ so that $L_D(s) \ne 0$ for $s \in \mathcal{R}_1(\sigma)$ and $\sigma - 1/2 \ge 1/\log|D|$. Put $\lambda = (4\sigma + 1)/6$, and let

$$\mathcal{R}(\sigma) = \{w : \lambda \le \Re e(w) \le 5/4, |\Im m(w)| \le (\log|D|)^2.$$

As in Davenport [6; §16], it follows that

$$\frac{L_D'}{L_D}(s) = \sum_{|\rho-s|<1} \frac{1}{s-\rho} + O(\log|D|) \ll (\log|D|)^2$$

for $s \in \mathcal{R}(\sigma)$. Now suppose that $\sigma \le s \le 5/4$. Then

$$\sum_{n=1}^{\infty} \left(\frac{D}{n}\right) \Lambda(n)n^{-s}e^{-n/x} = -\frac{1}{2\pi i}\int_{1-i\infty}^{1+i\infty} \frac{L_D'}{L_D}(s+w)\Gamma(w)x^w\, dw.$$

We replace the path of integration by the piecewise linear path with vertices $1-i\infty$, $1-i(\log|D|)^2$, $\lambda-s-i(\log|D|)^2$, $\lambda-s+i(\log|D|)^2$, $1+i(\log|D|)^2$, $1+i\infty$. The term $-\frac{L_D'}{L_D}(s)$ arises from the residue at $w = 0$. We note that for $\Re e(w) = \lambda - s$, $|\Im m(w)| \le (\log|D|)^2$ the integrand is

$$\ll \frac{e^{-|w|}}{|w|}x^{\lambda-\sigma}(\log|D|)^2 \ll e^{-|w|}(\log|D|)^{3-A(\sigma-1/2)/3} \ll e^{-|w|}(\log|D|)^{-1},$$

and hence this portion of the contour contributes an amount $\ll 1/\log|D|$. The rest of the contour contributes an amount

$$\ll x\exp(-(\log|D|)^2).$$

This is also $\ll 1/\log|D|$, since $x \le |D|$.

We now prove a lemma which we use in the proof of Corollary 3.

Lemma 3. *Let k be a non-negative integer. If $3 \le N \le X^{1/4}$ and $1/2 + 50(\log\log N)/\log N \le \sigma \le 1$ then*

$$\sum_{\substack{D\in\mathcal{Q}\\|D|\le X}} \left| L_D^{(k)}(\sigma) - \sum_{n=1}^{N} (-\log n)^k \left(\frac{D}{n}\right) n^{-\sigma}\right|^2$$

$$\ll XN^{1-2\sigma}(\log N)^{2k+1}(2\sigma-1)^{-2}.$$

Proof: Since

$$L_D^{(k)}(\sigma) = \sum_{n=1}^{\infty} (-\log n)^k \left(\frac{D}{n}\right) n^{-\sigma},$$

we may write

$$L_D^{(k)}(\sigma) - \sum_{n=1}^{N} (-\log n)^k \left(\frac{D}{n}\right) n^{-\sigma} = \sum_{N<n\le Y} + \sum_{Y<n} = \sum\nolimits_1 + \sum\nolimits_2$$

where $Y = X^{1/3}$. Let $S_D(y)$ be given by (3). By integrating by parts we see that

$$\sum\nolimits_2 \ll |S_D(Y)|(\log Y)^k Y^{-\sigma} + \int_Y^\infty |S_D(y)|(\log y)^k y^{-\sigma-1}\,dy = T_1 + T_2.$$

But Jutila [10] showed that

$$\sum_{\substack{D\in\mathcal{Q}\\|D|\le X}} |S_D(Y)|^2 \ll XY(\log X)^8 \tag{9}$$

uniformly for $X \ge 2$, $Y \ge 2$. Hence

$$\sum_{\substack{D\in\mathcal{Q}\\|D|\le X}} |T_1|^2 \ll XY^{1-2\sigma}(\log X)^{2k+8} \ll XN^{1-2\sigma}(\log N)^{2k}.$$

By the Cauchy-Schwarz inequality we see that

$$|T_2|^2 \ll \int_Y^\infty (\log y)^{2k} y^{-\sigma-1/2}\,dy \int_Y^\infty |S_D(y)|^2 y^{-\sigma-3/2}\,dy.$$

Here the first integral is $\ll Y^{1/2-\sigma}(\log Y)^{2k}(\sigma-12)^{-1}$. Hence by (9) it follows that

$$\sum_{\substack{D\in\mathcal{Q}\\|D|\le X}} |T_2|^2 \ll Y^{1/2-\sigma}(\log Y)^{2k}(\sigma-1/2)^{-1}\int_Y^\infty X(\log X)^8 y^{-\sigma-1/2}\,dy$$

$$\ll XY^{1-2\sigma}(\sigma-1/2)^{-2}(\log X)^{2k+8} \ll XN^{1-2\sigma}(\log N)^{2k}.$$

On combining these estimates, we find that

$$\sum_{\substack{D\in\mathcal{Q}\\|D|\le X}} \left|\sum\nolimits_2\right|^2 \ll XN^{1-2\sigma}(\log N)^{2k}.$$

On the other hand, from Lemma 1 we deduce that

$$\sum_{\substack{D\in\mathcal{Q}\\|D|\le X}} \left|\sum\nolimits_1\right|^2 \ll X \sum_{\substack{mn=\square\\N<m,n\le Y}} (\log m)^k(\log n)^k(mn)^{-\sigma}.$$

If mn is a square then there exists positive integers q,r,s such that $m=qr^2$ and $n=qs^2$. Hence the right hand side above is

$$\ll X\sum_{q\le Y}\left(\sum_{\substack{r\\N<qr^2}} (\log qr^2)^k(qr^2)^{-\sigma}\right)^2.$$

If $q\le N$ then r must satisfy the constraint $r\ge (N/q)^{1/2}$. For larger values of q there is no such condition. By treating these two situations separately we find that the above is

$$\ll XN^{1-2\sigma}(\log N)^{2k+1}(\sigma-1/2)^{-2},$$

and the proof is complete.

Lemma 4. *Let $\mathcal{F}$ be a finite set of prime numbers, and for $p\in\mathcal{F}$ let ϵ_p be given, $\epsilon_p=\pm1$ or 0. Let $\mathcal{D}(\epsilon)$ denote the set of those $d\in\mathcal{D}$ such that $\left(\frac{d}{p}\right)=\epsilon_p$ for all $p\in\mathcal{F}$. Then*

$$N_{\mathcal{D}(\epsilon)}(X)\sim(6/\pi^2)X\prod_{\substack{p\in\mathcal{F}\\\epsilon_p=\pm1}}\frac{p}{2(p+1)}\prod_{\substack{p\in\mathcal{F}\\\epsilon_p=0}}\frac{1}{(p+1)}.$$

Proof: By elementary techniques it is easy to show (e.g. see Cohen and Robinson [5]) that if a and q are given, with $\Delta=(a,q)$ squarefree, then

$$\sum_{\substack{n\le x\\n\equiv a(q)}}\mu(n)^2\sim(6/\pi^2)\frac{x}{q}\prod_{p|q/\Delta}\left(1-p^{-2}\right)^{-1}\prod_{\substack{p|q\\p\nmid q/\Delta}}\left(1+p^{-1}\right)^{-1}.$$

If p is an odd prime then $\left(\frac{d}{p}\right)=1$ (or -1) if and only if d lies in one of the $(p-1)/2$ quadratic residue (or nonresidue) classes $\pmod p$, and $\left(\frac{d}{p}\right)=0$ if and only if $p\mid d$. Also, $\left(\frac{d}{2}\right)=1$ for $d\equiv\pm1\pmod 8$, $\left(\frac{d}{2}\right)=-1$ for $d\equiv\pm3\pmod 8$, and $\left(\frac{d}{2}\right)=0$ when $2\mid d$. The stated result now follows by combining these observations, the definition of $\mathcal{D}$, and the above asymptotic estimate.

We now establish two probabilistic lemmas.

Lemma 5. *Suppose that for $r = 1, 2, 3, \ldots$ the random variables Z_{rn} are independent, where $1 \le n \le N_r$, and put*

$$Z_r = \sum_{n=1}^{N_r} Z_{rn}.$$

Suppose that $E(Z_{rn}) = 0$ for all n and r, and that $P(|Z_{rn}| \le c_n) = 1$ for all n and r, where $c_n \ge 0$ are constants such that

$$K = \sum_{n=1}^{\infty} c_n^3 < \infty.$$

Let

$$\sigma(r) = \left(\sum_{n=1}^{N_r} \mathrm{Var}(Z_{rn}) \right)^{1/2}$$

denote the standard deviation of Z_r, and suppose that $\sigma(r) \longrightarrow \infty$ as $r \longrightarrow \infty$. Then the distribution of the random variable $Z_r/\sigma(r)$ tends to the normal distribution with $\mu = 0$ and $\sigma = 1$ as $r \longrightarrow \infty$.

See Chung [4; Ch.7] for more general results in this direction.

Proof: The characteristic function of $Z_r/\sigma(r)$ is

$$\varphi_r(t) = E\left(\exp\left(itZ_r/\sigma(r)\right)\right) = \prod_{n=1}^{N_r} E\left(\exp\left(itZ_{rn}/\sigma(r)\right)\right).$$

Since $e^{iu} = 1 + iu - u^2/2 + O(|u|^3)$ uniformly for $|u| \le 1$, it follows that the multiplicand in the product above is

$$1 - \frac{1}{2}t^2 E(Z_{rn}^2)\sigma(r)^{-2} + O\left(|t|^3 c_n^3 \sigma(r)^{-3}\right)$$

$$= \exp\left(-\frac{1}{2}t^2 E(Z_{rn}^2)\sigma(r)^{-2} + O(|t|^3 c_n^3 \sigma(r)^{-3})\right)$$

uniformly for $|t| \le \sigma(r)K^{-1/3}$. Hence for such t we find that

$$\varphi_r(t) = \exp\left(-\frac{1}{2}t^2\right) + O\left(K|t|^3\sigma(r)^{-3}\right),$$

and the result follows by the method of characteristic functions.

Lemma 6. *let $\delta > 0$ and suppose that $Z_1, Z_2, \ldots, Z_R$ are independent random variables such that $P(Z_r > 0) \geq \delta$ and $P(Z_r < 0) \geq \delta$ for all r. Then*

$$P(S^-(Z_1, Z_2, \ldots, Z_R) \leq \frac{1}{5}\delta R) \ll e^{-\delta R/3}$$

uniformly in δ and R.

Proof: First we reduce to a simpler situation. Without loss of generality we may assume that our probability space is $[0,1]^R$, and that $Z_r(\mathbf{x})$ is a function of x_r alone, where $\mathbf{x} = (x_1, x_2, \ldots, x_R)$. For each r let $Y_r = Y_r(x_r)$ be defined so that if $Y_r < 0$ then $Z_r < 0$, if $Y_r > 0$ then $Z_r > 0$, and with the further property that $P(Y_r < 0) = P(Y_r > 0) = \delta$. Since

$$S^-(Z_1, Z_2, \ldots, Z_R) \geq S^-(Y_1, Y_2, \ldots, Y_R)$$

at all points of the probability space, it suffices to show that

$$P(S^-(Y_1, Y_2, \ldots, Y_R) \leq \frac{1}{5}\delta R) \ll e^{-\delta R/3}.$$

If $\delta R \leq 5$ then the above is trivial. Hence we may assume that $\delta R \geq 5$. We begin by computing $P(S^-(Y_1, Y_2, \ldots, Y_R) = k)$. Choose a subset $\mathcal{S}$ of n of the numbers $1, 2, \ldots, R$. There are $\binom{R}{n}$ such subsets, and $P(Y_r \neq 0 \iff r \in \mathcal{S}) = (2\delta)^n(1-2\delta)^{R-n}$. Suppose that $Y_r = \pm 1$ for $r \in \mathcal{S}$, $Y_r = 0$ otherwise. There are 2^n possible sequences of ± 1's . Of these, precisely $2\binom{n-1}{k}$ have exactly k changes of sign. Hence

$$P\left(S^-(Y_1, Y_2, \ldots, Y_R) = k\right) = \sum_{n=k+1}^{R} \binom{R}{n} (2\delta)^n (1-2\delta)^{R-n} 2 \binom{n-1}{k} 2^{-n}.$$

Let $m = n - k - 1$. Then the above is

$$2\binom{R}{k+1}\delta^{k+1} \sum_{m=0}^{R-k-1} \left(\frac{k+1}{m+k+1}\right) \binom{R-k-1}{m} \delta^m (1-2\delta)^{R-k-1-m}.$$

On replacing the first factor in the summand by 1 and using the binomial theorem, we see that the above is

$$\leq 2\binom{R}{k+1}\delta^{k+1}(1-\delta)^{R-k-1} = M_k,$$

say. Put $K = [\delta R/5]$. Then $K \geq 1$. If $k \leq K$ then $M_k/M_{k-1} \geq 2$. Thus the probability in question is $\leq \sum_{k \leq K} M_k \leq 2M_K$. Now $\binom{R}{K+1} \leq R^{K+1}/(K+1)!$ and $(K+1)! \geq (K/e)^{K+1}$. Hence

$$M_K \leq 2\left(\frac{eR\delta}{K(1-\delta)}\right)^{K+1} (1-\delta)^R.$$

Since $\delta \le 1/2$, we see that

$$\frac{eR\delta}{K(1-\delta)} \le 10e\frac{R\delta/5}{K} \le 10e\frac{K+1}{K}.$$

Also, $1-\delta \le e^{-\delta}$. Hence $M_K \ll (10e)^{K+1}e^{-\delta R} \ll (10e)^{\delta R/5}e^{-\delta R}$. But $(10e)^{1/5} \le e^{2/3}$, so we have the desired bound.

It is not hard to show that

$$E\left(S^-(Y_1, Y_2, \dots, Y_R)\right) = \delta R - 1/2 + O(e^{-\delta R}),$$

and that

$$\operatorname{Var}\left(S^-(Y_1, Y_2, \dots, Y_R)\right) = \delta(1-\delta)R - 1/4 + O(e^{-\delta R}).$$

Thus when δR is large the number of sign changes is usually $\sim \delta R$. Indeed, the distribution function of S^- resembles that of a binomial variable with parameters R, δ. Hence it is not surprising that

$$P\left(S^-(Y_1, Y_2, \dots, Y_R) < \frac{\delta R}{5}\right)$$

should be small.

We conclude this section by quoting a lemma of real analysis which we use in deriving Corollary 2. In its simplest form, this lemma is Descartes' rule of signs.

Lemma 7. *let f be a real-valued function defined on $\mathcal{R}$ which is Riemann-integrable on finite intervals, and suppose that the Laplace transform*

$$\mathcal{L}(s) = \int_{-\infty}^{\infty} f(x)e^{-sx}\,dx$$

converges for all $s > 0$. Then

$$S^-(f; -\infty, +\infty) \ge S^+(\mathcal{L}; 0, +\infty).$$

A proof of this may be found in Karlin [15; p.313]. See also Pólya-Szegö [18; vol 2, #33, 65,80].

3. Proof of the Theorem

Let $D \in \mathcal{Q}$. Taking a large value of R, we consider $\frac{L'_D}{L_D}(s)$ at the points $s = s_r = 1/2 + \exp(-4^r)$, $R_1 < r \le R$ where $R_1 = [R/5]$. We take $\sigma = s_R$, $A = 12/(\sigma - 1/2)$, and $x = (\log X)^A$. We assume that $X > X_1(R)$, that $X \le |D| \le 2X$, and that $D \notin \mathcal{E}(\sigma)$, so that the formula of Lemma 2 applies. Since $\Lambda(n) = 0$ if n is not a primepower, we may suppose that $n = p^k$. The contribution of $k \ge 3$ in this formula is $\ll 1$, uniformly for $s \ge 1/2$. The contribution of the terms $n = p^2$ is < 0 and $\ge -1/(2s - 1) + O(1)$. The sum over $n = p$ we break into three parts, according as $p \le u(s) = \exp((s - 1/2)^{-1/2})$, $u(s) < p \le v(s) = \exp((s - 1/2)^{-2})$, or $v(s) < p$. Let $\mathcal{E}_1(s, X)$ be the set of those $D \in \mathcal{Q} \bigcap [-X, X]$ for which

$$\left| \sum_{p \le u(s)} \left(\frac{D}{p}\right) (\log p) p^{-s} e^{-p/x} \right| > \frac{1}{6(s - 1/2)}.$$

By Lemma 1 we see that this sum, squared and summed over $D \in \mathcal{Q} \bigcap [X, 2X]$, is

$$\ll X \sum_{p \le u(s)} (\log p)^2/p \ll X(\log u(s))^2 \ll X/(s - 1/2)$$

provided that $X > X_2(s) = u(s)^2$. Hence

$$\operatorname{card}(\mathcal{E}_1(s, X)) \ll X(s - 1/2), \tag{10}$$

and it follows that

$$\operatorname{card}\left(\bigcup_{r=R_1+1}^{R} \mathcal{E}_1(s_r, X)\right) \ll X \sum_{r=R_1+1}^{R} (s_r - 1/2) \ll X e^{-R} \tag{11}$$

for $X \ge X_3(R)$. In the sum over $p > v(s)$, we note that the contribution of primes $p > x^2$ is

$$\ll \sum_{p > x^2} (\log p) p^{-1/2} e^{-p/x} \ll \sum_{n > x^2} e^{-n/x} \ll x e^{-x} \ll 1.$$

We apply Lemma 1 with $Y = x^2$ to the remaining range $v(s) < p \le x^2$, to see that

$$\sum_{\substack{D \in \mathcal{Q} \\ |D| \le 2X}} \left| \sum_{v(s) < p \le x^2} \left(\frac{D}{p}\right) (\log p) p^{-s} e^{-p/x} \right|^2 \ll X \sum_{p > v(s)} (\log p)^2 p^{-2s}.$$

Here the sum on the right is

$$\ll (s-1/2)^{-1}v(s)^{1-2s}(\log v(s))^2 \ll (s-1/2)^{-3}\exp(-2(s-1/2)^{-1}) \\ \ll \exp(-(s-1/2)^{-1}) \ll s-1/2.$$

Let $\mathcal{E}_2(s,X)$ be the set of $D \in \mathcal{Q} \cap [X,2X]$ for which

$$\left|\sum_{p>v(s)} \left(\frac{D}{p}\right)(\log p)p^{-s}e^{-p/x}\right| > \frac{1}{6(s-1/2)}.$$

If this inequality holds and s is near 1/2 then the corresponding sum with p restricted to the range $v(s) < p \leq x^2$ has modulus $\geq 1/(7(s-1/2))$. From the estimate above we deduce that (10) holds with $\mathcal{E}_1$ replaced by $\mathcal{E}_2$, and hence as in (11) we have

$$\operatorname{card}\left(\bigcup_{r=R_1+1}^{R} \mathcal{E}_2(s_r,X)\right) \ll Xe^{-R}. \tag{12}$$

We now treat the sum

$$\sum_{u(s)<p\leq v(s)} \left(\frac{D}{p}\right)(\log p)p^{-s}e^{-p/x}.$$

For p is this interval , $e^{-p/x} = 1 + O(v(s)/x)$. The contribution of this error term is $\ll v(s)^{3/2}/x$, which is $\ll 1$ when $s = s_r$, $R_1 < r \leq R$ and $X > X_4(R)$.

By the definition of the Legendre symbol, the definition of the set $\mathcal{Q}$, and the Chinese remainder theorem, it follows that the asymptotic distribution of the sum

$$K_D(s) = \sum_{u(s)<p\leq v(s)} \left(\frac{D}{p}\right)(\log p)p^{-s},$$

for $D \in \mathcal{Q}\bigcap[-X,X]$, $X \longrightarrow \infty$, is the same as the distribution function of the random variable

$$Z(s) = \sum_{u(s)<p\leq v(s)} (\log p)p^{-s}Z_p$$

where the Z_p are the independent random variables defined in (4). (If we worked instead with $d \in \mathcal{D}$ then the distribution of the individual variable would be slightly different, but the varibles would still be independent, as we see by Lemma 4.) Let

$$\rho(s) = \left(\sum_{u(s)<p\le v(s)} \frac{(p-1)(\log p)^2}{p^{2s+1}} \right)^{1/2}$$

be the standard deviation of $Z(s)$. By Lemma 5 we see that as $s \longrightarrow (1/2)^+$, the distribution function of $Z(s)/\rho(s)$ approaches the normal distribution with $\mu = 0$ and $\sigma = 1$. From the asymptotic estimates of Mertens we know that $\rho(s) \sim 1/(2s-1)$ as $s \longrightarrow (1/2)^+$. With $\Phi(x)$ given by (5), choose δ so that $0 < \delta < \Phi(-4) = 0.000031671\ldots$, say $\delta = 0.00003$. Since we are supposing that R is large, we have

$$P(Z(s_r) > 2/(s_r - 1/2)) \ge \delta, \qquad P(Z(s_r) < -2/(s_r - 1/2)) \ge \delta$$

for $R_1 < r \le R$. Put $B_r = 1$ if $Z(s_r) > 2/(s_r - 1/2)$, $B_r = -1$ if $Z(s_r) < -2/(s_r - 1/2)$, and $B_r = 0$ otherwise. Since the intervals $(u(s_r), v(s_r)]$ are disjoint, the variables $Z(s_r)$ are independent. Hence Lemma 6 applies to the B_r. Let

$$P_R = P(S^-(B_{R_1+1}, B_{R_1+2}, \ldots, B_R)) \le \delta(R - R_1)/5).$$

By Lemma 6 we see that $P_R \ll \exp(-\delta(R - R_1)/3)$.

For $D \in \mathcal{Q}$, $1/2 < s \le 1$, put $U_D(s) = -1,\ 0$, or 1 according as $K_D(s)$ lies in $(-\infty, -2/(s-1/2))$, $[-2/(s-1/2), 2/(s-1/2)]$, or $(2/(s-1/2), +\infty)$, respectively. Let $\mathcal{C}(R)$ denote the set of those $D \in \mathcal{Q}$ such that

$$S^-(U_D(s_{R_1+1}), U_D(s_{R_1+2}), \ldots, U_D(s_R)) \le \delta(R - R_1)/5.$$

Then for any given R, $N_{\mathcal{C}(R)}(X) \sim P_R$ as $X \longrightarrow \infty$. On combining this with Lemma 2, (11) and (12), we deduce that if $\epsilon > 0$ is given , then we may choose R so that for all large X we have both

$$\left|\frac{L'_D}{L_D}(s_r) + K_D(s_r)\right| < \frac{6}{7}(s_r - 1/2)^{-1}$$

for $R_1 < r \le R$, and also

$$S^-(U_D(s_{R_1+1}), U_D(s_{R_1+2}), \cdots, U_D(s_R)) > \delta(R - R_1)/5$$

for all but at most ϵX of the $D \in \mathcal{Q}$ for which $|D| \le X$. This gives the stated result, since $4\delta/25 > 1/300000$. To obtain the second part of the Theorem we argue similarly, but we use the second part of Lemma 1 instead of the first part.

4. Proof of the Corollaries

To derive Corollary 1 it suffices to note that if $L_D(s) > 0$ for $1/2 < s < 1$ then $S^-(L'_D, 1/2+\delta, 1) = S^-(\frac{L'_D}{L_D}, 1/2+\delta, 1)$, and that by Lemma 2, the set of $D \in \mathcal{Q}$ for which $L_D(s)$ has a zero in $(1/2+ ,1)$ has asymptotic density 0.

To derive Corollary 2 we first differentiate both sides of (2), to find that

$$L_D(s)\Gamma(s)\left(\frac{L'_D}{L_D}(s) + \frac{\Gamma'}{\Gamma}(s)\right) = \int_0^\infty F_D(e^{-x})(1-e^{-|D|x})^{-1}x^{s-1}(\log x)\,dx.$$

Suppose that δ is a small positive number. We may ignore those $D \in \mathcal{Q}$ for which $L_D(s)$ has a zero in the interval $(1/2+ ,1)$, since by Lemma 2 such D constitute a set of asymptotic density 0. Since $\frac{\Gamma'}{\Gamma}(s) \ll 1$ for $1/2 \le s \le 1$, it follows from the Theorem that the left hand side above has at least K changes of sign in the interval $(1/2+\delta, 1)$, for most D. Suppose that D is chosen so that the left hand side has at least K changes of sign. Then by Lemma 7, the integrand on the right hand side also has at least K changes of sign. The second and third factors are positive, and the fourth factor has only one change of sign (at $x = 1$), and hence $F_D(e^{-x})$ must have at least $K-1$ changes of sign for $0 < x < \infty$. That is , $F_D(z)$ has at least $K-1$ changes of sign for $0 < z < 1$.

Altered forms of these Corollaries may also be derived with $\mathcal{Q}$ replaced by $\mathcal{P}$.

We now derive Corollary 3. The existence of $\alpha(N)$ is assured by quadratic reciprocity. To demonstrate the existence of $\beta(N)$, we also appeal to Dirichlet's theorem on the uniform distribution of primes in arithmetic progressions. We note that if $S_D(x) \ge 0$ for $0 \le x \le N$ then

$$\begin{aligned}\sum_{n=1}^N \left(\frac{D}{n}\right) n^{-s} &= 1 + S_D(N)N^{-s} - 2^{-s} + s\int_2^N S_D(x)x^{-s-1}\,dx \\ &\ge 1 - 2^{-s} \ge 1 - 2^{-1/2}\end{aligned}$$

uniformly for $s \ge 1/2$. Similarly ,

$$\begin{aligned}\sum_{n=1}^N \left(\frac{D}{n}\right) n^{-s}\log n &= S_D(N)N^{-s}\log N + \int_1^N (s\log x - 1)S_D(x)x^{-s-1}\,dx \\ &\ge -\int_1^9 [x]x^{-3/2}\,dx > -9\end{aligned}$$

for $s \ge 1/2$. Let $s_r = 1/2 + \exp(-4^r)$, as in the Theorem, let $N_r = \exp((s_r - 1/2)^{-2})$, and let $\mathcal{G}_r$ be the collection of those ('good') $D \in \mathcal{Q}$ for which both

$$\left| L_D(s_r) - \sum_{n=1}^{N_r} \left(\frac{D}{n}\right) n^{-s_r} \right| \le 1/25$$

and

$$\left| L'_D(s_r) + \sum_{n=1}^{N_r} \left(\frac{D}{n}\right) n^{-s_r} \log n \right| \le 1.$$

Let $\mathcal{B}_r = \mathcal{Q} \setminus \mathcal{G}_r$ be the set of 'bad' D. If $S_D(n) \ge 0$ for $0 \le n \le N_r$ and $D \in \mathcal{G}_r$ then $L_D(s_r) \ge 1/4$ and $L'_D(s_r) \le 10$, so that

$$\frac{L'_D}{L_D}(s_r) \le 40. \tag{13}$$

Let R be large, and put $R_1 = [R/10^6]$. From Lema 3 we deduce that

$$\operatorname{card}\left(\bigcup_{R_1+1}^{R} \mathcal{B}_r \bigcap [-X, X] \right) \ll X \exp\left(-c \exp\left(\exp\left(R\right)\right)\right)$$

as $X \longrightarrow \infty$. But if (13) holds for all r in the interval $R_1 < r \le R$ then

$$S^-(A_D(s_{R_1+1}), A_D(s_{R_1+2}), \ldots, A_D(s_R)) = 0,$$

and hence

$$S^-(A_D(s_1), A_D(s_2), \ldots, A_D(s_R)) \le R/10^6.$$

By the Theorem, the set of such D has small density if R is large. This completes the proof of Corollary 3.

REFERENCES

[1] R.C. Baker and G. Harman, Unbalanced quadratic residues and non-residues, Proc. Camb. Philos. Soc. **98** (1975), 9–17.

[2] P.T. Bateman, G.B. Purdy and S. Wagstaff, Some numerical results on Fekete polynomials, Math. Comp. **29** (1975), 7–23.

[3] S.D. Chowla, Note on Dirichlet's L-fucnctions, Acta Arith. **1** (1936), 113–114.

[4] K.L Chung, *A course in probability theory*, Harcourt, Brace and World (New York), 1968.

[5] E. Cohen and R.L. Robinson, On the distribution of the k-free integers in residue classes, Acta Arith. **8** (1962/3), 283–293.

[6] H. Davenport, *Multiplicative Number Theory*, Second Edition, Springer–Verlag (New York), 1980, 178 pp.

[7] P.D.T.A. Elliott, On the mean value of $f(p)$, Proc. London Math. Soc. **(3) 21** (1970), 28–96.

[8] P.D.T.A. Elliott, On the distribution of the values of quadratic L-series in the half-plane $\sigma > 1/2$, Invent. Math. **21** (1973), 319–338.

[9] H. Heilbronn, On real characters, Acta Arith. **2** (1937), 212–213.

[10] M. Jutila, On character sums and class numbers, J. Number Theory **5** (1973), 203–214.

[11] M. Jutila, A density estimate for L-functions with a real character, Ann. Acad. Sci. Fenn. Ser. AI **508** (1972), 10 pp.

[12] M. Jutila, On the mean values of L-functions and short character sums with real characters, Acta Arith. **26** (1975), 405–410.

[13] M. Jutila, On the mean values of Dirichlet polynomials with real characters, Acta Arith. **27** (1975), 191–198.

[14] J. Kaczorowski, Some problems concerning roots of polynomials with Dirichlet characters as coefficients, in *Elementary and Analytic Theory of Numbers*, Henryk Iwaniec, ed., Banach Center Publ., 1985, pp. 333-337.

[15] S. Karlin, *Total Positivity*, Stanford University Press (Stanford), 1968.

[16] H.L. Montgomery and R.C. Vaughan, Mean values of character sums, Canadian J. Math. **31** (1979), 476–487.

[17] G. Pólya, Verschiedene Bemerkung zur Zahlentheorie, Jber. Deutsch. Math. Verein **28** (1919), 31–40.

[18] G. Pólya and G. Szegö, *Aufgaben und Lehrsätze aus der Analysis*, Vols I and II, Springer-Verlag (Berlin), 1964.

[19] D. Wolke, Eine Bemerkung über das Legendre-Symbol, Monat. Math. **77** (1973), 267–275.

R. C. Baker
Royal Holloway and Bedford New College
Egham
Surrey TW 20 OEX
England, U. K.

Hugh L. Montgomery
Department of Mathematics
University of Michigan
Ann Arbor, MI 48109-1003
USA

Elementary Proof of a Theorem of Bateman

MICHEL BALAZARD AND ABDELHAKIM SMATI

Dedicated to Paul T. Bateman

1. The distribution of values of the Euler totient function $\varphi(n)$ has been investigated from various points of view (*cf.* [7], [4], [2]) . In this paper, we shall study the asymptotic behaviour of the number $N(x)$ of those positive integers n which satisfy $\varphi(n) \le x$.

The best result up-to-date is the following

Theorem (Bateman 1972, [1]). *For every constant* $c < 1/\sqrt{2}$,

$$N(x) = A\,x + O_c(x\,e^{-c(\log x\ \log\log x)^{1/2}}), \tag{1}$$

where $A = \frac{\zeta(2)\zeta(3)}{\zeta(6)} = \prod_p (1 + \frac{1}{p(p-1)})$.

Bateman's proof is analytic: it starts from Perron's integral formula and uses a simple estimate of $|\zeta(s)|$ in the strip $0 < \Re(s) < 1$. As for every result involving only natural numbers and the elementary functions of real analysis, one may ask for a proof using only these elements, and neither complex variables nor Fourier analysis.

The first step in this direction is due to Dressler (*cf.* [3]) , who proved by elementary means in 1970 that $N(x) \sim Ax$. In 1984, Nicolas obtained the elementary error term $O(x/\log x)$ (*cf.* [6]). His starting point is the study of the weighted sum $\sum_{\varphi(n)\le x} \log\varphi(n)$ and lends itself to generalization. The idea is to estimate the sum $\sum_{\varphi(n)\le x} \log^k \varphi(n)$ in order to get the error term $O(x(\log x)^{-k})$ in (2).

This generalization has been studied, and completely worked out for $k = 2$, by the second author (*cf.* [8]). Nevertheless, the computations are so intricate that a new approach is needed to go further. During the conference, we presented an elementary proof of a result slightly weaker than (1), namely

$$N(x) = A\,x + O(x\,e^{-c_0(\log x)^{1/2}}), \tag{2}$$

where c_0 is some constant (positive, absolute and computable).

After our talk, G. Tenenbaum convinced us that obtaining Bateman's result by our method would demand only a few more lines of computation. As it seldom occurs that an elementary result reaches the degree of accuracy of the best known analytic one, we will follow Tenenbaum's advice and present a proof of (1) by elementary means.

2. Our starting point is the idea of Dressler. In his proof of $N(x) \sim Ax$, he approximated the Euler function by the truncated function

$$\varphi(n,y) = n \prod_{\substack{p \mid n \\ p \leq y}} (1 - \frac{1}{p}), \quad y \geq 2.$$

Denote by $N(x,y)$ the number of positive integers n such that $\varphi(n,y) \leq x$. We first give simple inequalities involving $N(x)$ and $N(x,y)$.

Lemma 1. *If x is large enough and $y > 3\log x$, then*

$$N(x,y) \leq N(x) \leq N(x(1 - \frac{3\log x}{y})^{-1},\ y). \tag{3}$$

Proof: The first inequality follows from $\varphi(n,y) \geq \varphi(n)$. For the second one, observe that the number $\omega(n)$ of prime divisors of n satisfies $\omega(n) \leq \log n / \log 2 \leq 2 \log n$ and that $(1-v)^\alpha \geq 1 - \alpha v$ if $\alpha \geq 1$ and $0 \leq v \leq 1$. Hence

$$\begin{aligned}\varphi(n) &= \varphi(n,y) \prod_{\substack{p \mid n \\ p > y}} (1 - \frac{1}{p}) \\ &\geq \varphi(n,y)(1 - \frac{1}{y})^{\omega(n)} \\ &\geq \varphi(n,y)(1 - \frac{2\log n}{y}).\end{aligned}$$

If $\varphi(n) \leq x$, the classical inequality $\varphi(n) \gg n(\log\log n)^{-1}$ shows that $n \ll x(\log\log x)$ and

$$x \geq \varphi(n) \geq \varphi(n,y)(1 - \frac{3\log x}{y})$$

if x is large enough.

3. In order to estimate $N(x,y)$, one writes $n = ab$ where, here and throughout this paper, a (resp. b) denotes a generic positive integer whose prime divisors p all satisfy $p \le y$ (resp. $p > y$) . One has $\varphi(n,y) = \varphi(a)b$. Since $b = 1$ or $b > y$, one gets

$$N(x,y) = \sum_{\varphi(a)\le x} 1 + \sum_{\varphi(a)\le x/y} \sum_{1<b\le x/\varphi(a)} 1 \tag{4}$$
$$= \sum_{\varphi(a)\le x/y} \sum_{1\le b\le x/\varphi(a)} 1 + O(\sum_{\varphi(a)\le x} 1).$$

The estimation of these sums depends on the following three lemmas.

Lemma 2. *Suppose y and k are real numbers so that $y \ge 2$ and $0 \le k \le \frac{1}{3}\log y$. Put $\sigma = 1 - k/\log y$. Then*

$$\sum_a \varphi(a)^{-\sigma} \ll \log y \, e^{O(e^k)}.$$

Proof: The sum $\sum_a \varphi(a)^{-\sigma}$ equals the Euler product

$$\prod_{p\le y}(1 + \frac{1}{(p-1)^\sigma}(1 + \frac{1}{p^\sigma} + \frac{1}{p^{2\sigma}} + \cdots))$$
$$= \prod_{p\le y}(1 + \frac{1}{(p-1)^\sigma} - \frac{1}{p^\sigma}) \prod_{p\le y}(1 + \frac{1}{p^\sigma} + \frac{1}{p^{2\sigma}} + \cdots)$$
$$\ll e^{\sum_{p\le y} p^{-\sigma}} \qquad \text{since } \sigma \ge \frac{2}{3}.$$

Now, since the function $(e^t - 1)/t$ increases with $t > 0$, we have

$$\sum_{p\le y} p^{-\sigma} = \sum_{p\le y} p^{-1} + \sum_{p\le y}(p^{1-\sigma} - 1)p^{-1}$$
$$\le \sum_{p\le y} p^{-1} + (e^k - 1)(\log y)^{-1} \sum_{p\le y} p^{-1}\log p$$
$$\le \log\log y + O(e^k),$$

and the Lemma follows.

Lemma 3. *Suppose $x \ge y \ge 2$ are real numbers and let $u = \log x/\log y$. Then*

$$\sum_{\varphi(a)\le x} 1 \le x \log y \, e^{-u\log u + O(u)},$$

provided that $u \le y^{1/3}$.

Proof: This is a typical application of the by now classical Rankin method. We have, for every positive σ,

$$\sum_{\varphi(a)\le x} 1 \le x^{\sigma} \sum_{a} \varphi(a)^{-\sigma}.$$

Choosing $\sigma = 1 - \log u/\log y$ and using Lemma 2 gives the result.

Lemma 4. *With the notations of Lemma 3 one has for every positive* ϵ

$$\sum_{b\le x} 1 = x \prod_{p\le y}(1-\frac{1}{p})(1+O_\epsilon(e^{-(1-\epsilon)u\log u}))$$

provided that $\log y > \sqrt{\log x}$ *(i.e.* $u < \log y$*).*

Proof: Although not stated explicitly by Halberstam and Richert, this is an easy consequence of their proof of the Fundamental Lemma of Brun's Sieve given in [5], pp. 82-83.

4. We now come back to (4). Suppose that x is large enough and that $\log y > \sqrt{\log x}$. Let ϵ be a positive real number. By Lemma 3, the error term in (4) is $O_\epsilon(x \log y\, e^{-(1-\epsilon)u\log u})$.

Moreover, by Lemma 4

$$\sum_{\varphi(a)\le x/y} \sum_{1\le b\le x/\varphi(a)} 1 \qquad (5)$$
$$= x\prod_{p\le y}(1-\frac{1}{p}) \sum_{\varphi(a)\le x/y} \frac{1}{\varphi(a)}(1+O_\epsilon(e^{-(1-\epsilon)u_a\log u_a})),$$

where $u_a = \log(x/\varphi(a))/\log y$.

The contribution of the main terms in (5) amounts to

$$x\prod_{p\le y}(1-\frac{1}{p})\left(\sum_{a}\frac{1}{\varphi(a)} - \sum_{\varphi(a)>x/y}\frac{1}{\varphi(a)}\right). \qquad (6)$$

By Rankin's method,

$$\sum_{\varphi(a)>x/y}\frac{1}{\varphi(a)} \le y^{\sigma}x^{-\sigma}\sum_{a}\frac{1}{\varphi(a)^{1-\sigma}} \quad (\sigma>0).$$

With $\sigma = \log u/\log y$ this is , by Lemma 2,

$$\ll ue^{-u\log u} \log y\ e^{O(u)}$$
$$\ll_\epsilon \log y\ e^{-(1-\epsilon)u\log u}.$$

Thus (6) can be written as

$$x\prod_{p\le y}(1+\frac{1}{p(p-1)}) + O_\epsilon(xe^{-(1-\epsilon)u\log u}).$$

We now turn to the contribution of the error term in (5). Observe that

$$u\log u - u_a\log u_a = \int_{u_a}^{u}(\log v + 1)dv$$
$$\le (\log u + 1)\frac{\log\varphi(a)}{\log y}.$$

Hence

$$\sum_{\varphi(a)\le\frac{x}{y}} \varphi(a)^{-1}e^{-(1-\epsilon)u_a\log u_a}$$
$$\le e^{-(1-\epsilon)u\log u}\sum_a \varphi(a)^{-1+\frac{\log u+1}{\log y}}$$
$$\ll_\epsilon \log y\ e^{-(1-2\epsilon)u\log u} \qquad \text{by Lemma 2.}$$

We summarize these computations in a Lemma.

Lemma 5. *Suppose $x \ge y > e^{\sqrt{\log x}}$ and x large enough. Then*

$$N(x,y) = x\prod_{p\le y}(1+\frac{1}{p(p-1)}) + O_\epsilon(x\log y\ e^{-(1-\epsilon)u\log u})$$

for every positive ϵ, where $u = \log x/\log y$.

5. We are now in a position to use Lemma 1. First, we observe that $\prod_{p\le y}(1+\frac{1}{p(p-1)}) = A + O(\frac{1}{y})$. Then, if x is large enough, $y > 4\log x$ and $\log y > [\log(x(1-\frac{3\log x}{y})^{-1})]^{1/2}$, Lemmas 1 and 5 give

$$N(x) = A\ x + O(x\ y^{-1}\log x) + O_\epsilon(x\ \log y\ e^{-(1-\epsilon)u\log u}).$$

We choose $y = e^{\sqrt{\frac{1}{2}\log x\ \log\log x}}$, so that

$$u\log u = (1+o(1))\sqrt{\frac{1}{2}\log x\ \log\log x},$$

and this completes the proof of (1).

In conclusion, let us point out that the prime number theorem is not required in our proof (neither in Bateman's proof, by the way). This is in contrast to [6] and [8], where the prime number theorem with remainder term is an essential tool.

REFERENCES

[1] P. T. Bateman, The distribution of values of Euler's function, Acta Arith. **21** (1972), 329–345.

[2] H.G. Diamond, The distribution of values of Euler's phi function, Proc. Symp. Pure Maths, AMS **24** (1973), 63–76.

[3] R.E. Dressler, A density which counts multiplicity, Pacific J. Math. **34** (1970), 371–378.

[4] P.Erdös, Some remarks on Euler's φ-function and some related problems, Bull. Amer. Math. Soc. **51** (1945), 540–544.

[5] H. Halberstam, H.-E. Richert, *Sieve Methods*, Academic Press, 1974.

[6] J.-L. Nicolas, Distribution des valeurs de la fonction d' Euler, L' Ens. Math. **30** (1984), 331–338.

[7] I.J.Schoenberg, Über die asymptotische Verteilung reeler Zahlen mod. 1, Math. Z. **28** (1928), 171–199.

[8] A. Smati, Répartition des valeurs de la fonction d' Euler, L' Ens. Math. **35** (1989), 61–76.

Michel Balazard and Abdelhakim Smati
Département de Mathématiques
Faculté des Sciences
123 Av. A. Thomas
87060 Limoges Cedex
France

The Prime k-Tuplets Conjecture on Average

ANTAL BALOG

Dedicated to Professor Paul Bateman on the occasion of his 70th birthday

1. Introduction.

The well-known twin prime conjecture states that there are infinitely many primes p such that $p+2$ is also a prime. Although the proof of this seemingly simple statement is hopeless at present many further connected conjectures exist. The conjecture in the title, for example, asks if k linear polynomials with suitable conditions on the coefficients represent simultaneously primes infinitely often. One can even ask how often this happens. In 1962 Bateman and Horn [2] gave a corresponding quantative conjecture with heuristic evidence. Before stating this conjecture we need to introduce some notations and conventions.

Let $\mathbf{a}$ and $\mathbf{b}$ be k-dimensional integer vectors, $x > 0$ a real number and $\pi(x;\mathbf{a},\mathbf{b})$ the number of integers such that $1 < \mathbf{a}n+\mathbf{b} \leq x$, $\mathbf{a}n+\mathbf{b}$ are primes. The last two conditions are understood to hold in each coordinates simultaneously. We keep this convention throughout the paper. For example, $(\mathbf{a},\mathbf{b}) = 1$ means that the corresponding coordinates are coprime. Having fixed $\mathbf{a}$ and $\mathbf{b}$, $\rho(p) = \rho(p;\mathbf{a},\mathbf{b})$ will denote the number of solutions of the congruence

$$(a_1n+b_1)(a_2n+b_2)\cdots(a_kn+b_k) \equiv 0 \pmod{p}. \tag{1.1}$$

The letter p with or without subscript will denote positive primes, $\mathbf{p}$ will denote a k-dimensional vector with prime coordiantes. The product on the left hand side of (1.1) will be abbreviated by $\prod(\mathbf{a}n+\mathbf{b})$.

We can now state the prime k-tuplets conjecture in the form given by Bateman and Horn. If $\rho(p) < p$ for every prime p (which implies, for example $(\mathbf{a},\mathbf{b}) = 1$) then

$$\pi(x;\mathbf{a},\mathbf{b}) \sim \prod_p \left(1-\frac{1}{p}\right)^{-k}\left(1-\frac{\rho(p)}{p}\right)\frac{x}{a\log^k x} \tag{1.2}$$

Written while the author was visiting the University of Georgia, Athens.

where a is the height of $\mathbf{a}$ (the maximal coordinate in modulus) and the product is extended over all primes. The product is convergent (and not zero) whenever $\rho(p) < p$ for all primes, that is whenever $\prod(\mathbf{a}n + \mathbf{b})$ has no fixed prime divisor. The classical twin prime conjecture corresponds to the case $\mathbf{a} = (1,1)$, $\mathbf{b} = (0,2)$, or in a slightly more general form to

$$\pi(x;(1,1),(0,h)) \sim 2 \prod_{p \neq 2} \frac{p(p-2)}{(p-1)^2} \prod_{\substack{p \mid h \\ p \neq 2}} \frac{p-1}{p-2} \frac{x}{\log^2 x} \tag{1.3}$$

for any even integer h. In 1947 Chudakov [3] proved that almost all even integers $h \leq x$ are the difference of two integers. It is implicit in his work that the relation (1.3) is true on average over h. To express this statement precisely we define the function (in a more general form for later use)

$$T(x;\mathbf{a},\mathbf{b}) = \sum_{1 < \mathbf{a}n + \mathbf{b} \leq x} \frac{1}{\log(a_1 n + b_1)\log(a_2 n + b_2) \cdots \log(a_k n + b_k)}.$$

One can easily see that $T(x;\mathbf{a},\mathbf{b})$ is the expected order of magnitude of $\pi(x;\mathbf{a},\mathbf{b})$. In fact

$$T(x;\mathbf{a},\mathbf{b}) \sim \frac{x}{a \log^k x}$$

for any fixed $\mathbf{a}$ and $\mathbf{b}$ as x tends to infinity. However $T(x;\mathbf{a},\mathbf{b})$ is the expected order of magnitude even uniformly in $\mathbf{a}$ and $\mathbf{b}$.

The above mentioned result of Chudakov states that

$$\sum_{2 \mid h} \left| \pi - 2 \prod_{p \neq 2} \frac{p(p-2)}{(p-1)^2} \prod_{\substack{p \mid h \\ p \neq 2}} \frac{p-1}{p-2} T \right|^2 \ll \frac{x^3}{\log^A x} \tag{1.4}$$

for any $A > 0$, where $\pi = \pi(x;(1,1),(0,h))$ and $T = T(x;(1,1),(0,h))$. Note that the sum over h is finite as for large h $\pi = T = 0$. Roughly speaking, for most even integers $h \leq x$ the number of prime pairs $1 < p \leq x$, $1 < p + h \leq x$ is equal to the expected number. Unfortunately this does not imply that $\pi(x;(1,1),(0,h))$ tends to infinity for some fixed h.

In 1961 Lavrik [5] extended this result to prime-twins in arithmetic progressions. To state this result we have to generalize our notation, and let $\pi(x;\mathbf{a},\mathbf{b};c,d)$ be the number of integers n in the residue class $c \pmod d$ such that $1 < \mathbf{a}n + \mathbf{b} \leq x$ and $\mathbf{a}n + \mathbf{b}$ are primes. His result is as follows: for any $A > 0$ and $B > 0$

$$\sum_{\substack{2 \mid h \\ (c+h,d)=1}} \left| \pi - \frac{2}{\varphi(d)} \prod_{p \neq 2} \frac{p(p-2)}{(p-1)^2} \prod_{\substack{p \mid hd \\ p \neq 2}} \frac{p-1}{p-2} T \right|^2 \ll \frac{x^3}{\log^A x} \tag{1.5}$$

uniformly for $d \leq \log^B x$ and $(c,d) = 1$. Here (and also in (1.6) below) $\pi = \pi(x;(1,1),(0,h);c,d)$ and $T = T(x;(1,1),(0,h))$. In the same paper he announced a generalization of this result to prime k-tuplets. The proof has appeared in [6].

Very recently Maier and Pomerance [7] extended (1.5) to a Bombieri-Vinogradov type theorem, namely

$$\sum_{d \leq x^{\delta}} \sum_{2 \mid h} \max_{\substack{(c,d)=1 \\ (c+h,d)=1}} \left| \pi - \frac{2}{\varphi(d)} \prod_{p \neq 2} \frac{p(p-2)}{(p-1)^2} \prod_{\substack{p \mid hd \\ p \neq 2}} \frac{p-1}{p-2} T \right| \ll \frac{x^2}{\log^A x} \tag{1.6}$$

for any $A > 0$, where $\delta > 0$ is some small computable constant. They established (1.6) to improve the size of the largest known gap between consecutive primes. One can also use (1.6) coupled with a lower bound sieve to deduce that infinitely often there are three primes and an almost prime in arithmetic progression. This was originally proved by Heath-Brown [4] in 1981. His rather different approach led him to the excellent approximation that the almost prime has at most two prime factors.

Our purpose is to prove (1.6) for prime k-tuplets. We are going to introduce one more piece of notation. Recall that $\rho(p) = \rho(p;\mathbf{a},\mathbf{b})$ is defined by the number of solutions of the congruence (1.1). We set

$$\sigma(\mathbf{a},\mathbf{b};c,d) = \frac{1}{d} \prod_{p \mid d} \frac{p}{p-\rho(p)} \prod_{p} \left(1 - \frac{1}{p}\right)^{-k} \left(1 - \frac{\rho(p)}{p}\right), \tag{1.7}$$

if $\rho(p) < p$ for all prime p and $(\mathbf{a}c+\mathbf{b}, d) = 1$, and $\sigma(\mathbf{a},\mathbf{b};c,d) = 0$ otherwise. That the infinite product in (1.7) is convergent has been proved in [2]. We can now state our main result.

Theorem. *Let $k \geq 1$ and b_k be fixed integers and* $\mathbf{a}$ *be a fixed k-dimensional integer vector. Let x be a real number with $x \geq |b_k|$ and let $Z = Z(b_k;x)$ be the set of k-dimensional integer vectors* $\mathbf{b}$ *such that the last coordiante of* $\mathbf{b}$ *is b_k and the set $\{n : 1 < \mathbf{a}n + \mathbf{b} \leq x\}$ is not empty. For any $A > 0$ there is a $B = B(A) > 0$ such that for any $D \leq x^{1/3}\log^{-B} x$ we have*

$$\sum_{d \leq D} \max_{c} \sum_{\mathbf{b} \in Z} |\pi(x;\mathbf{a},\mathbf{b};c,d) - \sigma(\mathbf{a},\mathbf{b};c,d)T(x;\mathbf{a},\mathbf{b})| \ll \frac{x^k}{\log^A x}. \tag{1.8}$$

The implied constant in the symbol $\ll$ depends at most on A and $\mathbf{a}$ but not on b_k. This will be important later. In other words, if we fix $\mathbf{a}$ and the last coordinate of $\mathbf{b}$ then we can prove the prime k-tuplets conjecture on average over the other coordinates of **b**. In addition we can do this in the form of a strong distribution theorem in residue classes.

The case $k = 1$ is essentially the Bombieri-Vinogradov theorem with a weaker D while the case $k = 2$ covers (1.6), the recent result of Maier and Pomerance. For $k \geq 3$ the result seems new. The full power of our Theorem is not clear at present; however it has already had several applications. We plan to return to a systematic discussion of the applications in a forthcoming paper. We mention here just some possibilities.

The result that there are infinitely many prime-triplets forming three term arithmetic progressions can be expressed by saying that there are infinitely many linear polynomials having prime values at three consecutive integers. Our Theorem enables us to generalize this to infinitely many polynomials of degree k having prime values at $2k+1$ consecutive integers. This corollary was discovered by Professor Andrew Granville.

In the next configurations all entries are primes and any three of them along an indicated line form a three term arithmetic progression.

```
      3                5 - 17 - 29
     / \               |    |    |
    7   13            23 - 53 - 83
   /      \            |    |    |
 11 - 17 - 23         41 - 89 - 137
```

As a consequence of our Theorem one can prove that there are infinitely many different "magic" triangles and squares with the above properties. One can also derive higher dimensional versions of these results. All of these applications are parts of a much grander design, namely the problem of solving systems of linear equations in many prime variables. This is the main objective of our forthcoming paper where we present the proof of the above results as well.

The proof of the Theorem is based on induction over k starting from the Bombieri-Vinogradov theorem. The main tool is a new version of the Hardy-Littlewood circle method that we will call the Weighted Circle Method. In section 2 we will explain how this method works and leads to a recursion formula for $\pi(x; \mathbf{a}, \mathbf{b}; c, d)$. In section 3 we shall complete the induction argument. In section 4 we introduce a certain weight function which is small on the "major arcs" where our generating function is possibly large. In this way the only information about the size of the generating function we need is an upper bound on the "minor arcs". Section 5 deals with this upper bound. We complete the proof in section 6 by calculating the "singular series".

The weighted circle method works in a more general context. Suppose we are interested in $r(n)$, the number of certain additive representations. By means of the circle method we can express $r(n)$ as a certain integral of some generating functions. The traditional way is to calculate this integral by giving asymptotic expansions for the generating functions. The use of our weight function changes the meaning of the integral to the difference between $r(n)$ and a certain average $\sum u(m)r(n-m)$ of $r(n)$. The weight function makes the integral small and it remains to calculate the above average. This is sometimes very simple if $r(n)$ is expected to behave regularly. However our method does not work if $r(n)$ behaves very wildly.

Acknowledgement: The author wishes to thank professors Carl Pomerance and Helmut Maier for proposing this problem and for stimulating discussions.

2. The Weighted Circle Method.

Let $k \geq 2$ be an integer, $\mathbf{a}$ be a fixed k-dimensional integer vector, b_k be a fixed integer and $Z = Z(b_k; x)$ be the set of k-dimensional vectors with the given b_k in the last coordinate as in the Theorem. We have to pay attention to uniformity in b_k as we will need this in the induction argument. For any k-dimensional integer vector $\mathbf{a}$, $\mathbf{b}$ or $\mathbf{p}$ we get $\tilde{\mathbf{a}}$, $\tilde{\mathbf{b}}$ or $\tilde{\mathbf{p}}$ by omitting the last coordinate. Let x be a sufficiently large real number, and c and d be integers. We define the functions

$$e(\alpha) = e^{2\pi i\alpha}, \qquad P(\alpha) = \sum_{1<p\leq x} e(\alpha p),$$

$$R(\alpha; c, d) = \sum_{\substack{1<a_k n+b_k=p_k\leq x \\ n\equiv c\ (d)}} e(\alpha n) = e\left(-\alpha\frac{b_k}{a_k}\right) \sum_{\substack{1<p_k\leq x \\ p_k\equiv a_k c+b_k\ (a_k d)}} e\left(\alpha\frac{p_k}{a_k}\right).$$

We can now define our weight function $W(\alpha)$ by

$$W(\alpha) = 1 - \sum_{q\leq Q}\frac{1}{qU}\sum_{m\leq qU}\left(\sum_{(f,q)=1} e\left(\frac{-fm}{q}\right)\right) e(\alpha m) = \sum_m w_m e(\alpha m). \tag{2.1}$$

The parameters U and Q here are positive integers that will be chosen later, and the variable f runs over a reduced system of residues $\mod q$.

We start with the integral

$$J = \sum_{d\leq D} d \max_c \int_0^1 \cdots \int_0^1 \left|R(-\tilde{\mathbf{a}}\boldsymbol{\alpha}; c, d)P(\alpha_1)\cdots P(\alpha_{k-1})W(-\tilde{\mathbf{a}}\boldsymbol{\alpha})\right|^2 d\boldsymbol{\alpha}, \tag{2.2}$$

where D is any number satisfying $1 \leq D \leq x^{1/3}\log^{-B} x$.

On the one hand we shall evaluate this integral by means of the $k-1$ dimensional Parseval identity; on the other hand we shall give an upper bound by means of upper bounds for $R(\alpha; c, d)$ and $W(\alpha)$.

On multiplying out we obtain

$$\begin{aligned} &R(-\tilde{\mathbf{a}}\boldsymbol{\alpha}; c, d)P(\alpha_1)\cdots P(\alpha_{k-1})W(-\tilde{\mathbf{a}}\boldsymbol{\alpha}) \\ &= \sum_{\substack{n\equiv c(d) \\ 1<a_k n+b_k=p_k\leq x}} \sum_{1<p_1\leq x} \cdots \sum_{1<p_{k-1}\leq x} \sum_m w_m e\left(\tilde{\mathbf{p}} - (n+m)\tilde{\mathbf{a}})\boldsymbol{\alpha}\right) \\ &= \sum_{b_1}\cdots\sum_{b_{k-1}} \left(\sum_n \sum_m \sum_{p_1} \cdots \sum_{p_{k-1}} w_m \right) e(\boldsymbol{\alpha}\tilde{\mathbf{b}}), \end{aligned}$$

where the variables of the summations satisfy

$$n \equiv c\ (d),\ 1 < \tilde{\mathbf{a}}(n+m) + \tilde{\mathbf{b}} = \tilde{\mathbf{p}} \leq x,\ 1 < a_k n + b_k = p_k \leq x. \tag{2.3}$$

The $(k-1)$-dimensional Parseval identity gives

$$\begin{aligned} J &= \sum_{d\leq D} d \max_c \sum_{\mathbf{b}\in Z} \left| \sum_n \sum_m \sum_{p_1} \cdots \sum_{p_{k-1}} w_m \right|^2 \\ &= \sum_{d\leq D} d \max_c \sum_{\mathbf{b}\in Z} \Bigg| \pi(x; \mathbf{a}, \mathbf{b}; c, d) \\ &\qquad - \sum_{q\leq Q} \frac{1}{qU} \sum_{m\leq qU} \sum_{(f,q)=1} e\left(-\frac{fm}{q}\right) \sum_n 1 \Bigg|^2, \end{aligned} \tag{2.4}$$

where the summation over n satisfies the conditions (2.3).

We will rearrange the summation over m and n into a summation over $l = n+m$ and n. Since the oscillating coefficient depends only on the residue class of $m \mod q$ we group the terms according to residue classes of l and $n \mod q$. The conditions (2.3) then lead to the conditions

$$l \equiv g(q),\ 1 < \tilde{\mathbf{a}}l + \tilde{\mathbf{b}} = \tilde{\mathbf{p}} \leq x, \tag{2.5}$$

$$n \equiv h(q),\ n \equiv c(d),\ 1 < a_k n + b_k = p_k \leq x,\ l - qU \leq n < l. \tag{2.6}$$

We can easily see that counting integers n satisfying (2.6) essentially amounts to counting primes in arithmetic progressions and in short intervals. The induction is based on the fact that the counting function of integers l satisfying (2.5) is $\pi(x; \tilde{\mathbf{a}}, \tilde{\mathbf{b}}; g, q)$ and $\tilde{\mathbf{a}}$, $\tilde{\mathbf{b}}$ are $(k-1)$-dimensional vectors. Note that the induction step leads to the problem of prime $(k-1)$-tuplets in residue classes that are different from the starting residue classes. Therefore the induction hypothesis must contain the equidistribution in residue classes. In other words, we cannot prove the special case $D = c = d = 1$ without proving the more general theorem.

Rearranging the last sum of (2.4) we arrive at

$$J = \sum_{d \le D} d \max_c \sum_{\mathbf{b} \in Z} \Bigg| \pi(x; \mathbf{a}, \mathbf{b}; c, d) - \sum_{q \le Q} \frac{1}{qU} \sum_g \sum_h \sum_f e\left(\frac{fh - fg}{q}\right) \sum_l \sum_n 1 \Bigg|^2 . \tag{2.7}$$

Here g and h run over a complete system of residues mod q while f runs over a reduced system of residues mod q, l satisfies (2.5) and n satisfies (2.6) . We can handle the sum over n (on average over d) by the Bombieri-Vinogradov theorem and the sum over l (on average over $\tilde{\mathbf{b}}$) by the induction hypothesis. We first have to choose U and Q. Thus we consider first the upper bound.

From the one dimensional Parseval identity we get

$$\begin{aligned} J &\le \int_0^1 \cdots \int_0^1 |P(\alpha_1) \cdots P(\alpha_{k-1})|^2 \sum_{d \le D} d \max_c |R(\tilde{\mathbf{a}}\boldsymbol{\alpha}; c, d) W(\tilde{\mathbf{a}}\boldsymbol{\alpha})|^2 \, d\boldsymbol{\alpha} \\ &\le \sup_\alpha |W(\alpha)|^2 \sum_{d \le D} d \max_c |R(\alpha; c, d)|^2 \\ &\quad \times \int_0^1 |P(\alpha_1)|^2 \, d\alpha_1 \cdots \int_0^1 |P(\alpha_{k-1})|^2 \, d\alpha_{k-1} \\ &\le \sup_\alpha |W(\alpha)|^2 \sum_{d \le D} d \max_c |R(\alpha; c, d)|^2 \frac{x^{k-1}}{\log^{k-1} x}. \end{aligned} \tag{2.8}$$

The upper bound for the supremum over α comes from the following two basic lemmata that we will prove in sections 4 and 5.

Lemma 1. *Let $W(\alpha)$ be defined by (2.1) where the positive integers U and Q satisfy $U \ge Q^{15}$. We have*

$$(i) \quad W(\alpha) \ll 1 \quad \text{and} \quad (ii) \quad W(\alpha) \ll \min_{v \le Q} U\|v\alpha\| + \frac{Q^{15}}{U}.$$

Lemma 2. *For any $A > 0$ there is a $B = B(A) > 0$ such that if*

$$D \le \frac{x^{1/3}}{\log^B x}, \quad \log^B x \le v \le \frac{x}{\log^B x}, \quad \|\alpha - \frac{u}{v}\| < v^{-2}$$

then

$$\sum_{d\le D} d \max_c |R(\alpha;c,d)|^2 \ll_{A,a_k} \frac{x^2}{\log^A x}.$$

Remarks: The constants implied by the symbols $\ll$ depend only on the indicated parameters; in particular, in Lemma 1 the implied constants are absolute. It is very important that Lemma 2 is uniform in b_k.

$||y||$ denotes the distance of y from the nearest integer.

Lemma 2 says that the generating function $R(\alpha;c,d)$ is small on the "minor arcs" around rational numbers with large denominator but only on average over d. The known uniform result (see [1]) would be too weak for our purposes.

The weight function $W(\alpha)$ is designed exclusively to be small on the "major arcs" around the rational numbers with small denominator where $R(\alpha;c,d)$ can be (but not necessarily is) large. This ensures that we have a good upper bound for the product $W(\alpha)R(\alpha;c,d)$ everywhere (on average over d), and no information about the behaviour of $R(\alpha;c,d)$ on the "major arcs" is needed.

Lemma 2 is responsible for the exponent 1/3 in the level of distribution of d in the Theorem. Any improvement on Lemma 2 automatically improves this exponent (up to 1/2).

One can consider Vinogradov's celebrated result [9] as the special case $D=1$ of Lemma 2. In fact our proof follows very closely Vaughan's proof [8] for Vinogradov's estimate.

As we have mentioned earlier we defer the proof of these lemmata to later sections, and we now return to the study of the integral J here.

For any $A>0$ we choose B according to Lemma 2, and we choose the parameters in the weight function as

$$Q = \left[1+\log^B x\right], \quad U = \left[\frac{x}{\log^{10B} x}\right]. \tag{2.0}$$

By the Dirichlet approximation theorem we can find for any α a rational number u/v such that

$$(u,v)=1, \quad 1\le v\le \frac{x}{Q}, \quad |\alpha-\frac{u}{v}|<\frac{Q}{vx}.$$

We divide the real numbers into two parts, the "major arcs" consisting of the real numbers α with $v\le Q$, and the "minor arcs" consisting of those with $Q<v\le x/Q$. As $Q/x\le 1/v$ we can apply Lemma 2 and Lemma 1 (i) on the "minor arcs" while we can apply Lemma 1 (ii) and a trivial

bound for $R(\alpha; c, d)$ on the "major arcs". We arrive at

$$\sup_{\alpha} |W(\alpha)|^2 \sum_{d \leq D} d \max_{c} |R(\alpha; c, d)|^2$$
$$\ll \frac{x^2}{\log^A x} + \frac{1}{\log^{18B} x} \sum_{d \leq D} d \left(1 + \frac{x}{d}\right)^2 \ll \frac{x^2}{\log^A x}.$$

Inserting this into (2.8) and (2.7) we get

$$\sum_{d \leq D} d \max_{c} \sum_{\mathbf{b} \in Z} \left| \pi(x; \mathbf{a}, \mathbf{b}; c, d) - \sum_{q \leq Q} \frac{1}{qU} \sum_{g} \sum_{h} \sum_{f} e\left(\frac{fh - fg}{q}\right) \sum_{l} \sum_{n} 1 \right|^2$$
$$\ll \frac{x^{k+1}}{\log^A x}. \tag{2.10}$$

We modify the conditions (2.5) and (2.6) slightly to

$$l \equiv g \ (q), \ 1 < \tilde{\mathbf{a}} l + \tilde{\mathbf{b}} = \tilde{\mathbf{p}} \leq x, \ 1 < a_k l + b_k \leq x, \tag{2.11}$$

$$n \equiv h \ (q), \ n \equiv c \ (d), \ a_k n + b_k = p_k, \ l - qU \leq n < l. \tag{2.12}$$

Changing the old conditions to the new ones introduces an error term which is bounded by the right hand side of (2.10). Thus in (2.10) we may assume that l satisfies (2.11) and n satisfies (2.12).

(2.10) is the result of the weighted circle method and the starting point for our further investigations.

3. The Induction Step.

We shall prove our Theorem by induction. For $k = 1$ the statement follows immediately from the Bombieri-Vinogradov theorem. (There is no average over $\mathbf{b}$ in this case.) We suppose that $k \geq 2$ and that the Theorem is true for $k-1$. Let $\mathbf{a}$ be a fixed k-dimensional integer vector and b_k be a fixed integer. We use all the notations and definitions introduced so far. Our starting point is (2.10). By the Cauchy-Schwarz inequality we deduce that

$$\sum_{d \leq D} \max_{c} \sum_{\mathbf{b} \in Z} \left| \pi(x; \mathbf{a}, \mathbf{b}; c, d) - \sum_{q \leq Q} \frac{1}{qU} \sum_{g} \sum_{h} \sum_{f} e\left(\frac{fh - fg}{q}\right) \sum_{l} \sum_{n} 1 \right|$$
$$\ll \frac{x^k}{\log^A x}. \tag{3.1}$$

Recall that g and h run over a complete system of residues modulo q, f runs over a reduced system of residues modulo q, l satisfies (2.11) and n satisfies (2.12). The sum over n is zero unless

$$(a_k, b_k) = (a_k c + b_k, d) = (a_k h + b_k, q) = 1, \; h \equiv c \mod (q, d), \tag{3.2}$$

in which case the conditions (2.12) are equivalent to

$$a_k l + b_k - a_k qU \leq p_k < a_k l + b_k, \; p_k \equiv a_k r + b_k \mod a_k [q, d],$$

where r is the unique solution modulo $[q, d]$ of the congruences $r \equiv h \ (q)$, $r \equiv c \ (d)$. We have

$$\sum_n \frac{1}{qU} = \frac{a_k}{\varphi(a_k[q,d]) \log(a_k l + b_k)} + O\left(\frac{qU}{\varphi(a_k q, d)(a_k l + b_k)}\right) + O\left(\frac{1}{qU} \max_{y \leq x} \max_r |E(y, r, a_k[q, d])|\right),$$

where the first error term comes from replacing the logarithmic integral by a single term, and $E(y, r, s)$ is the error term in the prime number formula for the arithmetic progression r modulo s. According to the induction hypothesis we have in a certain average sense

$$\sum_l 1 \sim \sigma(\tilde{\mathbf{a}}, \tilde{\mathbf{b}}; g, q) T(x; \tilde{\mathbf{a}}, \tilde{\mathbf{b}}),$$

where l satisifies the condition (2.5). However after summing over n we get a "smooth" weighting factor of l in (3.1), and the condition (2.11) is also slightly different. The expected main term is $\sigma(\tilde{\mathbf{a}}, \tilde{\mathbf{b}}; g, q) T(x; \mathbf{a}, \mathbf{b})$. Therefore we write

$$\sum_l \frac{1}{\log(a_k l + b_k)} = \sigma(\tilde{\mathbf{a}}, \tilde{\mathbf{b}}; g, q) T(x; \mathbf{a}, \mathbf{b}) + F(x; \mathbf{a}, \mathbf{b}; g, q),$$

Using this estimate in the second term of the expression enclosed in $|\cdots|$ in (3.1) we get

$$\begin{aligned} &\sum_{q \leq Q} \frac{1}{qU} \sum_g \sum_h \sum_f e\left(\frac{fg - fh}{q}\right) \sum_l \sum_n 1 \\ &= T(x; \mathbf{a}, \mathbf{b}) \sum_{q \leq Q} \frac{a_k}{\varphi(a_k[q,d])} \sum_g \sum_h \sum_f e\left(\frac{fh - fg}{q}\right) \sigma(\tilde{\mathbf{a}}, \tilde{\mathbf{b}}; g, q) \\ &\quad + O\left(\frac{Q^5 U}{\varphi(d)}\right) + O\left(\sum_{q \leq Q} \frac{xq}{U} \max_{y \leq x} \max_r |E(y, r, a_k[q, d])|\right) \\ &\quad + O\left(\sum_{q \leq Q} \frac{q^2}{\varphi(d)} \sum_g |F(x; \mathbf{a}, \mathbf{b}; g, q)|\right). \end{aligned}$$

First we consider the contribution of the error terms to the entire sum. In case of the first error term we have trivially

$$\sum_{d\le D}\sum_{\mathbf{b}\in Z}\frac{Q^5U}{\varphi(d)} \ll Q^5Ux^{k-1}\log x \ll \frac{x^k}{\log^A x},$$

where Z is the set of k-dimensional vectors $\mathbf{b}$ such that the last coordinate of $\mathbf{b}$ is b_k and the set $\{n : 1 < \mathbf{a}n + \mathbf{b} \le x\}$ is not empty. The second error term on average over $[q,d]$ can be estimated by the Bombieri- Vinogradov theorem:

$$\sum_{d\le D}\sum_{\mathbf{b}}\sum_{q\le Q}\frac{xq}{U}\max_{y\le x}\max_{r}|E(y,r,a_k[q,d])|$$
$$\ll \frac{x^kQ}{U}\sum_{n\le a_kQD}\tau(n)\max_{y\le x}\max_{r}|E(y,r,n)| \ll \frac{x^k}{\log^A x}.$$

The coefficients $\tau(n)$ here take into account the number of representations of n in the form $n = [q,d]$ and can be removed by an application of the Cauchy-Schwarz inequality.

The induction hypothesis enables us to handle the last error term on average over $\mathbf{b}$. Note that $|F(x;\mathbf{a},\mathbf{b};g,q)|$ is not exactly the error term in the theorem. We have to use partial summation or we need a slightly modified statement because of the "smooth" factor. We can state and prove the Theorem with any "smooth" function $s(n)$ contained in both $\pi(x;\mathbf{a},\mathbf{b};c,d)$ and $T(x;\mathbf{a},\mathbf{b})$, where "smooth" means differentiable with $s(t) = O(1)$, and $s'(t) = O(1/t)$. The only change is that $R(\alpha;c,d)$ must also contain $s(n)$. In this way we use the induction hypothesis for $s(n)/\log(an+b)$ which is again a "smooth" function. The "smooth" version of Lemma 2 can be deduced by partial summation. We leave the details to the reader. Note also that we do not use the full strength of the induction hypothesis because we are averaging over the last coordinate of $\tilde{\mathbf{b}}$ as well. However this is the reason for requiring the induction hypothesis to be uniform in b_k. We have

$$\sum_{d\le D}\sum_{\mathbf{b}\in Z}\sum_{q\le Q}\frac{q^2}{\varphi(d)}\sum_{g}|F(x;\mathbf{a},\mathbf{b};g,q)|$$
$$\ll Q^3\log x\sum_{b_{k-1}}\sum_{q\le Q}\max_{g}\sum_{\mathbf{b}}|F(x;\mathbf{a},\mathbf{b};g,q)| \ll \frac{x^k}{\log^A x}.$$

where $\mathbf{b}$ runs over Z with fixed last two coordinates. We arrive at

$$\sum_{d\le D}\max_{c}\sum_{\mathbf{b}\in Z}\Bigg|\pi(x;\mathbf{a},\mathbf{b};c,d) - T(x;\mathbf{a},\mathbf{b})\sum_{q\le Q}\frac{a_k}{\varphi(a_k[q,d])}$$
$$\times\sum_{g}\sum_{h}\sum_{f}e\left(\frac{fh-fg}{q}\right)\sigma(\tilde{\mathbf{a}},\tilde{\mathbf{b}};g,q)\Bigg| \ll \frac{x^k}{\log^A x}, \qquad (3.3)$$

where f, g and h run over a complete system of residues modulo q and satisfy the conditions

$$(f,q) = (\tilde{\mathbf{a}}g + \tilde{\mathbf{b}}, q) = (a_k h + b_k, q) = 1, \; h \equiv c \mod (q,d), \tag{3.4}$$

and where both $\pi(x;\mathbf{a},\mathbf{b};c,d)$ and the second term in the expression $|\cdots|$ on the left-hand side of (3.3) are zero unless

$$(\mathbf{a},\mathbf{b}) = (a_k c + b_k, d) = 1. \tag{3.5}$$

These conditions come from (3.2) and from considering terms with $\sigma(\tilde{\mathbf{a}},\tilde{\mathbf{b}};g,h) \neq 0$ only. Note that $\pi(x;\mathbf{a},\mathbf{b};c,d) = 0$ when $(\tilde{\mathbf{a}}c + \tilde{\mathbf{b}}, d) \neq 1$. In this case the second term in the expression $|\cdots|$ on the left hand side of (3.3) (the term involving the multiple sum) must be zero. Also we can define $\tilde{\rho}(p) = \rho(p;\tilde{\mathbf{a}},\tilde{\mathbf{b}})$ as the number of solutions of the congruence $\prod(\tilde{\mathbf{a}}n + \tilde{\mathbf{b}}) \equiv 0 \pmod p$. Obviously $\tilde{\rho}(p) = \rho(p)$ or $\tilde{\rho}(p) = \rho(p) - 1$, and $\tilde{\rho}(p) = p$ implies $\rho(p) = p$. But $\rho(p) = p$ can also happen when $\tilde{\rho}(p) < p$. Thus $\pi(x;\mathbf{a},\mathbf{b};c,d) = 0$ is possible when $\sigma(\tilde{\mathbf{a}},\tilde{\mathbf{b}};g,q) \neq 0$. This must be also reflected in the "singular series".

In section 6 we will evaluate this "singular series" by proving the following recursion formula.

Lemma 3. *Let* $\mathbf{a}$ *and* $\mathbf{b}$ *be fixed and define*

$$K = \prod(\mathbf{a}c + \mathbf{b}) = (a_1 c + b_1)\cdots(a_k c + b_k),$$

$$L = \prod(\tilde{\mathbf{a}}b_k - a_k\tilde{\mathbf{b}}) = (a_1 b_k - a_k b_1)\cdots(a_{k-1}b_k - a_k b_{k-1}).$$

We have uniformly in c, d *and* $\mathbf{b}$

$$\sum_{q\le Q} \frac{a_k}{\varphi(a_k[q,d])} \sum_g \sum_h \sum_f e\left(\frac{fh - fg}{q}\right) \sigma(\tilde{\mathbf{a}},\tilde{\mathbf{b}};g,q)$$
$$= \sigma(\mathbf{a},\mathbf{b};c,d) + O\left(\frac{((d,k))(k+1)^{\nu(Ld)}\log^{2k}(Q|\mathbf{a}||\mathbf{b}|)}{Q\varphi(d)}\right)$$

where the variables f, g *and* h *satisfy (3.5),* (n) *is the squarefree part of* n *(that is, the product of the distinct prime divisors of* n*),* $\nu(n)$ *is the number of distinct prime divisors of* n*, and* $|\mathbf{v}|$ *is the length of the vector* $\mathbf{v}$.

This lemma contains in some sense the arithmetic structure of the prime k-tuplets problem. Its proof is elegant and elementary but not very illuminating.

Substituting the estimate of Lemma 3 into (3.3) we get

$$\sum_{d\le D}\max_{c}\sum_{\mathbf{b}\in Z}|\pi(x;\mathbf{a},\mathbf{b};c,d)-T(x;\mathbf{a},\mathbf{b})\sigma(\mathbf{a},\mathbf{b};c,d)| \ll$$

$$\ll \sum_{d\le D}\sum_{\mathbf{b}}\frac{x((d,K))(k+1)^{\nu(Ld)}\log^k x}{Q\varphi(d)}+\frac{x^k}{\log^A x}.$$

Note that by (3.5) we have $(d,K)=(d,\tilde{K})$ where $\tilde{K}=\prod(\tilde{\mathbf{a}}c+\tilde{\mathbf{b}})$. The summation over d can be estimated the following way:

$$\sum_{d\le D}\frac{((d,\tilde{K}))(k+1)^{\nu(Ld)}}{\varphi(d)}\le (k+1)^{\nu(L)}\sum_{d\le D}\frac{(k+1)^{\nu(d)}}{\varphi(d)}\sum_{\delta\mid((d,\tilde{K}))}\varphi(\delta)$$

$$\le (k+1)^{\nu(L)}\sum_{\delta\mid(\tilde{K})}(k+1)^{\nu(\delta)}\sum_{d'\le D}\frac{(k+1)^{\nu(d')}}{\varphi(d')}$$

$$\ll (k+1)^{\nu(L)}(k+2)^{\nu(\tilde{K})}\log^k x.$$

By the Cauchy-Schwarz inequality we have

$$\frac{\log^{2k}x}{Q}\sum_{\mathbf{b}\in Z}(k+1)^{\nu(L)}(k+2)^{\nu(\tilde{K})}\le$$

$$\le \frac{\log^{2k}x}{Q}\left(\sum_{b_1}(k+1)^{2\nu(a_1c+b_1)}\sum_{b_1}(k+2)^{2\nu(a_1b_k-a_kb_1)}\right)^{1/2}\times\cdots\times$$

$$\times\left(\sum_{b_{k-1}}(k+1)^{2\nu(a_{k-1}c+b_{k-1})}\sum_{b_{k-1}}(k+2)^{2\nu(a_{k-1}b_k-a_kb_{k-1})}\right)^{1/2}\ll\frac{x^k}{\log^A x}.$$

This completes the proof of our Theorem. The remainder of the paper is devoted to the proof of our main lemmata.

4. The Weight Function.

Our goal in this section is to build up a trigonometric polynomial which is bounded (independently of the number of terms) and small around the rational numbers with small denominator. In fact, the result is given by (2.1). The notation here is completely independent of the other parts of the paper. Throughout this section the implied constants in the symbols $\ll$ are absolute unless specified otherwise. We start with a very simple observation.

Lemma 4. *Let $U > 0$ be an integer and*

$$U(\alpha) = 1 - \frac{1}{U} \sum_{l \le U} e(\alpha l). \tag{4.1}$$

We have (i) $U(\alpha) \ll 1$ and (ii) $U(\alpha) \ll U\|\alpha\|$.

Proof: (i) is trivial, and (ii) follows immediately from the estimate $e(l\alpha) = e(l\|\alpha\|) = 1 + O(l\|\alpha\|)$.

$U(\alpha)$ is a bounded trigonometric polynomial which is small around zero. For any real number r the polynomial $U(\alpha - r)$ is bounded and small around r. If we have a collection of real numbers then we may hope that the product of the associated polynomials will be small near these numbers. In fact, we have

Lemma 5. *Let $r_1, \cdots, r_R$ be real numbers with $\|r_i - r_j\| \ge \delta$ for all $i \ne j$, and let $U_1, \cdots, U_R$ be positive integers with $U_i \ge U > 0$ for all i. Let $V(\alpha)$ be defined by*

$$V(\alpha) = \prod_{i \le R} \left(1 - \frac{1}{U_i} \sum_{l \le U_i} e((\alpha - r_i)l) \right). \tag{4.2}$$

If $\log R < U\delta$ then (i) $V(\alpha) \ll 1$ and (ii) $V(\alpha) \ll \min_{i \le R} U_i \|\alpha - r_i\|$.

Proof: First we prove (i). We have, using the formula for the sum of the geometric series,

$$\begin{aligned} |V(\alpha)| &\le \prod_{i \le R} \left(1 + \frac{1}{U_i} \Big| \sum_{l \le U_i} e((\alpha - r_i)l) \Big| \right) \\ &\le \exp \left(\sum_{i \le R} \frac{1}{U_i} \Big| \sum_{l \le U_i} e((\alpha - r_i)l) \Big| \right) \\ &\ll \exp \left(\sum_{i \le R} \min \left(1, \frac{1}{U_i \|\alpha - r_i\|} \right) \right). \end{aligned}$$

For any fixed α the numbers $\alpha - r_i$ are spaced at least δ apart from each other. Therefore

$$|V(\alpha)| \ll \exp \left(1 + \sum_{j \le R} \frac{1}{U_j \delta} \right) \ll \exp \left(1 + \frac{\log R}{U\delta} \right) \ll 1,$$

which proves part (i) of the lemma. Part (ii) follows by selecting out one of the factors, estimating this factor by Lemma 4 and the remaining factors by part (i).

The last step is to show that the major part of the product comes from carrying out the multiplication and choosing 1 in each but at most one factors. This is the content of the next lemma.

Lemma 6. *Let $r_1, \cdots, r_R$ be real numbers with $\|r_i - r_j\| \geq \delta$ for all $i \neq j$ and $U_1, \cdots, U_R$ be positive integers with $QU \geq U_i \geq U > 0$ for all i. Let $W(\alpha)$ be defined by*

$$W(\alpha) = 1 - \sum_{i \leq R} \frac{1}{U_i} \sum_{l \leq U_i} e(-lr_i)e(\alpha l). \tag{4.3}$$

If $\delta^2 U > R^5 Q$ then (i) $W(\alpha) \ll 1$ and (ii) $W(\alpha) \ll \min_{i \leq R} U_i \|\alpha - r_i\| + R^5 Q/\delta^2 U$.

Proof: Let $V(\alpha)$ be defined by (4.2). We will show that under the given conditions we have

$$V(\alpha) - W(\alpha) \ll \frac{R^5 Q}{\delta^2 U}. \tag{4.4}$$

Lemma 6 then follows from Lemma 5.

We introduce some notation.

$$1 - \frac{1}{U_i} \sum_{l \leq U_i} e((\alpha - r_i)l) = \sum_l b_i(l) e(\alpha l),$$

$$b_i(l) = \begin{cases} 1 & \text{if } l = 0, \\ -\frac{1}{U_i} e(-lr_i) & \text{if } 1 \leq l \leq U_i, \\ 0 & \text{otherwise,} \end{cases}$$

$$V(\alpha) = \prod_{i \leq R} \left(\sum_l b_i(l) e(\alpha l) \right) = \sum_m a_m e(\alpha m),$$

$$a_m = \sum_{m = l_1 + \cdots + l_R} b_1(l_1) \cdots b_R(l_r).$$

In particular we have

$$a_0 = 1, \quad a_1 = \sum_{i \leq R} b_i(1), \quad a_m = \sum_{i \leq R} b_i(m) + s_m \quad \text{for } m \geq 2, \tag{4.5}$$

$$s_0 = s_1 = 0, \quad s_m = \sum_{m = l_1 + \cdots + l_R}^{*} b_1(l_1) \cdots b_R(l_R) \quad \text{for } m \geq 2, \tag{4.6}$$

where $\sum^*$ means that at least two summands l_i differ from zero.

Clearly we have

$$V(\alpha) - W(\alpha) = \sum_m s_m e(\alpha m), \tag{4.7}$$

and (4.4) will follow from an appropriate bound for the coefficients s_m. Since a_m is the Fourier coefficient of $V(\alpha)$ we get from Lemma 5 (i)

$$a_m = \int_0^1 V(\alpha)e(-\alpha m)\,d\alpha \ll 1. \tag{4.8}$$

Consider the expression (4.6) and group the terms according to the sum L of the two non-zero summands with largest indices $i < j$. For $m > 2$ we obtain the following important decomposition:

$$s_m = \sum_{1\le i<j\le R}\ \sum_{2\le L\le m}\ \sum_{1\le l<L} b_i(l)b_j(L-l) \sum_{m-L=l_1+\cdots+l_{i-1}} b_1(l_1)\cdots b_{i-1}(l_{i-1}). \tag{4.9}$$

The number of terms in the first sum is $\binom{R}{2} \ll R^2$. The last sum can be considered as a_{m-L} attached to the subsystem of real numbers $r_1, \cdots, r_{i-1}$, and (4.8) is applicable. The sum over l reduces to a sum of a geometric series that we can estimate easily. We have

$$\begin{aligned}\sum_{1\le l<L} b_i(l)b_j(L-l) &= \frac{1}{U_iU_j}\sum_l e(-lr_i)e(-(L-l)r_j) \\ &\ll \frac{1}{U_iU_j\|r_i - r_j\|} \ll \frac{1}{U_iU_j\delta},\end{aligned} \tag{4.10}$$

where in the second term l runs over an interval defined by the inequalities $1 \le l \le U_i$ and $1 \le L - l \le U_j$. In particular, the sum is empty unless $L \le U_i + U_j$. (4.9) and (4.10) give the bound $s_m \ll \sum_{1\le i<j\le R}(U_i + U_j)1/U_iU_j\delta \ll R^2/U\delta$. This bound is not strong enough for us, but together with (4.5) it enables us to improve (4.8) to

$$a_m \ll \sum_{i\le R}\frac{1}{U_i} + \frac{R^2}{U\delta} \ll \frac{R^2}{U\delta} \quad \text{for} \quad m \ge 1. \tag{4.11}$$

We can repeat the above argument using (4.11) instead of (4.8). This time we have to be more careful because (4.8) was true for every m but (4.11) is true for only $m \geq 1$. We therefore treat the term $L = m$ separately. Starting from (4.9) we get for $m \geq 2$

$$s_m = \sum_{1 \leq i < j \leq R} \left(\sum_{1 \leq l < m} b_i(l) b_j(L - l) \right.$$

$$\left. + \sum_{2 \leq L < m} \sum_{1 \leq l < L} b_i(l) b_j(L_l) \sum_{m - L = l_1 + \cdots + l_{i-1}} b_1(l_1) \cdots b_{i-1}(l_{i-1}) \right)$$

$$\ll \sum_{1 \leq i < j \leq R} \left(\frac{1}{U^2 \delta} + (U_i + U_j) \frac{1}{U_i U_j \delta} \frac{R^2}{U \delta} \right) \ll \frac{R^4}{U^2 \delta^2}.$$

(4.4) follows from this and (4.7) because the number of non-zero terms in (4.7) is at most $U_1 + \cdots + U_R \leq RQU$.

Lemma 1 is a special case of Lemma 6, corresponding to the choice $\{r_1, \cdots, r_R\} = \{a/q; 0 \leq a < q \leq Q, (a,q) = 1\}$, $U_i = U_{a/q} = qU$, $\delta \geq Q^{-2}$ and $R \leq Q^2$. There may be a shorter direct proof of Lemma 1, but the above approach shows clearly how the function $W(\alpha)$ arises.

5. The Generating Function.

This section is devoted to the proof of Lemma 2. Let a and b be fixed integers and let

$$R(\alpha; c, d) = e\left(-\alpha \frac{b}{a}\right) \sum_{\substack{1 < n \leq x \\ n \equiv ac + b \ (ad)}} \Lambda(n) e\left(\alpha \frac{n}{a}\right),$$

where $\Lambda(n)$ is the von Mangolt function. We want to prove that for any $A > 0$ there is a $B = B(A) > 0$ such that if

$$\log^B x \leq D \leq \frac{x^{1/3}}{\log^B x}, \ \log^B x \leq v \leq \frac{x}{\log^B x}, \ ||\alpha - \frac{u}{v}|| < v^{-2}$$

then

$$\sum_{d \leq D} d \max_c |R(\alpha; c, d)|^2 \ll_{A,a} \frac{x^2}{\log^A x}. \tag{5.1}$$

Lemma 2 follows from (5.1) by partial summation. Note that the bound (5.1) is uniform in b. This is important in the induction step in section 2. The symbol $m \sim M$ will stand for the inequality $M < m \leq M'$ for some $M' \leq 2M$.

For given d we choose c and c_d such that the maximum on the left hand side of (5.1) is attained at c and $c_d \equiv ac+b\ (ad)$. We may suppose that $(c_d, ad) = 1$. By a standard argument based on Vaughan's identity [8] the estimate (5.1) follows if we can prove

$$\sum_{d\sim D'} d \left| \sum_{\substack{mn\le x \\ mn\equiv c_d\ (ad)}} \sum_{m\sim M} a_m b_m e\left(\frac{\alpha mn}{a}\right) \right|^2 \ll \frac{x^2}{\log^A x} \tag{5.2}$$

for any $D' \le D$, for any coefficients $|a_m| \le 1$, $|b_n| \le 1$ and for any M satisfying

$$I: \quad M \le V^2, \text{ if } b_n = 1 \text{ for all } n,$$
$$II: V \le M \le \frac{X}{V}, \text{ otherwise,}$$

where V is a parameter to be chosen later. The left hand side of (5.2) will be denoted by I resp. II in the case I resp. II. We denote by $\bar{m}$ the solution of the congruence $m\bar{m} \equiv 1\ (ad)$ whenever $(m, ad) = 1$.

We have

$$I \le \sum_{d\sim D'} d \left(\sum_{\substack{m\sim M \\ (m,ad)=1}} \left| \sum_{\substack{n\le x/m \\ n\equiv \bar{m}c_d\ (ad)}} e\left(\frac{\alpha mn}{a}\right) \right| \right)^2$$

$$\ll x \sum_{d\sim D'} \sum_{\substack{m\sim M \\ (m,ad)=1}} \left| \sum_{\substack{n\le x/m \\ n\equiv \bar{m}c_d\ (ad)}} e\left(\frac{\alpha mn}{a}\right) \right|. \tag{5.3}$$

Writing $n = \bar{m}c_d + lad$ we can change the summation over n into summation over l where l runs over an interval of length at most x/amd and the summands form a geometric series. We have

$$\left| \sum_n e\left(\frac{\alpha mn}{a}\right) \right| = \left| \sum_l e(\alpha mld) \right| \ll \min\left(\frac{x}{MD'}, \|\alpha md\|^{-1}\right). \tag{5.4}$$

Thee basic tool in the estimates here is the following well-known estimate. If $\|\alpha - \frac{u}{v}\| < v^{-2}$ then

$$\sum_{n\le X} \min(Y, \|\alpha n\|^{-1}) \ll \frac{XY}{v} + (X+v)\log v.$$

A proof can be found in [7] but the result itself has been used already by Weyl and Vinogradov.

Combining this with the Cauchy-Schwarz inequality and known bounds for the mean square of the divisor function $\tau(n)$ we get

$$\sum_{n\leq X} \tau(n)\min(Y, ||\alpha n||^{-1}) \ll \left(\frac{XY}{v^{1/2}} + XY^{1/2} + (XYv)^{1/2}\right)\log^2(Xv). \tag{5.5}$$

Substituting (5.4) into (5.3) and then applying (5.5) we arrive at

$$\begin{aligned} I &\ll x \sum_{d\sim D'}\sum_{m\sim M} \min\left(\frac{x}{MD}, ||\alpha md||^{-1}\right) \\ &\ll x \sum_{n\sim MD'} \tau(n)\min\left(\frac{x}{MD}, ||\alpha n||^{-1}\right) \\ &\ll \left(\frac{x^2}{v^{1/2}} + x^{3/2}M^{1/2}D^{1/2} + x^{3/2}v^{1/2}\right)\log^2 x \ll \frac{x^2}{\log^A x}, \end{aligned}$$

provided that

$$I: \ MD \leq \frac{x}{\log^B x}. \tag{5.6}$$

In other words, (5.2) is proved in case I if (5.6) is satisfied.

Next we turn to (5.2) in case II. The idea is the same but technically more complicated as we have to get rid of the unknown coefficients. We start by subdividing the summation over m in (5.2) into residue classes mod ad. We obtain

$$II \leq \sum_{d\sim D'} d \left(\sum_{(f,ad)=1} \sum_{\substack{m\sim M \\ m\equiv f\ (ad)}} \left| \sum_{\substack{n\leq x/m \\ n\equiv \bar{f}c_d\ (ad)}} b_n e\left(\frac{\alpha mn}{a}\right)\right|\right)^2 .$$

Writing $n = \bar{f}c_d + lad$ and $b_n = b_{l,f,d}$ we can change the summation over n into a summation over l. The bound $n \leq x/m$ may be replaced by the bound $l \leq x/mad$ at the cost of an error term $O(1)$ in the inner sum. We arrive at

$$II \ll \sum_{d\sim D'} d \left(\sum_{(f,ad)=1} \sum_{\substack{m\sim M \\ m\equiv f\ (ad)}} \left| \sum_{l\le \frac{x}{mad}} b_{l,f,d} e(\alpha mld) \right| \right)^2 + M^2D^2$$

$$\ll \sum_{d\sim D'} Md \sum_{(f,ad)=1} \sum_{\substack{m\sim M \\ m\equiv f\ (ad)}} \left| \sum_{l\le x/mad} b_{l,f,d} e(\alpha mld) \right|^2 + M^2D^2$$

$$\ll \sum_{d\sim D'} Md \sum_{(f,ad)=1} \sum_{l,l'\le x/aMD'} \left| \sum_{m} e(\alpha m(l-l')d) \right| + M^2D^2,$$

where the variable m satisfies the conditions $m \sim M$, $m \equiv f\ (ad)$, $m \le x/adl$ and $m \le x/adl'$. Writing $m = f + kad$ we change the summation over m into a summation over k. At the cost of an error of order $O(1)$ we may write the conditions on k as $k \sim M/ad$, $k \le x/a^2d^2l$ and $k \le x/a^2d^2l'$. We have

$$II \ll \sum_{d\sim D'} Md \sum_{(f,ad)=1} \sum_{l\ne l'\le x/aMD'} \left| \sum_{k} e(\alpha k(l-l')ad^2) \right| + M^2D^2 + \frac{x^2D}{M} + xMD. \tag{5.7}$$

The last three terms have the desired size if we suppose that

$$II: \ D\log^B x \le M \le \frac{x}{D\log^B x}. \tag{5.8}$$

The summands are independent of f. Thus we can carry out the summation over f trivially. Depending on the size of D' we estimate the sum in (5.7) in two different ways. First we consider the inner sum as the sum of a geometric series. Setting $l - l' = h$ we obtain, by a standard argument,

$$II \ll \sum_{d\sim D'} d \sum_{h\le x/aMD'} \min\left(\frac{M}{d}, \|\alpha had^2\|^{-1}\right) + \frac{x^2}{\log^A x}$$

$$\ll xD' \sum_{n\le xD'/M} \tau(n) \min\left(\frac{M}{D'}, \|\alpha n\|^{-1}\right) + \frac{x^2}{\log^A x}$$

$$\le xD' \left(\frac{x}{v^{1/2}} + \frac{xD'^{1/2}}{M^{1/2}} + x^{1/2}v^{1/2} \right) \log^2 x + \frac{x^2}{\log^A x} \ll \frac{x^2}{\log^A x}$$

provided (5.8) holds and

$$D' \leq \log^{B/3} x. \tag{5.9}$$

If D' is large then we carry out first the summation over d. We start from (5.7) applying the Cauchy-Schwarz inequality. We suppose that (5.8) holds, but (5.9) is not satisfied, and we use these relations to simplify our expression in each step. We get

$$\begin{aligned}
II &\ll MD'^2 \sum_{d\sim D'} \sum_{l\neq l'\leq x/aMD'} \left|\sum_k e(\alpha k(l-l')ad^2)\right| + \frac{x^2}{\log^A x} \\
&\ll xD'^{3/2} \left(\sum_{d\sim D'} \sum_{l\neq l'\leq x/aMD'} \left|\sum_k e(\alpha k(l-l')ad^2)\right|^2\right)^{1/2} + \frac{x^2}{\log^A x} \\
&\ll xD'^{3/2} \left(\sum_{l\neq l'\leq x/aMD'} \sum_{k\neq k'\leq M/aD'} \left|\sum_d e(\alpha(k-k')(l-l')ad^2)\right|\right)^{1/2} \\
&\qquad + \frac{x^2}{\log^A x},
\end{aligned}$$

where d runs over an interval defined by the inequalities $d \sim D'$, $kad \sim M$, $k'ad \sim M$, $lka^2d^2 \leq x$, $lk'a^2d^2 \leq x$, $l'ka^2d^2 \leq x$, $l'k'a^2d^2 \leq x$. In any case the inner sum is at most as large as the maximum of the same type of sums with $d \sim D'$ replaced by $d \sim D''$ and the maximum is taken over all D'' satisfying $D'' \sim D'$. Writing $l - l' = h$, $k - k' = g$ and collecting the terms $n = gha$ we arrive at

$$\begin{aligned}
II &\ll x^{3/2}D'^{1/2} \left(\sum_{n\leq x/D'^2} \tau(n) \max_{D''\sim D'} \left|\sum_{d\sim D''} e(\alpha n d^2)\right|\right)^{1/2} + \frac{x^2}{\log^A x} \\
&\ll x^{7/4}\log x \left(\sum_{n\leq x/D'^2} \max_{D''\sim D'} \left|\sum_{d\sim D''} e(\alpha n d^2)\right|^2\right)^{1/4} + \frac{x^2}{\log^A x}.
\end{aligned} \tag{5.10}$$

We use a well-known linearization argument in bounding the inner sum.

$$
\begin{aligned}
\Big|\sum_{d\sim D''} e(\alpha n d^2)\Big|^2 &= \sum_{d,d'\sim D''} e(\alpha n(d^2-d'^2))\\
&= \sum_{d,d'\sim D''} e(\alpha n(d-d')(d+d'))\\
&\ll D'' + \sum_{h<D''}\Big|\sum_{d\sim D''} e(\alpha n h(h+2d))\Big|\\
&\ll D' + \sum_{h<2D'} \min(D', \|\alpha 2nh\|^{-1}).
\end{aligned}
$$

As a last step we substitute this back into (5.10) and then apply (5.5.) We arrive at

$$
\begin{aligned}
II &\ll x^{7/4}\log x \left(\sum_{n\le x/D'^2}\sum_{h<2D'} \min(D', \|\alpha 2nh\|^{-1})\right)^{1/4} + \frac{x^2}{\log^A x}\\
&\ll x^{7/4}\log x \left(\sum_{m\le 4x/D'} \tau(m)\min(D', \|\alpha m\|^{-1})\right)^{1/4} + \frac{x^2}{\log^A x}\\
&\ll x^{7/4}\log^2 x \left(\frac{x}{v^{1/2}} + \frac{x}{D'^{1/2}} + x^{1/2}v^{1/2}\right)^{1/4} + \frac{x^2}{\log^A x} \ll \frac{x^2}{\log^A x}.
\end{aligned}
$$

Comparing condition II and (5.3) a plausible choice for the parameter V is $V = D\log^B x$. This ensures that every arising bilinear form of type II satisfies (5.8). Finally we can easily verify that every bilinear form of type I satisfies (5.6) if $V^2 D \le x/\log^B x$ (that is $D \le x^{1/3}/\log^B x$).

As we have mentioned in section 2, this proof is direct generalization of Vaughan's proof [7]. An approach via the large sieve might possibly improve the exponent of the level of distribution to 1/2 in both Lemma 2 and the Theorem.

6. The "Singular Series".

In this section we will prove Lemma 3. We will use again the notations and definitions introduced in sections 1, 2 and 3. $\mathbf{a}$ and $\mathbf{b}$ are now fixed k-dimensional vectors, $\tilde{\mathbf{a}}$ and $\tilde{\mathbf{b}}$ are their $k-1$-dimensional projections without the last coordinate, c and d are fixed integers and $\rho(p;\mathbf{a},\mathbf{b})$, $\tilde\rho(p;\mathbf{a},\mathbf{b})$, $\sigma(\mathbf{a},\mathbf{b};c,d)$, K, L are as before. $Q>0$ is any real number. We can suppose (3.5), that is

$$(\mathbf{a},\mathbf{b}) = (a_k c + b_k, d) = 1. \tag{6.1}$$

We can also suppose that $\tilde{\rho}(p) < p$ for all primes p, for otherwise Lemma 3 is true with zero on both sides. For any integers q and $\Delta \mid q$ define

$$S(q; \mathbf{a}, \mathbf{b}; c, \Delta) = \sum_f \sum_g \sum_h e\left(\frac{fh - fg}{q}\right),$$

where f, g and h run over a complete system of residues mod q and satisfy the conditions

$$(f, q) = (\tilde{\mathbf{a}}g + \tilde{\mathbf{b}}, q) = (a_k h + b_k, q) = 1, \ h \equiv c\ (\Delta). \tag{6.2}$$

Note that under the conditions (6.1) and (6.2) $\sigma(\tilde{\mathbf{a}}, \tilde{\mathbf{b}}; g, q)$ is independent of g. We are going to prove Lemma 3, i. e.,

$$\prod_p \left(1 - \frac{1}{p}\right)^{-k+1} \left(1 - \frac{\tilde{\rho}(p)}{p}\right) \sum_{q \le Q} \frac{a_k}{q\varphi(a_k[q, d])} \prod_{p \mid q} \frac{p}{p - \tilde{\rho}(p)} S(q; \mathbf{a}, \mathbf{b}; c, (q, d))$$
$$= \sigma(\mathbf{a}, \mathbf{b}; c, d) + O\left(\frac{((d, k))(k+1)^{\nu(Ld)} \log^{2k}(Q|\mathbf{a}||\mathbf{b}|)}{Q\varphi(d)}\right) \tag{6.3}$$

uniformly in c, d and $\mathbf{b}$. Recall that (n) denotes the squarefree part of n.

First we show that $S(q; \mathbf{a}, \mathbf{b}; c, \Delta)$ is multiplicative in q.

Lemma 7. *If $\Delta \mid q = q'q''$, where $(q', q'') = 1$ then $\Delta = \Delta'\Delta''$, where $\Delta' \mid q'$ and $\Delta'' \mid q''$ and $S(q; \mathbf{a}, \mathbf{b}; c, \Delta) = S(q'; \mathbf{a}, \mathbf{b}; c, \Delta')S(q''; \mathbf{a}, \mathbf{b}; c, \Delta'')$.*

Proof: Let $\bar{q}'$ resp. $\bar{q}''$ be the inverse of q' resp. q'' modulo q'' resp. q', that is $q'\bar{q}' \equiv 1\ (q'')$ and $q''\bar{q}'' \equiv\ (q')$. We write $f = f'q'' + f''q'$, $g = g'q''\bar{q}'' + g''q'\bar{q}'$ and $h = h'q''\bar{q}'' + h''q'\bar{q}'$, where g', h' resp. g'', h'' are the residues of g, h modulo q' resp. q'' and f' resp. f'' are the residues of $f\bar{q}''$ modulo q' resp. of $f\bar{q}'$ modulo q''. f, g, and h run over a complete system of residues modulo q if and only if f', g' and h' run over a complete system of residues modulo q' and f'', g'' and h'' run over a complete system of residues modulo q''. Moreover

$$(f, q) = 1 \iff \begin{cases} (f', q') = 1, \\ (f'', q'') = 1, \end{cases}$$
$$(\tilde{\mathbf{a}}g + \tilde{\mathbf{b}}, q) = 1 \iff \begin{cases} (\tilde{\mathbf{a}}g'q''\bar{q}'' + \tilde{\mathbf{b}}, q') = (\tilde{\mathbf{a}}g' + \tilde{\mathbf{b}}, q') = 1, \\ (\tilde{\mathbf{a}}g''q'\bar{q}' + \tilde{\mathbf{b}}, q'') = (\tilde{\mathbf{a}}g'' + \tilde{\mathbf{b}}, q'') = 1, \end{cases}$$
$$(a_k h + b_k, q) = 1 \iff \begin{cases} (a_k h'q''\bar{q}'' + b_k, q') = (a_k h' + b_k, q') = 1, \\ (a_k h''q'\bar{q}' + b_k, q'') = (a_k h'' + b_k, q'') = 1, \end{cases}$$
$$h \equiv c\ (\Delta) \iff \begin{cases} h'q''\bar{q}'' \equiv h' \equiv c\ (\Delta'), \\ h''q'\bar{q}' \equiv h'' \equiv c\ (\Delta''). \end{cases}$$

Finally we have

$$\begin{aligned}
&\sum_f \sum_g \sum_h e\left(\frac{f}{q}(h-g)\right) \\
&= \sum_{f'} \sum_{f''} \sum_{g'} \sum_{g''} \sum_{h'} \sum_{h''} e\left(\frac{f'q''+f''q'}{q'q''}(h'q''\bar{q}''+h''q'\bar{q}'-g'q''\bar{q}''-g''q'\bar{q}')\right) \\
&= \sum_{f'} \sum_{g'} \sum_{h'} \sum_{f''} \sum_{g''} \sum_{h''} e\left(\frac{f'}{q'}(h'q''\bar{q}''-g'q''\bar{q}'')+\frac{f''}{q''}(h''q'\bar{q}'-g'q''\bar{q}'')\right) \\
&= \sum_{f'} \sum_{g'} \sum_{h'} e\left(\frac{f'}{q'}(h'-g')\right) \sum_{f''} \sum_{g''} \sum_{h''} e\left(\frac{f''}{q''}(h''-g'')\right).
\end{aligned}$$

Next we are going to calculate $S(q;\mathbf{a},\mathbf{b};c,\Delta)$ for prime powers.

Lemma 8. *Let $\alpha \geq \beta \geq 0$ be any integers and suppose that $(\mathbf{a},\mathbf{b})=1$. We have*

$$S(p^\alpha;\mathbf{a},\mathbf{b};c,p^\beta) = \begin{cases} p-\tilde{\rho}(p), & \text{if } \alpha=1,\ \beta=0,\ p|L,\ p\nmid a_k; \\ \tilde{\rho}(p)-p, & \text{if } \alpha=\beta=1,\ p|K,\ p\nmid a_kc+b_k; \\ -\tilde{\rho}(p), & \text{if } \alpha=1,\ \beta=0,\ p\nmid a_kL; \\ \tilde{\rho}(p), & \text{if } \alpha=\beta=1,\ p\nmid K; \\ 0 & \text{otherwise.} \end{cases}$$

Proof: First note that the last case in the statement of the lemma means either $\alpha=1$, $\beta=0$, $p|a_k$ or $\alpha=\beta=1$, $p|a_kc+b_k$ or $\alpha\geq 2$. Throughout the proof we omit the argument from $S(q;\mathbf{a},\mathbf{b};c,\Delta)$ that is we use the notation

$$S = \sum_{\substack{f\leq p^\alpha \\ p\nmid f}} \sum_{\substack{g\leq p^\alpha \\ p\nmid \tilde{\mathbf{a}}g+\tilde{\mathbf{b}}}} \sum_{\substack{h\leq p^\alpha \\ p\nmid \mathbf{a}_k h+b_k \\ h\equiv c\ (p^\beta)}} e\left(\frac{fh-fg}{p^\alpha}\right).$$

We consider five different cases. We will make frequent use of the well-known identities

$$\sum_{a\leq q} e\left(\frac{a}{q}\right)=0, \quad \sum_{\substack{a\leq q \\ (a,q)=1}} e\left(\frac{a}{q}\right)=\mu(q).$$

Case 1. $\beta=0$, $p|a_k$. In this case the sum over h extends over all $h\leq p^\alpha$ and is therefore zero. Hence $S=0$.

Case 2. $\beta = 0$, $p \nmid a_k$. Let $\bar{a}_k$ be defined by $a_k\bar{a}_k \equiv 1\ (p^\alpha)$. We have

$$\sum_{\substack{h\le p^\alpha \\ p\nmid a_k h+b_k}} e\left(\frac{fh}{p^\alpha}\right) = \sum_{\substack{h'\le p^\alpha \\ p\nmid h'+b_k}} e\left(\frac{f\bar{a}_k h'}{p^\alpha}\right)$$

$$= e\left(\frac{-f\bar{a}_k b_k}{p^\alpha}\right) \sum_{\substack{h'\le p^\alpha \\ p\nmid h'+b_k}} e\left(\frac{f\bar{a}_k}{p^\alpha}(h'+b_k)\right)$$

$$= e\left(\frac{-f\bar{a}_k b_k}{p^\alpha}\right) \sum_{\substack{h''\le p^\alpha \\ p\nmid h''}} e\left(\frac{f\bar{a}_k h''}{p^\alpha}\right) = e\left(\frac{-f\bar{a}_k b_k}{p^\alpha}\right)\mu(p^\alpha).$$

Thus $S = 0$ unless $\alpha = 1$ in which case

$$S = -\sum_{\substack{f\le p \\ p\nmid f}} \sum_{\substack{g\le p \\ p\nmid \tilde{\mathbf{a}}g+\tilde{\mathbf{b}}}} e\left(\frac{-f\bar{a}_k b_k - fg}{p}\right).$$

If we drop the condition $p\nmid \tilde{\mathbf{a}}g+\tilde{\mathbf{b}}$, then the double sum becomes zero. Thus, it suffices to consider those $g \le p$ that satisfy $\prod(\tilde{\mathbf{a}}g+\tilde{\mathbf{b}}) \equiv 0 \mod p$. There are exactly $\tilde{\rho}(p)$ such g. For any fixed g satisfying this condition the sum over f is either $p-1$ or -1 depending on whether $p|\bar{a}_k b_k + g$ or $p\nmid \bar{a}_k b_k + g$. If $p\nmid L$ then there is no g with $p|\bar{a}_k b_k + g$, while if $p|L$ then there is exactly one such g. Thus we arrive at

$$S = \sum_{\substack{g\le p \\ p|\prod(\tilde{\mathbf{a}}g+\tilde{\mathbf{b}})}} \sum_{\substack{f\le p \\ p\nmid f}} e\left(-(\bar{a}_k b_k + g)\frac{f}{g}\right)$$

$$= \sum_{\substack{g\le p \\ p|\prod(\tilde{\mathbf{a}}g+\mathbf{b})}} \begin{cases} -1, & \text{if } p\nmid a_k g + b_k \\ p-1, & \text{if } p|a_k g + b_k \end{cases}$$

$$= \begin{cases} -\tilde{\rho}(p), & \text{if } p\nmid L \\ p-\tilde{\rho}(p), & \text{if } p|L. \end{cases}$$

Case 3. $\alpha \ge \beta \ge 1$, $p|a_k c + b_k$. The sum over h is empty and thus $S = 0$.

Case 4. $\alpha > \beta \ge 1, p\nmid a_k c + b_k$. We have $p\nmid a_k h + b_k$ so that $S = 0$ again since

$$\sum_{\substack{h\le p^\alpha \\ h\equiv c\ (p^\alpha)}} e\left(\frac{fh}{p^\alpha}\right) = \sum_{h'\le p^{\alpha-\beta}} e\left(\frac{fc}{p^\alpha}\right) e\left(\frac{fh'}{p^{\alpha-\beta}}\right) = 0.$$

Case 5. $\alpha = \beta \geq 1$, $p \nmid a_k c + b_k$. In this case the sum over h extends over the value $h = c$. The argument then follows closely the argument of Case 2. If we drop the condition $p \nmid \tilde{\mathbf{a}}g + \tilde{\mathbf{b}}$, then the double sum becomes zero. Thus it suffices to consider those $g \leq p^\alpha$ that satisfy $\prod(\tilde{\mathbf{a}}g + \tilde{\mathbf{b}}) \equiv 0 \bmod p$. They are of the form $g = g' + g''p$, where $g' \leq p$, and satisfy $\prod(\tilde{\mathbf{a}}g' + \tilde{\mathbf{b}}) \equiv 0 \bmod p$ (there are exactly $\tilde{\rho}(p)$ such g'), and $g'' \leq p^{\alpha-1}$. The summation over g'' is zero unless $\alpha = 1$. We have

$$S = \sum_{\substack{f \leq p^\alpha \\ p \nmid f}} \sum_{\substack{g \leq p^\alpha \\ p \nmid \tilde{\mathbf{a}}g + \tilde{\mathbf{b}}}} e\left(\frac{fc - fg}{p^\alpha}\right) = - \sum_{\substack{g \leq p^\alpha \\ p \mid \prod(\tilde{\mathbf{a}}g + \tilde{\mathbf{b}})}} \sum_{\substack{f \leq p^\alpha \\ p \nmid f}} e\left(\frac{fc - fg}{p^\alpha}\right)$$

$$= - \sum_{\substack{g' \leq p \\ p \mid \prod(\tilde{\mathbf{a}}g' + \tilde{\mathbf{b}})}} \sum_{f \leq p^\alpha \; p \nmid f} \sum_{g'' \leq p^{\alpha-1}} e\left(\frac{fc - fg' - fg''p}{p^\alpha}\right)$$

$$= \begin{cases} 0, & \text{if } \alpha > 2 \\ - \displaystyle\sum_{\substack{g' \leq p \\ p \mid \prod(\tilde{\mathbf{a}}g' + \tilde{\mathbf{b}})}} \sum_{\substack{f \leq p \\ p \nmid f}} e\left(\frac{f(c - g')}{p}\right), & \text{if } \alpha = 1. \end{cases}$$

For any fixed g' the summation over f is either $p - 1$ or -1 depending on whether $c \equiv g'$ (p) or not. If $p \nmid K$ then there is no such g' while if $p | K$ then there is exactly one. We arrive at

$$S = - \sum_{\substack{g' \leq p \\ p \mid \prod(\tilde{\mathbf{a}}g' + \tilde{\mathbf{b}})}} \begin{cases} -1, & \text{if } c \not\equiv g' \ (p) \\ p - 1, & \text{if } c \equiv g' \ (p) \end{cases} = \begin{cases} \tilde{\rho}(p), & \text{if } p \nmid K \\ \tilde{\rho}(p) - p, & \text{if } p | K. \end{cases}$$

These five cases cover Lemma 8.

We are now in a position to prove (6.3). First of all we note that

$$\tilde{\rho}(p) = \begin{cases} \rho(p), & \text{if } p | a_k L, \\ \rho(p) - 1, & \text{if } p \nmid a_k L. \end{cases} \tag{6.4}$$

By Lemma 7 and Lemma 8 we can express $S(q; \mathbf{a}, \mathbf{b}; c, (q, d))$ exactly. Substituting the expression for $S(q; \mathbf{a}, \mathbf{b}; c, (q, d))$ into the left-hand side of (6.3) (abbreviated by Σ) we get

$$\Sigma = \frac{1}{d}\prod_{p|a_k d}\left(1-\frac{1}{p}\right)^{-1}\prod_{p}\left(1-\frac{1}{p}\right)^{-k+1}\left(1-\frac{\tilde{\rho}(p)}{p}\right)\sum_{q\le Q}\frac{(q,d)}{q^2}\prod_{\substack{p|q\\ p\nmid a_k d}}\frac{p}{p-1}$$

$$\times\prod_{p|q}\frac{p}{p-\tilde{\rho}(p)}\prod_{\substack{p|q\\ p\nmid d\\ p\nmid a_k\\ p|L}}(p-\tilde{\rho}(p))\prod_{\substack{p|q\\ p\nmid d\\ p\nmid a_k\\ p\nmid L}}(-\tilde{\rho}(p))\prod_{\substack{p|q\\ p|d\\ p\nmid a_k c+b_k\\ p|K}}(\tilde{\rho}(p)-p)\prod_{\substack{p|q\\ p|d\\ p\nmid K}}\tilde{\rho}(p)$$

$$=\frac{1}{d}\prod_{p|a_k d}\left(1-\frac{1}{p}\right)^{-1}\prod_{p}\left(1-\frac{1}{p}\right)^{-k+1}\left(1-\frac{\tilde{\rho}(p)}{p}\right)\sum_{q\le Q}\psi(q),$$

where

$$\psi(q)=\prod_{\substack{p|q\\ p|d\\ p|K}}(-1)\prod_{\substack{p|q\\ p|d\\ p\nmid K}}\frac{\tilde{\rho}(p)}{p-\tilde{\rho}(p)}\prod_{\substack{p|q\\ p\nmid a_k d\\ p|L}}\frac{1}{p-1}\prod_{\substack{p|q\\ p\nmid a_k d\\ p\nmid L}}\frac{-\tilde{\rho}(p)}{(p-1)(p-\tilde{\rho}(p))}$$

and in the summations over q we may assume

$$q \text{ is squarefree}, \ (q,a_k)|d, \ (q,d,a_k c+b_k)=1. \tag{6.5}$$

for otherwise $S(q;\mathbf{a},\mathbf{b};c,(q,d))=0$. The last condition in (6.5) is, in fact, guarranteed by (6.1).

We will show that the infinite sum $\sum\psi(q)$ is absolutely convergent. To this end we are looking for an upper bound for the tail of $\sum|\psi(q)|$ that is uniform in c, d and $\mathbf{b}$. Since $\tilde{\rho}(p)\le k-1$ we have the trivial bounds

$$\frac{\tilde{\rho}(p)}{(p-1)(p-\tilde{\rho}(p))}\le\frac{k}{p^2} \qquad (\text{for } p\ge k^2),$$

$$\psi(q)\ll((d,K))(q,Ld)\frac{k^{\nu(q)}}{q^2}$$

with an implied constant which depends at most on k. Thus

$$\sum_{q>Q}|\psi(q)|\ll\sum_{q>Q}((d,K))(q,Ld)\frac{k^{\nu(q)}}{q^2}\ll((d,k))\sum_{t|Ld}t\sum_{h>Q/t}\frac{k^{\nu(th)}}{t^2h^2}$$

$$\ll((d,K))\sum_{T|Ld}k^{\nu(t)}\frac{\log^k A}{Q}\ll((d,K))(k+1)^{\nu(Ld)}\frac{\log^k Q}{Q}. \tag{6.7}$$

Here we used the fact that the summations are only over squarefree numbers. If $p \nmid a_i b_j - a_j b_i$ for all possible pairs $i \neq j$ then $\tilde{\rho}(p) = k - 1$. This is certainly the case for $p > |\mathbf{a}||\mathbf{b}|$. By Mertens' prime number theorem we then have

$$\prod_p \left(1-\frac{1}{p}\right)^{-k+1}\left(1-\frac{\tilde{\rho}(p)}{p}\right)$$
$$\leq \prod_{p \leq |\mathbf{a}||\mathbf{b}|}\left(1-\frac{1}{p}\right)^{-k}\prod_{p>|\mathbf{a}||\mathbf{b}|}\left(1-\frac{1}{p}\right)^{-k+1}\left(1-\frac{k-1}{p}\right) \ll \log^k |\mathbf{a}||\mathbf{b}|. \tag{6.8}$$

Substituting (6.8) and (6.7) into (6.6) we arrive at

$$\Sigma = \frac{1}{d}\prod_{p|a_k d}\left(1-\frac{1}{p}\right)^{-1}\prod_p\left(1-\frac{1}{p}\right)^{-k+1}\left(1-\frac{\tilde{\rho}(p)}{p}\right)\sum_q \psi(q)$$
$$+O\left(\frac{((d,K))(k+1)^{\nu(Ld)}\log^{2k}(Q|\mathbf{a}||\mathbf{b}|)}{Q\varphi(d)}\right)$$
$$=\frac{1}{d}\prod_{p|a_k d}\left(1-\frac{1}{p}\right)^{-1}\prod_p\left(1-\frac{1}{p}\right)^{-k+1}\left(1-\frac{\tilde{\rho}(p)}{p}\right)\prod_p(1+\psi(p))$$
$$+O\left(\frac{((d,K))(k+1)^{\nu(Ld)}\log^{2k}(Q|\mathbf{a}||\mathbf{b}|)}{Q\varphi(d)}\right)$$
$$=\frac{1}{d}\prod_{p|a_k d}\left(1-\frac{1}{p}\right)^{-1}\prod_p\left(1-\frac{1}{p}\right)^{-k+1}\left(1-\frac{\tilde{\rho}(p)}{p}\right)\prod_{\substack{p|d\\ p\nmid K}}\frac{p}{p-\tilde{\rho}(p)}\prod_{\substack{p\nmid a_k d\\ p|L}}\frac{p}{p-1}$$
$$\times\prod_{\substack{p\nmid a_k d\\ p\nmid L}}\left(1-\frac{\tilde{\rho}(p)}{(p-1)(p-\tilde{\rho}(p))}\right)+O\left(\frac{((d,K))(k+1)^{\nu(Ld)}\log^{2k}(Q|\mathbf{a}||\mathbf{b}|)}{Q\varphi(d)}\right)$$
$$=\frac{1}{d}\prod_{p|d}\left(1-\frac{1}{p}\right)^{-1}\frac{p}{p-\tilde{\rho}(p)}\prod_{\substack{p|a_k\\ p\nmid d}}\left(1-\frac{1}{p}\right)^{-1}\prod_{\substack{p\nmid a_k L\\ p|L}}\left(1-\frac{1}{p}\right)^{-1}$$
$$\times\prod_{\substack{p\nmid a_k d\\ p\nmid L}}\left(1-\frac{\tilde{\rho}(p)}{(p-1)(p-\tilde{\rho}(p))}\right)\prod_p\left(1-\frac{1}{p}\right)^{-k+1}\left(\frac{p-\tilde{\rho}(p)}{p}\right)$$
$$+O\left(\frac{((d,K))(k+1)^{\nu(Ld)}\log^{2k}(Q|\mathbf{a}||\mathbf{b}|)}{Q\varphi(d)}\right)$$

if $(K,d)=1$. Otherwise the main term is zero. The main term is also zero if there is a prime p such that $p\nmid a_k dL$ and $\tilde{\rho}(p)=p-1$. By (6.4) we have

$\rho(p) = p$ in this case, and $\sigma(\mathbf{a}, \mathbf{b}; c, d)$ is also zero. Every prime p appears exactly twice on the right hand side, once in one of the first four products and a second time in the last product. For those primes that belong to the first product we can change $\tilde{\rho}(p)$ into $\rho(p)$ in both instances. For the primes that belong to the second or third product we have $\tilde{\rho}(p) = \rho(p)$, and we can change them. Finally if ρ belongs to the fourth product then $\tilde{\rho}(p) = \rho(p) - 1$ and we get that

$$\left(1 - \frac{1}{p}\right)\left(\frac{p - \tilde{\rho}(p)}{p}\right)\left(1 - \frac{\tilde{\rho}(p)}{(p-1)(p-\tilde{\rho}(p))}\right) = \left(1 - \frac{\rho(p)}{p}\right).$$

REFERENCES

[1] A. Balog, A. Perelli, Exponential sums over primes in an arithmetic progression, Proc. Amer. Math. Soc. **93** (1985), 578–581 MR 86b: 11053.

[2] P.T. Bateman, R. A. Horn, A heuristic asymptotic formula concerning the distribution of prime numbers, Math. Comp. **16** (1962), 363–367 MR 26: 6139.

[3] N.G. Chudakov, On Goldbach-Vinogradov's theorem, Ann. Math. **48** (1947), 515–545.

[2] D. R. Heath-Brown, Three primes and an almost-prime in arithmetic progression, J. London Math. Soc. **23** (1981), 396–414.

[5] A.F. Lavrik, The number of k-twin primes lying on an interval of a given length, Dokl. Acad. Nauk. SSSR **136** (1961), 281–283 (Russian); translated as Soviet Math. Dokl. **2** (1961), 52-55.

[6] A.F. Lavrik, On the theory of distribution of primes based on I.M.Vinogradov's method of trigonometric sums, Trudy Mat. Inst. Steklov **64** (1961), 90–125 (Russian).

[7] H. Maier, C. Pomerance, Unusually large gaps between consecutive primes.

[8] R.C. Vaughan, An elementary method in prime number theory, Acta Arithmetica **37** (1980), 111–115.

[9] I.M. Vinogradov, Represnetation of an odd number as a sum of three primes, Dokl. Acad. Nauk. SSSR **15** (1937), 169–172.

Antal Balog
Mathematical Institute of the
Hungarian Academy of Sciences
Budapest, Reáltanoda u. 13-15
Hungary-1053

On Arithmetic Functions Involving Consecutive Divisors

A. BALOG, P. ERDÖS, AND G. TENENBAUM

Dedicated to Paul Bateman

§ 1. Introduction.

This article is motivated by several questions posed in [3], which we can now at least partially answer.

Let $1 = d_1 < d_2 < \ldots < d_{\tau(n)} = n$ denote the increasing sequence of divisors of a general integer n. A quantitative measure of the growth of the d_i is provided by the arithmetic function

$$H(n) := \sum_{1 \le i < \tau(n)} (d_{i+1} - d_i)^{-1}.$$

It is established in [3] that

$$H(n) \ll \tau(n)(\log \tau(n))^{-\frac{1}{3}+\varepsilon} \qquad (n \ge 2) \tag{1.1}$$

holds for any fixed $\varepsilon > 0$ and that

$$\max_{n \le x} H(n) > \exp\left\{(\log x)^{\frac{1}{2}+o(1)}\right\} \qquad (x \to \infty). \tag{1.2}$$

Our first result is an upper bound for the left hand side of (1.2) in terms of the quantity

$$D(x) := \max_{n \le x} \tau(n) = 2^{(1+o(1))\frac{\log x}{\log_2 x}} \qquad (x \to \infty).$$

(Here and in the sequel we let $\log_k$ denote the k-fold iterated logarithm.)

Theorem 1. *Set* $c = \frac{5}{3} - \frac{\log 3}{\log 2} = 0.08170$. *Then we have*

$$\max_{n \le x} H(n) \le D(x)^{1-c+o(1)} \qquad (x \to \infty). \tag{1.3}$$

An analogous bound, with an unspecified constant c, has been independently obtained by Erdös and Sárközy, with a different method (unpublished). It emerges from (1.3) that (1.1) can be significantly improved in the case when $\tau(n)$ is "large". On the other hand, it follows from Theorem 9 of [3] that

$$H(n) \ll \tau(n)^{1-c} \log(1 + \omega(n)) \tag{1.4}$$

holds, with the same value of c, whenever n is squarefree. (We let $\omega(n)$ denote the number of distinct prime factors of n.) This leads to the conjecture that a bound of the type

$$H(n) \ll \tau(n)^{1-\delta} \tag{1.5}$$

with an absolute $\delta > 0$ could hold unconditionally. We haven't been able up to now to prove or disprove this hypothesis.

The lower bound (1.2) is probably not optimal, but it seems difficult to make a reasonable guess concerning the maximal order of $H(n)$. One trivially has

$$H(n) \ge \kappa(n) := \sum_{d(d+1) \mid n} 1 \tag{1.6}$$

and the function $\kappa(n)$ raises an interesting open problem. We certainly believe that

$$\kappa(n) \ll_\varepsilon \tau(n)^\varepsilon$$

holds for any $\varepsilon > 0$, but no upper estimate is actually available other than those which follow, via (1.6), from the results on $H(n)$. Erdös and Hall established in [2] the asymptotic inequality

$$\max_{n \le x} \kappa(n) > (\log x)^{\sqrt{e} + o(1)} \qquad (x \to \infty).$$

We can strengthen this estimate in the following way.

Theorem 2. *We have*

$$\max_{n \le x} \kappa(n) > (\log x)^{\frac{\log_3 x}{9 \log_4 x}} \qquad (x \to \infty). \tag{1.7}$$

This confirms a conjecture of Erdös. The analogous problem for the counting function of those divisors of the form $d(d+1)\dots(d+t-1)$ with fixed $t > 2$ seems much more difficult and we do not know in this case whether the maximal order exceeds an arbitrary power of $\log n$. Erdös and Hall prove in [2] that a power α_t is acceptable provided $\alpha_t < e^{1/t}$.

Our proof of Theorem 2 rests on an effective version of a result of Hildebrand [8] which is of independent interest. Let $P^+(n)$ (resp. $P^-(n)$) denote the largest (resp. the smallest) prime factor of n, with the convention $P^+(1) = 1$, $P^-(1) = +\infty$. Moreover, let us systematically put

$$u := \frac{\log x}{\log y} \qquad (x \geq y \geq 2).$$

The key to Theorem 2 is the following

Theorem 3. *The estimate*

$$\sum_{\substack{n \leq x \\ P^+(n(n+1)) \leq y}} 1 \gg x u^{-u^{7u}} \tag{1.8}$$

holds uniformly in the range

$$x \geq 3, \quad \max\{2, x^{\frac{8 \log_3 x}{\log_2 x}}\} \leq y \leq x. \tag{1.9}$$

Let $\rho(u)$ denote Dickman's function. It is known [9,10] that one has

$$\Psi(x,y) := \sum_{\substack{n \leq x \\ P^+(n) \leq y}} 1 \sim x\rho(u) \tag{1.10}$$

as x, y tend to infinity in the range

$$\exp\left\{(\log_2 x)^{\frac{5}{3}+\varepsilon}\right\} \leq y \leq x,$$

and the Riemann Hypothesis implies the persistence of (1.10) in any region of the type

$$(\log x)^{\xi(x)} \leq y \leq x \tag{1.11}$$

where $\xi(x) \to \infty$, see [7,13]. It is hence natural to conjecture that the left hand side of (1.8) is asymptotically $(1 + o(1))x\rho(u)^2$ when $x, y \to \infty$ in the range (1.11). Such a result would provide a strong measure of the multiplicative independence of n and $n + 1$, but seems at present very difficult, if not out of reach, even in a more modest region like $x^\varepsilon \leq y \leq x$.

In order to prove Theorem 3, we establish an elementary lower bound for the quantity

$$\Psi(x, y; a, q) := \text{card}\{n \leq x : P^+(n) \leq y, \ n \equiv a (\text{mod}\, q)\} \tag{1.12}$$

under the hypotheses

$$(\log x)^3 \le y \le x, \quad 1 \le q \le y^{1-\varepsilon},\ (a,q)=1.$$

This is the content of Lemma 3.2 below. In this context it is also natural to conjecture that

$$\Psi(x,y;a,q) \sim \frac{x}{q}\rho(u) \tag{1.13}$$

holds uniformly as $x, y \to \infty$ in the range (1.11) and $q = o(y)$. Fouvry and Tenenbaum have shown in [4] that (1.13) actually holds when

$$\exp\{c_0(\log_2 x)^2\} \le y \le x,\ 1 \le q \le e^{c_1\sqrt{\log y}} \tag{1.14}$$

where c_0, c_1 are absolute constants.

The bounds (1.2) and (1.3) summarize our knowledge on the maximal order of $H(n)$. The average behaviour of this function is given by the formula

$$\sum_{n\le x} H(n) = Bx + O\Big(x\frac{(\log_2 x)^3}{\log x}\Big), \tag{1.15}$$

proved in [3], sharpening an estimate of Ivič and De Koninck [12]. As for the normal behaviour, it is established in [3] that $H(n)$ has a distribution function. We are now able to provide some extra information.

Theorem 4. *The arithmetic function $H(n)$ has a distribution function which is everywhere continuous on the real line.*

We derive this result in Sect.4 from a theorem of Behrend concerning primitive sequences and an inequality proved in [3] which is essentially equivalent to (1.15).

§ 2. Proof of Theorem 1.

Let $\omega_1(n)$ denote the number of prime factors p of n such that $p^2 \nmid n$. From [3] (Th. 9) we have

$$H(n) \ll \tau(n) B_1^{\omega_1(n)} \log_2 n \qquad (n \ge 3) \tag{2.1}$$

with $B_1 = 3.2^{-\frac{5}{3}} = 0.94494$.

Lemma 2.1. *Set $\lambda := \frac{\log 3}{\log 4}$. We have*

$$\tau(n) \le D(n)^{\lambda + o(1)} 2^{(1-\lambda)\omega_1(n)} \qquad (n \to \infty). \tag{2.2}$$

Proof. Consider the canonical decomposition $n = ab$ where $a = \prod_{p\|n} p$. We have

$$\tau(n) = \tau(a)\tau(b) = 2^{\omega_1(n)}\tau(b). \tag{2.3}$$

Let t be a parameter chosen freely in the range $2 \le t \le b$. Then

$$\tau(b) = \prod_{p^\nu\|b}(1+\nu) \le \prod_{\substack{p^\nu\|b\\ p\le t}} \left(1+\frac{\log b}{\log 2}\right) \prod_{\substack{p^\nu\|b\\ p>t}} 2^{\lambda\nu}$$

where we have used the fact that $p^\nu\|b$ implies $\nu \ge 2$, whence $1+\nu \le 2^{\lambda\nu}$. If $b > 1$, it follows that

$$\tau(b) \le \left(1+\frac{\log b}{\log 2}\right)^{\pi(t)} 2^{\lambda\frac{\log b}{\log t}} \le D(b)^{\lambda+o(1)}$$

for the choice $t = \log b/(\log_2 b)^2$, where the quantity $o(1)$ above is defined for all $b > 1$ and is bounded for bounded b. Taking into account the easy estimate

$$D(a)D(b) \le D(ab)^{1+o(1)} \qquad (ab \to \infty)$$

we infer that

$$\tau(b) \le \left(\frac{D(n)}{D(a)}\right)^{\lambda+o(1)} \le (D(n)2^{-\omega(a)})^{\lambda+o(1)}.$$

Inserting this in (2.3), we get (2.2).

Theorem 1 is an immediate consequence of (2.2) and (2.1), in view of the estimate $\omega_1(n) \le \omega(n) \le (1+o(1))\frac{\log n}{\log_2 n}$.

§ 3. Proof of Theorem 3.

We use two auxiliary results. We say that an ordered set of integers $S = \{m_1 < m_2 < \ldots < m_R\}$ is *special* if we have

$$m_j - m_i = (m_i, m_j) \qquad (1 \le i < j \le R). \tag{3.1}$$

Lemma 3.1 (Heath–Brown, [6]). *There exists an absolute constant α such that, for every integer $R \ge 1$, there is a special set of R elements such that*

$$m_R \le R^{\alpha R^3}. \tag{3.2}$$

Lemma 3.2. *Let $0 < \varepsilon < 1$. With the notation (1.12) the estimate*

$$\Psi(x, y; a, q) \gg_\varepsilon \frac{x}{q} u^{-2u} \tag{3.3}$$

is uniformly valid under the conditions

$$x \geq 2, \ (\log x)^3 \leq y \leq x, \ 1 \leq q \leq y^{1-\epsilon}, (a,q) = 1. \tag{3.4}$$

Proof. Put $a_0 := 1$ if $q = 1$, $a_0 := a - q[a/q]$ otherwise. Then a_0 is counted by $\Psi(x, y; a, q)$, and this is always ≥ 1. We may therefore suppose without loss of generality that $x \geq x_0(\epsilon)$, whence $y \geq y_0(\epsilon)$.

Put $k := [u] - 1$, $\eta := \frac{1}{2}\varepsilon(1 - \frac{1}{6}\varepsilon)$. We obtain a lower bound for $\Psi(x, y; a, q)$ by counting all the integers n not exceeding x which have a representation in the form $n = mh\ell$ with the following conditions

(a) $p \mid m \Rightarrow p \in I_q := \{p : p \nmid q, \ y^{1-1/u} < p \leq y\}, \ \Omega(m) = k;$

(b) $p \mid h \Rightarrow p \in J_q := \{p : p \nmid q, \ y^{\eta} < p \leq y^{\frac{1}{2}\varepsilon}\};$

(c) $xy^{-1} \leq mh \leq xy^{\varepsilon - 1};$

(d) $\ell \equiv a\overline{m}\overline{h} \pmod q$.

Here and in the sequel, the letter p denotes exclusively a prime number. The symbols $\overline{m}, \overline{h}$ refer to the respective inverses of m, h modulo q.

When $1 \leq u \leq 2$, we have $m = 1$. Otherwise, I_q and J_q are disjoint. Thus, in any case $(m, h) = 1$. Furthermore, condition (c) implies

$$y^{1-\epsilon} \leq x/mh \leq y. \tag{3.5}$$

Since $l \leq x/mh$, it follows that the number of prime factors, counted with multiplicity, of (ℓ, mh) is at most $1/\eta$. But they must be chosen among the prime factors of n which belong to $]y^{\eta}, x]$ — and these are not more than $1 + [u/\eta]$ in number. Hence the total number of representations of a given n in the form $mh\ell$ is $\leq \binom{1+[u/\eta]}{[1/\eta]} \ll_{\epsilon} u^{1/\eta}$.

Now, inequality (3.5) shows that for fixed m, h there are at least

$$\left[\frac{x}{mhq}\right] \geq \frac{x}{2mhq}$$

values of ℓ satisfying (d). Hence we can write

$$\Psi(x, y; a, q) \gg_{\varepsilon} u^{-\frac{1}{\eta}} \frac{x}{q} \sum \frac{1}{m} \sum \frac{1}{h} \tag{3.6}$$

where, by convention, the letters m, h denote integers subjected to constraints (a), (b), (c).

For each m, put $T := \frac{2 \log(x/my)}{\varepsilon \log y}$; then

$$T \leq \frac{2}{\varepsilon}\Big\{u - 1 - k\big(1 - \frac{1}{u}\big)\Big\} \leq \frac{4}{\varepsilon},$$

and the length of the interval $\left[T/(1-\frac{1}{6}\varepsilon), T+2\right]$ is at least

$$2 - T\varepsilon/(6-\varepsilon) \geq 2 - 4/(6-\varepsilon) \geq 6/5 > 1.$$

It therefore contains an integer, say s. We restrict h to run through the products of s (not necessarily distinct) primes from J_q. We have in this circumstance

$$\frac{x}{my} \leq y^{\eta T/(1-\frac{1}{6}\varepsilon)} \leq y^{\eta s} \leq h \leq y^{\frac{1}{2}\varepsilon s} \leq y^{\frac{1}{2}\varepsilon(T+2)} \leq \frac{x}{my^{1-\varepsilon}}.$$

This shows that condition (c) is always fulfilled.

As $y \to \infty$, we have

$$\sum_{p\in J_q} \frac{1}{p} \geq \sum_{y^\eta < p \leq y^{\frac{1}{2}\varepsilon}} \frac{1}{p} - \frac{\omega(q)}{y^\eta} \geq \log\left(\frac{1}{1-\frac{1}{6}\varepsilon}\right) + o(1).$$

This sum is hence $\gg_\varepsilon 1$ for $y \geq y_0(\varepsilon)$. Since $s \ll_\varepsilon 1$, it follows that

$$\sum \frac{1}{h} \geq \frac{1}{s!}\Big(\sum_{p\in J_q} \frac{1}{p}\Big)^s \gg_\varepsilon 1. \tag{3.7}$$

It remains to estimate $\sum \frac{1}{m}$. We may plainly suppose that $u \geq 2$. We then have

$$\sum_{p\in I_q} \frac{1}{p} \geq \sum_{y^{1-\frac{1}{u}} < p \leq y} \frac{1}{p} - \frac{\omega(q)}{\sqrt{y}} = L + O\Big(\frac{\log y}{\sqrt{y}}\Big),$$

say. From the prime number theorem

$$L = \log\left(\frac{u}{u-1}\right) + O\Big(e^{-\sqrt{\log y}}\Big).$$

This implies

$$\sum_{p\in I_q} \frac{1}{p} \geq \frac{1}{2u} \tag{3.8}$$

provided $y_0(\varepsilon)$ is sufficiently large and $\log y \geq (\log_2 x)^3$. Moreover,for

$$(\log x)^3 \leq y \leq \exp\left\{(\log_2 x)^3\right\}$$

we have

$$1 + y^{-\frac{1}{3}} \leq y^{\frac{1}{u}} \leq 1 + o(1) \qquad (x \to \infty)$$

and Huxley's theorem [11] on the distribution of primes in short intervals gives the estimate $L \geq (1+o(1))\frac{1}{u}$. Thus (3.8) is again valid. For suitable $y_0(\varepsilon)$ we may therefore write

$$\sum \frac{1}{m} \geq \frac{1}{k!}\Big(\sum_{p\in I_q} \frac{1}{p}\Big)^k \gg \left(\frac{e}{2}\right)^u u^{-2u}.$$

Taking (3.6) and (3.7) into account, we readily obtain the required estimate.

Completion of proof of Theorem 3.

We may suppose x and u sufficiently large. Indeed the left hand side of (1.8) always counts $n = 1$, hence is ≥ 1, and is for fixed x a decreasing function of u.

Put $R := [(2u)^{2u}]$, $M := R^{2\alpha R^3}$. From Lemma 3.1, we can find R integers

$$m_1 < m_2 < \ldots < m_R \leq \sqrt{M}$$

satisfying (3.1). Let us now consider the sets of integers

$$T_i := \left\{t \leq \frac{x}{M} : P^+(m_i t + 1) \leq y\right\} \qquad (1 \leq i \leq R).$$

We have $|T_i| = \Psi((xm_i/M) + 1, y; 1, m_i)$ and can appeal to Lemma 3.2 to obtain a lower bound for this quantity. Indeed, by (1.9) we have for $x \geq x_0$

$$M \leq y, \tag{3.9}$$

whence

$$m_i \leq \sqrt{y}, \ \left(\log \frac{x}{\sqrt{M}}\right)^3 \leq y \leq \frac{x}{M}.$$

In order to check (3.9), it is sufficient to observe that $u \leq \frac{1}{8}\frac{\log_2 x}{\log_3 x}$, hence

$$\begin{aligned}\log M &\leq 2\alpha(2u)^{6u+1}\log(2u) \leq 2\alpha \log_3 x \exp\left\{(7/8)\log_2 x\right\}\\ &\leq (\log x)^{\frac{15}{16}} \leq \log y.\end{aligned}$$

With a suitable absolute constant C_0, we therefore have

$$|T_i| \geq C_0 \frac{x}{M} u^{-2u} \qquad (1 \leq i \leq R). \tag{3.10}$$

Furthermore, if $t \in T_i \cap T_j$ with $1 \leq i < j \leq R$, then

$$P^+(tm_i + 1) \leq y, \ P^+(tm_j + 1) \leq y$$

whence

$$P^+\left(t[m_i,m_j]+\frac{m_j}{(m_i,m_j)}\right)\le y,\ P^+\left(t[m_i,m_j]+\frac{m_i}{(m_i,m_j)}\right)\le y.$$

From (3.1), this last condition may be rewritten as $P^+(n(n+1))\le y$ for

$$n:=t[m_i,m_j]+\frac{m_i}{(m_i,m_j)}.$$

Let N denote the left hand side of (1.8). We deduce from the above reasoning that

$$|T_i\cap T_j|\le N \qquad (1\le i<j\le R).$$

The inclusion-exclusion principle then implies

$$\begin{aligned}\frac{x}{M}\ge\left|\bigcup_{i=1}^{R}T_i\right|&\ge\sum_{1\le i\le R}T_i-\sum_{1\le i<j\le R}|T_i\cap T_j|\\&\ge C_0\frac{x}{M}Ru^{-2u}-R^2N\end{aligned}$$

where we have taken (3.10) into account. It follows that

$$N\ge\frac{x}{MR^2}\{C_0Ru^{-2u}-1\}\gg xu^{-u^{7u}}$$

since $Ru^{-2u}\gg 2^{2u}$ and, for $u\ge u_0$,

$$MR^2\ll R^{2(\alpha+1)R^3}\ll(2u)^{2(\alpha+1)(2u)^{6u+1}}\ll u^{u^{7u}}.$$

This completes the proof of Theorem 3.

§ 4. Proof of Theorem 2.

Let y_0 be a sufficiently large constant, and suppose $y\ge y_0$. We put

$$x:=y^{\frac{\log_2 y}{8\log_3 y}}$$

so that (1.9) is satisfied. We also have

$$\log_2 y=8u\log_3 y>8u\log u$$

whence

$$y\ge e^{u^{8u}}. \tag{4.1}$$

We define, for each prime $p \le y$, the integer α_p by

$$y^2 < p^{\alpha_p} \le py^2$$

and we put

$$n := \prod_{p \le y} p^{\alpha_p}.$$

Plainly

$$\log n \asymp y. \tag{4.2}$$

Now, we have on the one hand, from Theorem 3,

$$\sum_{\substack{d \le x \\ P^+(d(d+1)) \le y}} 1 \gg xu^{-u^{7u}},$$

and on the other hand

$$\sum_{\substack{d \le x \\ P^+(d(d+1)) \le y \\ d(d+1) \nmid n}} 1 \le \sum_{d \le x} \sum_{\substack{p \le y \\ p^{\alpha_p+1} | d(d+1)}} 1 \le 2 \sum_{p \le y} \frac{x+1}{p^{\alpha_p+1}} \ll \frac{x}{y}.$$

Taking (4.1) into account, we get

$$\kappa(n) \gg xu^{-u^{7u}} \gg x^{\frac{9}{10}} \gg (\log n)^{\frac{\log_3 n}{9 \log_4 n}}.$$

This completes the proof.

§ 5. Proof of Theorem 4.

We use three lemmas. The first enunciates a property of primitive sequences — that is sequences no element of which divides any other — due to Behrend [1]. Another proof may be found in [5], Chap. V, Th. 6.

Lemma 5.1. *There exists an absolute constant K such that for every primitive sequence $\mathcal{A} \subseteq \mathbf{Z}^+$ and every $x \ge 3$ we have*

$$\sum_{\substack{a \le x \\ a \in \mathcal{A}}} \frac{1}{a} \le K \frac{\log x}{\sqrt{\log_2 x}}.$$

Lemma 5.2. *Let m, n be positive integers such that $m \mid n$. Then*

$$H(m) \le H(n). \tag{5.1}$$

Proof. The sequence of divisors of m is a subsequence of that of divisors of n. Consider two consecutive divisors of m, say d_j, d_{j+1}. The divisors of n in $[d_j, d_{j+1}]$ are $d_j = d_{j1} < d_{j2} < \ldots d_{jr} = d_{j+1}$ and we obviously have

$$(d_{j+1} - d_j)^{-1} \le \sum_{1 \le i < r} (d_{ji+1} - d_{ji})^{-1}.$$

Summing over j, $1 \le j < \tau(m)$, we obtain (5.1).

Lemma 5.3. *Let $y \geq 2$ and define, for every integer n,*

$$a_n := \prod_{\substack{p \leq y \\ p^\nu \| n}} p^\nu. \tag{5.2}$$

There exists an absolute constant C such that the inequalities

$$0 \leq H(n) - H(a_n) \leq y^{-1}(\log y)^4 \tag{5.3}$$

hold for all but at most $Cx(\log y)^{-1}$ integers $n \leq x$.

Proof. The left hand inequality follows from (5.1). The right hand inequality is implied by the estimate

$$\sum_{n \leq x} \{H(n) - H(a_n)\} \leq Cxy^{-1}(\log y)^3$$

established in [3], eq. (7.1).

We are now in a position to embark on the proof of Theorem 4.

We proceed by contradiction. If the required conclusion fails to hold, there is an $\alpha \geq 0$ and a $\delta > 0$ such that for every positive ε

$$\sum_{\substack{n \leq x \\ |H(n)-\alpha| \leq \frac{1}{2}\varepsilon}} 1 \geq \delta x \qquad (x > x_0(\varepsilon)). \tag{5.4}$$

From now on, we suppose that α and δ are given and agree that all the constants, implicit or explicit, may depend on these two quantities.

Put $y = \varepsilon^{-2}$ and define a_n by (5.2). From Lemma 5.3 and (5.4) it follows that, if ε is sufficiently small and $x_0(\varepsilon)$ is suitably chosen, then the inequality $|H(a_n) - \alpha| \leq \varepsilon$ holds, provided $x > x_0(\varepsilon)$, for at least $\frac{1}{2}\delta x$ integers $n \leq x$. Denote by $\mathcal{A} = \mathcal{A}(\varepsilon)$ the sequence of all integers a such that $|H(a) - \alpha| \leq \varepsilon$. By partial summation, the above property implies that

$$\log x \ll \sum_{\substack{n \leq x \\ a_n \in \mathcal{A}}} \frac{1}{n} \leq \sum_{\substack{P^+(a) \leq y \\ a \in \mathcal{A}}} \sum_{\substack{n \leq x \\ a_n = a}} \frac{1}{n}.$$

The inner sum is equal to

$$\sum_{\substack{b \leq x/a \\ P^-(b) > y}} \frac{1}{ab} \leq \frac{1}{a} \prod_{y < p \leq x} \Big(1 - \frac{1}{p}\Big)^{-1} \ll \frac{\log x}{a \log y}.$$

Hence

$$\sum_{\substack{P^+(a)\le y\\ a\in\mathcal{A}}} \frac{1}{a} \gg \log y. \tag{5.5}$$

Set $\theta := \frac{1}{\log y}$. For $t > 1$, we have

$$\sum_{\substack{a>y^t\\ P^+(a)\le y}} \frac{1}{a} \le \sum_{P^+(a)\le y} a^{\theta-1}y^{-\theta t} = e^{-t}\prod_{p\le y}(1-p^{\theta-1})^{-1}$$
$$\ll e^{-t}\log y.$$

Thus it follows from (5.5) that there exists a constant t such that

$$\sum_{\substack{a\le y^t\\ a\in\mathcal{A}}} \frac{1}{a} \gg \log y. \tag{5.6}$$

Lemma 5.1 enables us to deduce from (5.6) that $\mathcal{A}$ is not primitive for small ε. In this case there is at least one pair a, a' of elements of $\mathcal{A}$ such that

(i) $a \mid a',\ a' \le y^t$

(ii) $\alpha - \varepsilon \le H(a) \le H(a') \le \alpha + \varepsilon.$

We are going to show that the extra condition

(iii) $\exists p \mid \frac{a'}{a} :\ p \nmid a,\ p \le z := y^{\frac{1}{4}}.$

can also be imposed.

Indeed, suppose that (iii) fails to hold for all pairs a, a' satisfying (i) and (ii). Let $\mathcal{A}_0$ be the sequence of those elements of $\mathcal{A}$ which are divisible by no other element of $\mathcal{A}$. Then $\mathcal{A}_0$ is primitive — see e.g. [5], Chap. V, § 1. The hypothesis that (iii) never holds when (i) and (ii) are fulfilled implies that $\mathcal{A}\cap[1,y^t]$ is contained in the set

$$\{a_0 m : a_0 \in \mathcal{A}_0,\ p \mid m \Rightarrow p \mid a_0 \text{ or } p > z\}.$$

Hence

$$\sum_{\substack{a\le y^t\\ a\in\mathcal{A}}} \frac{1}{a} \le \sum_{\substack{a_0\le y^t\\ a_0\in\mathcal{A}_0}} \frac{1}{a_0} \prod_{\substack{p\mid a_0\\ p\le z}} \Big(1-\frac{1}{p}\Big)^{-1} \prod_{z<p\le y^t} \Big(1-\frac{1}{p}\Big)^{-1}$$
$$\ll \Big\{ \sum_{\substack{a_0\le y^t\\ a_0\in\mathcal{A}_0}} \frac{1}{a_0} \cdot \sum_{n\le y^t} \frac{n}{\varphi(n)^2} \Big\}^{\frac{1}{2}} \ll \frac{\log y}{(\log_2 y)^{\frac{1}{4}}}$$

by Lemma 5.1, since the second p-product is clearly bounded. This contradicts (5.6) for sufficiently small ε and in turn implies the existence of a, a' in $\mathcal{A}$ satisfying (i), (ii) and (iii).

For these a, a', denote by d_j, d_{j+1} the divisors of a such that $d_j < p < d_{j+1}$, with the convention that $d_{j+1} = d_{\tau(a)+1} = +\infty$ if $p > a$. We have

$$H(pa) - H(a) \geq \frac{1}{p - d_j} + \frac{1}{d_{j+1} - p} - \frac{1}{d_{j+1} - d_j} \geq \frac{1}{p}$$

whence

$$H(a') - H(a) \geq H(pa) - H(a) \geq \frac{1}{p} \geq \sqrt{\varepsilon}.$$

This is in contradiction to the definition of $\mathcal{A}$ when ε is small enough. The proof of Theorem 4 is thereby completed.

REFERENCES

[1] F. Behrend, On sequences of numbers not divisible one by another , *J. London Math. Soc.* **10** (1935), 42–44.

[2] P. Erdös and R.R. Hall, On some unconventional problems on the divisors of integers , *J. Austral. Math. Soc.* Ser. A **25** (1978), 479–485.

[3] P. Erdös et G. Tenenbaum, Sur les fonctions arithmétiques liées aux diviseurs consécutifs , *J. Number Theory* **31** (1989), 285–311.

[4] E. Fouvry et G. Tenenbaum, Entiers sans grand facteur premier en progressions arithmétiques , preprint.

[5] H. Halberstam and K.F. Roth, *Sequences* (1966, Oxford University Press; 2nd ed. 1983, Springer).

[6] D.R. Heath Brown, Consecutive almost primes , preprint.

[7] A. Hildebrand, Integers free of large prime factors and the Riemann Hypothesis , *Mathematika* **31** (1984), 258–271.

[8] A. Hildebrand, On a conjecture of Balog , *Proc. Amer. Math. Soc.* **95**, n° 4 (1985), 517-523.

[9] A. Hildebrand, On the number of positive integers $\leq x$ and free of prime factors $> y$, *J. Number Theory* **22** (1986), 289–307.

[10] A. Hildebrand and G. Tenenbaum, On integers free of large prime factors , *Trans. Amer. Math. Soc.* **296** (1986), 265–290.

[11] M. Huxley, On the difference between consecutive primes , *Inventiones Math.* **15** (1972), 164–170.

[12] A. Ivič and J.-M. De Koninck, On the distance between consecutive divisors of an integer , *Canad. Math. Bull.* **29** (2), (1986), 208–217.

[13] E. Saias, Sur le nombre des entiers sans grand facteur premier , *J. Number Theory* **32** (1989), 78–99.

Antal Balog and Paul Erdös
MTA-MKI
Budapest
Réaltanoda u. 13–15
H–1053 Hungary

G. Tenenbaum
Département de Mathématiques
Université de Nancy I
BP 239
54506 Vandœuvre Cedex
France

Small Zeros of Quadratic Forms Modulo p, II

TODD COCHRANE

Dedicated to Paul Bateman on his 70th birthday

Let $Q(\mathbf{x}) = Q(x_1, x_2, \dots, x_n)$ be a quadratic form with integer coefficients and p be an odd prime. Let $\mu = \mu(Q,p)$ be minimal such that there is a nonzero $\mathbf{x} \in \mathbf{Z}^n$ with $\max|x_i| \le \mu$ and

$$Q(\mathbf{x}) \equiv 0 \pmod{p}. \tag{1}$$

Heath-Brown [4] has shown that for $n \ge 4$, $\mu \ll p^{1/2}\log p$. Observing that any nonzero solution $\mathbf{x}$ of $x_1^2 + x_2^2 + \cdots + x_n^2 \equiv 0 \pmod{p}$ must satisfy $\max|x_i| \ge \frac{1}{\sqrt{n}} p^{1/2}$, the best result one can hope for is that $\mu \ll p^{1/2}$. This still has not been proven for a general Q.

When n is even one can use Minkowski's Theorem from the geometry of numbers to show $\mu < p^{1/2}$ for certain quadratic forms Q. Set

$$\Delta = \Delta_Q = \left(\frac{(-1)^{n/2}\det\ Q}{p}\right), \quad \text{(Legendre symbol)},$$

if $p \nmid \det Q$ and $\Delta = 0$ if $p | \det Q$. If $\Delta = 0$ or 1 then $\mu < p^{1/2}$; see [4, Theorem 2] for the case $n = 4$, and [2, Lemma 3, Theorem 2] for the general case. When $\Delta = -1$ the geometric method yields a weaker result than that of Heath-Brown mentioned above. This method was first used by Schinzel, Schlickewei and Schmidt [5] to obtain a small nonzero solution of (1) with a composite modulus. Geometric methods were also used in [3, Theorem 3] to show that if the number of variables is sufficiently large relative to p, specifically $n > 4\log_2 p + 3$, then $\mu < p^{1/2}$.

If Q is nonsingular (mod p) we define Q^* to be the quadratic form associated with the inverse (mod p) of the matrix representing Q, and set $\mu^* = \mu(Q^*, p)$. In Theorem 1 of [3] it was shown that for any quadratic form Q in an even number of variables $n \ge 4$ either $\mu \ll p^{1/2}$ or $\mu^* \ll p^{1/2}$. Here we obtain the stronger result

Theorem. *If $n \geq 4$ is even, then for any nonsingular quadratic form $Q(\mathbf{x})$ in n variables, $\mu\mu^* \ll p$. (The constant in the $\ll$ symbol can be taken as $2^{\frac{n^2}{4}+3n+1} n^{\frac{n}{4}+\frac{1}{2}}$).* ∎

This theorem is best possible in the sense that for a quadratic form such as $x_1^2 + x_2^2 + \cdots + x_n^2$, $\mu\mu^* \geq \frac{1}{n}p$. We wish to thank the referee of our paper [3] for suggesting the improvement in this theorem. The idea for the proof of the theorem, comes from Heath-Brown's paper [4].

Henceforth we shall assume that $Q(\mathbf{x})$ is a nonsingular quadratic form over $\mathbf{F}_p$, the finite field in p elements, where p is an odd prime. Let $e_p(\alpha) = e^{2\pi i\alpha/p}$, $\mathbf{x} \cdot \mathbf{y} = \sum_{i=1}^n x_i y_i$ and $\sum_{\mathbf{x}} = \sum_{\mathbf{x} \in \mathbf{F}_p^n}$. Let $V = V_Q$ denote the set of zeros of $Q(\mathbf{x})$ in $\mathbf{F}_p^n$. For $\mathbf{y} \in \mathbf{F}_p^n$, set

$$\phi(V, \mathbf{y}) = \begin{cases} \sum_{\mathbf{x} \in V} e_p(\mathbf{x} \cdot \mathbf{y}), & \text{for } \mathbf{y} \neq \mathbf{0} \\ |V| - p^{n-1}, & \text{for } \mathbf{y} = \mathbf{0}. \end{cases}$$

In Carlitz [1], it is shown that for even n,

$$\phi(V, \mathbf{y}) = \begin{cases} p^{n/2-1}(p-1)\Delta & \text{if } Q^*(\mathbf{y}) = 0 \\ -p^{n/2-1}\Delta & \text{if } Q^*(\mathbf{y}) \neq 0. \end{cases} \tag{2}$$

Let $B(M)$ denote the box of points $\mathbf{x}$ in $\mathbf{F}_p^n$ with $\max |x_i| \leq M$, where M is a positive integer less than $p/2$. (Here we have identified $\mathbf{F}_p$ with the set of representatives x in $\mathbf{Z}$ with $|x| < p/2$.) Let χ_B denote the characteristic function of B with Fourier expansion $\chi_B(\mathbf{x}) = \sum_{\mathbf{y}} a_B(\mathbf{y}) e_p(\mathbf{x} \cdot \mathbf{y})$. Then for $\mathbf{y} \in \mathbf{F}_p^n$,

$$a_B(\mathbf{y}) = p^{-n} \Pi_{i=1}^n \frac{\sin \pi m y_i/p}{\sin \pi y_i/p}, \tag{3}$$

where $m = 2M + 1$, and a term in the product is defined to be m if $y_i = 0$. Set $\chi_B * \chi_B(\mathbf{x}) \equiv \sum_{\mathbf{u}} \chi_B(\mathbf{u}) \chi_B(\mathbf{x} - \mathbf{u})$.

Proof of Theorem: If $\Delta_Q = 1$, then as we observed earlier, $\mu < p^{1/2}$ and $\mu^* < p^{1/2}$, and so the result is immediate. Thus we may suppose that $\Delta_Q = -1$. By Theorem 1 of [3] we have that either $\mu^* \ll p^{1/2}$ or $\mu \ll p^{1/2}$. Without loss of generality we suppose that the former holds; if the latter holds we simply interchange the roles of Q and Q^* in our proof. As shown in [3] we may assume that

$$\mu^* + 1 \leq 2^{3+\frac{n}{2}} \sqrt{n} p^{1/2}. \tag{4}$$

To show that $\mu \leq m$ it suffices to show that

$$\sum_{\substack{\mathbf{x} \in V \\ \mathbf{x} \neq \mathbf{0}}} \chi_B * \chi_B(\mathbf{x}) > 0. \tag{5}$$

Under the assumption that $\Delta = -1$, one deduces from (2) that

$$\sum_{\mathbf{x}\in V} \chi_B * \chi_B(\mathbf{x}) = \frac{m^{2n}}{p} + p^{\frac{n}{2}-1}m^n - p^{\frac{3n}{2}} \sum_{Q^*(\mathbf{y})=0} a_B^2(\mathbf{y}). \tag{6}$$

Let $\lambda = p/m\mu^*$ and assume that $\lambda^2 < 1/2n$. Then, using the fact that $Q^*(\mathbf{y}) \neq 0$ for any $\mathbf{y}$ with $0 < \max|y_i| < \mu^*$ one deduces from (3) that

$$\begin{aligned}\sum_{\substack{Q^*(\mathbf{y})=0\\ \mathbf{y}\neq\mathbf{0}}} a_B^2(\mathbf{y}) &\leq 2^{2n+1}n\lambda^2 m^{2n}(\mu^*+1)^n p^{-(2n+1)}\\ &\quad + 2^{n+1}n\lambda^2 m^{2n}p^{-(\frac{3}{2}n+1)}/3\\ &= A + B, \text{say.}\end{aligned} \tag{7}$$

See [3] for the details of (6) and (7). It follows from (6) and (7) that

$$\sum_{\substack{\mathbf{x}\in V\\ \mathbf{x}\neq\mathbf{0}}} \chi_B * \chi_B(\mathbf{x}) \geq \frac{m^{2n}}{p} + m^n\left(p^{\frac{n}{2}-1} - 1 - m^n p^{-n/2}\right) - p^{\frac{3n}{2}}(A+B),$$

and so (5) holds provided that $m^n < p^{n-1} - p^{n/2}$, $m^{2n}/p \geq 2p^{\frac{3n}{2}}A$ and $m^{2n}/p \geq 2p^{\frac{3n}{2}}B$. The latter two conditions are equivalent to

$$\lambda^2 \leq p^{n/2}/4n2^{2n}(\mu^*+1)^n \quad \text{and} \quad \lambda^2 \leq 3/4n2^n$$

respectively, and by (4) these two conditions hold if

$$\lambda^2 \leq \left(2^{\frac{n^2}{2}+5n+2}n^{\frac{n}{2}+1}\right)^{-1}.$$

Hence (5) holds if we take

$$m = \min\left\{2^{\frac{n^2}{4}+\frac{5}{2}n+1}n^{\frac{n}{4}+\frac{1}{2}}p/\mu^*,\ \ \frac{1}{2}\left(p^{n-1}-p^{n/2}\right)^{1/n}\right\}.$$

Thus $\mu\mu^* \leq m\mu^* \leq 2^{\frac{n^2}{4}+\frac{5}{2}n+1}n^{\frac{n}{4}+\frac{1}{2}}p$.

Remarks:

1. If we write $Q(\mathbf{x}) = \mathbf{x}A\mathbf{x}^T$, where A is a nonsingular symmetric matrix over $\mathbf{F}_p$, then $Q(\mathbf{x}) = 0$, for $\mathbf{x} \in \mathbf{F}_p^n$, if and only if $Q^*(\mathbf{x}A) = 0$. Thus, it follows from the Theorem that there exists a nonzero $\mathbf{x}$ with $\max|x_i| \ll p^{1/2}$ such that $Q(\mathbf{x}) = 0$ or $Q(\mathbf{x}A^{-1}) = 0$. In particular if the entries of A^{-1} are bounded by the positive constant α, then $\mu(Q,p) \leq \alpha c(n)p^{1/2}$, where $c(n)$ is a constant depending only on n.
2. Although the method discussed here yields no information on simultaneous small solutions of Q and Q^*, it follows from the geometric methods of [2, Remark after Theorem 2] that for any nonsingular form $Q(\mathbf{x})$ over $\mathbf{F}_p$ there exists a nonzero $\mathbf{x}$ with

$$Q(\mathbf{x}) = Q^*(\mathbf{x}) = 0 \quad \text{and} \quad \max|x_i| < p^{\frac{1}{2}+\frac{3}{2(n-1)}}.$$

REFERENCES

[1] L. Carlitz, Weighted quadratic partitions over a finite field, Can. J. of Math. **5** (1953), 317-323.

[2] T. Cochrane, Small solutions of congruences over algebraic number fields, Ill J. of Math. 31 **4** (1987), 618-625.

[3] T. Cochrane, Small zeros of quadratic forms modulo p, J. of Number Theory (to appear).

[4] D.R. Heath-Brown Small solutions of quadratic congruences, Glasgow Math. J. **27** (1985), 87-93.

[5] A. Schinzel, H.P. Schlickewei and W.M. Schmidt, Small solutions of quadratic congruences and small fractional parts of quadratic forms, Acta Arithmetica **37** (1980), 241-248.

Todd Cochrane
Mathematics Department
Kansas State University
Manhattan, KS 66506
USA

Zeros of Derivatives Of the Riemann Zeta-Function Near the Critical Line

J. B. CONREY AND A. GHOSH

To Paul Bateman on the occasion of his seventieth birthday

1. Introduction

The question of the horizontal distribution of the zeros of derivatives of Riemann's zeta-function is an interesting one in view of its connection with the Riemann Hypothesis. Indeed, Speiser [9] showed that the Riemann Hypothesis is equivalent to the assertion that no non-real zero of $\zeta'(s)$ is to the left of the critical line $\sigma = \Re s = 1/2$. Levinson and Montgomery [7] proved a quantitative version of this, namely that $\zeta(s)$ and $\zeta'(s)$ have essentially the same number of zeros to the left of $\sigma = 1/2$. More precisely, if $N_k(T)$ denotes the number of zeros of $\zeta^{(k)}(s)$ in the region $0 < t < T$, then

$$N_k(T) = \frac{T}{2\pi} \log \frac{T}{2\pi e} + O_k(\log T); \tag{1}$$

Montgomery and Levinson proved that up to a height T the difference between the number of zeros of ζ in $\sigma < 1/2$ and the number of zeros of ζ' there is $\ll \log T$. Moreover, they showed that $\zeta'(s)$ vanishes on $\sigma = 1/2$ only at a multiple zero of $\zeta(s)$ (hence probably never) and that

$$\sum_{\substack{0<\gamma_1<T\\ \beta_1>1/2}} (\beta_1 - 1/2) = \frac{T}{2\pi} \log\log \frac{T}{2\pi} + O(T)$$

Research supported in part by a grant from the NSF.

where $\rho_k = \beta_k + i\gamma_k$ denotes a zero of $\zeta^{(k)}(s)$ so that, on average at least, the zeros of $\zeta'(s)$ in $0 < t < T$ are a distance $(\log\log T)/(\log T)$ from the critical line. By contrast the consecutive ordinates of zeros of $\zeta(s)$ in $|t| < T$ differ by $\ll 1/(\log T)$ on average. Thus, zeros of ζ' are rather far from the critical line on average. These observations probably led Levinson to believe that $\zeta'(s)$ does not behave as "erratically" as $\zeta(s)$ in the immediate vicinity of the critical line (i.e. at a distance $\ll 1/(\log t)$ from the critical line) and $\zeta'(s)$ can be "mollified" or smoothed more efficiently near the critical line. Thus, he used Littlewood's lemma and an efficient mollifer to show that $\zeta'(s)$ does not have too many zeros to the left of $\sigma = 1/2$, whence the same is true of $\zeta(s)$. Of course, since the zeros of $\zeta(s)$ are symmetric about $\sigma = 1/2$, this implied that $\zeta(s)$ had zeros on the line $\sigma = 1/2$, specifically, at least 1/3 of the zeros of $\zeta(s)$ must be on $\sigma = 1/2$. This result was a quantitative improvement over Selberg's result that a positive proportion of zeros of $\zeta(s)$ are on the critical line. Of course the methods of Selberg and Levinson are different, but much of the success of Levinson's method should be attributed to the fact that a smoothing of $\zeta(s)$ on the critical line was replaced by a smoothing of $\zeta'(s)$ (near the critical line). Indeed, when smoothing (on $\sigma = 1/2$) with a Dirichlet polynomial

$$B(s) = \sum_{n \le T^\theta} b(n) n^{-s} \tag{2}$$

with $b(1) = 1$, $\theta < 1/2$, the best known result for $\zeta(s)$ is with

$$b(n) = \mu(n)(1 - (\log n)/(\theta \log T))$$

which leads to (as $\theta \to 1/2$)

$$\int_1^T |\zeta(1/2 + it)B(1/2 + it)|^2 \, dt \sim 3T, \tag{3}$$

while with $\zeta'(s)$ the same choice of B leads to

$$\int_2^T \left| \frac{\zeta'(1/2 + it)}{\log t} B(1/2 + it) \right|^2 \, dt \sim 4T/3. \tag{4}$$

In fact, a more elaborate choice of B allows the "4/3" in (4) to be replaced by

$$1/2 + \frac{\sqrt{3}}{3} \coth \frac{\sqrt{3}}{2} = 1.3255.... \tag{5}$$

It seems that $\zeta'(s)$ can be smoothed better than $\zeta(s)$ because the presence of zeros of $\zeta(s)$ on the critical line makes the smoothing more difficult.

We see more evidence for this relationship between good "smoothing" and absence of zeros when we consider zeros of higher derivatives of $\zeta(s)$. Thus, Levinson and Montgomery have shown that

$$2\pi \sum_{0<\gamma_k<T} (\beta_k - 1/2) = kT \log\log T + T(\frac{1}{2}\log 2 - k \log\log 2) - 2\pi k \mathrm{li}\left(\frac{T}{2\pi}\right) + O(\log T) \tag{6}$$

and that if the Riemann Hypothesis is true, then only finitely many of the ρ_k satisfy $\beta_k < 1/2$. Thus, on average in $0 < t < T$ the β_k are $1/2 + k(\log\log T)/(\log T)$. Of course the "average" situation may never take place. Nevertheless, there seems to be a definite migration of zeros of higher derivatives of ζ away from the critical line. (For an interesting chart on the location of zeros of $\zeta''(s)$ compared to zeros of $\zeta'(s)$, see Spira [10] where, for small ordinates, the ordinates of zeros of ζ' and ζ'' agree to a surprising degree, while the abscissa of a zero of ζ'' is larger than that of the "corresponding" zero of ζ'.) Thus, as k increases, the zeros of $\zeta^{(k)}(s)$ seem to move farther to the right of the critical line. As far as smoothing goes, we can show that with B as in (2) there is a choice of θ and $b(n)$ which leads to

$$\int_2^T \left| \frac{\zeta^{(k)}(1/2+it)}{\log^k t} B(1/2+it) \right|^2 dt \sim c_k T \tag{7}$$

where

$$c_k = 1/2 + \frac{\coth\left(\frac{k}{2}\sqrt{\frac{1+1/(2k)}{1-1/(2k)}}\right)}{2\sqrt{1-1/(4k^2)}} = 1 + O(1/k^2).$$

Thus, as k increases, $\zeta^{(k)}(s)$ can be smoothed more efficiently as well. (The presence of the $\log^{-k} t$ factor in this formula is inevitable because near the $1/2$ -line $\zeta^{(k)}(s)$ on average has an order of magnitude which is greater than that of $\zeta(s)$ by a factor of $\log^k t$.) We would like to know the precise horizontal distribution of zeros of $\zeta^{(k)}$. In particular, we would like to know whether in Levinson's method there is a loss due to the presence of zeros of ζ' in the region $\sigma < 1/2 + c/\log t$ for all $c > 0$. Unfortunately, we cannot answer this question. However, Theorem 2 below indicates that there probably is some loss. We would conclude that while Selberg's method cannot detect zeros on the critical line which have small gaps between them, Levinson's method cannot detect the zeros of ζ' too near the critical line (and we believe that such zeros exist).

In our statements k is fixed and $T \to \infty$. From Levinson and Montgomery [7] we can say that

(i) Almost all zeros of $\zeta^{(k)}(s)$ are in

$$1/2 - \frac{\phi(t)\log\log t}{\log t} \leq \sigma < 1/2 + \frac{\phi(t)\log\log t}{\log t}$$

where ϕ is any function which goes to infinity with t; on RH the lower bound 1/2 holds for $t > t_0$;

(ii) a positive proportion of the zeros of $\zeta^{(k)}(s)$ are in the region

$$\sigma < 1/2 + (k+\epsilon)\frac{\log\log T}{\log T}, \qquad 0 < t < T$$

for any $\epsilon > 0$; (this follows from (6) and (12))

(iii) there are $\gg_\epsilon T\log\log T$ zeros in the region

$$\sigma > 1/2 + (k-\epsilon)\frac{\log\log T}{\log T}, \qquad 0 < t < T$$

for any $\epsilon > 0$; (this also follows from (6) and (12)).

We add to these by proving

Theorem 1. *With the above notation:*

(a) *Almost all the zeros of* $\zeta^{(k)}(s)$ *are in the region*

$$\sigma > 1/2 - \frac{\phi(t)}{\log t}$$

for any $\phi(t)$ *which goes to infinity with* t;

(b) *for any* $c > 0$, *a positive proportion of zeros of* $\zeta^{(k)}(s)$ *are in the region*

$$\sigma \geq 1/2 + c/\log t;$$

(c) *assuming the Riemann Hypothesis, there are* $\gg_\epsilon T$ *zeros of*

$$\zeta^{(k)}(s)$$

in the region

$$1/2 \leq \sigma < 1/2 + \frac{(1+\epsilon)\log\log T}{\log T}, \qquad 0 < t < T$$

for any $\epsilon > 0$.

We remark that (a) and (c) give new information only when $k > 1$ while (b) is new for all $k \geq 1$. The first two results are a consequence of

Lemma 1. *Let T be large and $L = \log T$. Let*

$$G(s) = Q(\frac{1}{L}\frac{d}{ds})\zeta(s)$$

for some polynomial Q. Let B be as in (2) with

$$b(n) = \mu(n)P(1 - \frac{\log n}{\log y})$$

where P is real analytic with $P(0) = 0$ and $P(1) = 1$. Define

$$I = I(a, P, Q) := \frac{1}{T}\int_1^T |GB(a + it)|^2 \, dt.$$

Then for $0 < \theta < 1/2$ and $a = a(T)$ satisfying $|a - 1/2| = o(1)$ as $T \to \infty$ we have

$$I \sim \frac{T^{1-2a}Q(1)^2 + Q(0)^2}{2} + \theta \int_0^1 \int_0^1 (\frac{d}{dx}(T^{(1/2-a)x}Q(x)))^2 P(y)^2 \, dxdy$$
$$+ \frac{1}{\theta}\int_0^1 \int_0^1 T^{(1-2a)x} Q(x)^2 P'(y)^2 \, dxdy.$$

This result is essentially contained in Conrey [1].

The result (c) follows from

Theorem 2. *Assuming the Riemann Hypothesis,*

$$\sum_{0<\gamma_k<T} \chi(\rho_k) \sim \alpha_k \frac{T}{2\pi}$$

where $\chi(s) = 2(2\pi)^{s-1}\Gamma(1-s)\sin(\pi s/2)$ is the usual factor from the functional equation for $\zeta(s)$ and

$$\alpha_k = k + 1 - \sum_{\nu=1}^{k} e^{-z_\nu}$$

where the z_ν are roots of $f_k(z) = \sum_{j=0}^k \frac{z^j}{j!}$.

Remark. As a function of k we can show that $0 < \alpha_k \ll_\epsilon e^{-(b-\epsilon)k}$ for any $\epsilon > 0$ where $b = 1 - \log 2$ (see Conrey - Ghosh [4].) While (c) of Theorem 1 is all that we can conclude from Theorem 2, it seems that we can speculate more. The χ-function oscillates a lot – its argument at height t is essentially $t\log(t/2\pi e)$. However, the deduction of (c) ignores this fact altogether. Thus, it seems that the proper interpretation of Theorem 2 might be that a positive proportion of zeros of $\zeta^{(k)}$ are within $c/\log t$ of the critical line for any $c > 0$.

We will first show how to deduce the results (a) - (c) from Lemma 1 and Theorem 2 and then we will prove Theorem 2.

2. Deduction of results

As mentioned earlier, k is thought of as fixed. It is well known that

$$\chi(s) = \left(\frac{|t|}{2\pi}\right)^{1/2-\sigma} \exp(-it\log\frac{t}{2\pi e} + \pi/4)(1 + O(1/|t|)).$$

for $s = \sigma + it$. Then,

$$|\chi(s)| = \left(\frac{|t|}{2\pi}\right)^{1/2-\sigma} (1 + O(1/|t|)). \tag{8}$$

Thus, (c) follows directly from Theorem 2 and the theorem of Conrey - Ghosh [4] which gives the bound for α_k: for if $\sigma > 1/2 + ((1+\epsilon)\log\log t)/(\log t)$, then $|\chi(s)| \ll (\log|t|)^{-1-\epsilon}$.

To prove (a) and (b) we take

$$P(x) = \frac{\sinh\theta\Lambda x}{\sinh\theta\Lambda}$$

in Lemma 1 where if we let

$$v(x) = T^{(1/2-a)x}Q(x),$$

then Λ is defined by

$$\Lambda^2 = \frac{\int_0^1 v'(x)^2\,dx}{\int_0^1 v(x)^2\,dx}.$$

Then it is not hard to verify that

$$I = \frac{v(0)^2 + v(1)^2}{2} + \left(\int_0^1 v(x)^2\,dx \int_0^1 v'(x)^2\,dx\right)^{1/2} \coth\theta\Lambda. \tag{9}$$

We remark that (9) can be used to verify (3)-(5) and (7). Now take $Q(x) = x^k, k \geq 1$; then $v(0)^2 = 0$, $v(1)^2 = T^{1-2a}$, and if $a \neq 1/2$, then $\int_0^1 v(x)^2\,dx$

$$\begin{aligned}
&= \int_0^1 T^{(1-2a)x}x^{2k}\,dx \\
&= \frac{T^{(1-2a)}}{(1-2a)L}\left(1 - \frac{2k}{(1-2a)L} + \frac{2k(2k-1)}{((1-2a)L)^2} - + \cdots + \frac{(2k)!}{((1-2a)L)^{2k}}\right) \\
&\quad - \frac{(2k)!}{((1-2a)L)^{2k+1}}
\end{aligned}$$

where $L = \log T$. Thus

$$\int_0^1 v(x)^2\,dx = \begin{cases} \frac{T^{1-2a}}{(1-2a)L}(1+O(\frac{1}{(1-2a)L})) & \text{if } (1-2a)L \to \infty \\ \frac{-(2k)!}{((1-2a)L)^{2k+1}}(1+O(\frac{T^{1-2a}}{|1-2a|L})) & \text{if } (1-2a)L \to -\infty \\ \approx 1 & \text{if } |1-2a|L \ll 1; \end{cases}$$

the last formula follows from an integration by parts. Similarly,

$$\int_0^1 v'(x)^2\,dx = \int_0^1 T^{(1-2a)x} x^{2k-2}((1/2-a)Lx+k)^2\,dx$$

so that $\int_0^1 v'(x)^2\,dx$

$$= \begin{cases} \frac{T^{1-2a}}{4}(1-2a)L(1+O(\frac{1}{1-2a)L})) & \text{if } (1-2a)L \to \infty \\ \frac{-(2k)!}{4((1-2a)L)^{2k-1}}(1+O(\frac{T^{1-2a}}{|1-2a|L})) & \text{if } (1-2a)L \to -\infty \\ \approx 1 & \text{if } |1-2a|L \ll 1 \end{cases}$$

Thus,

$$I(a,x^k) = \begin{cases} T^{1-2a}(1+o(1)) & \text{if } (1-2a)L \to \infty \\ \frac{2(k)!}{2((1-2a)L^{2k}}(1+o(1)) & \text{if } (1-2a)L \to -\infty \\ \approx 1 & \text{if } |1-2a|L \ll 1. \end{cases} \tag{10}$$

Now let a be such that $|1/2-a| = o(1)$ as $T \to \infty$. We apply Littlewood's lemma to $\zeta^{(k)}(s)B(s)$ on the rectangle with vertices $a+i$, σ_k+i, σ_k+iT, $a+iT$ where $\sigma_k \ll_k 1$ is a number for which $\zeta^{(k)}(s)$ has no zeros in $\sigma > \sigma_k$. Now $B(s)$ is a Dirichlet polynomial with leading coefficient 1, bounded coefficients and length $\ll T^{1/2}$. Thus, in a completely standard way (see Levinson and Montgomery [7] Section 3 and Levinson [6] Section 1 for exact details) we obtain

$$2\pi \sum_{\substack{\beta_k > a \\ 0<\gamma_k<T}} (\beta_k - a) \le \int_2^T \log|\zeta^{(k)}B(a+it)|\,dt + T(a\log 2 - k\log\log 2) + O(\log T). \tag{11}$$

Now with $Q(x) = x^k$ we have $\zeta^{(k)}(s) = L^k G(s)$ with G as in Theorem 1. Then by the arithmetic mean-geometric mean inequality, the integral in (11) is

$$\le \frac{T}{2}\log\left(\frac{1}{T}\int_2^T |GB(a+iT)|^2\,dt\right)$$

so that

$$2\pi \sum_{\substack{\beta_k > a \\ 0<\gamma_k<T}} (\beta_k - a) \le kT \log\log T + \frac{T}{2}\log I(a, x^k)$$
$$+ T(a\log 2 - k\log\log 2) + O(\log T).$$

Then by (10), for $|a-1/2| = o(1)$, we have that

$$2\pi \sum_{\substack{\beta_k > a \\ 0<\gamma_k<T}} (\beta_k - a)$$

is

$$\le \begin{cases} kT\log\log T + T(1/2-a)\log T \\ \quad + T(a\log 2 - k\log\log 2 + O(T)) & \text{if } (1-2a)L \to \infty \\ kT\log\frac{1}{(2a-1)} + O(T) & \text{if } (1-2a)L \to -\infty \\ kT\log\log T + O(T) & \text{if } |1-2a|L \ll 1 \end{cases} \tag{12}$$

Next we note that using (1) and (6) we obtain

$$2\pi \sum_{\substack{\beta_k < a \\ 0<\gamma_k<T}} (a - \beta_k) = -kT\log\log T + 2\pi \sum_{\substack{\beta_k > a \\ 0<\gamma_k<T}} (\beta_k - a)$$
$$+ (a-1/2)T\log\frac{T}{2\pi e} - T(a\log 2 - k\log\log 2)$$
$$+ 2\pi k \operatorname{li}(\frac{T}{2\pi}) + O(\log T).$$

Combining this with (12) we get that

$$2\pi \sum_{\substack{\beta_k < a \\ 0<\gamma_k<T}} (a - \beta_k)$$

is

$$\le \begin{cases} kT\log\frac{1}{(2a-1)L} + (a-1/2)TL + O(T) & \text{if } (2a-1)L \to \infty \\ O(T) & \text{if } (1-2a)L \le C \end{cases} \tag{13}$$

for any fixed $C > 0$. Then, (a) follows in a straightforward way. Next we prove (b). Let $c > 0$ and suppose that almost all of the zeros of $\zeta^{(k)}(s)$ are in the region $\sigma < 1/2 + c/\log|t|, |t| \geq 2$. Then for $c' > c$ we have

$$\begin{aligned}
\sum_{\substack{\gamma_k \leq T \\ \beta_k < 1/2 + \frac{c'}{L}}} (1/2 + \frac{c'}{L} - \beta_k) &= \sum_{\substack{\gamma_k \leq T \\ \beta_k < 1/2 + \frac{c}{L}}} (1/2 + \frac{c'}{L} - \beta_k) \\
&\quad + O(L^{-1} \sum_{\substack{\beta_k < 1/2 + \frac{c'}{L} \\ \gamma_k \leq T}} 1) \\
&= \sum_{\substack{\gamma_k \leq T \\ \beta_k < 1/2 + c/L}} (1/2 + \frac{c'}{L} - \beta_k) + O(T) \\
&\geq \frac{c' - c}{L} \sum_{\substack{\gamma_k \leq T \\ \beta_k < 1/2 + c/L}} 1 + O(T) \\
&\geq (c' - c)\frac{T}{2\pi} - AT \qquad (14)
\end{aligned}$$

for some fixed $A \geq 0$. On the other hand, using (13) we see that the left hand side of (14) is

$$\leq \frac{kT}{2\pi} \log \frac{1}{c'} + \frac{c'T}{2\pi} + BT.$$

for some number B which is independent of T. This is a contradiction if c' is sufficiently large ($c' > e^{(c+A+B)/k}$); thus, (b) follows.

3. Proof of Theorem 2

In this section we assume the Riemann Hypothesis. The proof of Theorem 2 follows the lines of the proof in Conrey-Ghosh [3], so in some places we refer to that paper rather than give all the details. To begin with, we note that the complex poles of $\zeta^{(k+1)}(s)/\zeta^{(k)}(s)$ are in $\sigma \geq 1/2$, by Speiser's theorem if $k = 1$ and by (i) if $k > 1$. Thus, with T large and $U = TL^{-10}$,

$$S := \sum_{T < \gamma_k \leq T+U} \chi(\rho_k) = \frac{1}{2\pi i} \int_C \chi(s) \frac{\zeta^{(k+1)}(s)}{\zeta^{(k)}(s)}\, ds$$

where C is the positively oriented rectangle with vertices $\sigma_k + iT$, $\sigma_k + i(T + U)$, $1/2 - \delta + iT$, $1/2 - \delta + i(T + U)$ where $\sigma_k \geq \min\{3, 1 + \sup_{\rho_k} \beta_k\}$, δ is fixed with $0 < \delta < 1/8$ and where we assume that the horizontal sides of this rectangle are a distance $\gg L^{-1}$ from any zero of $\zeta^{(k)}(s)$. This last assumption entails no loss of generality since by (1) there are $\ll \log T$ zeros of $\zeta^{(k)}(1/2 + it)$ in an interval $(T, T + 1)$ so we only have to adjust T and U by an amount $\ll 1$ to justify the assumption and by (8) this involves an addition or deletion of $\ll \log T$ terms of size $\ll 1$. By (8) and the definition of σ_k the integrand is

$$\ll T^{-5/2}$$

for $s = \sigma_k + it$, $T \le t \le T + U$, while on the horizontal parts of the segment the integrand is

$$\ll L^2 T^\delta$$

by (8) and since $\zeta^{(k+1)}/\zeta^{(k)}(s) \ll L^2$ on the horizontal sides. (This can be proved in the case $k \ge 1$ exactly as for the case $k = 0$; see also equation (6.1) of Levinson and Montgomery [7].) Thus,

$$S = \frac{-1}{2\pi i} \int_{1/2-\delta+iT}^{1/2-\delta+i(T+U)} \chi(s) \frac{\zeta^{(k+1)}(s)}{\zeta^{(k)}(s)}\, ds + O(T^{1/2}L^2).$$

We make a change of variable $s \to 1 - s$ here and have

$$\overline{S} = \frac{1}{2\pi i} \int_{1/2+\delta+iT}^{1/2+\delta+i(T+U)} \chi(1-s) \frac{\zeta^{(k+1)}(1-s)}{\zeta^{(k)}(1-s)}\, ds + O(T^{1/2}L^2). \qquad (15)$$

Now we derive another expression for $\zeta^{(k+1)}/\zeta^{(k)}$. First of all,

$$\frac{\chi'}{\chi}(s) = -\log\frac{|t|}{2\pi} + O\left(\frac{1}{|t|}\right),$$

and

$$\left(\frac{d}{ds}\right)^n \frac{\chi'}{\chi}(s) \ll |t|^{-n}.$$

From these and the functional equation

$$\zeta(s) = \chi(s)\zeta(1-s)$$

it easily follows that for $\sigma \le 1/2$

$$(-1)^m \zeta^{(m)}(s) = \chi(s)(1 + O(1/|t|)) \left(\ell - \left(\frac{d}{ds}\right)\right)^m \zeta(1-s) \qquad (16)$$

where $\ell = \log\frac{|t|}{2\pi}$ (see Conrey [2], Lemma 2). Now let

$$G_k(s,z) = \left(z + \frac{d}{ds}\right)^k \zeta(s) = z^k\zeta(s) + kz^{k-1}\zeta'(s) + \cdots + \zeta^{(k)}(s).$$

Then using (16) in the numerator and denominator it is not hard to see that

$$\frac{\zeta^{(k+1)}}{\zeta^{(k)}}(1-s) = -(\ell + \frac{G_k'}{G_k}(s,\ell))(1+O(\frac{1}{|t|})) \tag{17}$$

where differentiation is with respect to s. (Use the relation $\binom{k+1}{j} = \binom{k}{j} + \binom{k}{j-1}$.) Next we observe that

$$\frac{G_k'}{G_k}(s,z) = \frac{\frac{\zeta'}{\zeta}(s) + \frac{k}{z}\frac{\zeta''}{\zeta}(s) + \cdots + \frac{1}{z^k}\frac{\zeta^{(k+1)}}{\zeta}(s)}{1 + \frac{k}{z}\frac{\zeta'}{\zeta}(s) + \cdots + \frac{1}{z^k}\frac{\zeta^{(k)}}{\zeta}(s)}.$$

Now assuming the Riemann Hypothesis it is not hard to show that

$$\frac{\zeta^{(j)}}{\zeta}(s) \ll (\log t)^{j+1-2\sigma}$$

uniformly for $1/2 < \sigma_0 \le \sigma \le \sigma_1 < 1, t \ge 2$. To prove this estimate one may proceed by Cauchy's theorem and induction starting from the case $j = 1$ which is well-known (see Titchmarsh [11], Theorem 14.55) For example, we see by Cauchy's theorem that

$$\frac{d}{ds}\frac{\zeta'}{\zeta}(s) = \frac{1}{2\pi i}\int_{|w-s|=\ell^{-1}} \frac{\zeta'/\zeta(w)}{(w-s)^2}\, ds \ll \ell^{3-2\sigma}$$

so that

$$\frac{\zeta''}{\zeta}(s) = \frac{d}{ds}\frac{\zeta'}{\zeta}(s) + (\frac{\zeta'}{\zeta}(s))^2 \ll \ell^{3-2\sigma} + \ell^{4-4\sigma} \ll \ell^{3-2\sigma}$$

for $1/2 < \sigma_0 \le \sigma \le \sigma_1 < 1$. To establish the case $j = 3$ we differentiate ζ''/ζ, and so on. We conclude that in the region $\sigma \ge 1/2+\delta, T \le t \le T+U$, $T \le \Re z \le T+U$, $\Im z \ll 1$, $|s-1| \gg 1$ there are no poles of G_k'/G_k and that

$$G_k'/G_k(s,z) = o(L) \tag{18}$$

uniformly. Then by Cauchy's Theorem

$$\frac{d}{dz}\frac{G_k'}{G_k}(s,z) \ll L^2 \tag{19}$$

there. Now it follows from (19) and the ordinary mean-value theorem of differential calculus that

$$\frac{G_k'}{G_k}(s,\ell) = \frac{G_k'}{G_k}(s,L) + O(L^{-8})$$

for $T \le t \le T+U, \sigma \ge 1/2+\delta$. Thus, by (17) and (18)

$$\frac{\zeta^{(k+1)}}{\zeta^{(k)}}(1-s) = -L - \frac{G_k'}{G_k}(s,L) + O(L^{-8})$$

for $T \le t \le T+U, \sigma \ge 1/2+\delta$. We insert this in (15) and obtain

$$\overline{S} = \frac{-1}{2\pi i}\int_{1/2+\delta+iT}^{1/2+\delta+i(T+U)} \chi(1-s)(L+\frac{G_k'}{G_k}(s,L))\,ds + O(T^\delta L^{-8}).$$

Then by Cauchy's theorem and the estimates (8) and (18) we have

$$\overline{S} = \frac{-1}{2\pi i}\int_{1+\delta+iT}^{1+\delta+i(T+U)} \chi(1-s)(L+\frac{G_k'}{G_k}(s,L))\,ds + O(T^{1/2+\delta}) \tag{20}$$

where $\delta > 0$ is still fixed. Next we expand $G_k'/G_k(s,L)$ into a Dirichlet series. Let

$$\alpha(s) = \frac{1}{L}\frac{\zeta'}{\zeta}(s) + \frac{k}{L^2}\frac{\zeta''}{\zeta}(s) + \cdots + \frac{1}{L^k}\frac{\zeta^{(k)}}{\zeta}(s).$$

Then

$$|\alpha(s)| \le C(\delta,k)L^{-1}$$

for $\sigma \ge 1+\delta$ and a positive constant $C = C(\delta,k)$. Thus, for T sufficiently large and $\sigma \le 1+\delta$,

$$(1+\alpha(s))^{-1} = 1 + \sum_{j=1}^{\infty}(-1)^j\alpha(s)^j = 1 + \sum_{j=1}^{J}(-1)^j\alpha(s)^j + O(T^{-1})$$

where $J = [2L/\log L]$. Now

$$\alpha(s) = \sum_{n=1}^{\infty}\frac{a(n,L)}{n^s}$$

where

$$|a(n,L)| \le C_1(\epsilon,k)n^\epsilon/L$$

for any $\epsilon > 0$ and some positive constant $C_1 = C_1(\epsilon,k)$. Thus,

$$1/G_k(s,L) = \sum_{n=1}^{\infty}\frac{b(n,L)}{n^s} + O(T^{-1}) \qquad (\sigma \ge 1+\delta) \tag{21}$$

where

$$|b(n,L)| \le n^{\epsilon} \sum_{j=1}^{J} \frac{C_1^j}{L^j} d_j(n).$$

Then by (8), (20), and (21),

$$\overline{S} = \frac{-1}{2\pi i} \int_{1+\delta+iT}^{1+\delta+i(T+U)} \chi(1-s)(L + \sum_{n=1}^{\infty} \frac{\beta(n,L)}{n^s})\, ds + O(T^{1/2+\delta} L)$$

where

$$G_k'(s,L) \sum_{n=1}^{\infty} \frac{b(n,L)}{n^s} = \sum_{n=1}^{\infty} \frac{\beta(n,L)}{n^s} \qquad (\sigma \ge 1+\delta). \tag{22}$$

Now

$$\sum_{n=1}^{\infty} \frac{|\beta(n,L)|}{n^{1+\delta}} \ll 1 \tag{23}$$

and according to some work of Karl Norton (unpublished),

$$d_j(n) \le n^{(\log j/(\log\log n)(1+o(1)))}$$

uniformly for $j \ll (\log n)/(\log\log n)$ so that for $T/2 < n < 3T/2$,

$$\sum_{j=1}^{J} (C_1/L)^j d_j(n) \le \sum_{j=1}^{J} (C_1/L)^j (3T/2)^{\frac{\log j}{\log\log T}(1+O(\frac{1}{L}))} \ll_\epsilon T^{\epsilon} \tag{24}$$

for any $\epsilon > 0$. Thus $|\beta(n,L)| \ll_\epsilon n^{\epsilon}$ for $n \approx T$. Then by (23), (24), and Lemmas 2 and 5 of Gonek [5],

$$\overline{S} = -\sum_{\frac{T}{2\pi} \le n \le \frac{T+U}{2\pi}} \beta(n,L) + O(T^{1/2+\delta} L). \tag{25}$$

Then by Perron's formula, (21), and (22),

$$\sum_{n \le x} \beta(n,L) = \frac{1}{2\pi i} \int_{1/2+\delta-iT}^{1/2+\delta+iT} \frac{G_k'}{G_k}(s,L) \frac{x^s}{s}\, ds + O((1+\frac{x}{T})T^{\delta})$$

for $x \ll T$. By Cauchy's theorem and (18),

$$\begin{aligned} \sum_{n \le x} \beta(n,L) &= \frac{1}{2\pi i} \int_{1/2+\delta-iT}^{1/2+\delta+iT} \frac{G_{k'}}{G_k}(s,L) \frac{x^s}{s}\, ds + O(\frac{xT^{\delta}}{T}) + \sum_R \\ &= \sum_R + O(\frac{xT^{\delta}}{T} + x^{1/2+\delta} L^2) \end{aligned} \tag{26}$$

where $\sum_R$ is the sum of the residues of the integrand at its poles in $|s-1| \le 1$. We now account for the poles of G'/G. Using the definition of G below (16), we see that G has a pole of order $k+1$ at $s=1$. Therefore, G_k'/G_k has a simple pole at $s=1$ with residue $-k-1$. Next, we apply the argument principle to $G_k(s,z)/(z^k\zeta(s))$ on the circle $|s-1|=1$. The estimate for $\zeta^{(k)}/\zeta$ given earlier shows that the total change in argument is 0. But $G_k(s,z)/(z^k\zeta(s))$ has a pole of order k at $s=1$ and no other poles whence $G_k(s,z)$ has k zeros (counting multiplicities) in $|s-1|\le 1$. Thus, G_k'/G_k has a simple pole at $s=1$ with residue $-k-1$ and simple poles at the zeros of

$$G_k(s,L) = L^k\zeta(s) + kL^{k-1}\zeta'(s) + \cdots + \zeta^{(k)}(s)$$

with residue equal to the multiplicity of the zero. In the neighborhood of $s=1$ we have

$$\zeta^{(j)}(s) = \frac{(-1)^j j!}{(s-1)^{j+1}} + O(1).$$

Thus

$$\begin{aligned}
G_k(s,L) &= \sum_{j=0}^{k} \binom{k}{j} \zeta^{(j)}(s) L^{k-j} \\
&= \sum_{j=0}^{k} \frac{k!}{j!(k-j)!} \left(\frac{j!(-1)^j}{(s-1)^{j+1}} + O(1) \right) L^{k-j} \\
&= \frac{(-1)^k k!}{(s-1)^{k+1}} \sum_{j=0}^{k} (-1)^j ((s-1)L)^j + O(\frac{1}{|s-1|^k}) \\
&= \frac{(-1)^k k!}{(s-1)^{k+1}} f_k((1-s)L) + O(|s-1|^{-k})
\end{aligned}$$

where f_k is as defined in the statement of Theorem 2. Denoting the zeros of $f_k(z)$ by z_ν, $1\le \nu \le k$, we see that the poles of $G_k'/G_k(s,L)$ are at

$$s_\nu = 1 - \frac{z_\nu}{L} + O_k\left(\frac{1}{L^2}\right).$$

Thus by (26),

$$\sum_{n\le x} \beta(n,L) = x(-k-1+\sum_{\nu=1}^{k} x^{s_\nu - 1}) + O(x^{1/2+\delta}L^2)$$

for $x \approx T$. Now it follows in a straightforward way that

$$S = (k+1-\sum_{\nu=1}^{k} e^{-z_\nu})\frac{U}{2\pi} + O(U/L)$$

which implies Theorem 2.

4. Conclusion

We remark that the techniques used in the proof of Theorem 2 can also be used to derive asymptotic formulae (on RH) for

$$\sum_{0<\gamma_k<T} \zeta^{(j)}(\rho_k)$$

for any positive integers j and k.

In the absence of precise knowledge of the horizontal distribution of zeros of derivatives of ζ we ask two questions which may be approachable: Let us use the notation

$$N_k^-(\sigma,T) := \#\{\rho_k : 0<\gamma_k \le T, \beta_k < \sigma\},$$
$$N_k^+(\sigma,T) := \#\{\rho_k : 0<\gamma_k \le T, \beta_k \ge \sigma\}.$$

Then

(α) does there exist a $c>0$ for which

$$N_k^+\left(1/2+\frac{c\log\log T}{\log T}, T\right) \gg N_k(T)?$$

(β) is there a $c>0$ for which

$$N_k^-\left(1/2+\frac{c}{\log T}, T\right) \gg N_k(T)?$$

REFERENCES

[1] J. B. Conrey, Zeros of derivatives of Riemann's xi-function on the critical line, J. Number Th. **16** (1983), 49-74.

[2] J. B. Conrey, The fourth moment of derivatives of the Riemann zeta-function, Quart. J. Math. Oxford (2) **39** (1988), 21-36.

[3] J. B. Conrey and A. Ghosh, A mean value theorem for the Riemann zeta-function at its relative extrema on the critical line, J. London Math. Soc. (2) **32** (1985), 193-202.

[4] J. B. Conrey and A. Ghosh, On the zeros of the Taylor polynomials associated with the exponential function, American Math. Monthly **95** (1988), 528-533.

[5] S. M. Gonek, Mean values of the Riemann zeta-function and its derivatives, Invent. Math. **75** (1984), 123-141.

[6] N. Levinson, More than one-third of zeros of Riemann's zeta-function are on $\sigma = 1/2$, Adv. in Math. **13** (1974), 383-436.

[7] N. Levinson and H. L. Montgomery, Zeros of the derivatives of the Riemann zeta-function, Acta Math. **133** (1974), 49-65.

[8] A. Selberg, On the zeros of Riemann's zeta-function, Skr. Norske Vid. Akad. Oslo **10** (1942), 1-59.

[9] A. Speiser, Geometrisches zur Riemannschen Zetafunktion, Math. Ann. **110** (1934), 514-521.

[10] R. Spira, Zero-free regions of $\zeta^{(k)}(s)$, J. London Math Soc. **40** (1965), 677-682.

[11] E. C. Titchmarsh, *The Theory of the Riemann Zeta-Function*, Oxford University Press, 1951.

J. B. Conrey and A. Ghosh
Department of Mathematics
Oklahoma State University
Stillwater, Oklahoma 74078

On some Exponential Sums

H. DABOUSSI

To Professor P. Bateman on his seventieth birthday

Let f be a multiplicative function, and let α be an irrational number. In this paper we want to estimate the exponential sum $\sum_{n\le x} f(n)e(n\alpha)$. If f is the constant multiplicative function 1 then trivially

$$\sum_{n\le x} 1(n)e(n\alpha) = \sum_{n\le x} e(n\alpha) = o(x);$$

in fact, the sum is bounded in this case. By a convolution argument, Wintner [10] proved that for any multiplicative function f satisfying $\sum_p \sum_{r\ge 1} |f(p^r) - f(p^{r-1})| p^{-r} < \infty$ (where here and in the sequel the letter p denotes a prime)

$$\sum_{n\le x} f(n)e(n\alpha) = o(x).$$

Such a function can be written as $f = h * 1$, where $*$ denotes the Dirichlet convolution and the function h satisfies $\sum |h(n)|/n < \infty$.

On the other hand, Davenport [4] proved that in the case of the Möbius function

$$\sum_{n\le x} \mu(n)e(n\alpha) = O(\frac{x}{\log^h x})$$

for any positive number h. His proof involves deep methods based on Vinogradov's work.

To obtain similar results for more general functions f requires a "Siegel–Walfisz estimate" for the sums

$$\sum_{\substack{n\le x \\ n\equiv l \bmod q}} f(n)$$

for $q \le \log^C x$ and the study of the corresponding Dirichlet series $L_f(s,\chi)$. This was accomplished by Dupain, Hall, and Tenenbaum [6] for certain classes of functions. They proved:

Theorem A. *For any fixed y satisfying* $0 < y < 2$,

$$\sum_{n\leq x} y^{\Omega(n)} e(n\alpha) = o\left(\sum_{n\leq x} y^{\Omega(n)}\right),$$

where $\Omega(n)$ *is the number of prime factors of n counted with multiplicity.*

In Davenport's case, the function $L_f(s,\chi)$ is the reciprocal of the ordinary L-function $L(s,\chi)$; in the Dupain–Hall–Tenenbaum case, it is essentially a yth power of $L(s,\chi)$. The approach of Davenport and Dupain–Hall–Tenenbaum seems to be hopeless for general multiplicative functions. However, using a different method based on the Túran–Kubilius inequality, I proved [**1**]:

Theorem B. *Let* f *be a multiplicative function satisfying* $|f(n)| \leq 1$. *Then*

$$\sum_{n\leq x} f(n)e(n\alpha) = o(x).$$

This has since been improved by many authors. The condition $|f(n)| \leq 1$ has been weakened by Delange [**5**], Indlekofer [**7**], Daboussi and Delange [**3**]. The best-possible error term has been obtained by Montgomery and Vaughan [**9**].

This result has applications to the Fourier analysis of multiplicative functions (see[**2**]), and also to the distribution modulo one of additive functions (Kátai [**8**]).

The purpose of this paper is to prove the following theorem.

Theorem 1. *Let* f *be a completely multiplicative function satisfying* $|f(p)| = y$ *for all primes* p *and* $0 < y < 2$. *Then*

$$\sum_{n<x} f(n)e(n\alpha) = o\left(\sum_{n\leq x} |f(n)|\right).$$

This answers, in part, a question of Dupain–Hall–Tenenbaum.

Notations: Given a finite set of primes E, we define two completely multiplicative functions $u = u_E$ and $v = v_E$ by

$$u(p^r) = \begin{cases} 1 & \text{if } p \notin E \\ 0 & \text{if } p \in E \end{cases}, \qquad v(p^r) = \begin{cases} 1 & \text{if } p \in E \\ 0 & \text{if } p \notin E \end{cases}$$

for any prime power p^r. For any positive integer k we let λ_k be the multiplicative function defined by

$$\lambda_k(p^r) = \begin{cases} 1 & \text{if } r \leq k-1 \\ 0 & \text{if } r > k-1 \end{cases}$$

for any prime power p^r. We set

$$M(x,f) = \sum_{n \le x} f(n)$$

and

$$M(x,f,\alpha) = \sum_{n \le x} f(n)e(n\alpha).$$

Some convolution identities. It is easy to see that for any integer n

$$(u * v)(n) = 1.$$

This implies, by the Möbius inversion formula,

$$u(n) = (v\mu * 1)(n). \tag{1}$$

One also gets

$$g(n) = (gu * gv)(n) \tag{2}$$

for any multiplicative function g, and

$$u(n)f(n) = (vf\mu * f)(n) \tag{3}$$

for any completely multiplicative function f.

We shall prove the following

Theorem 2. *Let f be a completely multiplicative function satisfying*

$$0 < a < |f(p)| < b < 2 \quad \text{for some fixed numbers } a \text{ and } b, \tag{4}$$

$$\sum_{n \le x} |f(n)| e(n\alpha) = o\left(\sum_{n \le x} |f(n)|\right) \quad \text{for any irrational } \alpha, \tag{5}$$

$$\sum_{n \le x/d} |f(n)| = \left(\frac{1}{d} + o(1)\right) \sum_{n \le x} |f(n)| \quad \text{for any fixed } d, \tag{6}$$

and such that for any constant $c > 0$ there exists a constant $C = C(c)$ with

$$\sum_{n \le x/d} |f(n)| < \frac{C}{d} \sum_{n \le x} |f(n)| \quad \text{for any } d < \log^c x. \tag{7}$$

Then

$$\sum_{n \le x} f(n)e(n\alpha) = o\left(\sum_{n \le x} |f(n)|\right) \quad \text{for any irrational } \alpha.$$

Theorem 1 is an immediate consequence of Theorem A, Theorem 2, and the classical estimate

$$\sum_{n\leq x}|f(n)| = \sum_{n\leq x} y^{\Omega(n)} = (c(y)+o(1))\, x(\log x)^{y-1},$$

which implies (6) and (7).

The hypothesis (4) implies that for any integer k the sum

$$\sum_{p}\frac{|f(p^k)|}{p^{k(1-\delta)}} \tag{8}$$

is finite for some δ depending only on b (more precisely, for any δ such that $b < 2^{1-\delta}$), and tends to zero as k tends to infinity. It also implies that

$$\sum_{p}\frac{2|f(p)|-|f(p)|^2}{p} = \infty,$$

which is equivalent to

$$\prod_{p<T}\left(1-\frac{|f(p)|}{p}\right)^2\left(1-\frac{|f(p)|^2}{p}\right)^{-1} \to 0 \quad \text{as } T\to\infty. \tag{9}$$

One also has

$$\frac{x}{\log x} \ll \sum_{n\leq x}|f(n)| \ll x\log x. \tag{10}$$

Our method of proof is different from that in our original paper; since the hypotheses of the theorem are trivially satisfied in the case when $|f(n)|=1$ we obtain a new proof of that result.

Lemma 1. *Let f satisfy the hypotheses of Theorem 2. Then*

$$M(x,u|f|) = (1+o(1))M(x,|f|)\prod_{p\in E}\left(1-\frac{|f(p)|}{p}\right), \tag{11}$$

$$M(x,u|f|,\alpha) = o(M(x,|f|) \quad \textit{for all irrational } \alpha. \tag{12}$$

Proof.: By (3) we have

$$u(n)|f(n)| = \sum_{d|n} v(d)\mu(d)|f(d)||f(n/d)|,$$

so

$$\sum_{n\leq x} u(n)|f(n)| = \sum_{d\leq x} v(d)\mu(d)|f(d)|M(x/d,|f|).$$

Now, $v(d)\mu(d) = 0$ unless d is a squarefree integer all of whose prime factors are in the set E. Since E is finite there are only finitely many such d, and so we get by (6)

$$\sum_{n\leq x} u(n)|f(n)| = (1+o(1))M(x,|f|)\sum_{d} \frac{v(d)\mu(d)|f(d)|}{d},$$

which proves (11).

We also have

$$\sum_{n\leq x} u(n)|f(n)|e(n\alpha) = \sum_{d\leq x} v(d)\mu(d)|f(d)|M(x/d,|f|,\alpha d),$$

where the set of d's to be considered is again finite. Since αd is irrational whenever α is irrational, an application of (5) then leads to (12).

It will be convenient to consider multiplicative functions g satisfying $g(p^r) = 0$ for $r \geq k$, for some fixed k and all primes p. This ensures that the set of integers d for which $v(d)g(d)$ is non-zero is finite.

Lemma 2. *Let f be completely multiplicative. Then, as k tends to infinity,*

$$\limsup_{x\to\infty} \frac{1}{M(x,|f|)} \sum_{n\leq x} |f(n)|(1-\lambda_k(n)) \to 0,$$

and consequently

$$\limsup_{x\to\infty} \frac{1}{M(x,|f|)} \left| \sum_{n\leq x} f(n)e(n\alpha) - \sum_{n\leq x} f(n)e(n\alpha)\lambda_k(n) \right| \to 0. \tag{13}$$

Proof: Define a multiplicative function h_k by

$$h_k(p^r) = \begin{cases} 0 & \text{if } r \neq k \\ -|f(p)|^k & \text{if } r = k \end{cases}$$

for every prime power p^r. It is easy to see that

$$|f|\lambda_k = h_k * |f|,$$

which gives

$$\sum_{n\leq x}|f(n)|(1-\lambda_k(n)) = -\sum_{1<n\leq x} h_k(n)\sum_{d\leq x/n}|f(d)|$$
$$= S_1 + S_2,$$

where in S_1, $1 < n \leq \log^c x$ and in S_2, $\log^c x < n \leq x$. (We will assume that c is a fixed number satisfying $c > 2/\delta$). By (7) we have

$$|S_1| \ll M(x,|f|)\sum_{n>1}\frac{|h_k(n)|}{n}.$$

We set

$$d_{k,\delta} = \sum_p |f(p^k)|p^{-k(1-\delta)}, \qquad d_k = d_{k,0},$$

and suppose that k is sufficiently large so that $d_{k,\delta}$ and d_k are less than one. Now $h_k(n) = 0$ unless $n = p_1^k \dots p_l^k$ with distinct primes p_j. So the contribution of those n with $\omega(n) = l$ to the sum $\sum_{n>1}|h_k(n)|/n$ is equal to

$$\sum_{p_1<\dots<p_l}\frac{f(p_1)^k \dots f(p_l)^k}{(p_1\dots p_l)^k},$$

which is less than $(d_k)^l$. Summing over l gives

$$\sum_{n>1}\frac{|h_k(n)|}{n} < \frac{d_k}{1-d_k},$$

and therefore

$$S_1 \ll M(x,|f|)\frac{d_k}{1-d_k}.$$

Since d_k tends to zero as k tends to infinity, it follows that $S_1 = o(M(x,|f|))$.

By (10) we have

$$S_2 \ll x\log x\sum_{n>\log^c x}\frac{|h_k(n)|}{n},$$

and using Rankin's trick we obtain

$$S_2 \ll x(\log x)^{1-c\delta}\sum_n |h_k(n)|n^{1-\delta}.$$

The sum on the right can be handled as before and is seen to be bounded by some absolute constant. Choosing c to be larger than $2/\delta$ we then obtain

$$S_2 = o\left(\frac{x}{\log x}\right) = o(M(x,|f|).$$

Proof of Theorem 2. We first consider the function $g = f\lambda_k$. Writing $g = gu * gv$, we have

$$\left|\sum_{n\le x} g(n)e(n\alpha)\right|^2 = \left|\sum_{n\le x} g(n)u(n) \sum_{d\le x/n} g(d)v(d)e(nd\alpha)\right|^2$$

$$\le \left(\sum_{n\le x} |f(n)|u(n)\left|\sum_{d\le x/n} g(d)v(d)e(nd\alpha)\right|\right)^2$$

$$\le \left(\sum_{n\le x} |f(n)|u(n)\right) \times$$

$$\times \left(\sum_{s,t\le x} g(s)v(s)g(t)v(t) \sum_{n\le x/\max(s,t)} u(n)|f(n)|e(n\alpha(t-s))\right),$$

using the inequality $|g(n)| \le |f(n)|$ and the Cauchy inequality. We divide both sides by $M(x,|f|)^2$ and let x tend to infinity. By the definition of g and v, $g(s)v(s) = 0$ for all but finitely many integers s, so we may take the limit inside the sum over s and t. Since by Lemma 1

$$\frac{M(x,u|f|)}{M(x,|f|)} \to \prod_{p\in E}\left(1 - \frac{|f(p)|}{p}\right)$$

and for $s \ne t$

$$\frac{M(x,u|f|,\alpha(t-s))}{M(x,|f|)} \to 0,$$

we see that the right hand side tends to

$$\prod_{p\in E}\left(1 - \frac{|f(p)|}{p}\right)^2 \sum_{d\ge 1} \frac{|g(d)|^2 v(d)}{d},$$

which is at most

$$\prod_{p\in E}\left(1 - \frac{|f(p)|}{p}\right)^2 \left(1 - \frac{|f(p)|^2}{p}\right)^{-1}.$$

We choose $E = \{p : p < T\}$ and let T tend to infinity. By (9) the right hand side then tends to zero and we obtain $M(x, f\lambda_k, \alpha) = o(M(x,|f|))$. In view of (13) this implies the desired relation $M(x,f,\alpha) = o(M(x,|f|))$ on letting k tend to infinity.

REFERENCES

[1] H. Daboussi, Fonctions multiplicatives presque périodiques B, Astérisque (Soc. Math. France), **24–25** (1975), 321–324.

[2] H. Daboussi, Caractérisation des fonctions multiplicatives p.p.B^{λ} à spectre non vide, Ann. Inst. Fourier **30** (1980), 141–166.

[3] H. Daboussi and H. Delange, On a class of multiplicative functions, Acta Scient. Math. **49** (1985), 143–149.

[4] H. Davenport, On some infinite series involving arithmetical functions, Quarterly J. Math. **8** (1937), 8–13; II, Quarterly J. Math. **8** (1937), 313-320.

[5] H. Delange, Generalization of Daboussi's theorem, *Topics in Classical Number Theory, Coll. Math. Soc. Janos Bolyai*, Budapest, 1981, pp. 305–318.

[6] Y. Dupain, R. R. Hall, and G. Tenenbaum, Sur l'équirépartition modulo 1 de certaines fonctions de diviseurs, J. London Math. Soc. **26** (1982), 397–411.

[7] K.-H. Indlekofer, Properties of uniformly summable multiplicative functions, Periodica Math. Hungarica **17** (1986), 143–161.

[8] I. Kátai, A remark on a theorem of H. Daboussi, Acta Math. Hung. (to appear).

[9] H. Montgomery and R. C. Vaughan, On exponential sums with multiplicative coefficients, Inventiones Math. 43 (1977), 69–82.

[10] A. Wintner, Number theoretical almost periodicities, Amer. J. Math. **67** (1945), 173–193.

H. Daboussi
Département de Mathématiques
Bâtiment 425
Université Paris-Sud
91405 Orsay Cedex
France

On the Integers n for which $\Omega(n)=k$

H. DELANGE

Dedicated to Paul Bateman for his seventieth birthday

1. Introduction.

We use the letter n to denote positive integers. $\Omega(n)$ is the number of prime factors in the factorization of n, counted with multiplicity.

We denote by $S(x,k)$ the set of $n \le x$ for which $\Omega(n)$ is equal to a given integer k, which may depend upon x. (This set is non-empty if and only if $0 \le k \le \log x/\log 2$.) We denote by $N(x,k)$ the number of elements of $S(x,k)$.

Given a prime p, we denote by $V_p(n)$ the exponent of p in the factorization of n (p-adic valuation of n).

We are concerned here with the following problems.

Problem 1. *Given a prime p study the distribution of the values of $V_p(n)$ on the set $S(x,k)$ for large x.*

Problem 2. *Given q distinct primes $p_1, p_2, \cdots, p_q$ study the distribution of the q-tuples $(V_{p_1}(n), V_{p_2}(n), \cdots, V_{p_q}(n))$ on the set $S(x,k)$ for large x.*

We can rephrase these problems using probabilistic terminology. For instance, given a prime p and a non-negative integer α, $\frac{1}{N(x,k)}\#\{n \in S(x,k) : V_p(n) = \alpha\}$ is the probability that $V_p(n) = \alpha$ if we choose at random an $n \in S(x,k)$, all these n having the same probability to be chosen. We will denote by Prob$(\ldots)$ the probability that an $n \in S(x,k)$ satisfies the condition, or the conditions, indicated inside the parentheses, i.e., $1/N(x,k)$ times the number of the n's $\in S(x,k)$ which satisfy the considered condition, or conditions.

It is known that the behavior of $N(x,k)$ as x tends to infinity is different depending on whether $k \le (2-\delta)\log\log x$ or $k \ge (2+\delta)\log\log x$ $(\delta > 0)$.

It was proved by Sathe [1] and A. Selberg [2] that we have uniformly for $1 \le k \le (2-\delta)\log\log x$ (with $0 < \delta < 2$)

$$N(x,k) = F\left(\frac{k-1}{\log\log x}\right)\frac{x(\log\log x)^{k-1}}{(k-1)!\log x}\left(1+O\left(\frac{1}{\log\log x}\right)\right), \quad (1)$$

where F is the meromorphic function defined by

$$F(z) = \frac{1}{\Gamma(z+1)}\prod_p \frac{\left(1-\frac{1}{p}\right)^z}{1-\frac{z}{p}},$$

whose poles are the primes. (Here and in the sequel the letter p is used to denote primes.)

Selberg noticed that, if $\delta > 0$ and $B > 2+\delta$, then we have uniformly for $(2+\delta)\log\log x \le k \le B\log\log x$

$$N(x,k) \sim C\frac{x\log x}{2^k},$$

where $C = \frac{1}{4}\prod_{p>2}\left(1+\frac{1}{p(p-2)}\right) =$ −residue of the pole of F at 2. Nicolas [3] extended the latter result in 1984.

The present author proved in 1983, but did not publish at that time, a result concerning the case when

$$2\log\log x - A\sqrt{\log\log x} \le k \le 2\log\log x + A\sqrt{\log\log x}.$$

Finally, Balazard proved in 1987 in his thesis [4] a formula which is valid on the whole range $1 \le k \le \frac{\log(x/3)}{\log 2}$ and from which the previous results can be derived.

We will consider here first the case when $k \le (2-\delta)\log\log x$ and then the case when $k \ge (2+\delta)\log\log x$. As it is natural to expect, it turns out that the results are quite different. In the first case no prime plays a special role, while in the second case the prime 2 does play a special role.

2.

It will be easier to study the probability that $V_p(n) \ge \alpha$ than the probability that $V_p(n) = \alpha$ (where α is a positive integer) since the condition "$V_p(n) \ge \alpha$" is equivalent to "$p^\alpha | n$". We will consider more generally the condition "$d|n$" , where d is a given positive integer. We will denote by $N_d(x,k)$ the number of the elements of $S(x,k)$ which are divisible by d. Of course $\Omega(d)$ has to be $\le k$.

Since the set of conditions "$n \leq x$, $\Omega(n) = k$, $d|n$" is equivalent to "$n = md$, $m \leq x/d$, $\Omega(m) = k - \Omega(d)$" we have

$$N_d(x,k) = N\left(\frac{x}{d}, k - \Omega(d)\right). \tag{2}$$

To obtain results for the p-adic valuations we will use the following formulas: Given a prime p and a non-negative integer α, we have

$$\mathrm{Prob}\left(V_p(n) = \alpha\right) = \mathrm{Prob}\left(p^\alpha|n\right) - \mathrm{Prob}\left(p^{\alpha+1}|n\right). \tag{3}$$

Given distinct primes $p_1, p_2, \cdots, p_q$ and non-negative integers $\alpha_1, \cdots, \alpha_q$, we have

$$\mathrm{Prob}\left(V_{p_j}(n) = \alpha_j \text{ for } j = 1, 2 \cdots, q\right) =$$

$$\sum_{\epsilon_1, \epsilon_2, \cdots, \epsilon_q = 0 \text{ or } 1} (-1)^{\epsilon_1 + \cdots + \epsilon_q} \mathrm{Prob}\left(p_1^{\alpha_1+\epsilon_1} p_2^{\alpha_2+\epsilon_2} \cdots p_q^{\alpha_q+\epsilon_q}/n\right). \tag{4}$$

Formula (3) is obvious.

Formula (4) is easily proved as follows. Let

$$\chi(n,d) = \begin{cases} 1 & \text{if } d|n \\ 0 & \text{otherwise.} \end{cases}$$

With a fixed n, $\chi(n,d)$ is multiplicative function of d. Given a prime p and a non-negative integer α, the characterictic function of the set of n's for which $V_p(n) = \alpha$ is $\chi(n,p^\alpha) - \chi(n,p^{\alpha+1})$. Therefore, given distinct primes $p_1, p_2, \cdots, p_q$ and non-negative integers $\alpha_1, \alpha_2, \cdots, \alpha_q$, the charateristic function of the set of n's such that $V_{p_j}(n) = \alpha_j$ for $j = 1, 2, \cdots, q$ is

$$\prod_{j=1}^{q} \left(\chi(n, p_j^{\alpha_j}) - \chi(n, p_j^{\alpha_j+1})\right),$$

which is equal to

$$\sum_{\epsilon_1, \epsilon_2, \cdots, \epsilon_q = 0 \text{ or } 1} (-1)^{\epsilon_1 + \cdots + \epsilon_q} \chi\left(n, p_1^{\alpha_1+\epsilon_1}\right) \chi\left(n, p_2^{\alpha_2+\epsilon_2}\right) \cdots \chi\left(n, p_q^{\alpha_q+\epsilon_q}\right),$$

which in turn is equal to

$$\sum_{\epsilon_1, \epsilon_2, \cdots, \epsilon_q = 0 \text{ or } 1} (-1)^{\epsilon_1 + \cdots + \epsilon_q} \chi\left(n, p_1^{\alpha_1+\epsilon_1} p_2^{\alpha_2+\epsilon_2} \cdots p_q^{\alpha_q+\epsilon_q}\right).$$

This yields a formula for the number of elements of $S(x,k)$ which satisfy $V_{p_j}(n) = \alpha_j$ for $j = 1, 2, \cdots, q$, and dividing by $N(x,k)$ this formula gives (4).

3. The case $k \le (2-\delta)\log\log x$

We will actually suppose throughout this section that

$$A\log\log x \le k \le (2-\delta)\log\log x \ (\delta \in]0,2[,\ 0 < A < 2-\delta).$$

Theorem 1. *Given $M>0$ and $\lambda \in]0,1[$ we have uniformly for $d \le (\log x)^M$ and $\Omega(d) \le \lambda k$*

$$\mathrm{Prob}(d|n) = \frac{1}{d}\left(\frac{k}{\log\log x}\right)^{\Omega(d)}\left(1+O\left(\frac{\Omega(d)^2}{\log\log x}\right)\right). \tag{5}$$

Proof: We may suppose that $d>1$, so that $\Omega(d)\ge 1$, for there is nothing to prove for $d=1$.

Formulas (1) and (2) give

$$N_d(x,k) = F\left(\frac{k-\Omega(d)-1}{\log\log\frac{x}{d}}\right)\frac{x}{d}\frac{(\log\log\frac{x}{d})^{k-\Omega(d)-1}}{(k-\Omega(d)-1)!\log\frac{x}{d}}(1+O\left(\frac{1}{\log\log\frac{x}{d}}\right)) \tag{6}$$

We have

$$\log\frac{x}{d} = \log x - \log d = \log x\left(1-\frac{\log d}{\log x}\right) = \log x\left(1+O\left(\frac{\log\log x}{\log x}\right)\right).$$

Therefore

$$\log\log\frac{x}{d} = \log\log x + O\left(\frac{\log\log x}{\log x}\right) = \log\log x\left(1+O\left(\frac{1}{\log x}\right)\right).$$

This allows us to replace $O\left(\frac{1}{\log\log\frac{x}{d}}\right)$ in (6) by $O\left(\frac{1}{\log\log x}\right)$. A simple calculation shows that

$$\frac{k-\Omega(d)-1}{\log\log\frac{x}{d}} - \frac{k-1}{\log\log x} = O\left(\frac{\Omega(d)}{\log\log x}\right).$$

Therefore

$$\begin{aligned} F\left(\frac{k-\Omega(d)-1}{\log\log\frac{x}{d}}\right) &= F\left(\frac{k-1}{\log\log x}\right) + O\left(\frac{\Omega(d)}{\log\log x}\right) \\ &= F\left(\frac{k-1}{\log\log x}\right)\left(1+O\left(\frac{\Omega(d)}{\log\log x}\right)\right). \end{aligned}$$

We also have

$$\begin{aligned} \left(\log\log\frac{x}{d}\right)^{k-\Omega(d)-1} &= (\log\log x)^{k-\Omega(d)-1}\left(1+O\left(\frac{k-\Omega(d)-1}{\log x}\right)\right) \\ &= (\log\log x)^{k-\Omega(d)-1}\left(1+O\left(\frac{\log\log x}{\log x}\right)\right) \end{aligned}$$

and it is easy to see that

$$\frac{1}{(k-\Omega(d)-1)!} = \frac{1}{(k-1)!}k^{\Omega(d)}\left(1+O\left(\frac{\Omega(d)^2}{\log\log x}\right)\right).$$

Using these relations in formula (6) we get

$$N_d(x,k) = F\left(\frac{k-1}{\log\log x}\right)\frac{x}{d}\frac{k^{\Omega(d)}(\log\log x)^{k-\Omega(d)-1}}{(k-1)!\log x}\left(1+O\left(\frac{\Omega(d)^2}{\log\log x}\right)\right). \tag{7}$$

Dividing (7) by (1) we obtain (5).

Remark: For fixed d and for k near $\log\log x$, $\mathrm{Prob}(d|n)$ is near $1/d$; that is the same as the probability that a positive integer $\leq x$ is divisible by d. This is consistent with the well known theorem of Hardy and Ramanujan on the *normal order* of $\Omega(n)$. In fact, the set of n's satisfying $1 < n \leq x$ is the union of the disjoint sets $S(x,k)$ where $1 \leq k \leq \log x/\log 2$, and most of these n's belong to an $S(x,k)$ where k is about $\log\log x$.

As an immediate consequence of Theorem 1 we have

Theorem 2. *For each prime p we have uniformly for all non-negative integers $\alpha \leq \lambda k$, where $0 < \lambda < 1$,*

$$\mathrm{Prob}\left(V_p(n) \geq \alpha\right) = \left(\frac{k}{p\log\log x}\right)^{\alpha}\left(1+O\left(\frac{\alpha^2}{\log\log x}\right)\right)$$

and

$$\mathrm{Prob}\left(V_p(n) = \alpha\right) = \left(\frac{k}{p\log\log x}\right)^{\alpha}\left(1-\frac{k}{p\log\log x}\right)\left(1+O\left(\frac{\alpha^2+1}{\log\log x}\right)\right)$$

The first formula is obtained by taking $d = p^{\alpha}$ in Theorem 1. The second formula follows from the first.

Remark: Given p and α, if k is near $2\log\log x$, then $\mathrm{Prob}(V_p(n) \geq \alpha)$ is near $(2/p)^{\alpha}$. In particular, for $p = 2$, $\mathrm{Prob}(V_p(n) \geq \alpha)$ is near 1, i.e., an element of $S(x,k)$ is very likely to be divisible by 2^{α}.

From Theorem 1 we also derive

Theorem 3. *Let $p_1, p_2, \cdots, p_q$ be given distinct primes and let $\alpha_1, \cdots, \alpha_q$ be non-negative integers. We have uniformly for $\alpha_1 + \alpha_2 + \cdots + \alpha_q \leq \lambda k$, where $0 < \lambda < 1$,*

$\mathrm{Prob}\left(V_{p_j}(n) = \alpha_j \text{ for } j = 1, 2, \cdots, q\right) =$

$$\left(\prod_{j=1}^{q}\left(\frac{k}{p_j\log\log x}\right)^{\alpha_j}\left(1-\frac{k}{p_j\log\log x}\right)\right)\left(1+O\left(\frac{(\alpha_1+\cdots+\alpha_q)^2+1}{\log\log x}\right)\right).$$

To prove this result we use formula (4) and Theorem 1 with

$$d = p_1^{\alpha_1+\epsilon_1}p_2^{\alpha_2+\epsilon_2}\cdots p_q^{\alpha_q+\epsilon_q}.$$

4. The case $k \geq (2+\delta)\log\log x$.

Here we need the work of Balazard.

4.1 Balazard starts with a formula due to Halász. Consider the arithmetic function ψ defined by $\psi(m) = 2^{-\Omega(m)}m$. This function is completely multiplicative. We have $\psi(m) \geq 1$ for all m and $\psi(m) = 1$ if m is a power of 2. Further, $\psi(m)$ tends to infinity as m tends to infinity through odd values. For, if m is odd, then $m \geq 3^{\Omega(m)}$ and therefore $\Omega(m) \leq \log m/\log 3$, which yields $\psi(m) \geq m^{1-\log 2/\log 3}$. So, given $y > 0$, there are only finitely many odd m's for which $\psi(m) \leq y$. However, ψ is not non-decreasing on odd numbers (*e.g.* $\psi(25) = 6.25$ and $\psi(27) = 3.375$).

The formula of Halász is

$$N(x,k) = \sum_{\substack{m \text{ odd} \\ \psi(m)\leq x/2^k \\ \Omega(m)\leq k}} 1.$$

This follows from the fact that there is a one-to-one correspondence between $S(x,k)$ and the set of those odd positive integers m which satisfy $\psi(m) \leq x/2^k$ and $\Omega(m) \leq k$. It is obtained by associating to each $n \in S(x,k)$ the integer $m = 2^{-V_2(n)}n$. This m is odd, $\Omega(m) = k - V_2(n) \leq k$ and $\psi(m) = 2^{-k}n \leq x/2^k$. If m is an odd positive integer satisfying $\psi(m) \leq x/2^k$ and $\Omega(m) \leq k$, then m is the image of a unique element of $S(x,k)$, namely $n = 2^{k-\Omega(m)}m$. Note that, if m is the odd integer associated to $n \in S(x,k)$, then $V_2(n) = k - \Omega(m)$.

In the following we will set

$$T(y,k) = \sum_{\substack{m \text{ odd} \\ \psi(m)\leq y \\ \Omega(m)\leq k}} 1.$$

Halász's formula can be written as

$$N(x,k) = T\left(\frac{x}{2^k}, k\right).$$

Balazard proves a general formula for $T(y,k)$ which gives $N(x,k)$ by taking $y = x/2^k$. This formula holds for $k \geq 2$ and $y \geq 3$. Before stating it we have to introduce the following notations:

$$P_q(x) = \sum_{j=0}^{q} \frac{x^j}{j!},$$

$$Q(\lambda) = \lambda\log\lambda - \lambda + 1 \text{ for } \lambda \geq 1,$$

$$R_\lambda(y) = (\log\log y)^{-\frac{1}{2}}(\log y)^{-2Q(\lambda)} \text{ for } \lambda \geq 1 \text{ and } y \geq 3.$$

(Note that $Q(1) = 0$ and Q is increasing for $\lambda > 1$ and that we have always $R_\lambda(y) = O(1/\log\log y)$.)

Balazard's formula is the following:

Given any real $B < 3/2$ we have uniformly for $k \geq 2$ and $y \geq 3$

$$T(y,k) = f(2r(y,k))\frac{y}{\log y}P_{k-1}(2\log\log y) \\ \times \left(1 + O\left(\min\left((\log\log y)^{-1}, R_\lambda(y)\right)\right)\right),$$

where

$$f(z) = (2-z)F(z) = \frac{2^{1-z}}{\Gamma(z+1)}\prod_{p>2}\frac{\left(1-\frac{1}{p}\right)^z}{1-\frac{z}{p}},$$

(so that $f(2) = C$),

$$r(y,k) = \frac{P_{k-2}(2\log\log y)}{P_{k-1}(2\log\log y)}$$

and

$$\lambda = \max\left(1, \min\left(B, \frac{k}{2\log\log y}\right)\right).$$

4.2 From now on we will suppose that

$$(2+\delta)\log\log x \leq k \leq \frac{\log(x/3)}{\log 2},$$

where $0 < \delta < 1$. It will be understood that $y = x/2^k$ (≥ 3). If, instead of $k \leq \log(x/3)/\log 2$, we suppose that $k \leq \lambda \log x/\log 2$, where $\lambda < 1$, this implies $\log\log y - \log\log x + O(1)$, and it is easy to see that this permits us to replace y by x in the statements of the theorems below.

4.3 We will have to use the following properties of the polynomials P_q.

(a) We have

$$e^{-X}P_q(X) = G\left(\frac{q-X}{\sqrt{X}}\right) + O\left(\frac{1}{\sqrt{X}}\right)$$

uniformly for $q \geq 0$ and $X \geq 1$, where

$$G(t) = \frac{1}{\sqrt{2\pi}}\int_{-\infty}^{t} e^{-u^2/2}du.$$

This is well known.

(b) Given $\rho > 1$ we have uniformly for $X > 0$ and $q \geq \rho X$

$$P_{q-1}(X) = e^X \left(1 + O\left(X^{-\frac{1}{2}} e^{-XQ(\rho)}\right)\right)$$

and

$$\frac{P_{q-2}(X)}{P_{q-1}(X)} = 1 + O\left(X^{-\frac{1}{2}} e^{-XQ(\rho)}\right).$$

The first formula follows from $P_{q-1}(X) = e^X(1 - \sum_{j=q}^{\infty} \frac{e^{-X}X^j}{j!})$ by majorizing the series by a geometric series and using the fact that $q! > q^{q+1/2}e^{-q}\sqrt{2\pi}$. The second formula follows from

$$\frac{P_{q-2}(X)}{P_{q-1}(X)} = 1 - \frac{X^{q-1}/(q-1)!}{P_{q-1}(X)}.$$

(c) Given $A > 0$ we have uniformly for $q \geq X - A\sqrt{X}$

$$\frac{P_{q-1}(X)}{P_q(X)} = 1 + O\left(\frac{1}{\sqrt{X}}\right).$$

This follows from $\frac{P_{q-1}(X)}{P_q(X)} = 1 - \frac{X^q/q!}{P_q(X)}$ and (a).

4.4 Theorem 4. *Let μ be a real number satisfying $1 < \mu < 1 + \delta/2$, let η be any positive number and d a positive integer. We have for $\Omega(d) \leq k - 2\mu \log\log y$ and $\log \psi(d) \leq \eta(\log y)^{1-2Q(\mu)}(\log\log y)^{-1/2}$*

$$\text{Prob}(d|n) = \frac{1}{\psi(d)}\left(1 + O(R_\mu(y))\right).$$

Proof: We may suppose $y \geq y_0$ for some given y_0, for it is easy to see that the result holds for $y < y_0$.

We have $k \geq (2+\delta)\log\log y > 2\mu \log\log y$. Taking $B = \mu$, Balazard's formula gives

$$N(x,d) = f(2r(y,k))\frac{y}{\log y}P_{k-1}(2\log\log y)(1 + O(R_\mu(y))).$$

Moreover, by (b) above we have $r(y,k) = 1 + O(R_\mu(y))$, which implies

$$f(2r(y,k)) = f(2) + O(R_\mu(y)) = C(1 + O(R_\mu(y)))$$

and

$$P_{k-1}(2\log\log y) = (\log y)^2(1 + O(R_\mu(y))).$$

We thus see that we have

$$N(x,k) = Cy\log y(1 + O(R_\mu(y))). \tag{8}$$

We know that $N_d(x,k) = N(x/d, k - \Omega(d))$. By Halász's formula this is equal to

$$T\left(\frac{x}{2^{k-\Omega(d)}d}, k - \Omega(d)\right) = T\left(\frac{y}{\psi(d)}, k - \Omega(d)\right).$$

Set $y/\psi(d) = w$ (which is ≥ 3 provided that y_0 has been chosen large enough). Since $k - \Omega(d) \geq 2\mu \log\log y \geq 2\mu\log\log w$, Balazard's formula gives

$$N_d(x,k) = f(2r(w, k - \Omega(d)))\frac{w}{\log w}P_{k-\Omega(d)-1}(2\log\log w)(1 + O(R_\mu(w))).$$

We have

$$\begin{aligned}\log w = \log y - \log\psi(d) &= \log y\left(1 - \frac{\log\psi(d)}{\log y}\right)\\ &= \log y(1 + O(R_\mu(y))).\end{aligned}$$

It follows first that $R_\mu(w) = O(R_\mu(y))$, so that $O(R_\mu(w))$ may be replaced by $O(R_\mu(y))$. Further, since $k - \Omega(d) \geq 2\mu\log\log w$ it follows, again by (b), that

$$r(w, k - \Omega(d)) = 1 + O(R_\mu(w)) = 1 + O(R_\mu(y)),$$

which yields $f(2r(w, k - \omega(d))) = C(1 + O(R_\mu(y)))$. We have also by (b)

$$\begin{aligned}P_{k-\Omega(d)-1}(2\log\log w) &= (\log w)^2(1 + O(R_\mu(w)))\\ &= (\log w)^2(1 + O(R_\mu(y)))\end{aligned}$$

Combining these estimates we get

$$\begin{aligned}N_d(x,k) &= Cw\log w(1 + O(R_\mu(y)))\\ &= C\frac{y}{\psi(d)}\log y(1 + O(R_\mu(y))).\end{aligned}$$

This with (8) gives the desired result.

4.5. From Theorem 4 we derive

Theorem 5. *Let p be any odd prime, and let $\delta' \in \,]0,\delta[$. Let α be a non-negative integer. We have uniformly for $\alpha \le \delta' \log\log y$*

$$\operatorname{Prob}(V_p(n) \ge \alpha) = \left(\frac{2}{p}\right)^{\alpha} (1 + O(R_\mu(y)))$$

and

$$\operatorname{Prob}(V_p(n) = \alpha) = \left(\frac{2}{p}\right)^{\alpha} (1 - \frac{2}{p})(1 + O(R_\mu(y))),$$

where $\mu = 1 + \frac{\delta-\delta'}{2}$.

For the proof we take $d = p^\alpha$ in Theorem 4. We have

$$k - \Omega(d) = k - \alpha \ge (2 + \delta - \delta') \log\log y = 2\mu \log\log y,$$

whence $\Omega(d) \le k - 2\mu \log\log y$. Moreover,

$$\log \psi(d) = \alpha \log \frac{p}{2} \le \delta' \log \frac{p}{2} \log\log y,$$

which is much smaller than required. This gives the first formula.

For the second one we apply Theorem 4 to $d = p^\alpha$ and to $d = p^{\alpha+1}$, and use formula (3).

We also have

Theorem 6. *Let $p_1, p_2, \cdots, p_q$ be distinct odd primes, $\delta' \in \,]0,\delta[$, and let $\alpha_1, \alpha_2, \cdots, \alpha_q$ be non-negative integers. We have uniformly for $\alpha_1 + \cdots + \alpha_q \le \delta' \log\log y$*

$$\operatorname{Prob}\left(V_{p_j}(n) = \alpha_j \text{ for } j = 1, 2, \cdots, q\right)$$
$$= \left(\prod_{j=1}^{q} \left(\frac{2}{p_j}\right)^{\alpha_j} \left(1 - \frac{2}{p_j}\right)\right) (1 + O(R_\mu(y))),$$

where $\mu = 1 + \frac{\delta-\delta'}{2}$.

We obtain this result by applying Theorem 4 to $d = p_1^{\alpha_1+\epsilon_1} \cdots p_q^{\alpha_q+\epsilon_q}$, where $\epsilon_i = 0$ or 1, and applying formula (4).

4.6. As $\psi(2^\alpha) = 1$, Theorem 4 with $d = 2^\alpha$ gives $\operatorname{Prob}(V_2(n) \ge \alpha) = 1 + O(R_\mu(y))$ uniformly for $\alpha \le \delta' \log\log y$ $(0 < \delta' < \delta)$, where $\mu = 1 + \frac{\delta-\delta'}{2}$. We can prove a result of a different kind.

Theorem 7. *Given $A > 0$ we have uniformly for $t \leq A$*

$$\mathrm{Prob}(V_2(n) \leq k - 2\log\log y + t\sqrt{2\log\log y}) = G(t) + O\left(\frac{1}{\sqrt{\log\log y}}\right),$$

where $G(t) = \frac{1}{\sqrt{2\pi}}\int_{-\infty}^{t} e^{-u^2/2}\,du$.

Proof: The formula to be proved is equivalent to

$$\begin{aligned}\mathrm{Prob}(V_2(n) > k - 2\log\log y + t\sqrt{2\log\log y}) &= 1 - G(t) + O\left(\frac{1}{\sqrt{\log\log y}}\right)\\ &= G(-t) + O\left(\frac{1}{\sqrt{\log\log y}}\right). \qquad (9)\end{aligned}$$

Let $\alpha = [k - 2\log\log y + t\sqrt{2\log\log y}] + 1$. The condition $V_2(n) > k - 2\log\log y + t\sqrt{2\log\log y}$ is equivalent to $V_2(n) \geq \alpha$. So the left-hand side of (9) is $N_{2^\alpha}(x,k)/N(x,k)$.

As $\psi(2^\alpha) = 1$, $N_{2^\alpha}(x,k) = T(y, k-\alpha)$. Balazard's formula gives

$$T(y, k-\alpha) = f(2r(y,k-\alpha))\frac{y}{\log y}P_{k-\alpha-1}(2\log\log y)(1 + O(\frac{1}{\log\log y})).$$

We have $k - \alpha - 1 = 2\log\log y - u\sqrt{2\log\log y}$, where $u = t + O(1/\sqrt{\log\log y})$. Now, $t \leq A$ implies $u \leq$ some constant.

By (c) of §4.3, $r(y, k-\alpha) = 1 + O(1/\sqrt{\log\log y})$, so that $f(2r(y,k-\alpha)) = C(1 + O(1/\sqrt{\log\log y}))$, and

$$\begin{aligned}P_{k-\alpha-1}(2\log\log y) &= (\log y)^2\left(G(-u) + O\left(\frac{1}{\sqrt{\log\log y}}\right)\right)\\ &= (\log y)^2\left(G(-t) + O\left(\frac{1}{\sqrt{\log\log y}}\right)\right)\\ &= (\log y)^2 G(-t)\left(1 + O\left(\frac{1}{\sqrt{\log\log y}}\right)\right).\end{aligned}$$

Therefore

$$N_{2^\alpha}(x,k) = T(y,k-\alpha) = Cy\log y G(-t)(1 + O(1/\sqrt{\log\log y})).$$

By (8) we have

$$N(x,k) = Cy\log y(1 + O(1/\sqrt{\log\log y})) \quad (for\ R_\mu(y) = O(1/\sqrt{\log\log y})).$$

The last two relations give the desired result.

4.6.1 Theorem 7 is obviously equivalent to

Theorem 7′. *Write $n = 2^{\alpha}m$, where m is odd ($m = 2^{-V_2(n)}n$). We have uniformly for $t \leq A$*

$$\operatorname{Prob}(\Omega(m) \geq 2\log\log y - t\sqrt{2\log\log y}) = G(t) + O\left(\frac{1}{\sqrt{\log\log y}}\right)$$

(because $\Omega(m) = k - V_2(n)$).

Now, as we have remarked earlier (see §4.2), if we suppose $(2+\delta)\log\log x \leq k \leq \lambda \log x/\log 2$, where $\lambda \in]0,1[$ is given , then we may replace $\log\log y$ by $\log\log x$. It follows that, for every positive A,

$$\lim_{x\to\infty} \operatorname{Prob}(|\Omega(m) - 2\log\log x| \leq A\sqrt{2\log\log x}) = \frac{1}{\sqrt{2\pi}}\int_{-A}^{A} e^{-u^2/2}du,$$

and therefore, for every positive η,

$$\operatorname{Prob}(|\Omega(m) - 2\log\log x| > 2\eta\log\log x) = o(1) \qquad \text{as } x \to \infty.$$

It is easy to deduce that, for every positive ϵ,

$$\operatorname{Prob}(m > e \text{ and } |\Omega(m) - 2\log\log m| \geq \epsilon\log\log m) = o(1).$$

(It is useful to notice that the number of $n \in S(x,k)$ for which $m \leq \sqrt{x}$ is $\leq \sqrt{x}$.) We can therefore say that, on the set $S(x,k)$, $\Omega(m)$ has *normal order* $2\log\log m$.

4.7 Theorem 7 does not give a good evaluation of $\operatorname{Prob}(V_2(n) = \alpha)$ for α near $k - 2\log\log y$. It gives only

$$\operatorname{Prob}(V_2(n) = \alpha) = O\left(\frac{1}{\sqrt{\log\log y}}\right).$$

By another method we can prove

Theorem 8. *Given $A > 0$ we have uniformly for $|\alpha - (k - 2\log\log y)| \leq A\sqrt{2\log\log y}$*

$$\operatorname{Prob}(V_2(n) = \alpha) =$$
$$\frac{1}{\sqrt{4\pi\log\log y}}\exp\left(-\frac{(\alpha-(k-2\log\log y))^2}{4\log\log y}\right)\left(1+O\left(\frac{1}{\sqrt{\log\log y}}\right)\right).$$

Proof: We may suppose $y \geq y_0$ for some given y_0, for it is easy to see that the result holds for $y < y_0$.

We have seen in §4.1 that there is a one-to-one correspondence between the set $S(x,k)$ and the set of those odd positive integers m which satisfy $\psi(m) \le y$ and $\Omega(m) \le k$, and that $V_2(n) = k - \Omega(m)$ where m is the odd integer associated to n. It follows that

$$\#\{n \in S(x,k) : V_2(n) = \alpha\} = \sum_{\substack{m \text{ odd} \\ \psi(m)\le y \\ \Omega(m)=k-\alpha}} 1.$$

Following Balazard we start with the formula

$$\sum_{\substack{m \text{ odd} \\ \psi(m)\le y}} z^{\Omega(m)} = zf(2z)y(\log y)^{2z-1} + O(y \log\log y(\log y)^{2\Re e\, z-2}),$$

which holds uniformly for $|z| \le R < 3/2$. But now we argue as Selberg. We see that, given $\epsilon \in]0,3[$, we have uniformly for $1 \le k-\alpha \le (3-\epsilon)\log\log y$

$$\sum_{\substack{m \text{ odd} \\ \psi(m)\le y \\ \Omega(m)=k-\alpha}} 1 = f\left(\frac{k-\alpha-1}{\log\log y}\right)\frac{y(2\log\log y)^{k-\alpha-1}}{(k-\alpha-1)!\log y}\left(1+O\left(\frac{1}{\log\log y}\right)\right).$$

The hypothesis on α gives

$$\left|\frac{k-\alpha}{\log\log y} - 2\right| \le \frac{A\sqrt{2}}{\sqrt{\log\log y}}.$$

It follows that $\frac{k-\alpha}{\log\log y} \le 3-\epsilon$, where $\epsilon \in\ ]0,1[$ is given, provided that y_0 has been chosen large enough. Moreover

$$f\left(\frac{k-\alpha-1}{\log\log y}\right) = C\left(1+O\left(1/\sqrt{\log\log y}\right)\right).$$

This gives

$$\#\{n \in S(x,k) : V_2(n) = \alpha\} \\ = C\frac{y}{\log y}\frac{(2\log\log y)^{k-\alpha-1}}{(k-\alpha-1)!}\left(1+O\left(\frac{1}{\sqrt{\log\log y}}\right)\right).$$

As $N(x,k) = Cy\log y(1+O(1/\log\log y))$ it follows that

$$\text{Prob}(V_2(n)=\alpha) = \frac{1}{(\log y)^2}\frac{(2\log\log y)^{k-\alpha-1}}{(k-\alpha-1)!}\left(1+O\left(\frac{1}{\sqrt{\log\log y}}\right)\right).$$

As $\frac{k-\alpha}{2\log\log y} = 1 + O(1/\sqrt{\log\log y})$ we have

$$\frac{(2\log\log y)^{k-\alpha-1}}{(k-\alpha-1)!} = \frac{(2\log\log y)^{k-\alpha}}{(k-\alpha)!}\left(1+O\left(\frac{1}{\sqrt{\log\log y}}\right)\right).$$

Setting $\alpha = k - 2\log\log y + t\sqrt{2\log\log y}$, so that $|t| \leq A$, and using Stirling's formula we arrive at

$$\begin{aligned}&\operatorname{Prob}(V_2(n) = \alpha)\\&\quad= \frac{1}{\sqrt{4\pi\log\log y}}\exp\left(-\frac{t^2}{4\log\log y}\right)\left(1+O\left(\frac{1}{\sqrt{\log\log y}}\right)\right),\end{aligned}$$

which is the desired result.

REFERENCES

[1] L.G.Sathe, On a problem of Hardy on the distribution of integers having a prescribed number of prime factors, I, II, III, IV, J. Indian Math. Soc. **17** (1953), 63–141; **18** (1954), 27–81.

[2] A. Selberg, Note on a paper by L.G. Sathe, J. Indian Math. Soc. **18** (1954), 83–87.

[3] J.L. Nicolas, Sur la distribution des entiers ayant une quantité fixée de facteurs premiers, Acta Arith. **44** (1984), 191–200.

[4] M. Balazard, Sur la répartition des valeurs de certaines fonctions arithmétiques additives, Thèse, Université de Limoges, 1987.

H. Delange
Département de Mathématiques
Université de Paris-Sud
91405 Orsay
France

A Boundary Value Problem for a Pair of Differential Delay Equations Related to Sieve Theory, I

H. DIAMOND, H. HALBERSTAM
AND H.-E. RICHERT

Dedicated to Paul T. Bateman on the occasion of his retirement

1. Introduction

We showed in [**DHR 1**] how to construct sieves of dimension (or *sifting density*) $\kappa > 1$, on the assumption that the following result is true:

Theorem 0. *Let $\kappa \geq 1$ be given, and let $\sigma = \sigma_\kappa$ be the continuous solution of the differential-difference problem*

$$\begin{cases} u^{-\kappa}\sigma(u) = A^{-1}, \quad 0 < u \leq 2, \quad A = A_\kappa = (2e^\gamma)^\kappa\Gamma(\kappa+1), \\ (u^{-\kappa}\sigma(u))' = -\kappa u^{-\kappa-1}\sigma(u-2), \quad 2 < u, \end{cases} \tag{1.1}$$

where γ is Euler's constant and Γ is Euler's gamma function. Then there exist numbers $\alpha = \alpha_\kappa$, $\beta = \beta_\kappa$ satisfying $\alpha \geq \beta \geq 2$ such that the simultaneous differential-difference system

$$\begin{cases} (i) & F(u) = 1/\sigma(u), & 0 < u \leq \alpha, \\ (ii) & f(u) = 0, & 0 < u \leq \beta, \\ (iii) & (u^\kappa F(u))' = \kappa u^{\kappa-1} f(u-1), & \alpha < u, \\ (iv) & (u^\kappa f(u))' = \kappa u^{\kappa-1} F(u-1), & \beta < u \end{cases} \tag{1.2}$$

Research supported in part by grants from the National Science Foundation and the Research Board of the University of Illinois

has continuous solutions $F = F_\kappa$ and $f = f_\kappa$ with the properties

$$F(u) = 1 + O(e^{-u}), \quad f(u) = 1 + O(e^{-u}), \tag{1.3}$$

$$F(u) \quad \textit{decreases monotonically towards} \quad 1 \quad \textit{as} \quad u \to \infty \tag{1.4}$$

$$f(u) \quad \textit{increases monotonically towards} \quad 1 \quad \textit{as} \quad u \to \infty. \tag{1.5}$$

The case $\kappa = 1$ (with $\alpha_1 = \beta_1 = 2$) of Theorem 0—corresponding to the so-called linear sieve—is well-known and two distinct accounts ([**JR**] and [**HR**], [**I**]) exist in the literature. We shall not deal with it again here but *assume from now on that*

$$\kappa > 1. \tag{1.6}$$

The proof of Theorem 0 is complicated and will occupy several papers bearing the same title. In this, the first of the series, we prove subject to (1.6) that if (1.2) possesses solutions $F = F_\kappa$, $f = f_\kappa$ satisfying (1.3), (1.4) and (1.5) then, *necessarily*,

$$\alpha_\kappa > \beta_\kappa \quad \text{when} \quad \kappa > 1. \tag{1.7}$$

Although Theorem 0 arose in the context of sieves, it deals really with a boundary value problem for a linked pair of differential-difference equations with retarded argument and will, we hope, be of some interest also in that context.

We suppose that

$$\alpha_\kappa > 1, \quad \beta_\kappa > 1 \tag{1.8}$$

and, in view of (1.7), assume from now on that[1]

$$\alpha_\kappa \le \beta_\kappa. \tag{1.9}$$

Our aim is to show that *the system* (1.2) *with* (1.9) *has no solutions that satisfy all three of* (1.3), (1.4) *and* (1.5). In the course of achieving this aim we shall lay the technical foundations of our method, especially in an extended Appendix consisting of several sections, each dealing with one or more of the auxiliary functions that are connected with the theorem.

[1] We shall have occasion subsequently to quote results from this article that are independent of (1.9). We put a star next to each displayed relation that depends on (1.9) to provide a warning that it is conditional.

1. Necessary implications of (1.9)

An account of relevant properties of the Ankeny-Onishi function $\sigma_\kappa(u)$ (see (1.1)) is given in the Appendix, in Section Aσ, and we shall quote freely from there as the need arises. Suffice it to say here that $\sigma(u)$ is positive and strictly increasing in u for $u > 0$ by (1.1) and tends exponentially fast to 1 as $u \to \infty$ (σ8).

We form the functions

$$P(u) := F(u) + f(u), \quad Q(u) := F(u) - f(u). \tag{2.1}$$

Since F, f must satisfy (1.4) and (1.5) we may take it from now on that

$$Q(u) > 0 \quad \text{if} \quad u > 0; \tag{2.2}$$

and similarly, in view of (1.3), that

$$Q(u) \ll e^{-u} \tag{2.3}$$

and that

$$P(u) = 2 + O(e^{-u}). \tag{2.4}$$

We shall now restate (1.2) in terms of P and Q, and on the basis of (1.9). First note that by (1.2), (ii), $P(u) = Q(u) = F(u)$ when $0 < u \leq \beta$; more specifically,

$$P(u) = Q(u) = \frac{1}{\sigma(u)}, \quad 0 < u \leq \alpha \tag*{(2.5)*}$$

by (1.2), (i), and

$$P(u) = Q(u) = \frac{\alpha^\kappa}{\sigma(\alpha)} u^{-\kappa}, \quad \alpha \leq u \leq \beta \tag*{(2.6)*}$$

on integrating (1.2), (iii), from α to u and using (ii) and (i). Then (1.2), (iii) and (iv), with $u > \beta$ may be rephrased as

$$(u^\kappa P(u))' = \kappa u^{\kappa-1} P(u-1), \quad u > \beta, \tag*{(2.7)*}$$

and

$$(u^\kappa Q(u))' = -\kappa u^{\kappa-1} Q(u-1), \quad u > \beta. \tag*{(2.8)*}$$

Alternative versions of these equations are

$$uP'(u) = -\kappa P(u) + \kappa P(u-1), \quad u > \beta, \tag*{(2.9)*}$$

and

$$uQ'(u) = -\kappa Q(u) - \kappa Q(u-1), \quad u > \beta. \tag*{(2.10)*}$$

Following Iwaniec [I] (although the Laplace transform approach of Rawsthorne [R] leads to the same conclusions), we introduce auxiliary functions $p = p_\kappa$ and $q = q_\kappa$ which satisfy the adjoint differential equations

$$(up(u))' = \kappa p(u) - \kappa p(u+1), \quad u > 0, \tag{2.11}$$

and

$$(uq(u))' = \kappa q(u) + \kappa q(u+1), \quad u > 0, \tag{2.12}$$

and the normalized polynomial-like boundary conditions

$$p_\kappa(u) \sim u^{-1}, \quad q_\kappa(u) \sim u^{2\kappa-1}, \quad u \to \infty. \tag{2.13}$$

In contrast to P and Q, we have explicit representations for the p and q functions. These auxiliary functions are important here because the two "inner product" expressions

$$up(u)P(u) + \kappa \int_{u-1}^{u} P(t)p(t+1)dt$$

and

$$uq(u)Q(u) - \kappa \int_{u-1}^{u} Q(t)q(t+1)dt$$

are each *constant* for $u > \beta_\kappa$ (cf. [I], (5–3)). We determine these constants by letting $u \to \infty$ and using (2.3), (2.4), and (2.13). We find that

$$up(u)P(u) + \kappa \int_{u-1}^{u} P(t)p(t+1)dt = 2, \quad u \geq \beta, \tag{2.14}^*$$

and

$$uq(u)Q(u) - \kappa \int_{u-1}^{u} Q(t)q(t+1)dt = 0, \quad u \geq \beta. \tag{2.15}^*$$

(Although (2.14) and (2.15) are valid only for $u > \beta$ in the first instance, their truth at $u = \beta$ follows by continuity.)

An account of the auxiliary functions $p_\kappa(u)$ and $q_\kappa(u)$ is given in sections Ap and Aq—for example, the asymptotic statements in (2.13) are given in (p4) and (q3) respectively. Iwaniec showed that $p_\kappa(u)$ is positive and decreasing in u and that $q_\kappa(u)$ has fewer than 2κ real zeros. The largest real zero $\rho = \rho_\kappa$ of $q_\kappa(u)$ plays a major rôle throughout this investigation; for the moment we note only that (q5)

$$\rho_\kappa > \kappa > 1.$$

Then we can deduce at once that

$$\beta_\kappa > \rho_\kappa. \tag{2.16}^*$$

For suppose on the contrary that $\beta \leq \rho$. Then (2.15) may be invoked at $u = \rho$ and there gives

$$\int_{\rho-1}^{\rho} Q(t)q(t+1)dt = 0, \quad *$$

obviously a contradiction by (2.2) since $Q > 0$ and $q(u) > 0$ if $u > \rho$, by (q4).

We now come to a crucial stage of this account: we interpret the 'orthogonality' relations (2.14) and (2.15) at $u = \beta$ with the aid of (2.5) and (2.6). The moment we come to do so, however, we are forced to distinguish (as Rawsthorne [**R**] was) between two cases:

Case I: * $\alpha_\kappa \leq \beta_\kappa - 1$

and

Case II: * $\beta_\kappa - 1 < \alpha_\kappa \leq \beta_\kappa$.

We deal first with Case I; here it is convenient to begin with (2.15) at $u = \beta$, which states, using (2.6), that

$$\beta^{1-\kappa}q(\beta)\frac{\alpha^\kappa}{\sigma(\alpha)} = \kappa \int_{\beta-1}^{\beta} \frac{\alpha^\kappa}{\sigma(\alpha)} \cdot t^{-\kappa}q(t+1)dt. \quad *$$

Since (2.12) may be restated as

$$(u^{1-\kappa}q(u))' = \kappa u^{-\kappa}q(u+1) \tag{2.12$'$}$$

the integral on the right may be evaluated and we arrive immediately at

$$q(\beta - 1) = 0. \quad *$$

Thus $\beta - 1$ is a zero of $q(u)$; and, indeed, we claim that

$$\beta = 1 + \rho \quad \text{in Case I.} \tag{2.17}^*$$

Otherwise, if $\rho' = \rho'_\kappa$ denotes the next–to largest real zero of $q(u)$, then $\beta - 1 \leq \rho'$. But, by (q6), $\rho' < \rho - 1$ and then we arrive at $\beta < \rho$ contradicting (2.16).

Next, still in Case I, set $u = \beta$ in (2.14) and again use (2.6) to obtain

$$\beta^{1-\kappa}p(\beta)\frac{\alpha^\kappa}{\sigma(\alpha)} + \kappa \int_{\beta-1}^{\beta} \frac{\alpha^\kappa}{\sigma(\alpha)} t^{-\kappa}p(t+1)dt = 2. \quad *$$

Since (2.11) may be rewritten in the form

$$(u^{1-\kappa}p(u))' = -\kappa u^{-\kappa}p(u+1) \tag{2.11'}$$

we obtain after integration and substitution from (2.17)

$$\frac{\alpha^\kappa}{\sigma(\alpha)}\rho^{1-\kappa}p(\rho) = 2. \qquad *$$

This and (2.17) are the necessary conditions in Case I. By (1.1), $u^\kappa/\sigma(u)$ increases with u and therefore

$$\alpha^\kappa/\sigma(\alpha) \le (\beta-1)^\kappa/\sigma(\beta-1) = \rho^\kappa/\sigma(\rho)$$

by (2.17). It follows that

$$\rho p(\rho)/\sigma(\rho) \ge 2 \quad \textit{is necessary in Case I.} \tag{2.18*}$$

In §4 we shall show that this inequality does not hold.

We now turn to Case II, when $\beta - 1 < \alpha \le \beta$, and proceed as before, starting with (2.14) this time at $u = \beta$ and applying both (2.5) and (2.6):

$$\begin{aligned}\beta^{1-\kappa}p(\beta)\frac{\alpha^\kappa}{\sigma(\alpha)} &+ \kappa\int_{\beta-1}^{\alpha}\frac{p(t+1)}{\sigma(t)}dt \\ &+ \kappa\int_{\alpha}^{\beta}\frac{\alpha^\kappa}{\sigma(\alpha)}\cdot t^{-\kappa}p(t+1)dt = 2 \qquad *\end{aligned}$$

which becomes, after using (2.11′) in the second integral,

$$\frac{\alpha p(\alpha)}{\sigma(\alpha)} + \kappa\int_{\beta-1}^{\alpha}\frac{p(t+1)}{\sigma(t)}dt = 2. \tag{2.19*}$$

In the same way (2.15) at $u = \beta$ in combination with (2.5), (2.6) and (2.12′) leads to

$$\frac{\alpha q(\alpha)}{\sigma(\alpha)} - \kappa\int_{\beta-1}^{\alpha}\frac{q(t+1)}{\sigma(t)}dt = 0. \tag{2.20*}$$

Thus (2.19) *and* (2.20) *are necessary conditions in Case* II.

However, as in Case I we take matters further. First comes a useful deducation from (2.20): since $\beta > \rho$ by (2.16), the integral in (2.20) is positive and hence so is $q(\alpha)$; but $\alpha > \beta - 1 > \rho - 1 \ge \rho'$ (and $q(u) < 0$ for $\rho' < u < \rho$ by Lemma q2) whence even

$$\alpha > \rho \quad \text{in Case II.} \tag{2.21*}$$

Next, write the integrand in (2.20) in the form

$$\frac{t^\kappa}{\sigma(t)}\kappa t^{-\kappa} q(t+1)$$

and integrate by parts on the basis of (2.12′) and (1.1). We obtain

$$\begin{aligned} 0 &= \frac{(\beta-1)q(\beta-1)}{\sigma(\beta-1)} + \int_{\beta-1}^{\alpha} t^{1-\kappa} q(t) d\left(\frac{t^\kappa}{\sigma(t)}\right) \\ &= \frac{(\beta-1)q(\beta-1)}{\sigma(\beta-1)} + \kappa \int_{\beta-1}^{\alpha} \frac{q(t)\sigma(t-2)}{\sigma^2(t)} dt. \end{aligned} \qquad (2.22)^*$$

Now *suppose if possible that* $\alpha \le 2$ *in Case* II. Then the integral in (2.22) vanishes and the same argument that led in Case I to (2.17) here also gives

$$\beta = 1 + \rho \quad \text{in Case II when} \quad \alpha \le 2. \qquad (2.23)^*$$

At the same time, when $\alpha \le 2$ equation (2.19) using (the first line of) (1.1) reads

$$2 = A\alpha^{1-\kappa} p(\alpha) + A\kappa \int_{\beta-1}^{\alpha} t^{-\kappa} p(t+1) dt = A(\beta-1)^{1-\kappa} p(\beta-1) \qquad *$$

by (2.11′). Hence, by (2.23), $A\rho^{1-\kappa}p(\rho) = 2$; and since $\rho < \alpha \le 2$ from (2.21) we obtain by (1.1) that (cf. (2.18))

$$\rho p(\rho)/\sigma(\rho) = 2 \quad \textit{is necessary in Case II with} \quad \alpha \le 2. \qquad (2.24)^*$$

In §4 we shall show that this equality does not hold.

There remains Case II with $\alpha > 2$. Here we claim that

$$\beta < 1 + \rho; \qquad *$$

for if, on the contrary, $\beta - 1 \ge \rho$ the integral in (2.22) is positive and (2.22) cannot be true. Hence in Case II with $\alpha > 2$ the numbers α, β and ρ stand in the following relationship:

$$\beta - 1 < \rho < \alpha \le \beta < \rho + 1, \quad 2 < \alpha \quad \textit{in Case II}. \qquad (2.25)^*$$

Before going further in this part of Case II we have to introduce two functions which play a crucial rôle in this and later papers. We do this in the next section, where also some of their basic properties are derived and their place in the present investigation is established.

3. The functions Π and χ

For $u > 0$ and $v > 1$ define

$$\Pi(u,v) = \Pi_\kappa(u,v) := \frac{up(u)}{\sigma(u)} + \kappa \int_{v-1}^{u} \frac{p(t+1)}{\sigma(t)} dt \tag{3.1}$$

and

$$\chi(u,v) = \chi_\kappa(u,v) := \frac{up(u)}{\sigma(u)} - \kappa \int_{v-1}^{u} \frac{q(t+1)}{\sigma(t)} dt. \tag{3.2}$$

Integration by parts (as in going from (2.20) to (2.22)) on the basis of (2.11′), (2.12′) and (1.1) leads to alternative versions

$$\Pi_\kappa(u,v) = \frac{(v-1)p(v-1)}{\sigma(v-1)} + \kappa \int_{v-1}^{u} \frac{p(t)\sigma(t-2)}{\sigma^2(t)} dt \tag{3.3}$$

and

$$\chi_\kappa(u,v) = \frac{(v-1)q(v-1)}{\sigma(v-1)} + \kappa \int_{v-1}^{u} \frac{q(t)\sigma(t-2)}{\sigma^2(t)} dt. \tag{3.4}$$

Then (2.19) and (2.20) assert that

$$\Pi_\kappa(\alpha_\kappa, \beta_\kappa) = 2, \quad \chi_\kappa(\alpha_\kappa, \beta_\kappa) = 0 \text{ are necessary in Case II.} \tag{3.5*}$$

By (3.3) and (3.4)

$$\frac{\partial}{\partial u}\Pi(u,v) = \kappa \frac{p(u)\sigma(u-2)}{\sigma^2(u)} > 0 \quad \text{if} \quad u > 2 \tag{3.6}$$

and

$$\frac{\partial}{\partial u}\chi(u,v) = \kappa \frac{q(u)\sigma(u-2)}{\sigma^2(u)} > 0 \quad \text{if} \quad u > \max(2,\rho); \tag{3.7}$$

it follows in particular that each of $\Pi(u,\beta)$, $\chi(u,\beta)$ increases (strictly) with u for $u > \max(2,\rho)$ and we deduce from (3.5) by (2.25) that

$$\Pi(\beta,\beta) \geq 2 \quad \text{and} \quad \chi(\beta,\beta) \geq 0 \quad \text{in Case II with} \quad \alpha > 2. \tag{3.8*}$$

Next define

$$\begin{aligned} \Pi(u) &= \Pi_\kappa(u) := \Pi_\kappa(u,u), \\ \chi(u) &= \chi_\kappa(u) := \chi_\kappa(u,u) \quad (u > 1); \end{aligned} \tag{3.9}$$

by (3.1) and (3.2), (3.6) and (3.7),

$$\begin{aligned} \Pi'(u) &= \frac{\partial}{\partial u}\Pi(u,v)|_{v=u} - \kappa \frac{p(u)}{\sigma(u-1)} \\ &= \kappa p(u) \left\{ \frac{\sigma(u-2)}{\sigma^2(u)} - \frac{1}{\sigma(u-1)} \right\} < 0 \quad (u > 1) \end{aligned} \tag{3.10}$$

and

$$\begin{aligned}\chi'(u) &= \frac{\partial}{\partial u}\chi(u,v)|_{v=u} + \kappa\frac{q(u)}{\sigma(u-1)} \\ &= \kappa q(u)\left\{\frac{\sigma(u-2)}{\sigma^2(u)} + \frac{1}{\sigma(u-1)}\right\} > 0 \quad (u > \rho).\end{aligned} \tag{3.11}$$

It follows that

$$\Pi(u) \quad \textit{is strictly decreasing in} \quad u > 1 \tag{3.12}$$

and

$$\chi(u) \quad \textit{is strictly increasing in} \quad u > \rho. \tag{3.13}$$

Moreover, by (1.1), the positivity of p and σ, (2.13), and ($\sigma 8$) we have

$$\lim_{u\to 1+0} \Pi(u) = +\infty, \quad \lim_{u\to\infty} \Pi(u) = 1, \tag{3.14}$$

so that by (3.12) *the equation*

$$\Pi_\kappa(u) = 2$$

possesses a unique root, to be denoted by $z_\Pi = z_\Pi(\kappa)$, *located to the right of* 1.

Similarly, by (3.2), (3.4), and (2.13)

$$\chi_\kappa(\rho_\kappa) < 0 \quad \text{and} \quad \lim_{u\to\infty} \chi_\kappa(u) = +\infty,$$

so that, by (3.13), *the equation*

$$\chi_\kappa(u) = 0$$

possesses a root, to be denoted by $z_\chi = z_\chi(\kappa)$, *which is unique on the interval* (ρ_κ, ∞).

We may now deduce from (3.8), (3.12) and (3.13) that

$$z_\chi(\kappa)(\leq \beta_\kappa) \leq z_\Pi(\kappa) \quad \textit{in Case II with } \alpha_\kappa > 2. \tag{3.16)*$$

We shall prove that the three necessary conditions (2.18), (2.24) and (3.16) are false; so that *Theorem 0 with* $\kappa > 1$ *can hold only with* $\alpha_\kappa > \beta_\kappa$. The proof for all $\kappa > 1$ is given in the next section and makes essential use of computer calculations.

4. Disproof of (2.18), (2.24), and (3.16) for all $\kappa > 1$

We now establish the inequality $z_\Pi(\kappa) < z_\chi(\kappa)$ for $\kappa > 1$. The principal tools are a lower estimate for z_χ of Grupp [G] and a monotonicity result for Π.

Here we shall have to make use of numerical data. Calculations of σ, ρ, q and other functions were made independently in Ulm and Urbana and are in full agreement. The programs of Wheeler for computing $p_\kappa(u)$ and $\sigma_\kappa(u)$ were designed to guarantee a relative error of less than 10^{-14}. For p this accuracy is in the range $1 \le \kappa \le 15$ and $1 \le u \le 50$; for σ the range is $\kappa > 0$ and $0 \le u \le 6$. Since p and σ are smaller than 1 in these ranges, the error estimates are also absolute. Wheeler's methods are described in his Illinois Ph.D. Thesis [W]. His data agree with those of te Riele [te R] to the ten decimal places given by te Riele.

Define

$$\nu(\kappa) := \begin{cases} 3\kappa - 1.4, & \kappa \ge 2.4 \\ 3\kappa - 1.45, & 1.5 \le \kappa < 2.4 \\ 3\kappa - 1.4, & 1.44 \le \kappa < 1.5 \\ 2 + 2.25(\kappa - 1), & 1.05 \le \kappa < 1.44 \\ 2 + 2.48(\kappa - 1), & 1 \le \kappa < 1.05 \end{cases} \tag{4.1}$$

(note that $\nu(\kappa) > 2$ if $\kappa > 1$).

Grupp [G] (Theorem 5) proved that

$$\max(2, \rho_\kappa + \frac{1}{2}) < z_\chi(\kappa) < \rho_\kappa + 1 \tag{4.2}$$

and

$$\nu(\kappa) < z_\chi(\kappa), \quad \kappa > 1. \tag{4.3}$$

We shall now prove what is, in effect, the principal result of this paper:

Theorem 1. *We have for every $\kappa > 1$ that*

$$\Pi_\kappa(\nu(\kappa)) < 2,$$

and consequently that

$$z_\Pi(\kappa) < \nu(\kappa) < z_\chi(\kappa), \quad \kappa > 1.$$

This result disproves (3.16). By the preceding inequalities $z_\Pi(\kappa) < z_\chi(\kappa) < \rho + 1$, and it follows from the monotonicity of $\Pi(u)$ and from (3.3) (with $u = v = \rho + 1 > 2$) that

$$2 > \Pi(\rho + 1) > \rho p(\rho)/\sigma(\rho).$$

Hence

Corollary. *For* $\kappa > 1$, $\rho_\kappa p_\kappa(\rho_\kappa)/\sigma_\kappa(\rho_\kappa) < 2$.

This disproves both (2.18) and (2.24).

The rest of this section is devoted to a proof of Theorem 1. Since $\Pi_1(\nu(1)) = \Pi_1(2) = A_1 p_1(1) = 2$ by (p8), Theorem 1 would follow at once if we could prove that $\Pi_\kappa(\nu(\kappa))$ is strictly decreasing in κ. Unfortunately we have failed so far to prove this conjecture, amply supported though it is by the data.

Therefore we shall proceed by devising upper estimates for $\Pi_\kappa(u)$ that are good enough for the various ranges of values of κ indicated by the definition of $\nu(\kappa)$ (see (4.1)). Since $\Pi_1(\nu(1)) = 2$ we may expect the smallest κ's to give the most trouble. It comes, under the circumstances, as a pleasant surprise that we can establish Theorem 1 in three stages.

(a) $\kappa \geq 2.4$. Here $\nu(\kappa) = 3\kappa - 1.4 \geq 2\kappa + 1$, and we cite from **[GR]**, (6.15),

$$z_\Pi(\kappa) < 2\kappa + 1, \quad \kappa > 1.$$

By (4.3), this proves our result in case (a).

(b) $1.05 \leq \kappa < 2.4$. We begin with a lemma establishing an upper estimate of $\Pi_\kappa(u)$ that is very well suited to take advantage of numerical information. The lemma is not elegant in form; nevertheless we shall see that, together with six sets of numerical evaluations, it suffices to account for all but a small neighborhood of $\kappa = 1$, of the remaining values of κ.

Lemma 4.1. *For all κ satisfying $(1 \leq)\kappa_1 \leq \kappa \leq \kappa_2$ and all constants b (independent of κ) satisfying $0 \leq b \leq \frac{1}{2}$, we have uniformly*

$$\begin{aligned}\Pi_\kappa(2\kappa + b) < \frac{1}{\sigma_{\kappa_2}(2\kappa_2 + b)} &\left\{(2\kappa_1 + b)p_{\kappa_1}(2\kappa_1 + b) + \frac{\kappa_2}{4} p_{\kappa_2}(2\kappa_2 + b + 1)\right\} \\ &+ \frac{\kappa_2}{8}\frac{(2\kappa_1)^{\kappa_1}}{\sigma_{\kappa_2}(2\kappa_2)} \left\{ \frac{3p_{\kappa_2}\left(2\kappa_2 + b + \frac{1}{2}\right)}{\left(2\kappa_1 + b - \frac{1}{2}\right)^{\kappa_1}} \right. \\ &+ p_{\kappa_2}\left(2\kappa_2 + b + \frac{1}{4}\right)\left(\frac{2}{\left(2\kappa_1 + b - \frac{3}{4}\right)^{\kappa_1}} \right. \\ &\left.\left. + \frac{1}{(2\kappa_1 + b - 1)^{\kappa_1}} \frac{2\kappa_1 + b + \frac{1}{4}}{2\kappa_1 + b}\right)\right\}.\end{aligned}$$

Proof: Let

$$u = 2\kappa + b, \quad 0 \le b \le \frac{1}{2} \tag{4.4}$$

We have

$$\Pi_\kappa(u) = \frac{up(u)}{\sigma(u)} + \kappa \int_{u-1}^{u} \frac{p(t+1)}{\sigma(t)} dt$$

from (3.9) and (3.1) (with $v = u$). We split up the range of integration into intervals $(u-1, u-\frac{3}{4})$, $(u-\frac{3}{4}, u-\frac{1}{2})$, and $(u-\frac{1}{2}, u)$ and invoke on each interval the convexity of $p(t+1)/\sigma(t)$ (established in Lemma p1). We then obtain

$$\Pi(u) < \frac{up(u)}{\sigma(u)} + \frac{1}{4}\frac{\kappa p(u+1)}{\sigma(u)} + \frac{3}{8}\frac{\kappa p\left(u+\frac{1}{2}\right)}{\sigma\left(u-\frac{1}{2}\right)} + \frac{1}{4}\frac{\kappa p\left(u+\frac{1}{4}\right)}{\sigma\left(u-\frac{3}{4}\right)} + \frac{1}{8}\frac{\kappa p(u)}{\sigma(u-1)}.$$

By (p7) $up(u)$ is increasing in u and, in particular, $up(u) < (u+\frac{1}{4})p(u+\frac{1}{4})$; hence

$$\Pi(u) - \frac{up(u)}{\sigma(u)} - \frac{1}{4}\frac{\kappa p(u+1)}{\sigma(u)} < \frac{3}{8}\frac{\kappa p\left(u+\frac{1}{2}\right)}{\sigma\left(u-\frac{1}{2}\right)} + \frac{\kappa}{8} p\left(u+\frac{1}{4}\right)\left(\frac{2}{\sigma\left(u-\frac{3}{4}\right)} + \frac{1}{\sigma(u-1)}\frac{u+\frac{1}{4}}{u}\right).$$

Next, for each factor $1/\sigma(t)$ on the right use the inequality

$$\frac{1}{\sigma_\kappa(t)} \le \left(\frac{2\kappa}{t}\right)^\kappa \frac{1}{\sigma_\kappa(2\kappa)} \quad \text{if} \quad t \le 2\kappa,$$

valid because $t^\kappa/\sigma_\kappa(t)$ is increasing in $t \ge 0$ (see (1.1)), to obtain

$$\Pi_\kappa(u) < \frac{up_\kappa(u)}{\sigma_\kappa(u)} + \frac{1}{4}\frac{\kappa p_\kappa(u+1)}{\sigma_\kappa(u)} + \frac{3}{8}\left(\frac{2\kappa}{u-\frac{1}{2}}\right)^\kappa \frac{\kappa p_\kappa\left(u+\frac{1}{2}\right)}{\sigma_\kappa(2\kappa)} + \frac{1}{8}\frac{\kappa p_\kappa\left(u+\frac{1}{4}\right)}{\sigma_\kappa(2\kappa)}\left(2\left(\frac{2\kappa}{u-\frac{3}{4}}\right)^\kappa + \left(\frac{2\kappa}{u-1}\right)^\kappa \frac{u+\frac{1}{4}}{u}\right).$$

To deduce the Lemma from this inequality we note that (with u as in (4.4)) each of $1/\sigma_\kappa(u)$ and $1/\sigma_\kappa(2\kappa)$ is increasing in κ by **[GR]** (Theorem 4); $up_\kappa(u)$ is decreasing in κ by Lemma p2; each of $\kappa p_\kappa(u+1)$, $\kappa p_\kappa(u+\frac{1}{2})$, $\kappa p_\kappa(u+\frac{1}{4})$ is increasing in κ by Lemma p3; and each of $(\frac{2\kappa}{u-\frac{1}{2}})^\kappa$, $(\frac{2\kappa}{u-\frac{3}{4}})^\kappa$, $(\frac{2\kappa}{u-1})^\kappa$, $\frac{u+\frac{1}{4}}{u}$ is decreasing in κ. ∎

We now apply Lemma 4.1 six times to cover the range from $\kappa = 1.05$ to $\kappa = 2.4$:

(b_1) $1.56 \le \kappa < 2.4$. Here $\nu(\kappa) = 3\kappa - 1.45 \ge 2\kappa + 0.11$ and therefore

$$\Pi_\kappa(\nu(\kappa)) \le \Pi_\kappa(2\kappa + 0.11).$$

Apply Lemma 4.1 with $\kappa_1 = 1.56$, $\kappa_2 = 2.4$ and $b = 0.11$. Six data-points are required:

$\sigma_{\kappa_2}(2\kappa_2)$ 0.54067...	$\sigma_{\kappa_2}(2\kappa_2 + b)$ 0.56004...	$p_{\kappa_1}(2\kappa_1 + b)$ 0.21555...
$p_{\kappa_2}\left(2\kappa_2 + b + \frac{1}{4}\right)$ 0.13497...	$p_{\kappa_2}\left(2\kappa_2 + b + \frac{1}{2}\right)$ 0.13048...	$p_{\kappa_2}(2\kappa_2 + b + 1)$ 0.12236...

In each case the cited decimal is truncated; the last digit should be increased by 1 for an upper estimate. When these six numbers, suitably rounded, are substituted in Lemma 4.1 we obtain

$$\Pi_\kappa(\nu(\kappa)) < 1.99241.$$

(b_2) $1.44 \le \kappa < 1.56$. Here $\nu(\kappa) = 3\kappa - 1.45$, $1.5 \le \kappa < 1.56$, and $\nu(\kappa) = 3\kappa - 1.4$, $1.44 \le \kappa < 1.5$, so that $\nu(\kappa) \ge 2\kappa + 0.04$ over the whole range. Therefore

$$\Pi_\kappa(\nu(\kappa)) \le \Pi_\kappa(2\kappa + 0.04)$$

and we apply Lemma 4.1 with $\kappa_1 = 1.44$, $\kappa_2 = 1.56$ and $b = 0.04$, to obtain (for the data see the Table below)

$$\Pi_\kappa(\nu(\kappa)) < 1.99756.$$

Similarly, for the remainder of the range, $\nu(\kappa) = 2.25\kappa - 0.25$; therefore, in

(b_3) $1.185 \le \kappa < 1.44$ take $b = 0.04625$, to obtain

$$\Pi_\kappa(\nu(\kappa)) \le \Pi_\kappa(2\kappa + 0.04625) < 1.99867;$$

(b_4) $1.093 \le \kappa < 1.185$ take $b = 0.02325$, to obtain

$$\Pi_\kappa(\nu(\kappa)) \le \Pi_\kappa(2\kappa + 0.02325) < 1.99895;$$

(b_5) $1.06 \le \kappa < 1.093$ take $b = 0.015$, to obtain

$$\Pi_\kappa(\nu(\kappa)) \le \Pi_\kappa(2\kappa + 0.015) < 1.99991;$$

and, finally for

(b_6) $1.05 \le \kappa < 1.06$, take $b = 1/80$, to obtain (working with six digit accuracy this time)

$$\Pi_\kappa(\nu(\kappa)) < \Pi_\kappa(2\kappa + 1/80) < 1.999989.$$

This proves the Theorem for the entire (b)-range of κ.

	$\sigma_{\kappa_2}(2\kappa_2)$	$\sigma_{\kappa_2}(2\kappa_2 + b)$	$p_{\kappa_1}(2\kappa_1 + b)$
b_2	0.55067...	0.55928...	0.23763...
b_3	0.55298...	0.56329...	0.28961...
b_4	0.55862...	0.56437...	0.31701...
b_5	0.56048...	0.56439...	0.32815...
b_6	0.560996...	0.564326...	0.331684...

	$p_{\kappa_2}\left(2\kappa_2 + b + \frac{1}{4}\right)$	$p_{\kappa_2}\left(2\kappa_2 + b + \frac{1}{2}\right)$	$p_{\kappa_2}(2\kappa_2 + b + 1)$
b_2	0.20727...	0.19679...	0.17879...
b_3	0.22358...	0.21142...	0.19076...
b_4	0.27112...	0.25338...	0.22420...
b_5	0.29361...	0.27289...	0.23930...
b_6	0.302574...	0.280598...	0.245196...

TABLE for proof of Theorem 1, (b).
(for b_6 six-figure accuracy was required)

(c) $1 < \kappa < 1.05$. We come to the last and most delicate part of the κ-range. After (4.1) we let

$$u = u_\kappa = 2 + a(\kappa - 1), \quad a = 2.48,$$

and note that

$$2 < u < 2.124.$$

Since $u-1<2$ we may write (3.3) with $v=u$ as

$$\begin{aligned}\Pi(u) &= \frac{(u-1)p(u-1)}{\sigma(u-1)} + \kappa \int_2^u \frac{p(t)\sigma(t-2)}{\sigma^2(t)}dt \\ &= A(u-1)^{1-\kappa}p(u-1) + \frac{\kappa}{A}\int_2^u \frac{p(t)(t-2)^\kappa}{\sigma^2(t)}dt\end{aligned}$$

by (1.1), so that

$$\Pi(u) = A\left\{(u-1)^{1-\kappa}p(u-1) + \frac{\kappa(\kappa-1)^{\kappa+1}}{A^2}\int_0^a \frac{p(2+(\kappa-1)s)}{\sigma^2(2+(\kappa-1)s)}s^\kappa ds\right\}.$$

In the integral on the right arguments of p and σ lie between 2 and 2.124 so that their values do not change drastically over the range of integration. With the important factor $(\kappa-1)^{\kappa+1}$ outside, we estimate the integrand: $p_\kappa(2+(\kappa-1)s) \le p_\kappa(2)$ by (p5) and $\sigma_\kappa(2+(\kappa-1)s) \ge \sigma_\kappa(2) = 2^\kappa/A_\kappa$ by (σ7), whence

$$\Pi(u) \le A\left\{(u-1)^{1-\kappa}p(u-1) + \frac{2\kappa}{\kappa+1}2p(2)(\frac{a}{4})^{\kappa+1}(\kappa-1)^{\kappa+1}\right\}.$$

The factor

$$\frac{2\kappa}{\kappa+1}(\frac{a}{4})^{\kappa+1}2p_\kappa(2) < \frac{2.1}{2.05}(0.62)^2 2p_\kappa(2) < 0.4(2p_\kappa(2)) < 0.4$$

by (p3); hence

$$\Pi_\kappa(u) < A_\kappa\left\{(u-1)^{1-\kappa}p_\kappa(u-1) + \frac{2}{5}(\kappa-1)^{\kappa+1}\right\}.$$

Define

$$\hat{\Pi}_\kappa(u) := A_\kappa\left\{(u-1)^{1-\kappa}p_\kappa(u-1) + \frac{2}{5}(\kappa-1)^{\kappa+1}\right\}; \qquad (4.5)$$

we check easily that $\lim_{\kappa\to 1+0}\hat{\Pi}_\kappa(u) = 2$, since $p_1(1) = e^{-\gamma}$ by (p8). Thus if $\hat{\Pi}_\kappa(u_\kappa)$ is decreasing in κ, then

$$\Pi_\kappa(u_\kappa) = \Pi_\kappa(\nu(\kappa)) < 2, \quad 1 < \kappa < 1.05,$$

follows at once. Therefore it now suffices to prove that

$$\frac{d}{d\kappa}\hat{\Pi}_\kappa(u_\kappa) < 0, \qquad (4.6)$$

and the next lemma is a first and critical step in that direction.

Lemma 4.2. *Let*

$$u = a\kappa - b \geq 1, \quad b \geq 0.$$

Then

$$\frac{d}{d\kappa}(u^{1-\kappa}p_\kappa(u)) \leq -u^{1-\kappa}\left\{(1+\log u)p_\kappa(u) + \frac{b}{u}p_\kappa(u+1)\right\} < 0.$$

Proof: From (p1), the definition of p_κ,

$$\frac{d}{d\kappa}(u^{1-\kappa}p_\kappa(u))$$
$$= -u^{1-\kappa}\int_0^\infty e^{-ux-\kappa g(x)}\left\{\log u + \frac{a(\kappa-1)}{u} + ax + g(x)\right\}dx,$$

where g is defined in Appendix Ag. Since $g(x) \geq 1 - e^{-x}$ for $x \geq 0$ by (g3) we obtain by using (p1) again,

$$\frac{d}{d\kappa}(u^{1-\kappa}p_\kappa(u))$$
$$\leq -u^{1-\kappa}\left\{(\log u + \frac{a(\kappa-1)}{u} + 1)p_\kappa(u) - ap'_\kappa(u) - p_\kappa(u+1)\right\},$$

and this gives the stated conclusion after an application of (2.11). ∎

We now apply Lemma 4.2 with $u-1$ in place of u and $b = a-1$ so $u - 1 = a\kappa - (a-1)$. Although A_κ increases with κ it turns out that

$$\frac{d}{d\kappa}(A_\kappa(u-1)^{1-\kappa}p_\kappa(u-1)) < 0. \tag{4.7}$$

Indeed, we have

$$\frac{d}{d\kappa}(A_\kappa(u-1)^{1-\kappa}p_\kappa(u-1))$$
$$= (u-1)^{1-\kappa}p_\kappa(u-1)A_\kappa(\gamma + \log 2 + \psi(\kappa+1))$$
$$+ A_\kappa\frac{d}{d\kappa}((u-1)^{1-\kappa}p_\kappa(u-1))$$
$$\leq A_\kappa(u-1)^{1-\kappa}\Big\{(\gamma + \log 2 + \psi(\kappa+1))p_\kappa(u-1)$$
$$- (1+\log(u-1))p_\kappa(u-1) - \frac{a-1}{u-1}p_\kappa(u)\Big\}$$

by Lemma 4.2, so that

$$\frac{d}{d\kappa}(A_\kappa(u-1)^{1-\kappa}p_\kappa(u-1)) \leq A_\kappa(u-1)^{1-\kappa}\{\phi_1(\kappa)p_\kappa(u-1) - \phi_2(\kappa)\} \tag{4.8}$$

where

$$\phi_1(\kappa) = \gamma + \log 2 - 1 + \psi(\kappa+1) - \log(u-1)$$

and

$$\phi_2(\kappa) = \frac{a-1}{u-1} p_\kappa(u).$$

We have, by (ψ5),

$$\frac{d}{d\kappa}\phi_1(\kappa) = \psi'(\kappa+1) - \frac{a}{u-1} < \frac{1}{\kappa + \frac{1}{2}} - \frac{a}{a\kappa - a + 1}$$
$$= \frac{-\frac{3}{2} + \frac{1}{a}}{(\kappa + \frac{1}{2})(\kappa - \frac{a-1}{a})} < 0;$$

thus $0 < \phi_1(1.05) < \phi_1(\kappa) < \phi_1(1) = \log 2$ by (ψ2). Since $p_\kappa(u-1)$ is positive and also decreases in κ (by differentiating (p1)) it follows that $\phi_1(\kappa)p_\kappa(u-1)$ is decreasing in κ and therefore by (p8)

$$\phi_1(\kappa)p_\kappa(u-1) < \phi_1(1)p_1(1) = e^{-\gamma}\log 2. \tag{4.9}$$

Similarly $\phi_2(\kappa)$ is decreasing in κ and therefore

$$-\phi_2(\kappa) < -\phi_2(1.05) = -\frac{1.48}{1.124} p_{1.05}(2.124). \tag{4.10}$$

Substituting from (4.9) and (4.10) in (4.8) we obtain

$$\begin{aligned}&\frac{d}{d\kappa}(A_\kappa(u-1)^{1-\kappa}p_\kappa(u-1))\\ &< -A_\kappa(u-1)^{1-\kappa}\left\{1.31672 p_{1.05}(2.124) - e^{-\gamma}\log 2\right\}\\ &< -(0.04581)A_\kappa(u-1)^{1-\kappa}\end{aligned}$$

since $p_{1.05}(2.124) = 0.33036\ldots$ by a numerical evaluation. Hence by (4.5)

$$\frac{d}{d\kappa}\hat{\Pi}_\kappa(u) < -(0.04581)A_\kappa(u-1)^{1-\kappa} + \frac{2}{5}\frac{d}{d\kappa}A_\kappa(\kappa-1)^{\kappa+1}. \tag{4.11}$$

It is easy to check that

$$\frac{d}{d\kappa}A_\kappa(\kappa-1)^{\kappa+1} = A_\kappa(\kappa-1)^\kappa\phi_3(\kappa),$$

where

$$\phi_3(\kappa) = (\kappa - 1)\left\{\log 2 + \gamma + \psi(\kappa + 1) + \log(\kappa - 1) + \frac{\kappa + 1}{\kappa - 1}\right\}.$$

Now $(\kappa - 1)^\kappa$ is positive, increases with κ, and A_κ and ϕ_3 are positive, so

$$\frac{d}{d\kappa} A_\kappa(\kappa - 1)^{\kappa+1} \leq (.05)^{1.05} A_\kappa \phi_3(\kappa).$$

Also

$$\phi_3''(\kappa) = \frac{1}{\kappa - 1} + 2\psi'(\kappa + 1) + (\kappa - 1)\psi''(\kappa + 1) > 0,$$

so that $\phi_3(\kappa)$ is convex and is maximal at an end point of $1 \leq \kappa \leq 1.05$. Hence $\phi_3(\kappa)$ assumes its largest value at $\kappa = 1.05$, and this maximum is[2] less than 1.98646. Thus, by (4.11),

$$\begin{aligned}\frac{d}{d\kappa}\hat{\Pi}_\kappa(u) &< -A_\kappa \left\{(0.04581)(u - 1)^{1-\kappa} - 0.03421\right\} \\ &< -A_\kappa \left\{0.04554 - 0.03421\right\} < 0,\end{aligned}$$

and (4.6) is true. This completes the proof of Theorem 1. ∎

We remark in conclusion that the proof of Theorem 1 is characteristic of the methods developed in this series of papers: with, usually, a multiplicity of diverse functions simultaneously in play, we weld together arguments using monotonicity (with respect to u or κ) or convexity (when available) or both, with numerical data making the 'joins'. While such procedures are not always pretty, we believe that the interplay between theoretical estimates and numerical information is novel (at least in this context) and interesting.

Appendix

Here we define three families of special functions that occur in our study and establish some of their properties. Also, we list for convenience several basic properties of two more familiar functions—the Euler gamma function and the exponential integral.

Aψ. *Derivatives of* $\log\Gamma$.

We list the following elementary facts about derivatives of $\log\Gamma$, quoting from [**AS**].

$$\psi(z) := \Gamma'(z)/\Gamma(z) \qquad \text{[AS], 6.3.1} \qquad (\psi 1)$$

$$\psi(2) := 1 - \gamma \qquad \text{[AS], 6.3.2} \qquad (\psi 2)$$

[2] $\psi(2.05) = 1/1.05 + \psi(1.05) = 0.45453\ldots$ from [**AS**].

$$\psi(z+1) = \psi(z) + 1/z \qquad \textbf{[AS]}, 6.3.5 \tag{ψ3}$$

$$\psi'(w) = \sum_{\nu=0}^{\infty}(w+\nu)^{-2}, \ \psi''(w) = -2\sum_{\nu=0}^{\infty}(w+\nu)^{-3} \qquad \textbf{[AS]}, 6.4.10 \tag{ψ4}$$

Also, we have, by (ψ4),

$$\psi'(w) < w^{-2} + (w+1/2)^{-1} < (w-1/2)^{-1}. \tag{ψ5}$$

Indeed, by convexity

$$(w+\nu)^{-2} < \int_{w+\nu-1/2}^{w+\nu+1/2} t^{-2}dt,$$

so that

$$\sum_{\nu=1}^{\infty}(w+\nu)^{-2} < \int_{w+1/2}^{\infty} t^{-2}dt = \frac{1}{w+1/2}.$$

Ag. *The exponential integral.*

For $z \in \mathbf{C}$, define $g(z)$, the exponential integral, by

$$g(z) := \int_0^z (1-e^{-t})t^{-1}dt. \tag{g1}$$

This is an entire function of z. (In [**AS**], §5.1, it is denoted by Ein(z).) For $x > 0$ we have

$$g(x) = \log x + \gamma + \int_x^{\infty} e^{-t}t^{-1}dt, \qquad \textbf{[AS]}, 5.1.39. \tag{g2}$$

Since $t^{-1}(1-e^{-t})$ decreases, it follows that

$$g(x) \geq 1 - e^{-x}, \quad x \geq 0. \tag{g3}$$

Also, we have

$$g(x) \leq (x+1-e^{-x})/2, \quad x \geq 0. \tag{g4}$$

Indeed, if we set

$$\phi(x) = x + 1 - e^{-x} - 2g(x),$$

then $\phi(0) = \phi'(0) = 0$ and $\phi''(x) > 0$, so $\phi(x) \geq 0$ for $x \geq 0$.

Aσ. *The sigma function.*

$\sigma_\kappa(u)$ (briefly: σ), defined in (1.1) provides the upper bound function for our sieve in the initial range $0 < u \le \alpha_\kappa$ (cf. (1.2)). This family of functions was introduced by Ankeny and Onishi [**AO**]. It is convenient to extend the definition of σ to $\mathbf{R}$ as the continuous solution of the difference differential equation

$$\sigma_\kappa(u) = 0, \quad u \le 0, \tag{σ1}$$

$$u^{-\kappa}\sigma_\kappa(u) = A_\kappa^{-1}, \quad 0 < u \le 2, \tag{σ2}$$

where

$$A_\kappa = (2e^\gamma)^\kappa\Gamma(\kappa+1), \tag{σ3}$$

and

$$(u^{-\kappa}\sigma_\kappa(u))' = -\kappa u^{-\kappa-1}\sigma_\kappa(u-2), \quad u > 2, \tag{σ4}$$

or, equivalently,

$$u\sigma_\kappa'(u) = \kappa\sigma_\kappa(u) - \kappa\sigma_\kappa(u-2), \quad u > 2. \tag{σ5}$$

Actually, the last two equations are valid for $0 < u \le 2$ also, by (σ2). *In what follows we shall assume that* $\kappa > 1$, though several of the results hold also for $\kappa = 1$.

Ankeny and Onishi [**AO**], pp. 38-40 and Theorem 2.3[3], proved that,

$$\sigma(u) > 0, \quad u > 0, \tag{σ6}$$

$$\sigma'(u) > 0, \quad u > 0, \tag{σ7}$$

and

$$\sigma(u) = 1 + O(e^{-u/2}), \quad u \to \infty. \tag{σ8}$$

We record also the relation

$$u\sigma''(u) = (\kappa-1)\sigma'(u) - \kappa\sigma'(u-2), \quad u > 0, \tag{σ9}$$

and we quote from [**GR**], (6.14), that

$$\sigma_\kappa(2\kappa) > \frac{1}{2}, \quad \kappa > 1. \tag{σ10}$$

[3]Note that [**AO**] uses the functions

$$F_\kappa(u/2) = e^{\gamma\kappa}\Gamma(\kappa)(1-\sigma_\kappa(u)), \ \tau_\kappa(u/2) = 2e^{\gamma\kappa}\Gamma(\kappa)\sigma_\kappa'(u)$$

and the letter α instead of κ.

Lemma $\sigma 1$. *Let $\kappa > 1$. Then $(\sigma'(u)/\sigma(u))' < 0$, $u > 0$. Moreover, $u\sigma'(u)/\sigma(u)$ decreases in u for $u > 0$ and is strictly decreasing for $u > 2$.*

Proof: Differentiation and ($\sigma 5$) and ($\sigma 9$) give

$$\begin{aligned} u(\frac{\sigma'}{\sigma}(u))' &= u\sigma(u)^{-2}\left\{\sigma(u)\sigma''(u) - \sigma'(u)^2\right\} \\ &= \sigma(u)^{-2}\left\{-\sigma(u)\sigma'(u) - \kappa\sigma(u)\sigma'(u-2) + \kappa\sigma'(u)\sigma(u-2)\right\} \\ &= -\frac{\sigma'}{\sigma}(u) + \kappa\frac{\sigma(u-2)}{\sigma(u)}\left\{\frac{\sigma'}{\sigma}(u) - \frac{\sigma'}{\sigma}(u-2)\right\} \\ &= -\frac{\sigma'}{\sigma}(u) + \kappa\frac{\sigma(u-2)}{\sigma(u)}\int_{u-2}^{u}(\frac{\sigma'}{\sigma}(t))'dt. \end{aligned}$$

The integral expression is valid because $\sigma'(u)/\sigma(u)$ has a continuous derivative for $0 < u < \infty$.

Now $(\sigma'(u)/\sigma(u))' = -\kappa/u^2 < 0$ for $0 < u \leq 2$. If the first statement of the lemma were false, there would be a least number, say v, $v > 2$, such that $(\sigma'/\sigma)'(v) = 0$ and $(\sigma'/\sigma)' < 0$ on $(0, v)$. If we take $u = v$ in the formula for $u(\sigma'/\sigma)'(u)$, the left side is zero and the right side is negative, which is impossible. This proves the first statement.

If we rewrite the formula as

$$\left(u\frac{\sigma'}{\sigma}(u)\right)' = \kappa\frac{\sigma(u-2)}{\sigma(u)}\int_{u-2}^{u}\left(\frac{\sigma'}{\sigma}(t)\right)' dt$$

the lemma now follows at once. ∎

Corollary $\sigma 1$. *For $u > 0$ we have $(1/\sigma(u))'' > 0$.*

Proof: We have

$$\left(\frac{1}{\sigma(u)}\right)'' = -\left(\frac{\sigma'}{\sigma}\cdot\frac{1}{\sigma}\right)' = -\left(\frac{\sigma'}{\sigma}\right)'\cdot\frac{1}{\sigma} - \left(\frac{\sigma'}{\sigma}\right)\left(-\frac{\sigma'}{\sigma^2}\right) > 0.$$

Ap. *The p function.*

For each $\kappa \geq 1$ we define p_κ (briefly: p) following Iwaniec ([**I**], (2.6)) by

$$p_\kappa(u) = \int_0^\infty e^{-ux-\kappa g(x)}dx, \quad u > 0, \qquad \text{(p1)}$$

where g is given in (g1). Iwaniec also proved ([**I**], p. 196) that $p_\kappa(u)$ satisfies

$$(up(u))' = \kappa p(u) - \kappa p(u+1), \quad u > 0, \qquad \text{(p2)}$$

and

$$up(u) + \kappa \int_u^{u+1} p(t)dt = 1, \quad u > 0, \tag{p3}$$

and that $p(u) \ll u^{-1}$, as $u \to \infty$, so that by (p3)

$$p(u) \sim u^{-1}, \quad u \to \infty. \tag{p4}$$

It follows at once from (p1) that

$$sgn\ p^{(\nu)}(u) = (-1)^\nu, \quad u > 0 \quad \nu = 0, 1, 2.... \tag{p5}$$

In particular, p is positive, decreasing, and convex on the positive reals. Because p is decreasing, (p2) implies that

$$up(u) \quad \text{is increasing,} \quad u > 0. \tag{p6}$$

Also we have

$$(u + \kappa)p(u) \quad \text{is decreasing,} \quad u > 0. \tag{p7}$$

Indeed, by (p2), the mean value theorem, and the convexity of p we have

$$(up(u))' = -\kappa(p(u+1) - p(u)) = -\kappa p'(u + \delta) < -\kappa p'(u).$$

We need one specific value of p:

$$p_1(1) = \int_0^\infty e^{-x-g(x)}dx = \int_0^\infty d(xe^{-g(x)}) = e^{-\gamma}, \tag{p8}$$

since $\log x - g(x) \to -\gamma$ at ∞ by (g2).

Lemma p1. *$p(t+1)/\sigma(t)$ is convex in $\{t : t > 0\}$.*

Proof: We have

$$\left\{\frac{p(t+1)}{\sigma(t)}\right\}'' = p''(t+1) \cdot \frac{1}{\sigma(t)} - 2p'(t+1)\frac{\sigma'}{\sigma^2}(t) + p(t+1)\left\{\frac{1}{\sigma(t)}\right\}'' > 0,$$

since p, $-p'$, $p'' > 0$ by (p5) and σ, $\sigma' > 0$ by (σ7) and $(1/\sigma)'' > 0$ by Corollary σ1. ∎

In order to obtain estimates that are uniform in κ, we need to know how p_κ and some related functions vary with changes in κ.

Lemma p2. *Let $u = a\kappa + b$, $a > 0$, $b \geq 0$. Then $up_\kappa(u)$ is decreasing in κ.*

Proof: We have

$$\frac{d}{d\kappa}(up_\kappa(u)) = a(up_\kappa(u))' + \frac{\partial}{\partial\kappa}(up_\kappa(u))$$
$$= a\kappa(p(u) - p(u+1)) - u\int_0^\infty e^{-ux-\kappa g(x)} g(x)dx.$$

Since $g(x) \geq 1 - e^{-x}$ by (g3), the last term is at most

$$-u\int_0^\infty e^{-ux-\kappa g(x)}(1-e^{-x})dx = -u(p(u) - p(u+1)).$$

It follows that

$$\frac{d}{d\kappa}(up_\kappa(u)) \leq -b(p_\kappa(u) - p_\kappa(u+1)) \leq 0. \quad \blacksquare$$

Lemma p3. *Let $u = a\kappa + b$, $a > 0$, $a + 1 > b$, and $ab \geq 1/2$. Then $\kappa p_k(u)$ is increasing in κ.*

Proof: Since $g(x) \leq \frac{1}{2}(1 + x - e^{-x})$ by (g4), we have

$$\frac{d}{d\kappa}(\kappa p_\kappa(u)) = p_\kappa(u) - \kappa\int_0^\infty e^{-ux-\kappa g(x)}(ax + g(x))dx$$
$$\geq p_\kappa(u) - \kappa\int_0^\infty e^{-ux-\kappa g(x)}\left(ax + \frac{1}{2}(x + 1 - e^{-x})\right)dx$$
$$= p_\kappa(u)(1 - \kappa/2) + \left(a + \frac{1}{2}\right)\kappa p_\kappa'(u) + \kappa p_\kappa(u+1)/2$$
$$= p_\kappa(u)\left\{1 - \frac{\kappa}{2} + \frac{\left(a + \frac{1}{2}\right)\kappa(\kappa - 1)}{u}\right\} - p_\kappa(u+1)\left\{\frac{\left(a + \frac{1}{2}\right)\kappa^2}{u} - \frac{\kappa}{2}\right\}$$

by (p2). Since the quantity in the last bracket is positive, we may use the estimate

$$-p_\kappa(u+1) \geq -\frac{u+\kappa}{u+\kappa+1}p_\kappa(u),$$

which follows from (p7), and conclude that

$$u(u+\kappa+1)\frac{d}{d\kappa}(\kappa p_\kappa(u)) \geq p_\kappa(u)\left\{\kappa\left(ab - \frac{1}{2}\right) + b(b+1)\right\} > 0. \quad \blacksquare$$

Aq. *The q function.*

For each number $\kappa > 1$, q_κ (briefly: q) is defined ([I], (5.4)) by

$$q_\kappa(u) = \frac{\Gamma(2\kappa)}{2\pi i}\int_C z^{-2\kappa}e^{uz+\kappa g(-z)}dz \tag{q1}$$

where g is given in Appendix Ag, $z^{-2\kappa} = exp\{-2\kappa \log z\}$, and C is the path from $-\infty$ back to $-\infty$ which surrounds the negative real axis in the positive sense.

Iwaniec proved ([I], pp. 182-183) that this function satisfies

$$(uq(u))' = \kappa q(u) + \kappa q(u+1), \quad u > 0, \tag{q2}$$

and

$$q(u) \sim u^{2\kappa-1}, \quad u \to \infty. \tag{q3}$$

Moreover ([I], pp. 184-185)

Lemma q1. *$q_\kappa(u)$ has a largest real zero, say ρ_κ (briefly: ρ), which is simple, and we have*

$$q(u) > 0, \quad q'(u) > 0, \quad u > \rho \tag{q4}$$

and

$$\rho_\kappa > \kappa. \tag{q5}$$

Further we quote from Rawthorne ([R], p. 92).

Lemma q2. *Let ρ'_κ (briefly: ρ') denote the second largest positive zero of q. If it exists, we have*

$$\rho' < \rho - 1, \tag{q6}$$

and q is negative in (ρ', ρ) (if ρ' does not exist, $q(u) < 0$ in $0 < u < \rho$).

Proof: Rewrite (q2) in the form

$$(u^{1-\kappa}q(u))' = \kappa u^{-\kappa}q(u+1).$$

Let v satisfy $\rho - 1 \le v < \rho$, and integrate the preceding formula to give

$$-v^{1-\kappa}q(v) = \kappa\int_v^\rho u^{-\kappa}q(u+1)du > 0$$

since $q(u+1) > 0$ for $u > \rho - 1$. It follows that $\rho' < \rho - 1$ and $q(v) < 0$ for $\rho' < v < \rho$. ∎

REFERENCES

[AO] N. C. Ankeny and H. Onishi, The general sieve, Acta Arith. **10** (1964), 31-62.

[AS] M. Abramowitz and I. Stegun, Handbook of Mathematical Functions. Dover, N.Y., 1965.

[DHR1] H. G. Diamond, H. Halberstam, and H.-E. Richert, Combinatorial sieves of dimension exceeding one, J. Number Theory **28** (1988), 306-346.

[DHR2] H. G. Diamond, H. Halberstam, and H.-E. Richert, Sieve auxiliary functions, Banff Conference, Proc. Canadian Number Theory Assn. (1988). To appear.

[G] F. Grupp, On zeros of functions satisfying certain difference-differential equations, Acta. Arith **51** (1988), 247-268.

[GR] F. Grupp and H.-E. Richert, Notes on functions conected with the sieve, Analysis **8** (1988), 1-23.

[HR] H. Halberstam and H.-E. Richert, *Sieve Methods*, Academic Press, London, 1974.

[I] H. Iwaniec, Rosser's sieve, Acta Arith. **36** (1980), 171-202.

[JR] W. B. Jurkat and H.-E. Richert, An improvement of Selberg's sieve method I, Acta Arith. **11** (1965), 217-240.

[R] D. A. Rawsthorne, Improvements in the small sieve estimate of Selberg by iteration, Thesis, University of Illinois, Urbana, 1980.

[te R] H. J. J. te Riele, Numerical solution of two coupled nonlinear equations related to the limits of Buchstab's iteration sieve, Afd. Numer. Wisk. **86** (1980), Stichting Math. Centrum, Amsterdam. 15pp.

[W] F. S. Wheeler, On two differential difference equations arising in analytic number theory, Thesis, University of Illinois, Urbana, 1988.

H. Diamond and H. Halberstam
Department of Mathematics
University of Illinois
1409 West Green Street
Urbana, Illinois 61801

H.-E. Richert
Mathematik III
Universität Ulm
Albert-Einstein-Allee 11
7900 Ulm, West Germany

Some Remarks about Multiplicative Functions of Modulus ≤ 1

P.D.T.A. ELLIOTT

Dedicated to P. T. Bateman

1. In this paper g will denote a complex-valued multiplicative function which satisfies $|g(n)| \leq 1$ for all positive integers n.

I begin with a statement of

Theorem 0. *Let $1 \leq w_0 \leq x$. There is a real τ, $|\tau| \leq (\log x)^{1/19}$, so that*

$$\sum_{n \leq x/w} g(n) = w^{-1-i\tau} \sum_{n \leq x} g(n) + O\left(\frac{x}{w}\left(\frac{\log 2w_0}{\log 2x}\right)^{1/19}\right)$$

uniformly for $1 \leq w \leq w_0$. If g is real-valued, then we may set $\tau = 0$. The implied constant is absolute.

The constant 1/19 can be improved, but in its present form the method can be expected to lead to an exponent less than 1. The parameter τ depends upon both g and x, but its dependence upon x is in fact rather weak, so that a result which is uniform over a range $N \leq x \leq N^{\beta}$, for any fixed $\beta > 0$, could be obtained.

A detailed proof of this theorem will appear in the Proceedings of the Royal Dutch Academy [3]. Here I shall give some applications.

2. It follows straightaway from the theorem that

$$\left|\sum_{n \leq x/w} g(n)\right| = \frac{1}{w}\left|\sum_{n \leq x} g(n)\right| + O\left(\frac{x}{w}\left(\frac{\log 2w}{\log 2x}\right)^{1/19}\right) \qquad (1)$$

Partially funded by N.S.F. contract DMS-8722913

uniformly for $1 \le w \le x$ and all g. This much improves a result of Hildebrand [6] obtained by a different method. In particular, it enables one to extend character sum estimates, such as those of Burgess [1], on a sliding scale.

Let χ be a non-principal character of order m to a prime-modulus p, and ρ one of the values it can assume. Let $t(p)$ be the least positive integer n for which $\chi(n) = \rho$.

Theorem 1. *There is a positive absolute constant c so that for each fixed* $\varepsilon > 0$

$$t(p) \ll p^{\beta+\varepsilon} \quad \textit{with} \quad \beta = \frac{1}{4}\left(1 - \frac{c}{m^{19}}\right).$$

Proof. Since

$$m^{-1}\sum_{k=1}^{m} (\bar{\chi}(n)\rho)^k = \begin{cases} 1 \text{ if } \chi(n) = \rho, \\ 0 \text{ otherwise,} \end{cases}$$

$$\sum_{\substack{n\le y \\ \chi(n)=\rho}} 1 - \frac{1}{m}[y] = m^{-1}\sum_{k=1}^{m-1} \rho^k \sum_{n\le y} (\bar{\chi}(n))^k .$$

For $y \ge p^{1/4+\varepsilon}$ the above mentioned estimate of Burgess shows that a typical character sum is $O\left(p^{1-\delta}y\right)$ for some positive δ depending only upon ε. If we replace y by y/w and appeal to Theorem 0, the character sums do not exceed

$$m^{-1}\sum_{k=1}^{m} \frac{1}{w}\left|\sum_{n\le y}(\bar{\chi}(n))^k\right| + O\left(\frac{x}{w}\left(\frac{\log 2w}{\log 2x}\right)^{1/19}\right) .$$

This is less that $[yw^{-1}]m^{-1}$ for w of the form $\exp(c_1 m^{-19}\log y)$, and $y = p^{1/4+\varepsilon}$.

It is clear from this example that amelioration of the exponent $1/19$ in Theorem 0 would be worthwhile.

It is interesting to compare the bound for $t(p)$ given in Theorem 1 with the estimate $\beta = 1/2 - \eta$ of Davenport and Erdös [2]. The value $1/2$ rather than $1/4$ in their result comes from an application of the Pólya-Vinogradov inequality rather than the above estimate of Burgess. The constant is $\delta_\nu^2/2(2k+1)$, where ν denotes the number of distinct prime factors of k, and the sequence δ_j is defined inductively by $\delta_1 = 1/(k+1)$, $\delta_{s+1} = \delta_s^2/(2k^2)$. They regard this value of η as "very small", and go on to say "but it is difficult to see how one can obtain a reasonably good result without making some assumptions about the arithmetical nature of k." No such assumptions are made in Theorem 1.

Let $S(x) = \sum_{n\leq x} g(n)$.

Suppose now that g is real-valued. Then for $0 < \delta < 1$, $\delta x \leq y \leq x/\delta$, say,

$$\sum_{n\leq x} g(n+y) = S(x+y) - S(y) = S(x) + O\Big(x(\log x)^{-1/19}\Big),$$

the implied constant depending only upon δ. In particular

$$|S(x)|^2 \leq x^{-1} \sum_{\delta x\leq y\leq x/\delta} \Big|\sum_{n\leq x} g(n+y)\Big|^2 + O\left(x^2(\log x)^{-2/19}\right).$$

If $g = \chi$, a real non-principal character (mod p), and $1 \leq x < p$, then we may introduce the very wasteful step of increasing the outer summation over y to run over a complete set of residue class representatives (mod p). For our extravagance the resulting multiple sum can be evaluated:

$$\sum_{y=0}^{p-1} \Big|\sum_{n\leq x} \chi(n+y)\Big|^2 = \sum_{n_1\leq x}\sum_{n_2\leq x}\sum_{y=0}^{p-1} \chi(n_1+y)\chi(n_2+y) = p[x] - [x]^2 ,$$

and we gain the estimate

$$\sum_{n\leq x} \chi(n) \ll p^{1/2} + x(\log x)^{-1/19} .$$

This is not spectacular, but for x below $p^{1/2}\log p$ or so it already improves upon the well-known Pólya-Vinogradov inequality.

Let $\theta > 0$. A more careful version of this argument which relates $S(x+y) - S(y)$ first to $S(y)$, and then to $S(x)$, employing y in the range $x(\log x)^\theta \leq y \leq 2x(\log x)^\theta$, yields

$$\sum_{n\leq x} \chi(n) \ll (p(\log x)^{-\theta})^{1/2} + x(\log x)^\theta \left(\frac{\log\log x}{\log x}\right)^{1/19} .$$

In particular

$$\sum_{n\leq x} \chi(n) \ll p^{1/2}(\log x)^{-1/40} + x(\log x)^{-1/400}$$

uniformly in p and $x \geq 2$.

3. Returning to the general relation of Theorem 0, multiplying by $w^{i\tau}$ and summing over the interval $1 \le w \le w_0$ gives

$$\sum_{w \le w_0} \sum_{n \le x/w} g(n) w^{i\tau} = S(x)(\log w_0 + O(1)) + O\left(x \log w_0 \left(\frac{\log 2w_0}{\log 2x} \right)^{1/19} \right).$$

The order of summation in the double sum can be inverted, and the resulting inner sum estimated:

$$\sum_{n \le x} g(n) \sum_{w \le \min(w_0, x/n)} w^{i\tau} = \sum_{x/w_0 < n \le x} g(n) \frac{(x/n)^{1+i\tau}}{1+i\tau} + O(x) .$$

We have reached

Theorem 2. *In the notation of Theorem 0*

$$\sum_{n \le x} g(n) = \frac{x^{1+i\tau}}{(1+i\tau)\log w_0} \sum_{x/w_0 < n \le x} \frac{g(n)}{n^{1+i\tau}} + O\left(\frac{x}{\log 2w_0} + x \left(\frac{\log 2w_0}{\log 2x} \right)^{1/19} \right).$$

This reduces the study of the mean-value of g to that of $g(n)n^{-1-i\tau}$, which is generally thought to be an easier problem.

As an example, suppose g to be real and define the Dirichlet convolution $h = 1 * g$. Then $h(p) = g(p) + 1 \ge 0$ for all primes p, and $|h(p^k)| \le 2$. Easy elementary arguments (for example Elliott [5] Chapter 1) give on the one hand

$$\sum_{n \le x} h(n) \le \sum_{n \le x} |h(n)| \le x \exp\left(\sum_{p \le x} \frac{|h(p)| - 1}{p} \right) \ll x \log x e^{-\Delta}$$

with

$$\Delta = \sum_{p \le x} \frac{1 - g(p)}{p} .$$

On the other hand, from its property as a convolution

$$\sum_{n \le x} h(n) = \sum_{m \le x} g(m) \left[\frac{x}{m} \right] = x \sum_{m \le x} \frac{g(m)}{m} + O(x) ,$$

so that together

$$\sum_{m \le x} \frac{g(m)}{m} \ll 1 + e^{-\Delta} \log x. \tag{2}$$

A simple modification of this argument with

$$\sum_{x/w_0 < m \leq x} \frac{g(m)}{m} = \frac{1}{x} \sum_{n \leq x} h(n) - \frac{w_0}{x} \sum_{n \leq x/w_0} h(n) + O(1)$$

shows that for $w_0 \leq x^{1/2}$ the restriction $x/w_0 < m$ can be adjoined to the sum in (2), for then

$$\sum_{x/w_0 < p \leq x} \frac{|h(p)| - 1}{p} \ll 1 + \log\left(\frac{\log x}{\log x/w_0}\right) \ll 1 .$$

It follows from Theorem 2 that

$$\sum_{n \leq x} g(n) \ll \frac{x}{\log w_0} e^{-\Delta} \log x + \frac{x}{\log w_0} + x\left(\frac{\log 2w_0}{\log 2x}\right)^{1/19} .$$

Choosing w_0 favorably we have established

Theorem 3. *For real g*

$$\sum_{n \leq x} g(n) \ll x \exp\left(-\frac{1}{40} \sum_{p \leq x} \frac{(1 - g(p))}{p}\right).$$

Of course the constant $1/40$ is far from best, but the proof is not too messy, the result is 'clean', and the method lends itself to generalizations allowing $|g(p)| > 1$.

4. My last application is to Probabilistic Number Theory. Let $f(n)$ be a real-valued additive function and assume that the distribution function

$$F_x(z) = \nu_x(n;\ f(n) - \alpha(x) \leq z\beta(x))$$

which counts the frequency of those integers in the interval $1 \leq n \leq x$ for which the inequality $f(n) - \alpha(x) \leq z\beta(x)$ is satisfied, converges weakly to a proper law, as $x \to \infty$.

The function $\beta(x)$ must then satisfy some growth restrictions, which are not well understood. However, in the course of proving that all proper limit laws are continuous, Timofeev [7] showed that (in our present notation) $\beta(x/w)/\beta(x) \to 1$ certainly holds if $\log w/\log x \to 0$, $x \to \infty$. This last result could also be deduced by the arguments in Chapter 17 of Elliott [4], but in neither case can the proof be considered simple. I show here how to deduce it rapidly from Theorem 0.

Define the multiplicative function $g(n) = \exp(it\, f(n)/\beta(x))$, t real. Let $\phi(t)$ denote the characteristic function of the limit law. Then our hypothesis can be expressed in the form

$$[x]^{-1} \exp\Big(-it\, \alpha(x)/\beta(x)\Big) \sum_{n \le x} g(n) = \int_{-\infty}^{\infty} e^{itz} dF_x(z) \to \phi(t), \; x \to \infty,$$

uniformly on compact t-sets. It follows from Theorem 0 that if $\log w/\log x \to 0$, then

$$\left|\phi\Big(t\beta(x/w)/\beta(x)\Big)\right| - |\phi(t)| \to 0, \; x \to \infty. \tag{3}$$

Suppose now that for some unbounded (increasing) sequence of values x_j, with corresponding w_j, we have $\beta(x_j/w_j)/\beta(x_j) \to \theta \neq 1$. Without loss of generality $0 \le \theta < 1$. It follows from (3) that for all real t, $|\phi(t)| = |\phi(t\theta)|$. Arguing by induction $|\phi(t)| = |\phi(t\theta^k)|$ for all positive integers k and real t, so that $|\phi(t)| = \lim_{k\to\infty} |\phi(t\theta^k)| = |\phi(0)| = 1$. The limit law ϕ must be improper, contrary to assumption.

REFERENCES

[1] Burgess, D.A., On character sums and primitive roots. Proc. London Math. Soc. (3) **12** (1962), 179-192.
[2] Davenport, H., Erdös, P., The distribution of quadratic and higher residues. Publ. Math. Debrecen **2** (1952), 252-265.
[3] Elliott, P.D.T.A., Extrapolating the mean values of multiplicative functions. To appear in Indigationes Math. = Proc. Kon. Ned. Akad. Wetensch.
[4] Elliott, P.D.T.A., *Probablistic Number Theory II: Central Limit Theorems.* Grund. der math. Wiss., **240**, Springer-Verlag, Berlin, Heidelberg, New York, 1980.
[5] Elliott, P.D.T.A., *Arithmetic Functions and Integer Products.* Grundl. der math. Wiss., **272**, Springer-Verlag, New York, Berlin, Heidelberg, Tokyo, 1985.
[6] Hildebrand, A., A note on Burgess' character sum estimate. C.R. Acad. Royale Canada, **VIII** (1986), 35-37.
[7] Timofeev, N.M., On the convergence to the discontinuous distribution. Liet. Mat. Rinkinys = Litovsk Mat. Sb. **22** (1982), 159-171.

P.D.T.A. Elliott
Dept. of Mathematics
University of Colorado
Boulder, CO 80309

On the Normal Behavior of the Iterates Of some Arithmetic Functions

P. ERDÖS, A. GRANVILLE[1], C. POMERANCE[2],
AND C. SPIRO

Dedicated to our friend, colleague and teacher, Paul Bateman

Abstract

Let $\varphi_1(n) = \varphi(n)$ where φ is Euler's function, let $\varphi_2(n) = \varphi(\varphi(n))$, etc. We prove several theorems about the normal order of $\varphi_k(n)$ and state some open problems. In particular, we show that the normal order of $\varphi_k(n)/\varphi_{k+1}(n)$ is $ke^\gamma \log\log\log n$ where γ is Euler's constant. We also show that there is some positive constant c such that for all n, but for a set of asymptotic density 0 , there is some k with $\varphi_k(n)$ divisible by every prime up to $(\log n)^c$. With $k(n)$ the first subscript k with $\varphi_k(n) = 1$, we show, conditional on a certain form of the Elliott-Halberstam conjecture, that there is some positive constant α such that $k(n)$ has normal order $\alpha \log n$. Let $s(n) = \sigma(n) - n$ where σ is the sum of the divisors function, let $s_2(n) = s(s(n))$, etc. We prove that $s_2(n)/s(n) = s(n)/n + o(1)$ on a set of asymptotic density 1 and conjecture the same is true for $s_{k+1}(n)/s_k(n)$ for any fixed k .

§1. Introduction

Let $\varphi(n) = \varphi_1(n)$ denote Euler's phi-function and if $\varphi_{k-1}(n)$ has already been defined, let $\varphi_k(n) = \varphi(\varphi_{k-1}(n))$. If $n > 1$, then $n > \varphi(n)$. Thus the sequence $n, \varphi_1(n), \varphi_2(n), \cdots$ is strictly decreasing until it reaches 1 when it becomes constant. Let $k(n) = k$ be the least number such that $\varphi_k(n) = 1$. Further, let $k(1) = 1$.

[1] Supported in part by an NSERC grant
[2] Supported in part by an NSF grant

Note that if $n = 2^j$, then $k(n) = j = (\log n)/\log 2$. Also if $n = 2 \cdot 3^j$, then $k(n) = j + 1 = \lceil(\log n)/\log 3\rceil$ where $\lceil x\rceil$ denotes the least integer $\geq x$. It turns out that these two examples essentially demonstrate the extreme behavior of $k(n)$, for as Pillai [14] showed in 1929,

$$\lceil(\log n)/\log 3\rceil \leq k(n) \leq \lceil(\log n)/\log 2\rceil \tag{1.1}$$

for all n . Further, by considering numbers n of the form 2^a3^b it is easy to see that the set of numbers of the form $k(n)/\log n$ is dense in $[1/\log 3, 1/\log 2]$. What is still in doubt about $k(n)$ is its average and normal behavior. We conjecture that there is some constant α such that $k(n) \sim \alpha \log n$ on a set of asymptotic density 1 . If this is true, then (1.1) immediately would imply that

$$\frac{1}{x}\sum_{n\leq x} k(n) \sim \alpha \log x.$$

The function $k(n)$ possesses more algebraic structure than is immediately apparent from its definition. Shapiro [16] has shown that the function $g(n) := k(n) - 1$ is additive and in fact satisfies the stronger relation

$$g(mn) = g(m) + g(n) + \epsilon_{(m,n)}$$

for all natural numbers m, n where $\epsilon_{(m,n)}$ is 0 unless (m,n) is even in which case $\epsilon_{(m,n)} = 1$.

Let $F(n)$ denote the number of even terms of the sequence

$$n,\ \varphi(n),\ \varphi_2(n),\ \cdots .$$

Then $F(n) = k(n)$ for n even and $F(n) = k(n) - 1$ for n odd. It is not hard to show (we leave this for the reader) that the function $F(n)$ is completely additive; that is, $F(mn) = F(m) + F(n)$ for all natural numbers m, n . Note that $F(2) = 1$ and for p an odd prime, $F(p) = F(p-1)$. So in fact, we have an alternative definition of F that does not have anything to do with iterating the phi-function. Namely, F is the completely additive function which is defined inductively on the primes as follows:

$$F(p) = \begin{cases} 1, & \text{if } p = 2 \\ F(p-1), & \text{if } p > 2. \end{cases}$$

Thus our conjectures on the normal and average orders of $k(n)$ can be equivalently put in terms of the normal and average orders of the function $F(n)$. Using this translation of the problem, we are able to prove these conjectures conditionally on a certain form of the Elliott-Halberstam conjecture. This conjecture states that for any A,

$$\sum_{k\le Q}\max_{(a,k)=1}\max_{x'\le x}\left|\pi(x';k,a)-\frac{\pi(x')}{\varphi(k)}\right|\ll_A\frac{x}{\log^A x},\tag{1.2}$$

where $\pi(x;k,a)$ denotes the number of primes $p\le x$ with $p\equiv a \pmod k$, where $\pi(x)$ is the number of primes $p\le x$, and where Q is some function of the form $x^{1-o(1)}$. This conjecture for $Q=x/\log^B x$, which was the original conjecture of Elliott and Halberstam, was recently disproved in [4], while in [5] the conjecture is disproved for $Q=x/\exp\{c(\log\log x)^2/\log\log\log x\}$ for some positive constant c. But presumably, if $Q=x^{1-\epsilon(x)}$ and $\epsilon(x)$ tends to 0 slowly enough, then (1.2) holds. In section 2 below, we show that $F(n)$ (and thus $k(n)$) possesses normal and average order $\alpha\log n$ provided (1.2) holds for $Q=x^{1-\epsilon(x)}$ with $\epsilon(x)=(\log\log x)^{-2}$. Further we can weaken (1.2) by deleting the double max (letting $x'=x$ and $a=1$), by restricting k to integers with at most two prime factors, and taking $A=2$.

Short of proving our conjecture on the normal order of $k(n)$ unconditionally, there are still many interesting questions about the normal behavior of the functions $\varphi_k(n)$. In 1928, Schoenberg [15] showed that $n/\varphi(n)$ has a distribution function. That is, $D_\varphi(u)$, defined as the asymptotic density of the set of n with $n/\varphi(n)\le u$, exists for every u. In addition, $D_\varphi(u)$ is continuous and strictly increasing on $[1,\infty)$, with asymptotic limit 1.

It turns out that the situation for the higher iterates of φ is much simpler. We show below that the normal order of $\varphi_k(n)/\varphi_{k+1}(n)$ is

$$ke^\gamma\log\log\log n,$$

where γ is Euler's constant, for each fixed $k\ge 1$. In fact, this result continues to hold true if k is allowed to tend to infinity at a modest rate. (For fixed k, this result was stated without proof in [7].)

As a corollary, we have that the set

$$\{n: n/\varphi_{k+1}(n)\le uk!e^{k\gamma}(\log\log\log n)^k\}$$

has asymptotic density $D_\varphi(u)$ for every integer $k\ge 0$ and for every real number u.

It is well known that the maximal order of $n/\varphi(n)$ is $e^\gamma\log\log n$ but that very few integers n have $n/\varphi(n)$ this order of magnitude. We show below the existence of a positive constant c such that

$$\max_k \varphi_k(n)/\varphi_{k+1}(n)>c\log\log n$$

holds for a set of n of asymptotic density 1 . In fact a stronger result is true. We show the existence of a positive constant c' such that the set of n for which there is a k with $\varphi_k(n)$ divisible by every prime up to $(\log n)^{c'}$ has asymptotic density 1 .

The following two conjectures are perhaps tractable, but so far have resisted our efforts. We define

$$\Phi(n) = n \prod_{k \geq 1} \varphi_k(n).$$

Conjecture 1. *For each prime p, let $N(x,p)$ denote the number of $n \leq x$ with $p \mid \Phi(n)$. Then for every $\epsilon > 0$, $N(x,p) = o(x)$ uniformly in the region $p > (\log x)^{1+\epsilon}$ and $N(x,p) \sim x$ uniformly in the region $p < (\log x)^{1-\epsilon}$.*

Conjecture 2. *For each $\epsilon > 0$, the upper asymptotic density of the set of n with the property that the largest prime factor of $\varphi_k(n)$ exceeds n^ϵ tends to 0 as $k \to \infty$.*

Concerning Conjecture 1, we show below that for every n , the number of distinct prime factors of $\Phi(n)$ is at most $\lceil (\log n)/\log 2 \rceil$. Thus for each $\epsilon > 0$ and all $x \geq x_0(\epsilon)$, there is no $n \leq x$ with $\Phi(n)$ divisible by every prime $p \leq (\log x)^{1+\epsilon}$. However, we not only cannot prove the first assertion in Conjecture 1 for every $\epsilon > 0$, we cannot prove it for any specific choice of ϵ , even for very large choices. From our theorem mentioned above on $\varphi_k(n)$ being divisible by every prime up to $(\log n)^{c'}$, it follows that if $0 < c < c'$, then $N(x,p) \sim x$ uniformly for $p < (\log x)^c$. The second assertion in Conjecture 1 has both stronger and weaker versions that may be worth stating. The stronger version is that for each $\epsilon > 0$, there is a set $S_\epsilon(x)$ of integers $n \leq x$ of cardinality $o_\epsilon(x)$ such that if $n \leq x$, $n \notin S_\epsilon(x)$, then $\Phi(n)$ is divisible by every prime $p \leq (\log x)^{1-\epsilon}$. From the above mentioned theorem, this is true for all $\epsilon < 1 - c'$. The weaker version is that

$$\sum_{\substack{p < \log n \\ p \nmid \Phi(n)}} 1/p = o(1)$$

on a set of n of asymptotic density 1. Perhaps this is tractable. Note that from the above comments, we have

$$\sum_{\substack{p > \log n \\ p \mid \Phi(n)}} 1/p \to 0 \quad \text{as } n \to \infty.$$

By using sieve methods, we can prove Conjecture 2 for $\epsilon > 2/3$. We do not give the proof here.

The sum of the divisors function $\sigma(n)$ resembles in many ways Euler's function $\varphi(n)$. Yet it seems very difficult to prove anything non-trivial about the sequence of k-fold iterates $\sigma_k(n)$. For example, consider the following statements:

(i) for every $n > 1$, $\sigma_{k+1}(n)/\sigma_k(n) \to 1$ as $k \to \infty$;
(ii) for every $n > 1$, $\sigma_{k+1}(n)/\sigma_k(n) \to \infty$ as $k \to \infty$;
(iii) for every $n > 1$, $\sigma_k(n)^{1/k} \to \infty$ as $k \to \infty$;
(iv) for every $n > 1$, there is some k with $n \mid \sigma_k(n)$;
(v) for every n, $m > 1$, there is some k with $m \mid \sigma_k(n)$;
(vi) for every n, $m > 1$, there are some k, ℓ, with $\sigma_k(m) = \sigma_\ell(n)$.

We can neither prove nor disprove any of these statements.

Let $s(n) = \sigma(n) - n$ and let $s_k(n)$ be the k-fold iterate of s at n. In [8], the first author stated the following: For each $\epsilon > 0$ and k, the set of n with

$$\left| \frac{s(n)}{n} - \frac{s_{j+1}(n)}{s_j(n)} \right| < \epsilon \quad \text{for } j = 1, 2, \cdots, k$$

has asymptotic density 1. This result is "half proved" in [8]. Namely, it is shown that the set of n with

$$\frac{s_{j+1}(n)}{s_j(n)} > \frac{s(n)}{n} - \epsilon \quad \text{for } j = 1, 2, \cdots, k$$

has asymptotic density 1. The other half of the statement is claimed, but no argument is given. The first author now wishes to retract this claim and state the following as an open problem.

Conjecture 3. *For each $\epsilon > 0$ and k, the set of n with*

$$\frac{s_{j+1}(n)}{s_j(n)} < \frac{s(n)}{n} + \epsilon \quad \text{for } j = 1, ,\cdots, k$$

has asymptotic density 1.

In section 5 we give a proof of Conjecture 3 in the case $k = 1$. We also show that the full Conjecture 3 would be implied by the following conjecture.

Conjecture 4. *If $\mathcal{A}$ is a set of natural numbers of positive upper density, then $s(\mathcal{A}) = \{s(n) : n \in \mathcal{A}\}$ also has positive upper density.*

Note that it is possible for $s(\mathcal{A})$ to have positive density when $\mathcal{A}$ has density 0. For example, if $p \neq q$ are primes, then $s(pq) = p + q + 1$.

While the set of integers of the form pq has asymptotic density 0 , the set of integers of the form $p+q+1$ with p, q distinct primes has asymptotic density 1/2 . This follows from work on the "exceptional set" in Goldbach's conjecture. In fact, a more complicated version of this idea gives that the set of $s_k(pq)$ has lower asymptotic density at least 1/2 for any fixed k. We show this in section 5.

Suppose for every K there is a number C_K such that for any m there are at most C_K numbers $n \leq Km$ with $s(n) = m$. We are not sure whether we believe this hypothesis and in fact it may be possible to disprove it. We note though that it implies Conjecture 4.

In some sense, the paper [8] was motivated by a problem of H. W. Lenstra, Jr. [12] to show that for each k , there is an n with

$$n < s(n) < s_2(n) < \cdots < s_k(n). \tag{1.2}$$

Let α be the asymptotic density of the set of n with $n < s(n)$. Then $\alpha > 0$ and the correct half of [8] shows that for each k, (1.2) holds for a set of n of asymptotic density. That is, if the first inequality in (1.2) holds, then almost certainly all of the inequalities in (1.2) hold. Thus [8] provides a very strong solution to Lenstra's problem. The third author wishes to acknowledge a conversation with Lenstra in which the difficulty in the proof of the other half of [8] was discovered.

In [9], the first and third authors prove a theorem on the normal number of prime factors of $\varphi(n)$. Abdelhakim Smati has pointed out to us an error in the proof of Lemma 2.2 in this paper and another minor error. We correct these errors below in the last section.

Throughout the paper the letters p, q, r will always denote primes.

§2. The average and normal order of $F(n)$

Most of the results in this section are conditional on certain suitably strong versions of the Elliott-Halberstam conjecture. Before we state our results we define a few terms.

Definition. *We say a positive, continuous function $\epsilon(x)$ defined on $(1,\infty)$ is acceptable if*

(i) $\epsilon(x)\log x$ is eventually increasing and $\to \infty$ as $x \to \infty$;

(ii) for some $\delta > 0$, $\epsilon(x)(\log\log x)^{1+\delta}$ is eventually decreasing.

Some examples of acceptable functions are

$$\epsilon(x) = (\log\log 3x)^{-2},$$
$$\epsilon(x) = (\log x)^{-1/2},$$
$$\epsilon(x) = \exp\left((\log\log 3x)^{1/2}\right)/\log x.$$

Consider the two statements:

$$\sum_{p\leq x^{1-\epsilon(x)}} \left|\pi(x;p,1)-\frac{\pi(x)}{p-1}\right| \ll \epsilon(x)\pi(x), \tag{A_ϵ}$$

$$\sum_{\substack{m\leq x^{1-\epsilon(x)}\\ \Omega(m)\leq 2}} \left|\pi(x;m,1)-\frac{\pi(x)}{\varphi(m)}\right| \ll \epsilon(x)\pi(x). \tag{B_ϵ}$$

Here the function $\Omega(m)$ counts the total number of prime factors of m with multiplicity, so that the statement B_ϵ implies the statement A_ϵ.

We now state the principal results of this section. Please note that if $\epsilon(x)$ is an acceptable function, then $\epsilon(x)\log\log x = o(1)$.

Theorem 2.1. *If A_ϵ holds for some acceptable function $\epsilon(x)$, then there is some positive constant α such that*

$$\frac{1}{x}\sum_{n\leq x} F(n) = \alpha\log x + O(\epsilon(x)\log x\log\log x). \tag{2.1}$$

Theorem 2.2. *If B_ϵ holds for some acceptable function $\epsilon(x)$ and if α is the constant of Theorem 2.1, then*

$$\frac{1}{x}\sum_{n\leq x}(F(n)-\alpha\log n)^2 \ll \epsilon(x)\log^2 x\log\log x.$$

In particular, $F(n)$ has normal order $\alpha\log n$.

Corollary 2.3. *If $\epsilon(x)$ is an acceptable function of the form $(\log x)^{-1+o(1)}$ and if B_ϵ holds, then for each $\delta>0$, the set of n with*

$$|F(n)-\alpha\log n| < (\log n)^{1/2+\delta}$$

has asymptotic density 1.

The implied constants in Theorems 2.1, 2.2 depend, respectively, on the implied constants in A_ϵ, B_ϵ and on which specific function $\epsilon(x)$ is used. Thus if one had A_ϵ with an explicit constant for some explicit $\epsilon(x)$, say $\epsilon(x)=(\log\log 3x)^{-2}$, then the constant α would be effectively computable.

We begin the proof of Theorem 2.1 with an unconditional result.

Lemma 2.4. *For any function $\epsilon(x)$ with $x^{1/2} \le x^{1-\epsilon(x)} \le (1-\delta)x$ for x large and $\delta > 0$ some constant, we have*

$$\frac{1}{\pi(x)}\sum_{p\le x} F(p) - \sum_{p\le x}\frac{F(p)}{p} \ll \epsilon(x)\log x + \frac{\log^2 x}{x}\sum_{p\le x^{1-\epsilon(x)}}\left|\pi(x;p,1)-\frac{\pi(x)}{p-1}\right|.$$

Proof: From the definition of F we have

$$\begin{aligned}\sum_{p\le x} F(p) &= 1 + \sum_{3\le p\le x} F(p) = 1 + \sum_{3\le p\le x} F(p-1)\\ &= 1 + \sum_{3\le p\le x}\sum_{q^a|p-1} F(q) = 1 + \sum_{q^a\le x} F(q)\pi(x;q^a,1).\end{aligned}$$

Thus

$$\sum_{p\le x} F(p) - \pi(x)\sum_{p\le x}\frac{F(p)}{p} = \sum\nolimits_1 + \sum\nolimits_2 + \sum\nolimits_3, \tag{2.3}$$

where

$$\sum\nolimits_1 = 1 + \sum_{\substack{p^a\le x\\ a\ge 2}} F(p)\pi(x;p^a,1),$$

$$\sum\nolimits_2 = \sum_{p\le x^{1-\epsilon(x)}} F(p)\left(\pi(x;p,1) - \frac{\pi(x)}{p}\right),$$

$$\sum\nolimits_3 = \sum_{x^{1-\epsilon(x)}<p\le x} F(p)\left(\pi(x;p,1) - \frac{\pi(x)}{p}\right).$$

We have (using $F(p) \ll \log p$)

$$\begin{aligned}\sum\nolimits_1 &\ll 1 + \sum_{\substack{p^a\le x^{1/3}\\ a\ge 2}} (\log p)\left(\pi(x;p^a,1) - \frac{\pi(x)}{\varphi(p^a)}\right)\\ &\quad + \sum_{\substack{p^a\le x^{1/3}\\ a\ge 2}} (\log p)\frac{\pi(x)}{\varphi(p^a)} + \sum_{\substack{x^{1/3}<p^a\le x\\ a\ge 2}} (\log p)\pi(x;p^a,1)\\ &\ll \frac{x}{\log^2 x} + \frac{x}{\log x} + \sum_{\substack{p^a>x^{1/3}\\ a\ge 2}}\frac{x\log p}{p^a} \ll \frac{x}{\log x},\end{aligned} \tag{2.4}$$

where we used the Bombieri-Vinogradov theorem for the first sum over $p^a \le x^{1/3}$. In addition, we have

$$\sum\nolimits_2 \ll \log x \sum_{p\le x^{1-\epsilon(x)}}\left|\pi(x;p,1) - \frac{\pi(x)}{p-1}\right| + \frac{x}{\log x}. \tag{2.5}$$

For $\sum_3$, we have

$$\sum\nolimits_3 \ll \log x \sum_{x^{1-\epsilon(x)}<p\le x} \pi(x;p,1) + \pi(x) \sum_{x^{1-\epsilon(x)}<p\le x} \frac{\log p}{p}$$
$$\ll \log x \sum_{x^{1-\epsilon(x)}<p\le x} \pi(x;p,1) + \pi(x)\epsilon(x)\log x. \tag{2.6}$$

We estimate the sum on the right of (2.6) using Brun's method as follows:

$$\sum_{x^{1-\epsilon(x)}<p\le x} \pi(x;p,1) = \sum_{x^{1-\epsilon(x)}<p\le x} \sum_{\substack{q\le x \\ q\equiv 1 \ (\mathrm{mod}\ p)}} 1$$
$$\le \sum_{m<x^{\epsilon(x)}} \sum_{\substack{p\le x/m \\ pm+1 \text{ is prime}}} 1 \ll \sum_{m<x^{\epsilon(x)}} \frac{m}{\varphi(m)} \frac{x/m}{\log^2(x/m)}$$
$$\ll \frac{x}{\log^2 x} \sum_{m<x^{\epsilon(x)}} \frac{1}{\varphi(m)} \ll \frac{\epsilon(x)x}{\log x}. \tag{2.7}$$

Putting this estimate in (2.6) and assembling (2.3)-(2.6), we obtain the lemma.

Corollary 2.5. *If $\epsilon(x)$ is some function that satisfies the hypothesis of Lemma 2.4 and if A_ϵ holds, then*

$$\frac{1}{\pi(x)} \sum_{p\le x} F(p) - \sum_{p\le x} \frac{F(p)}{p} \ll \epsilon(x)\log x.$$

Proof of Theorem 2.1: We unconditionally have

$$\frac{1}{x}\sum_{n\le x} F(n) = \sum_{p\le x} \frac{F(p)}{p} + O(1). \tag{2.8}$$

Indeed, using $F(p) \ll \log p$, we have

$$\frac{1}{x}\sum_{n\le x} F(n) = \frac{1}{x}\sum_{n\le x}\sum_{p^a|n} F(p) = \frac{1}{x}\sum_{p^a\le x} F(p)\left[\frac{x}{p^a}\right]$$
$$= \sum_{p^a\le x} \frac{F(p)}{p^a} + O(1) = \sum_{p\le x} \frac{F(p)}{p} + O(1).$$

Corollary 2.5 gives a (conditional) connection between $\sum_{p\le x} F(p)/p$ and $\sum_{p\le x} F(p)$. There is another (unconditional) connection which comes from partial summation. Let

$$R(x) := \frac{1}{x}\sum_{p\le x} F(p).$$

Then

$$\sum_{p\le x}\frac{F(p)}{p}=R(x)+\int_2^x\frac{1}{t}R(t)\,dt. \tag{2.9}$$

Assume now that $\epsilon(x)$ is an acceptable function and that A_ϵ holds. Then from Corollary 2.5 and (2.9) we have

$$R(x)\log x=R(x)+\int_2^x\frac{1}{t}R(t)\,dt+O(\epsilon(x)\log x),$$

so that

$$R(x)=\frac{1}{\log x}\int_2^x\frac{1}{t}R(t)\,dt+O(\epsilon(x)), \tag{2.10}$$

using $R(x)\ll 1$.

Let

$$V(x):=\frac{1}{\log x}\int_2^x\frac{1}{t}R(t)\,dt.$$

Since $R(x)$ is continuous but for a discrete set of jump discontinuities it follows that $V(x)$ is continuous, differentiable where $R(x)$ is continuous and satisfies

$$V(x)=\int_2^x V'(t)\,dt. \tag{2.11}$$

But at points where $R(x)$ is continuous , we have

$$\begin{aligned}V'(x)&=\frac{R(x)}{x\log x}-\frac{1}{x\log^2 x}\int_2^x\frac{1}{t}R(t)\,dt\\ &=\frac{1}{x\log x}(R(x)-V(x))\ll\frac{\epsilon(x)}{x\log x},\end{aligned} \tag{2.12}$$

by (2.10).

Note that by the definition of acceptable function, we have

$$\int_2^x\frac{\epsilon(t)}{t\log t}\,dt<\infty.$$

Thus by (2.11) and (2.12), we have that

$$\alpha:=\int_2^\infty V'(t)\,dt=\lim_{x\to\infty}V(x)$$

exists and is positive.

We define now

$$\bar\epsilon(x):=\int_x^\infty\frac{\epsilon(t)}{t\log t}\,dt \tag{2.13}$$

and note that from the definition of acceptable function we have

$$\begin{aligned}\epsilon(x) &= \int_x^\infty \frac{\epsilon(x)\log x}{t\log^2 t}\,dt \le \int_x^\infty \frac{\epsilon(t)\log t}{t\log^2 t}\,dt = \bar{\epsilon}(x)\\ &= \int_x^\infty \frac{\epsilon(t)(\log\log t)^{1+\delta}}{t\log t(\log\log t)^{1+\delta}}\,dt \le \int_x^\infty \frac{\epsilon(x)(\log\log x)^{1+\delta}}{t\log t(\log\log t)^{1+\delta}}\,dt\\ &= \frac{1}{\delta}\epsilon(x)\log\log x \end{aligned} \tag{2.14}$$

for some $\delta > 0$ and all sufficiently large x. From (2.10)-(2.14), we have

$$R(x) = \alpha + O(\bar{\epsilon}(x)).$$

Putting this estimate and (2.14) into (2.9) gives

$$\sum_{p\le x}\frac{F(p)}{p} = \alpha\log x + O(\bar{\epsilon}(x)\log x) = \alpha\log x + O(\epsilon(x)\log x\log\log x). \tag{2.15}$$

Thus (2.1) follows from (2.8) and (2.15).

We now turn our attention to the proof of Theorem 2.2. We shall prove this result by Turán's method (see Elliott [3], vol. II, p.112). In particular, let

$$E(x) := \frac{1}{x}\sum_{n\le x} F(n).$$

Thus (2.2) follows directly from (2.1) and the assertion

$$\frac{1}{x}\sum_{n\le x}(F(n) - E(x))^2 \ll \epsilon(x)\log^2 x\log\log x. \tag{2.16}$$

But

$$\begin{aligned}\frac{1}{x}\sum_{n\le x}(F(n) - E(x))^2 &= \frac{1}{x}\sum_{n\le x}F(n)^2 - \frac{2E(x)}{x}\sum_{n\le x}F(n) + \frac{[x]}{x}E(x)^2\\ &= \frac{1}{x}\sum_{n\le x}F(n)^2 - E(x)^2 + O(\frac{\log^2 x}{x}).\end{aligned}$$

Thus (2.16) follows from (2.1) and the assertion

$$\frac{1}{x}\sum_{n\le x}F(n)^2 = \alpha^2\log^2 x + O(\epsilon(x)\log^2 x\log\log x). \tag{2.17}$$

We have thus reduced Theorem 2.2 to proving (2.17) (under the hypothesis of Theorem 2.2) .

We now turn to the sum in (2.17) . We have

$$\sum_{n\le x} F(n)^2 = \sum_{n\le x}\Bigg(\sum_{p^a\mid n} F(p)\Bigg)^2 = \sum_{p^a,q^b\le x} F(p)F(q) \sum_{\substack{n\le x\\ p^a\mid n,\ q^b\mid n}} 1$$

$$= \sum_{p\le x} F(p)^2\left[\frac{x}{p}\right] + 2\sum_{\substack{pq\le x\\ p<q}} F(p)F(q)\left[\frac{x}{pq}\right] + \sum\nolimits_1 + \sum\nolimits_2, \qquad (2.18)$$

where

$$\sum\nolimits_1 = \sum_{\substack{p^a,p^b\le x\\ a+b\ge 3}} F(p)^2\left[\frac{x}{p^{\max\{a,b\}}}\right]$$

$$\ll x\sum_{\substack{p^a\\ a\ge 2}}\frac{F(p)^2}{p^a}\sum_{b\le(\log x)/\log p} 1$$

$$\ll x(\log x)\sum_{\substack{p^a\\ a\ge 2}}\frac{\log p}{p^a} \ll x\log x$$

and

$$\sum\nolimits_2 = 2\sum_{\substack{p^aq^b\le x\\ p<q\\ a+b\ge 3}} F(p)F(q)\left[\frac{x}{p^aq^b}\right]$$

$$\ll x\sum_{p^a\le x}\frac{F(p)}{p^a}\sum_{\substack{q^b\\ b\ge 2}}\frac{F(q)}{q^b} \ll x\sum_{p^a\le x}\frac{F(p)}{p^a} \ll x\log x.$$

Further note that removing the brackets on the right of (2.18) introduces an error of at most $O(x\log x)$. Thus

$$\frac{1}{x}\sum_{n\le x} F(n)^2 = \sum_{p\le x}\frac{F(p)^2}{p} + 2\sum_{\substack{pq\le x\\ p<q}}\frac{F(p)F(q)}{pq} + O(\log x). \qquad (2.19)$$

We thus will have (2.17) and Theorem 2.2 from (2.14), (2.19) and the following result.

Proposition 2.6. *Under the hypothesis of Theorem 2.2 we have*

$$\sum_{p\le x}\frac{F(p)^2}{p}=\frac{1}{2}\alpha^2\log^2 x+O(\epsilon(x)\log^2 x\log\log x), \tag{2.20}$$

$$2\sum_{\substack{pq\le x\\ p<q}}\frac{F(p)F(q)}{pq}=\frac{1}{2}\alpha^2\log^2 x+O(\bar\epsilon(x)\log^2 x) \tag{2.21}$$

where $\bar\epsilon(x)$ is defined in (2.13).

Proof: We begin with the proof of (2.21) which is easier and actually used in the proof of (2.20). First note that from $\epsilon(x)\le\bar\epsilon(x)$ for large x (see (2.14)), we have

$$\frac{d}{dx}\left(\bar\epsilon(x)\log x\right)=\frac{\bar\epsilon(x)-\epsilon(x)}{x}\ge 0$$

for large x , so that $\bar\epsilon(x)\log x$ is eventually increasing. Thus from (2.15) we have

$$\begin{aligned}\sum_{\substack{pq\le x\\ p<q}}\frac{F(p)F(q)}{pq}&=\sum_{p\le\sqrt{x}}\frac{F(q)}{q}\sum_{p<q\le x/p}\frac{F(p)}{p}\\
&=\sum_{p\le\sqrt{x}}\frac{F(p)}{p}\left(\alpha\log\frac{x}{p}-\alpha\log p+O\left(\bar\epsilon\left(\frac{x}{p}\right)\log\left(\frac{x}{p}\right)\right)\right.\\
&\qquad\left.+O\left(\bar\epsilon(p)\log p\right)\right)\\
&=\alpha\log x\sum_{p\le\sqrt{x}}\frac{F(p)}{p}-2\alpha\sum_{p\le\sqrt{x}}\frac{F(p)\log p}{p}\\
&\qquad+O\left(\bar\epsilon(x)\log x\sum_{p\le\sqrt{x}}\frac{F(p)}{p}\right)\\
&=2\alpha\int_2^{\sqrt{x}}\frac{1}{t}\sum_{p\le t}\frac{F(p)}{p}\,dt+O\left(\bar\epsilon(x)\log^2 x\right).\end{aligned} \tag{2.22}$$

By (2.15), the integral is

$$\begin{aligned}2\alpha\int_2^{\sqrt{x}}&\frac{\alpha\log t}{t}\,dt+O\left(\int_2^{\sqrt{x}}\frac{\bar\epsilon(t)\log t}{t}\,dt\right)\\
&=\alpha^2\log^2\sqrt{x}+O\left(\bar\epsilon(x)\log x\int_2^{\sqrt{x}}\frac{dt}{t}\right)\\
&=\frac{1}{4}\alpha^2\log^2 x+O\left(\bar\epsilon(x)\log^2 x\right),\end{aligned}$$

so (2.22) gives (2.21).

We now turn to the proof of (2.20). By partial summation, we have

$$\sum_{p\le x}\frac{F(p)^2}{p}=\frac{1}{x}\sum_{p\le x}F(p)^2+\int_2^x\frac{1}{t^2}\sum_{p\le t}F(p)^2\,dt. \tag{2.23}$$

We now expand $\sum_{p\le x}F(p)^2$. We have

$$\begin{aligned}\sum_{p\le x}F(p)^2&=1+\sum_{3\le p\le x}F(p-1)^2=1+\sum_{3\le p\le x}\left(\sum_{q^a\mid p-1}F(q)\right)^2\\&=1+\sum_{p^a,q^b\le x}F(p)F(q)\pi(x;[p^a,q^b],1)\\&=1+\sum_{p\le x}F(p)^2\pi(x;p,1)+2\sum_{\substack{pq\le x\\p<q}}F(p)F(q)\pi(x;pq,1)\\&\quad+\sum_{\substack{p^a,q^b\le x\\a+b\ge 3}}F(p)F(q)\pi(x;[p^a,q^b],1).\end{aligned} \tag{2.24}$$

We have

$$\begin{aligned}&\sum_{p\le x}F(p)^2\pi(x;p,1)\\&=\sum_{p\le x}F(p)^2\frac{\pi(x)}{p-1}+\sum_{p\le x}F(p)^2\left(\pi(x;p,1)-\frac{\pi(x)}{p-1}\right)\\&=\pi(x)\sum_{p\le x}\frac{F(p)^2}{p-1}+O\left(\log^2 x\sum_{p\le x^{1-\epsilon(x)}}\left|\pi(x;p,1)-\frac{\pi(x)}{p-1}\right|\right)\\&\quad+O\left(\log^2 x\sum_{x^{1-\epsilon(x)}<p\le x}\pi(x;p,1)\right)+O\left(\pi(x)\sum_{x^{1-\epsilon(x)}<p\le x}\frac{\log^2 p}{p}\right)\\&=\frac{x}{\log x}\sum_{p\le x}\frac{F(p)^2}{p}+O\left(\epsilon(x)x\log x\right),\end{aligned} \tag{2.25}$$

using hypothesis A_ϵ and (2.7). Next, we have using hypothesis B_ϵ,

$$\sum_{\substack{pq\le x\\p<q}}F(p)F(q)\pi(x;pq,1)$$

$$= \pi(x) \sum_{\substack{pq \le x \\ p<q}} \frac{F(p)F(q)}{\varphi(pq)} + O\left(\log^2 x \sum_{pq \le x^{1-\epsilon(x)}} \left| \pi(x;,pq,1) - \frac{\pi(x)}{\varphi(pq)} \right| \right)$$

$$+ O\left(\log x \sum_{\substack{x^{1-\epsilon(x)} < pq \le x \\ p<q}} (\log p)\pi(x;pq,1) \right)$$

$$+ O\left(\pi(x) \sum_{x^{1-\epsilon(x)} < pq \le x} \frac{\log p \log q}{pq} \right)$$

$$= \frac{x}{\log x} \sum_{\substack{pq \le x \\ p<q}} \frac{F(p)F(q)}{pq} + O\left(\frac{x}{\log x} \sum_{\substack{pq \le x \\ p<q}} \frac{\log p \log q}{p^2 q} \right) + O\left(\epsilon(x) x \log x\right)$$

$$+ O\left(\log x \sum_{m < x^{\epsilon(x)}} \sum_{p \le \sqrt{\frac{x}{m}}} \log p \sum_{\substack{q \le \frac{x}{pm} \\ qpm+1 \text{ prime}}} 1 \right)$$

$$+ O\left(\frac{x}{\log x} \sum_{p \le \sqrt{x}} \frac{\log p}{p} \sum_{\frac{x^{1-\epsilon(x)}}{p} < q \le \frac{x}{p}} \frac{\log q}{q} \right)$$

$$= \frac{x}{\log x} \sum_{\substack{pq \le x \\ p<q}} \frac{F(p)F(q)}{pq} + O\left(\epsilon(x) x \log x\right) + O\left(\frac{x}{\log x} \sum_{m < x^{\epsilon(x)}} \sum_{p \le \sqrt{\frac{x}{m}}} \frac{\log p}{p\varphi(m)} \right)$$

$$= \frac{x}{\log x} \sum_{\substack{pq \le x \\ p<q}} \frac{F(p)F(q)}{pq} + O\left(\epsilon(x) x \log x\right). \tag{2.26}$$

For the last term in (2.24) we have the estimate

$$\ll \sum_{\substack{p^a \le x \\ a \ge 2}} \log p \sum_{\substack{r \le x \\ r \equiv 1 \ (\mathrm{mod}\ p^a)}} \sum_{q^b \mid r-1} \log q$$

$$= \sum_{\substack{p^a \le \sqrt{x} \\ a \ge 2}} \log p \sum_{\substack{r \le x \\ r \equiv 1 \ (\mathrm{mod}\ p^a)}} \log r + \sum_{\substack{\sqrt{x} < p^a \le x \\ a \ge 2}} \log p \sum_{\substack{r \le x \\ r \equiv 1 \ (\mathrm{mod}\ p^a)}} \log r$$

$$= x \sum_{\substack{p^a \le \sqrt{x} \\ a \ge 2}} \frac{\log p}{p^a} + x \log x \sum_{\substack{\sqrt{x} < p^a \le x \\ a \ge 2}} \frac{\log p}{p^a} \ll x,$$

using the Brun-Titchmarsh inequality. Putting this estimate, (2.25) and (2.26) into (2.24) we get

$$\frac{\log x}{x}\sum_{p\le x}F(p)^2=\sum_{p\le x}\frac{F(p)^2}{p}+2\sum_{\substack{pq\le x\\ p<q}}\frac{F(p)F(q)}{pq}+O\left(\epsilon(x)\log^2 x\right). \quad (2.27)$$

Now using this estimate with (2.14), (2.21) and (2.23), we get

$$\frac{\log x}{x}\sum_{p\le x}F(p)^2=\frac{1}{2}\alpha^2\log^2 x+\int_2^x\frac{1}{t^2}\sum_{p\le t}F(p)^2\,dt+O(\bar{\epsilon}(x)\log^2 x),$$

so that if

$$R_2(x):=\frac{1}{x}\sum_{p\le x}F(p)^2,$$

then we have

$$R_2(x)=\frac{1}{2}\alpha^2\log x+\frac{1}{\log x}\int_2^x\frac{1}{t}R_2(t)\,dt+O(\bar{\epsilon}(x)\log x). \quad (2.28)$$

Let

$$V_2(x):=\frac{1}{\log x}\int_2^x\frac{1}{t}R_2(t)\,dt.$$

As in the proof of Theorem 2.1, we have

$$V_2(x)=\int_2^x V_2'(t)\,dt \quad (2.29)$$

and

$$V_2'(x)=\frac{1}{x\log x}(R_2(x)-V_2(x))=\frac{\alpha^2}{2x}+O\left(\frac{\bar{\epsilon}(x)}{x}\right).$$

Thus from (2.29), we have

$$V_2(x)=\frac{1}{2}\alpha^2\log x+O\left(1+\int_2^x\frac{\bar{\epsilon}(t)}{t}\,dt\right).$$

But for large x

$$\begin{aligned}\int_2^x\frac{\bar{\epsilon}(t)}{t}\,dt&=\bar{\epsilon}(x)\log x-\int_2^x\bar{\epsilon}'(t)\log t\,dt\\&=\bar{\epsilon}(x)\log x+\int_2^x\frac{\epsilon(t)\log t}{t\log t}\,dt\\&\le\bar{\epsilon}(x)\log x+\epsilon(x)\log x\int_2^x\frac{dt}{t\log t}\\&=\bar{\epsilon}(x)\log x+\epsilon(x)\log x(\log\log x-\log\log 2)\\&\ll\epsilon(x)\log x\log\log x\end{aligned}$$

by (2.14), so that

$$V_2(x) = \frac{1}{2}\alpha^2 \log x + O(\epsilon(x) \log x \log\log x).$$

Thus from (2.28), we get

$$\frac{1}{x}\sum_{p\le x} F(p)^2 = \alpha^2 \log x + O(\epsilon(x) \log x \log\log x).$$

Finally, using this and (2.21) in (2.27) gives (2.20).

REMARKS: With a little more care, the right side of (2.2) can be replaced with

$$\bar{\epsilon}(x)\log^2 x + \log x \int_2^x \frac{\epsilon(t)}{t} dt\ .$$

For some choices of acceptable functions $\epsilon(x)$, this expression is $O(\epsilon(x)\log^2 x)$, which is smaller than the right side of (2.2) by a $\log\log x$ factor. For example, we would have this for $\epsilon(x) = (\log x)^{-\delta}$ for some fixed $\delta, 0 < \delta < 1$.

For each prime q , define a completely additive function $F_q(n)$ by inductively defining its values on the primes as follows:

$$F_q(p) = \begin{cases} 0, & \text{if } p < q \\ 1, & \text{if } p = q \\ F_q(p-1), & \text{if } p > q. \end{cases}$$

Thus $F_2(n) = F(n)$. The functions $F_q(n)$ have the following connection with the iterated phi-function:

$$F_q(n) = \#\{j \ge 0 :\ q \mid \varphi_j(n)\},$$

where we interpret $\varphi_0(n) = n$. We have already seen this for $q = 2$ in the Introduction.

Theorems 2.1 and 2.2 hold for the functions F_q for each q with corresponding constants α_q (with $\alpha_2 = \alpha$) , except that we are not sure that $\alpha_q > 0$ for $q > 2$. This, in fact, can be proved assuming hypothesis A_ϵ holds for $\epsilon(x) = (\log x)^{c-1}$ for some c with $0 < c < c_{10}$ where c_{10} is the constant of Theorem 4.5 below. Indeed, if $\varphi_j(n)$ is divisible by every prime up to $(\log n)^{c_{10}}$ and if n is large, then $\varphi_{j+1}(n)$ is divisible by q^k where

$$k > \frac{(\log n)^{c_{10}}}{qc_{10}\log\log n}.$$

Thus Theorem 4.5 implies $F_q(n) \gg (\log n)^{c_{10}}/\log\log n$ on a set of asymptotic density 1. However, this is incompatible with (2.1) if $\alpha_q = 0$, $\epsilon(x) = (\log x)^{c-1}$.

Let $v_p(n)$ denote the exponent on p in the prime factorization of n. Note that for any natural number m we have

$$v_p(\varphi(m)) - v_p(m) = \begin{cases} -1 + \sum_{q|m} v_p(q-1), & \text{if } v_p(m) > 0 \\ \sum_{q|m} v_p(q-1), & \text{if } v_p(m) = 0. \end{cases}$$

Let $k = k(n)$. Then

$$\begin{aligned} 0 = v_p(\varphi_k(n)) &= v_p(n) + \sum_{i=1}^{k} (v_p(\varphi_i(n)) - v_p(\varphi_{i-1}(n))) \\ &= v_p(n) - \sum_{\substack{i \geq 0 \\ v_p(\varphi_i(n))>0}} 1 + \sum_{i\geq 0} \sum_{q \,|\, \varphi_i(n)} v_p(q-1) \\ &= v_p(n) - F_p(n) + \sum_q v_p(q-1) F_q(n); \end{aligned}$$

that is, for every prime p and every natural number n , we have

$$F_p(n) = v_p(n) + \sum_q v_p(q-1) F_q(n), \tag{2.30}$$

where the sum is over all primes q .

We can generate another pretty identity involving the functions F_p via the elementary relation

$$\log m - \log \varphi(m) = \sum_{p \,|\, m} \log \frac{p}{p-1}.$$

We have

$$\begin{aligned} \log n = \sum_{i\geq 0} (\log \varphi_i(n) - \log \varphi_{i+1}(n)) &= \sum_{i\geq 0} \sum_{p \,|\, \varphi_i(n)} \log \frac{p}{p-1} \\ &= \sum_p F_p(n) \log \frac{p}{p-1}. \end{aligned} \tag{2.31}$$

Using (2.30) with $p = 2$ to eliminate $F_2(n)$ in (2.31), we have

$$\log n = v_2(n)\log 2 + F_3(n)\log 3 + F_5(n)\log 5 + F_7(n)\log\frac{7}{3} + \cdots \tag{2.32}$$

where the general term on the right is $F_p(n)\log\frac{p}{(p-1)_3}$ for $p \geq 3$ and where $(p-1)_2$ is the largest odd divisor of $p-1$. We can now use (2.30) to eliminate $F_3(n)$ in (2.32) and continuing, if we eliminate all $F_p(n)$ for $p \leq q$, we obtain the identity (valid for all n and q):

$$\log n = \sum_{p\leq q} v_p(n)\log p + \sum_{p>q} F_p(n)\log\frac{p}{(p-1)_q}, \tag{2.33}$$

where $(p-1)_q$ denotes the largest divisor of $p-1$ not divisible by any prime up to and including q.

Since for every $p > 2$ we have $(p-1)_q = 1$ for some $q < p$, a corollary of (2.33) is the theorem

$$F_p(n) \leq \log n/\log p \tag{2.34}$$

for all n and all $p > 2$. From (2.31), this inequality holds for $p = 2$ as well. Note that if $n = p^k$, then $F_p(n) = k = \log n/\log p$, so (2.34) is best possible.

Suppose now that A_ϵ holds for some acceptable function $\epsilon(x)$. Then each of the numbers α_p exists and an immediate corollary of (2.34) is that

$$\alpha_p \leq 1/\log p \tag{2.35}$$

for each p. In particular, $\lim_{p\to\infty}\alpha_p = 0$. Further, (2.30) implies that

$$\alpha_p \geq \sum_{q\leq p_0} v_p(q-1)\alpha_q$$

for any prime p_0. Letting $p_0 \to \infty$, we obtain

$$\alpha_p \geq \sum_q v_p(q-1)\alpha_q \tag{2.36}$$

for every p. The case $p = 2$ shows that $\sum \alpha_p$ converges. Similarly, using (2.31) and (2.33) we get

$$1 \geq \sum_p \alpha_p \log\frac{p}{p-1},$$

$$1 \geq \sum_{p>q} \alpha_p \log\frac{p}{(p-1)_q} \tag{2.37}$$

for every q. Thus if infinitely many p have $\alpha_p > 0$, we have strict inequality in (2.35) for every p.

Assume now that $0 < c < c_{10}$ and that A_ϵ holds for $\epsilon(x) = (\log x)^{c-1}$. We've seen that this then implies each $\alpha_p > 0$. Thus (2.36) and Dirichlet's theorem on primes in an arithmetic progression imply we have $\alpha_p > \alpha_q$ for all primes p, q with $q \equiv 1 \pmod p$. We conjecture that we have $\alpha_p > \alpha_q$ whenever $q > p$.

We can prove that we have equality in the first statement in (2.37) as follows. By (2.34), we have

$$\frac{1}{\log n} \sum_{p>p_0} F_p(n) \log \frac{p}{p-1} \leq \sum_{p>p_0} \frac{\log(p/(p-1))}{\log p} \to 0 \text{ as } p_0 \to \infty. \quad (2.38)$$

But for any p_0, we have by (2.31)

$$\begin{aligned} 1 &= \frac{1}{[x]} \sum_{n \leq x} \frac{1}{\log n} \sum_{p \leq p_0} F_p(n) \log \frac{p}{p-1} + \frac{1}{[x]} \sum_{n \leq x} \frac{1}{\log n} \sum_{p > p_0} F_p(n) \log \frac{p}{p-1} \\ &= \sum_{p \leq p_0} \alpha_p \log \frac{p}{p-1} + o(1) + \frac{1}{[x]} \sum_{n \leq x} \frac{1}{\log n} \sum_{p > p_0} F_p(n) \log \frac{p}{p-1} \end{aligned}$$

as $x \to \infty$. But from (2.38) we can make the last expression as small as we please uniformly for every x by taking p_0 large enough. Thus

$$1 = \sum_p \alpha_p \log \frac{p}{p-1}.$$

We conjecture we also have equality in (2.36) and in the second statement of (2.37).

§3. Results on the sum of the reciprocals of primes

From a theorem of Landau (for example, see Davenport [2], p. 94) there is a positive constant c_0 with the following property. Let $\mathcal{E}(c_0)$ denote the set of natural numbers n for which there is a real primitive character χ mod n for which $L(s, \chi)$ has a real root $\beta \geq 1 - c_0/\log n$. Then $1 \notin \mathcal{E}(c_0)$ and for any x there is at most one member n of $\mathcal{E}(c_0)$ between x and x^2.

Lemma 3.1. *There are positive absolute constants $c_1 \leq 1, c_2 > 1$ such that if $n > 1$ is a natural number with n not divisible by any member of $\mathcal{E}(c_0)$, then*

$$\sum_{\substack{p \leq x \\ p \equiv 1 \pmod n}}' \frac{1}{p} \geq \frac{c_1}{\varphi(n)} (\log \log x - \log \log n)$$

for all $x \geq n^{c_2}$, where $\sum'$ signifies that the sum is over primes not in $\mathcal{E}(c_0)$.

Proof: This result follows from the proof of Linnik's theorem given in Section 6 of Bombieri [1]. In particular, from this proof, if c_2 is sufficiently large, then

$$\sum_{\substack{p \leq t \\ p \equiv 1 \pmod n}}' \log p > \frac{t}{2\varphi(n)}$$

for any $t \geq n^{c_2/2}$. Then if $x \geq n^{c_2}$,

$$\begin{aligned}\sum_{\substack{p\leq x\\ p\equiv 1 \pmod n}}{}' \frac{1}{p} &\geq \int_n^x \frac{1}{t^2\log t} \sum_{\substack{p\leq t\\ p\equiv 1 \pmod n}}{}' \log p\, dt \\ &\geq \frac{1}{2\varphi(n)}\int_{n^{c_2/2}}^x \frac{dt}{t\log t} \\ &= \frac{1}{2\varphi(n)}(\log\log x - \log\log(n^{c_2/2})) \\ &\geq \frac{c_1}{\varphi(n)}(\log\log x - \log\log n)\end{aligned}$$

for $c_1 \leq (\log 2)/(2\log c_2)$.

Lemma 3.2. *Suppose $\mathcal{S}$ is a set of primes. For any x, let*

$$S_1 = \sum_{p\in\mathcal{S}} \sum_{\substack{q\leq x\\ q\equiv 1 \pmod p}}{}' \frac{1}{q}, \quad S_2 = \sum_{\substack{p,p'\in\mathcal{S}\\ p<p'}} \sum_{\substack{q\leq x\\ q\equiv 1 \pmod{pp'}}}{}' \frac{1}{q}.$$

If $q \leq x$ is prime, let a_q denote the number of prime factors of $q-1$ that are in $\mathcal{S}$. If $S_1 > 0$, then

$$\sum_{\substack{q\leq x\\ a_q>0}} \frac{1}{q} \geq \frac{S_1^2}{2S_2+S_1}.$$

Proof: This is just the Cauchy-Schwarz inequality. In fact,

$$\begin{aligned}S_1 = \sum_{q\leq x}{}' \frac{a_q}{q} = \sum_{q\leq x}{}' \frac{1}{\sqrt{q}}\cdot\frac{a_q}{\sqrt{q}} &\leq \left(\sum_{\substack{q\leq x\\ a_q>0}}{}' \frac{1}{q}\right)^{1/2} \left(\sum_{q\leq x}{}' \frac{a_q^2}{q}\right)^{1/2} \\ &= \left(\sum_{\substack{q\leq x\\ a_q>0}}{}' \frac{1}{q}\right)^{1/2} (2S_2+S_1)^{1/2},\end{aligned}$$

since $a_q^2 = 2\binom{a_q}{2} + a_q$.

Lemma 3.3. *Suppose $y \geq 3$ and $\mathcal{S}$ is a set of primes such that if $p \in \mathcal{S}$ then $p \leq y$ and $p \notin \mathcal{E}(c_0)$. There is an absolute positive constant c_3 such that if $x \geq y^{c_2}$, then*

$$\sum_{\substack{q\leq x\\ a_q>0}}{}' \frac{1}{q} \geq \min\left\{\frac{c_1^2}{16c_3}\frac{(\log\log x - \log\log y)^2}{\log\log x}, \frac{c_1}{4}(\log\log x - \log\log y)\sum_{p\in\mathcal{S}}\frac{1}{p}\right\}$$

where a_q is defined in Lemma 3.2.

Proof: The lemma is clearly true if $2 \in \mathcal{S}$ or if $\mathcal{S} = \emptyset$, so assume $2 \notin \mathcal{S}$ and $\mathcal{S} \neq \emptyset$. Using the notation of Lemma 3.2, we have

$$S_1 \geq \frac{1}{2} c_1 (\log\log x - \log\log y) \sum_{p \in \mathcal{S}} \frac{1}{p}$$

from Lemma 3.1. Also, using partial summation and the Brun-Titchmarsh inequality we have for some absolute constant $c_3 \geq 1$,

$$\sum_{\substack{q \leq x \\ q \equiv 1 \pmod n}} \frac{1}{q} \leq \frac{c_3}{\varphi(n)} \log\log x \tag{3.1}$$

for any natural number n and any $x \geq 3$. Thus

$$S_2 \leq c_3 \log\log x \sum_{\substack{p,p' \in \mathcal{S} \\ p<p'}} \frac{1}{(p-1)(p'-1)} \leq c_3 \log\log x \left(\sum_{p \in \mathcal{S}} \frac{1}{p} \right)^2 =: S_2',$$

since $2 \notin \mathcal{S}$. Thus from Lemma 3.2, we have

$$\sum_{\substack{q \leq x \\ a_q > 0}}{}' \frac{1}{p} \geq \frac{S_1^2}{S_1 + 2S_2} \geq \frac{S_1^2}{S_1 + 2S_2'} \geq \min\left\{ \frac{S_1^2}{4S_2'}, \frac{1}{2} S_1 \right\}$$

and our conclusion follows.

If k, n are natural numbers, let

$$S_k'(x,n) = \sum_{\substack{p \leq x \\ n \mid \varphi_k(p)}}{}' \frac{1}{p}$$

where again the dash means that $p \notin \mathcal{E}(c_0)$.

Theorem 3.4. *There are absolute constants $0 < c_4,\ c_5,\ c_6 \leq 1$ such that for any A and $x \geq x_0(A)$ we have*

$$S_k'(x,n) \geq \min\left\{ c_4 \log\log x, \frac{1}{\varphi(n)} \left(\frac{c_5 \log\log x}{k} \right)^k \right\}$$

for all $n \le (\log x)^A$ *and* $k \le c_6 \log\log x$.

Proof: Fix an arbitrary number A and assume $n \le (\log x)^A$. If $y \ge \exp\left((\log x)^{1/3}\right)$, then by partial summation and the Siegel-Walfisz theorem, provided $x \ge x_0(A)$, we have

$$S_1'(y,n) \ge \int_{\exp((\log x)^{1/6})}^{\exp((\log x)^{1/3})} \frac{1}{t^2} \sum_{\substack{p \le t \\ p \equiv 1 \pmod n}} 1\, dt \ge \frac{1}{7\varphi(n)} \log\log x. \tag{3.2}$$

By letting $y = x$ in (3.2) we have the theorem for $k = 1$.

Suppose now $k = 2$. Let $\mathcal{S}$ be the set of primes $p \le \exp((\log x)^{1/3})$ for which $p \equiv 1 \pmod n$ and $p \notin \mathcal{E}(c_0)$. Then in the notation of Lemma 3.2, we have

$$S_2'(x,n) \ge \sum_{\substack{q \le x \\ a_q > 0}}{}' \frac{1}{q}.$$

From Lemma 3.3 and (3.2) with $y = \exp((\log x)^{1/3})$ we have

$$S_2'(x,n) \ge \min\left\{ \frac{c_1^2}{36c_3} \log\log x, \frac{c_1}{42\varphi(n)} (\log\log x)^2 \right\},$$

which gives the theorem for $k = 2$.

Now let $k = 3$. Let

$$S_i' = S_i'\left(\exp((\log x)^{i/3}), n\right) \quad \text{for } i = 1,\ 2,\ 3.$$

Then from Lemma 3.3 we have

$$S_2' \ge \min\left\{ \frac{c_1^2}{96c_3} \log\log x, \frac{c_1}{12} (\log\log x) S_1' \right\},$$

$$\begin{aligned} S_3' &\ge \min\left\{ \frac{c_1^2}{144c_3} \log\log x, \frac{c_1}{12} (\log\log x) S_2' \right\} \\ &\ge \min\left\{ \frac{c_1^2}{144c_3} \log\log x, \frac{c_1^3}{1152c_3} (\log\log x)^2, \frac{c_1^2}{144} (\log\log x)^2 S_1' \right\}. \end{aligned}$$

Since $S_1' \ge \frac{1}{7\varphi(n)} \log\log x$ by (3.2), we have our theorem for $k = 3$.

Suppose now $k \ge 4$. Let

$$y_j = \exp\left((\log x)^{\frac{1}{3} + \frac{j}{3(k-3)}}\right) \quad \text{for } j = 0,\ 1,\ \cdots,\ k-3.$$

If c_6 is sufficiently small, then $k \leq c_6 \log\log x$ implies that $y_j \geq y_{j-1}^{c_2}$ for $j = 1, \cdots, k-3$. Note that

$$\log\log y_j - \log\log y_{j-1} = \frac{1}{3(k-3)} \log\log x.$$

Thus from Lemma 3.3 we have for $j = 1, \cdots, k-3$

$$S'_{j+1}(y_j, n) \geq \min\left\{\frac{c_1^2 \log\log x}{96c_3(k-3)^2}, \frac{c_1 \log\log x}{12(k-3)} S'_j(y_{j-1}, n)\right\}. \tag{3.3}$$

The min is the first term if and only if

$$S'_j(y_{j-1}, n) \geq \frac{c_1}{8c_3(k-3)}. \tag{3.4}$$

We shall choose c_6 so small that we also have $c_6 \leq c_1/12$. Then $k-3 \leq (c_1/12)\log\log x$, so that

$$\frac{c_1^2 \log\log x}{96c_3(k-3)^2} \geq \frac{c_1}{8c_3(k-3)}. \tag{3.5}$$

Thus if $0 < j < k-3$ and the min in (3.3) is the first term, then (3.5) implies that

$$S'_{j+1}(y_j, n) \geq \frac{c_1}{8c_3(k-3)}$$

and so (3.4) implies the same is true when j is replaced with $j+1$; i.e., the min in (3.3) is again the first term. Thus by iterating (3.3), we have

$$S'_{k-2}(y_{k-3}, n) \geq \min\left\{\frac{c_1^2 \log\log x}{96c_3(k-3)^2}, \left(\frac{c_1 \log\log x}{12(k-3)}\right)^{k-3} S'_1(y_0, n)\right\}. \tag{3.6}$$

Note that from (3.2) we have

$$S'_1(y_0, n) \geq \frac{1}{7\varphi(n)} \log\log x. \tag{3.7}$$

Note also that $y_{k-3} = \exp((\log x)^{2/3})$. Thus from Lemma 3.3, we have

$$S'_{k-1} := S'_{k-1}\left(\exp((\log x)^{5/6}),\ n\right) \geq$$
$$\min\left\{\frac{c_1^2}{480c_3} \log\log x,\ \frac{c_1}{24}(\log\log x) S'_{k-2}(y_{k-3}, n)\right\},$$

$$S_k'(x,n) \geq \min\left\{\frac{c_1^2}{576c_3}\log\log x,\ \frac{c_1}{24}(\log\log x)S_{k-1}'\right\}.$$

Thus from (3.6) and (3.7)

$$S_{k-1}' \geq \min\left\{\frac{c_1^2}{480c_3}\log\log x, \frac{c_1^3(\log\log x)^2}{2304c_3(k-3)^2}, \left(\frac{c_1\log\log x}{12(k-3)}\right)^{k-2}\frac{\log\log x}{14\varphi(n)}\right\},$$

so that

$$S_k'(x,n) \geq \min\left\{\frac{c_1^2}{576c_3}\log\log x,\ \frac{c_1^3}{11520c_3}(\log\log x)^2,\ \frac{c_1^4(\log\log x)^3}{55296c_3(k-3)^2},\right.$$
$$\left.\left(\frac{c_1\log\log x}{12(k-3)}\right)^{k-1}\frac{\log\log x}{28\varphi(n)}\right\}$$

Thus our theorem holds with

$$c_4 = \min\left\{\frac{c_1^2}{576c_3}, \frac{c_1^4}{55296c_3c_6^2}\right\},\ c_5 = c_1/15$$

if $x \geq x_0$.

Theorem 3.5. *If c_3 is the constant in (3.1), we have*

$$\sum_{\substack{n\leq x\\ p\,|\,\varphi_k(n)}} 1 \leq \frac{x}{p}(2c_3\log\log x)^k$$

for every odd prime p, for every $k \geq 0$ and for all x with $\log\log x \geq 2/c_3$. (We define $\varphi_0(n) = n$.)

Proof: The theorem holds for $k = 0$ since $\varphi_0(n) = n$. Suppose $k \geq 0$ and the theorem holds for k. If $p\,|\,\varphi_{k+1}(n)$ then either $p^2\,|\,\varphi_k(n)$ or there is some prime $q\,|\,\varphi_k(n)$ with $q \equiv 1 \pmod p$. Thus

$$\sum_{\substack{n\leq x\\ p\,|\,\varphi_{k+1}(n)}} 1 \leq \sum_{\substack{n\leq x\\ p^2\,|\,\varphi_k(n)}} 1 + \sum_{q\equiv 1 \pmod p}\ \sum_{\substack{n\leq x\\ q\,|\,\varphi_k(n)}} 1$$
$$\leq x(2c_3\log\log x)^k\left(\frac{1}{p} + \sum_{\substack{q\leq x\\ q\equiv 1 \pmod p}}\frac{1}{q}\right)$$

by the induction hypothesis, the fact that $p^2 \mid \varphi_k(n)$ implies $p|\varphi_k(n)$ and the observation that if $n \le x$ and $q \mid \varphi_k(n)$, then $q \le x$. Using (3.1) to estimate the remaining sum we have

$$\begin{aligned}\sum_{\substack{n\le x\\ p\mid\varphi_{k+1}(n)}} 1 &\le x(2c_3\log\log x)^k\left(\frac{1}{p}+\frac{1}{p-1}c_3\log\log x\right)\\ &\le \frac{x}{p}(2c_3\log\log x)^k\left(1+\frac{3}{2}c_3\log\log x\right)\\ &\le \frac{x}{p}(2c_3\log\log x)^{k+1}.\end{aligned}$$

REMARK: If we let $S_k(x,p)$ denote the sum of $1/q$ for primes $q \le x$ with $p \mid \varphi_k(q)$, then by essentially the same proof we have

$$S_k(x,p) \le \frac{1}{p}(2c_3\log\log x)^k$$

for the same set of p, k, x as in Theorem 3.5. Although this result will not be of use to us it is interesting to compare it with Theorem 3.4 in the case $n = p$.

§4. More on the iterated phi-function

Using the constants c_3, c_5 of the preceding section, let

$$\alpha_k(x) = (c_5k^{-1}\log\log x)^k, \quad \beta_k(x) = (2c_3\log\log x)^k.$$

Let

$$f_k(n,x) = \sum_{\substack{p\le(\log\log x)^k\\ p\nmid\varphi_k(n)}}\frac{1}{p} + \sum_{\substack{p>(\log\log x)^k\\ p\mid\varphi_k(n)}}\frac{1}{p}.$$

Thus if $n \le x$, $f_k(n,x)$ measures in some sense how far $\varphi_k(n)/\varphi_{k+1}(n)$ is from

$$\prod_{p\le(\log\log x)^k}(1-1/p)^{-1}.$$

Theorem 4.1. *There is an absolute constant c_7 such that*

$$\frac{1}{x}\sum_{n\le x} f_k(n,x) \le c_7(\log k)/(\log\log\log x - \log k)$$

for all $x \geq x_0$, $1 \leq k < \log\log x$.

Proof: We have $\alpha_k(x) \leq (\log\log x)^k \leq \beta_k(x)$ for all $k \geq 1$, $x \geq 3$. Thus

$$\begin{aligned}\sum_{n\leq x} f_k(n,x) &= \sum_{p\leq(\log\log x)^k} \frac{1}{p} \sum_{\substack{n\leq x\\ p\nmid \varphi_k(n)}} 1 + \sum_{p>(\log\log x)^k} \frac{1}{p} \sum_{\substack{n\leq x\\ p\mid \varphi_k(n)}} 1\\ &\leq \sum_{p\leq\alpha_k(x)} \frac{1}{p} \sum_{\substack{n\leq x\\ p\nmid \varphi_k(n)}} 1 + \sum_{\alpha_k(x)<p\leq\beta_k(x)} \frac{1}{p}\sum_{n\leq x} 1 + \sum_{p>\beta_k(x)} \frac{1}{p} \sum_{\substack{n\leq x\\ p\mid \varphi_k(n)}} 1\\ &= S_1 + S_2 + S_3, \quad \text{say.} \end{aligned} \tag{4.1}$$

If $p\nmid\varphi_k(n)$, then n is not divisible by any prime q with $p|\varphi_k(q)$. Thus by Brun's method (see Halberstam-Richert [11]) we have

$$\begin{aligned}\sum_{\substack{n\leq x\\ p\nmid \varphi_k(n)}} 1 \ll x \prod_{\substack{q\leq x\\ p\mid\varphi_k(q)}} (1-\frac{1}{q}) \ll x\exp\left(-\sum_{\substack{q\leq x\\ p\mid\varphi_k(q)}} \frac{1}{q}\right)\\ \leq x\exp\left(-S_k'(x,p)\right)\end{aligned}$$

uniformly for all x, p, k, where $S_k'(x,p)$ is defined in section 3. Let

$$\alpha_k'(x) = \alpha_k(x)/(c_4 \log\log x).$$

Thus by Theorem 3.4, there is an absolute constant c_8 such that

$$\sum_{\substack{n\leq x\\ p\nmid \varphi_k(n)}} 1 \leq \begin{cases} c_8 x e^{-\alpha_k(x)/(p-1)}, & \text{if } p > \alpha_k'(x)+1\\ c_8 x(\log x)^{-c_4}, & \text{if } p \leq \alpha_k'(x)+1\end{cases} \tag{4.2}$$

for all $x \geq x_0$, $p \leq (\log x)^2$, $k \leq c_6 \log\log x$.

The theorem holds trivially if $k \gg \log\log x$, so assume $k/\log\log x \leq \min\{\frac{1}{2}c_5, c_6\}$. Since for any k, $\alpha_k(x) \leq (\log x)^{c_5/e}$, (4.2) implies

$$\begin{aligned}S_1 &= \sum_{p\leq\alpha_k(x)} \frac{1}{p} \sum_{\substack{n\leq x\\ p\nmid \varphi_k(n)}} 1\\ &\leq c_8 x(\log x)^{-c_4} \sum_{p\leq\alpha_k'(x)+1} \frac{1}{p} + c_8 x \sum_{\alpha_k'(x)<p\leq\alpha_k(x)} \frac{1}{p} e^{-\alpha_k(x)/(p-1)}\\ &\ll x/\log\alpha_k(x) \ll x/(\log\log\log x - \log k).\end{aligned} \tag{4.3}$$

Since we are assuming $k \le \frac{1}{2}c_5 \log\log x$, we have

$$S_2 \le x \sum_{\alpha_k(x)<p\le\beta_k(x)} \frac{1}{p} = x\,(\log\log\beta_k(x) - \log\log\alpha_k(x) + O(1/\log\alpha_k(x)))$$
$$\ll x(\log k)/(\log\log\log x - \log k) \tag{4.4}$$

uniformly in k.

For S_3 we use Theorem 3.5 to estimate the inner sum. We have

$$S_3 \le x\beta_k(x) \sum_{p>\beta_k(x)} \frac{1}{p^2} \ll x/\log\beta_k(x) \le x/\log\log\log x.$$

Assembling this estimate, (4.1), (4.3) and (4.4) we have the theorem.

Theorem 4.2. *Let $\epsilon(x) > 0$ tend to 0 arbitrarily slowly as $x \to \infty$. If $k \le (\log\log x)^{\epsilon(x)}$, then the normal order of $\varphi_k(n)/\varphi_{k+1}(n)$ for $n \le x$ is $ke^\gamma \log\log\log x$.*

Proof: Let $\delta > 0$ be arbitrary. Let x be large and let $k \le (\log\log x)^{\epsilon(x)}$. From Theorem 4.1, the average value of $f_k(n,x)$ for $n \le x$ is $O(\epsilon(x))$. Thus if $x \ge x_0(\delta)$, $f_k(n,x) < \delta$ for at least $(1-\delta)x$ values of $n \le x$. But

$$\frac{\varphi_k(n)}{\varphi_{k+1}(n)} = \left(\prod_{p\le(\log\log x)^k}\left(1-\frac{1}{p}\right)^{-1}\right)\left(\prod_{\substack{p\le(\log\log x)^k\\ p\nmid\varphi_k(n)}}\left(1-\frac{1}{p}\right)\right)$$
$$\left(\prod_{\substack{p>(\log\log x)^k\\ p\mid\varphi_k(n)}}\left(1-\frac{1}{p}\right)^{-1}\right),$$

so that

$$\log\left(\frac{\varphi_k(n)}{\varphi_{k+1}(n)}\prod_{p\le(\log\log x)^k}\left(1-\frac{1}{p}\right)\right) \ll f_k(n,x).$$

Thus, for at least $(1-\delta)x$ values of $n \le x$

$$\frac{\varphi_k(n)}{\varphi_{k+1}(n)} = (1+O(\delta))ke^\gamma\log\log\log x.$$

Theorem 4.3. *Let $\epsilon(x) > 0$ tend to 0 arbitrarily slowly as $x \to \infty$. Then if $k \le \epsilon(x) \log\log\log x / \log\log\log\log x$, the normal order of $\varphi(n)/\varphi_{k+1}(n)$ for $n \le x$ is $k! e^{k\gamma} (\log\log\log x)^k$.*

Proof: From the proof of Theorem 4.2, the number of $n \le x$ for which

$$\left| \log \left(\frac{\varphi_j(n)}{\varphi_{j+1}(n)} (je^{\gamma} \log\log\log x)^{-1} \right) \right| \le \frac{\log j}{\log\log\log x}$$

fails is $O\left(\frac{x \log j}{\log\log\log x}\right)$ uniformly for any $j \le k$. Summing for $j = 1, \cdots, k$ we have that

$$\left| \log \left(\frac{\varphi(n)}{\varphi_k(n)} \left(k! e^{k\gamma} (\log\log\log x)^k\right)^{-1} \right) \right| \le \epsilon(x)$$

but for at most $O(\epsilon(x)x)$ integers $n \le x$. Since $\epsilon(x) \to 0$, we have our theorem.

Theorem 4.4. *There is an absolute constant $c_9 > 0$ such that if $1 \le k \le c_9 \log\log x$, then the number of $n \le x$ for which*

$$\frac{\varphi_k(n)}{\varphi_{k+1}(n)} > k(\log\log\log x - \log k) \tag{4.5}$$

fails is $O(xk^{-1}(\log\log\log x - \log k)^{-1})$. In particular

$$\max_k \frac{\varphi_k(n)}{\varphi_{k+1}(n)} \gg \log\log n$$

for a set of n of asymptotic density 1.

Proof: As in (4.3), if $c_9 > 0$ is small enough, then

$$\sum_{n \le x} \sum_{\substack{p \le \alpha_k(x) \\ p \nmid \varphi_k(n)}} \frac{1}{p} = \sum_{p \le \alpha_k(x)} \frac{1}{p} \sum_{\substack{n \le x \\ p \nmid \varphi_k(n)}} 1 \ll x / \log \alpha_k(x)$$
$$\ll xk^{-1}(\log\log\log x - \log k)^{-1}$$

uniformly for all $k \le c_9 \log\log x$. Thus but for at most $O(xk^{-1}(\log\log\log x - \log k)^{-1})$ exceptional values of $n \le x$, we have

$$\prod_{\substack{p \le \alpha_k(x) \\ p \nmid \varphi_k(n)}} \left(1 - \frac{1}{p}\right) \ge \frac{3}{4}.$$

For those values of n we have

$$\frac{\varphi_k(n)}{\varphi_{k+1}(n)} \geq \prod_{p\leq\alpha_k(x)}\left(1-\frac{1}{p}\right)^{-1}\prod_{\substack{p\leq\alpha_k(x)\\ p\nmid\varphi_k(n)}}\left(1-\frac{1}{p}\right)$$
$$\geq \frac{3}{4}e^\gamma \log\alpha_k(x) - 1$$
$$> k(\log\log\log x - \log k)$$

provided x is sufficiently large and $c_9 \leq \frac{1}{2}c_5$. This proves the theorem.

Theorem 4.5. *There is a positive absolute constant c_{10} such that the set of natural numbers n, for which there is some k with $\varphi_k(n)$ divisible by every prime up to $(\log n)^{c_{10}}$, has asymptotic density 1 .*

Proof: There is a positive absolute constant c_{11} such that if $k = [c_{11} \log\log x]$, then $\alpha_k(x) > (\log x)^{c_{11}}$. Then by (4.2) we have

$$\sum_{\substack{n\leq x\\ p\nmid\varphi_k(n)}} 1 \leq c_8 x(\log x)^{-c_4}$$

for all $x \geq x_0$, primes $p \leq (\log x)^{c_{11}/2}$, $k = [c_{11}\log\log x]$. Let $c_{10} = \min\{c_4/2, c_{11}/2\}$. Then

$$\sum_{p\leq(\log x)^{c_{10}}}\ \sum_{\substack{n\leq x\\ p\nmid\varphi_k(n)}} 1 < c_8 x(\log x)^{-c_4/2}.$$

Thus but for at most $c_8x(\log x)^{-c_4/2}$ exceptional integers $n \leq x$ we have $p\,|\,\varphi_k(n)$ for every prime $p \leq (\log x)^{c_{10}}$ if $k = [c_{11}\log\log x]$ and $x \geq x_0$. This proves the theorem.

In contrast to Theorem 4.5 we give the following result. The proof is not an application of the theorems in section 3, but rather follows from the easy identity (2.30). Let $\nu(m)$ denote the number of distinct prime factors of m .

Theorem 4.6. *Let $\Phi(n) = n\prod_{k=1}^\infty \varphi_k(n)$. Then for all n, $\nu(\Phi(n)) \leq \lceil(\log n)/\log 2\rceil$. In particular, for all n there is some prime $p \ll \log n \log\log n$ with $p\nmid\Phi(n)$.*

Proof: For any $n > 1$ we have, using (2.30) with $p = 2$ and (1.1),

$$\nu(\Phi(n)) = 1 + \sum_{\substack{q>2\\ q\,|\,\Phi(n)}} 1 \leq 1 + \sum_{q>2} F_q(n) \leq \begin{cases} 1 + F(n), & n \text{ odd} \\ F(n), & n \text{ even}\end{cases}$$
$$= k(n) \leq \lceil(\log n)/\log 2\rceil.$$

§5. Aliquot sequences

Let $s(n) = \sigma(n) - n$, where σ is the sum of the divisors function. Let $s_1(n) = s(n)$, $s_2(n) = s(s_1(n))$, etc. What is now known as the Catalan-Dickson conjecture is that for any n , the "aliquot sequence" n, $s_1(n)$, $s_2(n)$, $\cdots$ eventually terminates at 0 or is eventually periodic. The least n for which this conjecture is in doubt is 276. Guy and Selfridge [10] instead conjecture that for infinitely many n the aliquot sequence beginning with n tends to ∞. The function $s(n)$ has been studied since antiquity when numbers were classified as perfect, abundant or deficient depending on whether $s(n) = n$, $s(n) > n$ or $s(n) < n$, respectively.

As discussed in the Introduction, the first author proved in [8] that for each $\epsilon > 0$ and k , the set of n for which

$$\frac{s_{j+1}(n)}{s_j(n)} > \frac{s(n)}{n} - \epsilon \text{ for } j = 1, \cdots, k \tag{5.1}$$

has asymptotic density 1 . Further, he claimed that similar methods would show that

$$\frac{s_{j+1}(n)}{s_j(n)} < \frac{s(n)}{n} + \epsilon \text{ for } j = 1, \cdots, k$$

for a set of n of asymptotic density 1. This claim of a proof is now retracted but we still remain convinced of the truth of this statement; it is our Conjecture 3 in section 1. We now give a proof of the case $k = 1$.

Theorem 5.1. *For each $\epsilon > 0$, the set of n with*

$$\frac{s_2(n)}{s(n)} < \frac{s(n)}{n} + \epsilon \tag{5.2}$$

has asymptotic density 1 .

Proof: Let $1 > \delta > 0$ be arbitrary. We shall show that for all large x, the number of $n \leq x$ for which (5.2) fails is at most $c\delta x$ for some absolute constant c.

Let $P(n)$ denote the largest prime factor of n. If $\eta > 0$ is sufficiently small, then the number of $n \leq x$ for which

$$P(n) > x^\eta, \quad P(n)^2 \nmid n \tag{5.3}$$

fails is at most δx for all large x . This result follows from either sieve methods or work of Dickman and others on the distribution of integers n with no large prime factors. Fix such a number η.

Since $\sum_{n \leq x} \sigma(n)/n \ll x$, there is a number B so large that the number of $n \leq x$ for which

$$\sigma(n)/n \leq B \tag{5.4}$$

fails is at most δx for all large x. Fix such a number B.

If $\alpha > 0$, say that an integer n is *α-primitive* if $s(n)/n \geq \alpha$ and if $d \mid n$, $d < n$, then $s(d)/d < \alpha$ (also called a primitive $(1+\alpha)$-abundant number).

Let α be a rational number with $0 < \alpha_1 < 1/2$, $\alpha_1 \leq \epsilon/4B$. Also let α_2 be a rational number with $0 < \alpha_2 < \alpha_1\eta/24$. Since α_1, α_2 are rational, it follows from the proof in [6] for the case $\alpha = 1$, that

$$\sum\nolimits^{(1)} 1/a < \infty, \quad \sum\nolimits^{(2)} 1/a < \infty,$$

where for $i = 1, 2$, $\sum^{(i)}$ denotes a sum over α_i-primitive numbers. Since $a/\varphi(a)$ is bounded if a is α_i-primitive, it follows that there is a number T so large that

$$\sum_{a \geq T}^{(1)} 1/a < \delta, \quad \sum_{a \geq T}^{(2)} 1/\varphi(a) < \delta\eta. \tag{5.5}$$

Also assume T is so large that

$$T > \frac{1}{\alpha_2} + 1, \quad \prod_{p \geq T} \frac{p^2}{p^2 - 1} < 1 + \frac{1}{2}\alpha_2. \tag{5.6}$$

If $n > 1$ is an integer, factor n as $n_1 n_2$ and $s(n)$ as $N_1 N_2$ where every prime factor of $n_1 N_1$ is less than T and every prime factor of $n_2 N_2$ is at least T. It follows from the work in [8] that but for a set of n of asymptotic density 0 , we have

$$n_1 = N_1. \tag{5.7}$$

The idea of the proof is that but for a set of n of asymptotic density 0 , the number n_1 is not too large, say $n_1 < (\log\log n)^{1/2}/\prod_{p<T} p$. For these n , there is almost certainly a prime $q \| n$ with

$$q \equiv 1 \pmod{n_1 \prod_{p<T} p}.$$

Then but for a set of n of asymptotic density 0 , we have $n_1 \prod_{p<T} p \mid \sigma(n)$. For these n we have $n_1 | s(n)$ and $(\prod_{p<T} p, s(n)/n_1) = 1$, i.e. (5.7) holds.

The number of $n \leq x$ with n_2 divisible by an α_1-primitive number a is at most

$$\sum_{(a, \prod_{p<T} p) = 1}^{(1)} \left[\frac{x}{a}\right] \leq x \sum_{a \geq T}^{(1)} \frac{1}{a} < \delta x$$

by (5.5). Thus but for at most δx exceptional values of $n \leq x$, we have

$$\sigma(n_2)/n_2 < 1 + \alpha_1. \tag{5.8}$$

Suppose now that (5.2) fails for n. By adding 1 to both sides, we get

$$\frac{\sigma(s(n))}{s(n)} \geq \frac{\sigma(n)}{n} + \epsilon,$$

so that from (5.7) and (5.4)

$$\frac{\sigma(N_2)}{N_2} \geq \frac{\sigma(N_2)/N_2}{\sigma(n_2)/n_2} = \frac{\sigma(s(n))/s(n)}{\sigma(n)/n}$$
$$\geq 1 + \frac{\epsilon}{\sigma(n)/n} \geq 1 + \frac{\epsilon}{B} \geq 1 + 4\alpha_1.$$

Factor N_2 as $N_3 N_4$ where every prime in N_3 also divides n and $(N_4, n) = 1$. If $N_3 = \prod p_i^{\beta_i}$, where $p_i \geq T$ are distinct primes and each $\beta_i \geq 1$, then

$$\frac{\sigma(N_3)}{N_3} = \prod \frac{p_i - p_i^{-\beta_i}}{p_i - 1} < \prod \frac{p_i}{p_i - 1}$$
$$\leq \left(\prod \frac{p_i}{p_i - 1} \cdot \frac{p_i}{p_i + 1} \right) \frac{\sigma(n_2)}{n_2}$$

since each $p_i \mid n_2$. Then from (5.6) and (5.8) we have

$$\frac{\sigma(N_3)}{N_3} < \left(\prod_{p \geq T} \frac{p^2}{p^2 - 1} \right) \frac{\sigma(n_2)}{n_2} < \left(1 + \frac{1}{2}\alpha_2\right)(1 + \alpha_1) < 1 + 2\alpha_1.$$

Thus

$$\frac{\sigma(N_4)}{N_4} = \frac{\sigma(N_2)/N_2}{\sigma(N_3)/N_3} > \frac{1 + 4\alpha_1}{1 + 2\alpha_1} > 1 + \alpha_1,$$

so $s(n)$ is divisible by an α_1 -primitive number a_1 not divisible by any prime below T and with $(a_1, n) = 1$.

We now show that any α_1-primitive number a_1 which is not divisible by any primes below T must have an α_2-primitive divisor a_2 with $a_2 \leq a_1^{\eta/2}$. Indeed, let the distinct prime factors of a_1 be $q_1, \cdots, q_t$, where

$$T \leq q_1 < \cdots < q_t.$$

Let $a_0 = q_1 q_2 \cdots q_{[\eta t/2]}$. Then

$$a_0 \leq (q_1 \cdots q_t)^{[\eta t/2]/t} \leq a_1^{\eta/2},$$

so it is sufficient to show $\sigma(a_0)/a_0 \geq 1 + \alpha_2$, for this will guarantee it having an α_2-primitive divisor a_2.

Note that

$$[\eta t/2] \geq \eta t/3,$$

since if not, we have $t < 6/\eta$, which implies by (5.6)

$$\begin{aligned} 1+\alpha_1 \leq \frac{\sigma(a_1)}{a_1} &< \prod_{i\leq t} \frac{q_i}{q_i-1} \\ &< \left(1+\frac{1}{T-1}\right)^{6/\eta} < (1+\alpha_2)^{6/\eta} \\ &< \left(1+\frac{\alpha_1\eta}{24}\right)^{6/\eta} < 1+\frac{1}{2}\alpha_1, \end{aligned}$$

a contradiction. Thus from (5.6),

$$\begin{aligned} \frac{\sigma(a_0)}{a_0} = \prod_{i\leq[\frac{\eta t}{2}]} \frac{q_i+1}{q_i} &> \left(\prod_{i\leq[\frac{\eta t}{2}]} \frac{q_i}{q_i-1}\right) \prod_{p\geq T} \frac{p^2-1}{p^2} \\ &> \left(\prod_{i\leq t} \frac{q_i}{q_i-1}\right)^{[\eta t/2]/t} \left(1+\frac{1}{2}\alpha_2\right)^{-1} \\ &> (1+\alpha_1)^{\eta/3}\left(1+\frac{1}{2}\alpha_2\right)^{-1} \\ &> \left(1+\frac{24\alpha_2}{\eta}\right)^{\eta/3}\left(1+\frac{1}{2}\alpha_2\right)^{-1} > 1+\alpha_2. \end{aligned}$$

We have seen above, but for $O(\delta x)$ integers $n \leq x$, if $n \leq x$ does not satisfy (5.2), then $s(n)$ is divisible by an α_1-primitive number a_1 with $(a_1, n) = 1$ and a_1 not divisible by any prime below T and further that (5.3) holds. Thus such an n must have $s(n)$ divisible by an α_2-primitive number a_2 with $(a_2, n) = 1$, with a_2 not divisible by any prime below T and with

$$a_2 \leq a_1^{\eta/2} \leq s(n)^{\eta/2} < x^{2\eta/3}$$

for x large. For such an n, we factor it as mp where $p = P(n)$. From (5.3), $m < x^{1-\eta}$, $p\nmid m$. Consider the α_2-primitive number a_2 just discovered dividing $s(n)$. We have $s(n) = p(\sigma(m) - m) + \sigma(m)$, so that

$$p(\sigma(m)-m) \equiv -\sigma(m) \pmod{a_2}. \tag{5.9}$$

Since $(a_2, pm) = 1$ we have $(a_2, \sigma(m)) = 1$ so that there is a certain residue class $c(m, a_2)$ mod a_2 such that if p, m, a_2 satisfy (5.9), then $p \equiv c(m, a_2)$

mod a_2. Thus but for $O(\delta x)$ integers, the number of $n \le x$ which do not satisfy (5.2) is at most

$$\begin{aligned}
&\sum_{x^{2\eta/3}\ge a_2\ge T}^{(2)} \sum_{m<x^{1-\eta}} \sum_{\substack{p\le x/m \\ p\equiv c(m,a) \pmod{a_2}}} 1 \\
&\ll \sum_{x^{2\eta/3}\ge a_2\ge T}^{(2)} \sum_{m<x^{1-\eta}} \frac{x}{\varphi(a_2)m\log(x/a_2m)} \\
&\ll \frac{x}{\eta}\sum_{a_2\ge T}\frac{1}{\varphi(a_2)} < \delta x,
\end{aligned}$$

where we used the Brun-Titchmarsh theorem for the first inequality and (5.5) for the last.

Theorem 5.2. *Conjecture 4 implies Conjecture 3.*

Proof: Let k be a natural number. Let $T = T(n)$ tend to infinity very slowly, say $T(n)$ is the $3k$-fold iterated logarithm. For $j = 1, \cdots, k$, factor $s_j(n) = m_j n_j$ where every prime factor of m_j is less than T and every prime factor of n_j is at least T. We analogously factor $n = m_0 n_0$. In the same way as (5.7) is established, the set of n for which

$$m_0 = m_1 = \cdots = m_k \tag{5.10}$$

fails has asymptotic density 0 . Indeed, this is essentially established in [8].

By a simple averaging argument one can show that the set of n for which

$$\sum_{\substack{p\mid n\\ p\ge T}} \frac{1}{p-1} < \frac{1}{T}$$

fails has asymptotic density 0 . Indeed, the average value of the sum is $\sim (T\log T)^{-1}$. But

$$\log\frac{\sigma(n_0)}{n_0} < \log\left(\prod_{\substack{p\mid n\\ p\ge T}}\left(1+\frac{1}{p-1}\right)\right) < \sum_{\substack{p\mid n\\ p\ge T}}\frac{1}{p-1}.$$

Thus, but for a set of n of asymptotic density 0 , we have

$$\sigma(n_j)/n_j < e^{1/T} \quad \text{for } j = 0,\ 1,\ \cdots,\ k, \tag{5.11}$$

using Conjecture 4 in the form: if $\mathcal{A}$ has an asymptotic density 0, then $s^{-1}(\mathcal{A})$ has asymptotic density 0.

By the same argument involved with (5.4), we have that the set of n for which

$$\frac{\sigma(m_0)}{m_0} < \log T \tag{5.12}$$

fails has asymptotic density 0 . Then from (5.10) and (5.11), for $j \le k$ we have

$$\begin{aligned}\frac{s_{j+1}(n)}{s_j(n)} - \frac{s(n)}{n} &= \frac{\sigma(m_0)}{m_0}\left(\frac{\sigma(n_j)}{n_j} - \frac{\sigma(n_0)}{n_0}\right)\\ &< (\log T)(e^{1/T} - 1) \ll (\log T)/T = o(1),\end{aligned}$$

which gives Conjecture 3.

REMARK. Note that (5.10), the case $j = 0$ of (5.11) (which does not require Conjecture 4) and (5.12) immediately give

$$\begin{aligned}\frac{s(n)}{n} - \frac{s_{j+1}(n)}{s_j(n)} &= \frac{\sigma(m_0)}{m_0}\left(\frac{\sigma(n_0)}{n_0} - \frac{\sigma(n_j)}{n_j}\right)\\ &< (\log T)(e^{1/T} - 1) = o(1).\end{aligned}$$

That is, (5.1) holds for all n, but for a set of asymptotic density 0, the principal result of [8].

Theorem 5.3. *Let $S_k(x)$ denote the number of odd numbers $m \le x$ not in the range of the function s_k. There is a positive number δ_0 such that*

$$S_k(x) \ll x^{1-\delta_0}$$

uniformly for all natural numbers k and $x > 0$.

Proof: Let $E(x,y)$ denote the number of odd integers $n \le x$ with $r(n) \le y$, where $r(n)$ is the number of representations of n in the form $1+p+q$ where $p < q$ are primes. Since

$$s(pq) = 1 + p + q,$$

it follows that for any $y \ge 0$

$$S_1(x) \le E(x,y). \tag{5.13}$$

We now prove that for any natural number k and any $y > 0$,

$$S_{k+1}(x) \le \frac{S_k(x^2)}{y} + E(x,y). \tag{5.14}$$

Let S_j denote the set of odd numbers not in the range of s_j. Suppose $n \in S_{k+1}$. Consider the $r(n)$ representations

$$n = 1 + p_i + q_i, \quad i = 1, \cdots, r(n)$$

where $p_i < q_i$ are primes. Then all of the numbers $p_i q_i$ are in S_k, for if $p_i q_i = s_k(m)$ for some m, then $n = s_{k+1}(m)$, contradicting $n \in S_{k+1}$. Note that the integers $p_i q_i$ are distinct and each $p_i q_i < n^2$. Moreover if p'_j, q'_j are associated with n' and $n \neq n'$, then $p_i q_i \neq p'_j q'_j$. Thus

$$\begin{aligned} S_{k+1}(x) &= \#\{n \leq x : n \in S_{k+1}, r(n) \leq y\} + \#\{n \leq x : n \in S_{k+1}, r(n) > y\} \\ &\leq \#\{n \leq x : \ r(n) \leq y\} + y^{-1} \cdot \#\{m \leq x^2 : \ m \in S_k\} \\ &= E(x,y) + y^{-1} S_k(x^2), \end{aligned}$$

which is (5.14).

Next we show there is some $\delta_1 > 0$, $B > 0$ such that

$$E(x,y) \leq By \log^{38} x \tag{5.15}$$

for all $x \geq 2$, $y \geq x^{1-\delta_1}$. This result follows from the proof in Montgomery and Vaughan [13]. To see this, let $E_0(x,y)$ denote the number of odd numbers n with $x/2 < n \leq x$ and $r(n) \leq y$. Then from the proof in [13], we have

$$E_0(x, x^{1-\frac{3}{2}\delta} \log^{-3} x) \ll x^{1-2\delta} \log^{35} x$$

uniformly for $\delta \leq \delta_0$ for some $\delta_0 > 0$. Let $z = x^{3\delta/2}$. Then for i such that $2^i \leq z$,

$$\begin{aligned} E_0\left(2^{-i}x, \frac{x}{z\log^3 x}\right) &\leq E_0\left(2^{-i}x, \frac{2^{-i}x}{2^{-i}z\log^3(2^{-i}x)}\right) \\ &\ll \frac{2^{-i}x}{(2^{-i}z)^{4/3}} \log^{35}(2^{-i}x) \leq 2^{i/3}\frac{x}{z^{4/3}} \log^{35} x. \end{aligned}$$

Let j be such that $2^j \leq z < 2^{j+1}$. Then

$$\begin{aligned} E\left(x, \frac{x}{z\log^3 x}\right) &= E\left(2^{-(j+1)}x, \frac{x}{z\log^3 x}\right) + \sum_{i=0}^{j} E_0\left(2^{-i}x, \frac{x}{z\log^3 x}\right) \\ &\ll 2^{-(j+1)}x + \sum_{i=0}^{j} 2^{i/3}\frac{x}{z^{4/3}} \log^{35} x \\ &\ll \frac{x}{z} \log^{35} x. \end{aligned}$$

Thus letting $y = xz^{-1}\log^{-3} x$, we have (5.15) for $y \geq x^{1-3\delta_0/2}\log^3 x$. Letting $\delta_1 = 5\delta_0/4$, we have (5.15) for $y \geq x^{1-\delta_1}$.

Suppose we know that for some specific $k \geq 1$, there is some constant $C(k) \geq B$ with

$$S_k(x) \leq C(k)x^{1-\delta_1}\log^{38} x \tag{5.16}$$

for all $x \geq e$. Then letting $y = 2^{19}(C(k)/B)^{1/2}x^{1-\delta_1}$ and using (5.14) and (5.15) we have

$$S_{k+1}(x) \leq C(k+1)x^{1-\delta_1}\log^{38} x,$$

where

$$C(k+1) := 2^{20}(C(k)B)^{1/2}. \tag{5.17}$$

Since we have (5.16) for $k = 1$ and $C(1) = B$ by (5.13) and (5.15), we thus have it for all k where $C(k)$ is inductively defined by (5.17). Note that $C(k) < 2^{40}B$ for all k. In addition, since $\delta_1 = 5\delta_0/4$, we have our theorem.

§6. Corrections for an earlier paper

In [9], the first and third authors considered the normal number of prime factors of $\varphi(n)$. The principal result is that this normal order is $\frac{1}{2}(\log\log n)^2$ and there is a Gaussian distribution with standard deviation $\frac{1}{\sqrt{3}}(\log\log n)^{3/2}$. It has been pointed out to us by Abdelhakim Smati that there is an error in the proof of Lemma 2.2 of this paper. We now give a (hopefully) correct proof of this result.

Let $\Omega_y(n)$ denote the number of prime factors $p \leq y$ of n counted with multiplicity. Lemma 2.1 of [9] gives the average order for $\Omega_y(p-1)$ for p prime:

$$\sum_{p\leq x}\Omega_y(p-1) = \frac{x\log\log y}{\log x} + O\left(\frac{x}{\log x}\right) \tag{6.1}$$

uniformly for $3 \leq y \leq x$. Lemma 2.2 estimates the square mean.

"Lemma 2.2". *If $3 \leq y \leq x$, then*

$$\sum_{p\leq x}\Omega_y(p-1)^2 = \frac{x(\log\log y)^2}{\log x} + O\left(\frac{x\log\log y}{\log x}\right)$$

where the implied constant is uniform.

Proof: Let u range over the integers with exactly 2 distinct prime factors, neither exceeding y. Then

$$\begin{aligned}\sum_{p\leq x}\Omega_y(p-1)^2 &= \sum_{p\leq x}\sum_{\substack{q^a\|p-1\\ q\leq y}} a^2 + 2\sum_{p\leq x}\sum_{u\,|\,p-1} 1\\ &= S_3 + S_4,\end{aligned}$$

say. (In [9], the expression for S_4 is wrong.)

As in [9], we get

$$S_3 = O\left(\frac{x \log\log y}{\log x}\right)$$

using (6.1) and the Brun-Titchmarsh inequality.

For S_4, we write

$$S_4 = S_{4,1} + S_{4,2}$$

where in $S_{4,1}$ neither prime power in u exceeds $x^{1/6}$ and is $S_{4,2}$ at least one prime power in u exceeds $x^{1/6}$. We have

$$\begin{aligned} S_{4,1} &= 2 \sum_{\substack{1<q^a,\, r^b \le x^{1/6} \\ q<r\le y}} \pi(x; q^a r^b, 1) \\ &= 2\mathrm{li}(x) \sum_{\substack{1<q^a,\, r^b \le x^{1/6} \\ q,r\le y}} \frac{1}{\varphi(q^a r^b)} + O\left(\frac{x}{\log^2 x}\right) \\ &= \frac{x(\log\log y)^2}{\log x} + O\left(\frac{x \log\log y}{\log x}\right) \end{aligned}$$

using the Bombieri-Vinogradov theorem and a simple calculation.

For $S_{4,2}$ we have

$$S_{4,2} \ll \sum_{p\le x} \Omega_y(p-1) \ll \frac{x \log\log y}{\log x},$$

using (6.1). This, together with our estimates for S_3 and $S_{4,1}$ completes the proof.

A. Smati also points out that the three cases on p. 350 of [9] for $p^2 \mid \varphi(n)$, $p > y$ (where now $y = (\log\log x)^2$) do not exhaust all possibilities. This is fixed by changing (i) to (i') $p^2 \mid n$. The number of $n \le x$ in this case is at most $\sum_{p>y} x/p^2 = o(x/y) = o(x)$.

We are grateful to A. Smati for pointing these difficulties out to us.

REFERENCES

[1] E. Bombieri, Le grand crible dans la théorie analytique des nombres, Astérisque **18** (1974), 1–87.

[2] H. Davenport, *Multiplicative Number Theory , 2nd edition*, Springer Verlag, New York, 1980.

[3] P.D.T.A. Elliott, *Probabilistic Number Theory , vols. I, II*, Springer Verlag, New York, 1980.

[4] J. Friedlander and A. Granville, Limitations to the equi-distribution of primes I, Ann. Math. **129** (1989), 363–382.

[5] J. Friedlander, A. Granville, A. Hildebrand and H. Maier, Oscillation theorems for primes in arithmetic progressions and for sifting functions, preprint..

[6] P. Erdös, On the density of the abundant numbers, J. London Math. Soc. **9** (1934), 278–282.

[7] P. Erdös, Some remarks on the iterates of the φ and σ functions, Colloq. Math. **17** (1967), 195–202.

[8] P. Erdös, On asymptotic properties of aliquot sequences, Math. Comp. **30** (1976), 641–645.

[9] P. Erdös and C. Pomerance, On the normal number of prime factors of $\varphi(n)$, Rocky Mountain Math. J. **15** (1985), 343–352.

[10] R. K. Guy and J. L. Selfridge, What drives an aliquot sequence?, Math. Comp. **29** (1975), 101–107; Corrigendum, Math. Comp. **34** (1980), 319–321.

[11] H. Halberstam and H.-E. Richert, *Sieve Methods*, Academic Press, London, 1974.

[12] H. W. Lenstra, Jr., Problem 6064, American Math. Monthly **82** (1975), p. 1016; Solution by the proposer, **84** (1977), p. 580.

[13] H. L. Montgomery and R. C. Vaughan, The exceptional set in Goldbach's problem, Acta Arith. **27** (1975), 353–370.

[14] S. S. Pillai, On a function connected with $\varphi(n)$, Bull. A.M.S. **35** (1929), 837-841.

[15] I. J. Schoenberg, Über die asymptotische Verteilung reeller Zahlen mod 1, Math. Z. **28** (1928), 171–200.

[16] H. Shapiro, An arithmetic function arising from the φ-function, American Math. Monthly **50** (1943), 18–30.

Paul Erdös
Mathematical Institute
Hungarian Academy of Sciences
Reáltanoda u. 13-15
Budapest, Hungary

Carl Pomerance
Department of Mathematics
University of Georgia
Athens, Georgia 30602

Andrew Granville
School of Mathematics
Institute for Advanced Study
Princeton, NJ 08540

Claudia Spiro
Department of Mathematics
M. I. T.
Cambridge, Massachusetts 02139

On the Number of Partitions of n Without a Given Subsum, II

P. ERDÖS, J. L. NICOLAS, AND A. SÁRKÖZY

Dedicated to Professor Paul T. Bateman for his seventieth birthday

Abstract

Let $R(n,a)$ denote the number of unrestricted partitions of n whose subsums are all different of a, and $Q(n,a)$ the number of unequal partitions (i.e. each part is allowed to occur at most once) with the same property. In a preceding paper, we considered $R(n,a)$ and $Q(n,a)$ for $a \le \lambda_1\sqrt{n}$, where λ_1 is a small constant. Here we study the case $a \ge \lambda_2\sqrt{n}$. The behaviour of these quantities depends on the size of a, but also on the size of $s(a)$, the smallest positive integer which does not divide a.

1. Introduction

Let us denote by $p(n)$ the number of unrestricted partitions of n, by $r(n,m)$ the number of partitions of n whose parts are at least m, and by $R(n,a)$ the number of those partitions

$$n = n_1 + \cdots + n_t \quad (n_1 \le \cdots \le n_t)$$

of n which do not represent a, i.e. whose subsums $n_{i_1} + \cdots + n_{i_j}$ are all different from a.

We shall consider also partitions of n into distinct parts. In that case the above notations will be change to $q(n)$, $\rho(n,m)$ and $Q(n,a)$.

In [5] J. Dixmier considered $R(n,a)$ when a is fixed, and in [7], we studied $R(n,a)$ and $Q(n,a)$ when $a < \lambda_1\sqrt{n}$, where λ_1 is a small constant. Here we shall consider the case $\lambda_2\sqrt{n} \le a \le n/2$, where λ_2 is a large constant. (Since $R(n,a) = R(n,n-a)$ and $Q(n,a) = Q(n,n-a)$, we may suppose that $a \le n/2$.) We shall prove:

Research partially supported by Hungarian National Foundation for Scientific Research, grant no. 1811, and by C.N.R.S., Greco Calcul Formel and PRC Math.-Info.

Theorem 1. *For $n > n_0$ and*

$$10^{18}\sqrt{n} \le a \le n^{5/7}, \tag{1.1}$$

we have

$$Q(n,a) \le q([n/2]) \exp(5 \cdot 10^3 a^{-1/3} n^{2/3} \log(a^{1/3} n^{-1/6})) \tag{1.2}$$

and

$$R(n,a) \le p([n/2]) \exp(5 \cdot 10^3 a^{-1/3} n^{2/3} \log(a^{1/3} n^{-1/6})) \tag{1.3}$$

where $[x]$ denotes the integral part of x.

Theorem 2. *For $n > n_0$ and*

$$n^{5/7} < a \le n/2 \tag{1.4}$$

we have

$$Q(n,a) \le q([n/2]) \exp(n^{1/2-1/30}) \tag{1.5}$$

and

$$R(n,a) \le p([n/2]) \exp(n^{1/2-1/30}). \tag{1.6}$$

It follows from Theorems 1 and 2 that

Corollary. *If $a = a(n)$ is such that $a/\sqrt{n} \to \infty$ and $a \le n/2$, we have $Q(n,a) = (q([n/2]))^{1+o(1)}$ and $R(n,a) = (p([n/2]))^{1+o(1)}$.*

Theorem 3. *Let $s(a)$ denote the smallest positive integer which does not divide a. For $n \ge (2500)^2$, $s(a) \ge 40000$ and*

$$\frac{7}{100} n^{1/2}(s(a))^{3/2} \le a \le \frac{1}{40} n(s(a))^{-1} \tag{1.7}$$

we have

$$Q(n,a) < \exp(201 n^{1/2}(s(a))^{-1/2} \log(s(a))) \tag{1.8}$$

and

$$R(n,a) < \exp(301 n^{1/2} s(a)^{-1/2} \log(s(a))). \tag{1.9}$$

Remark: By Lemma 1 below, (1.7) holds for all a's such that

$$(7/10)n^{1/2}(\log n)^{3/2} \le a \le (1/200)n(\log n)^{-1}.$$

To give lower bounds for $R(n,a)$ and $Q(n,a)$, first we note that if a is odd and n is even, then multiplying the parts of a partition of $n/2$ by 2 we get a partition of n whose subsums are all even and thus different from a. Hence

$$R(n,a) \ge p([n/2]) \tag{1.10}$$

and

$$Q(n,a) \ge q([n/2]) \tag{1.11}$$

which show the exponent $1+o(1)$ in the above corollary to be best possible.

This argument can be extended, and yields:

Theorem 4. *Let $h = h(n,a,m)$ be given by $h = 0$ if $m|n$ and*

$$h \equiv n \pmod{m}, \quad \text{if} \quad a < h \leq a + m.$$

Then

$$Q(n,a) \geq q\left(\frac{n - h(n,a,s(a))}{s(a)}\right) \tag{1.12}$$

and

$$R(n,a) \geq p\left(\frac{n - h(n,a,s(a))}{s(a)}\right). \tag{1.13}$$

Proof: When $s(a)$ divides n, we consider all the partitions of $n/s(a)$, and we multiply their parts by $s(a)$. In this way we obtain a partition of n whose subsums are all divisible by $s(a)$ and they cannot be equal to a.

When $s(a)$ does not divide n, let us set $h = h(n,a,s(a))$. We consider all the partitions of $(n-h)/s(a)$, we multiply their parts by $s(a)$, and we complete them with a part equal to h to obtain a partition of n. As $h > a$, the subsums of such a partition are all different from a.

Taking into account the results of Hardy and Ramanujan (cf. [9]):

$$\begin{aligned} p(n) &\sim \frac{1}{4\sqrt{3}n} \exp\left(\pi\sqrt{\frac{2}{3}}\sqrt{n}\right), \\ q(n) &\sim \frac{1}{4(3n^3)^{1/4}} \exp\left(\pi\sqrt{\frac{n}{3}}\right), \end{aligned} \tag{1.14}$$

we observe that for $a = o(n)$, the upper bounds given in Theorem 3 for $\log Q(n,a)$ and $\log R(n,a)$ are of the same order of magnitude as the lower bounds given in Theorem 4 (apart from a factor $\log s(a)$). This shows that the behavior of $Q(n,a)$ and $R(n,a)$ depends on the arithmetical structure of a if a is large.

In [7], we gave bounds for $R(n,a)$, and a lower bound for $Q(n,a)$ when a is $< \lambda_1\sqrt{n}$. Here we will prove the following upper bound:

Theorem 5. *For $a \leq \frac{3}{5}\sqrt{n}$ and n large enough, we have*

$$Q(n,a) \leq q(n) \exp\left(-a \log\frac{2}{\sqrt{3}} + \frac{\pi a^2}{8\sqrt{3}\sqrt{n}}\right). \tag{1.15}$$

The proof follows the same principle as in [7] for unrestricted partitions: if a partition π does not represent a, then i and $a-i$ cannot belong simultaneously to π. So, for every i, $1 \leq i < a/2$, there are three possibilities: $i \in \pi$ and $a-i \notin \pi$, $i \notin \pi$ and $a-i \in \pi$, $i \notin \pi$ and $a-i \notin \pi$. When a is even, and $i = a/2$, there are only two possibilities. Therefore, the number of possible sets $\mathcal{A}$ of parts $< a$ is at most $3^{a/2}$. For such a set $\mathcal{A}$, there are $\rho(n - \sum_{x \subset \mathcal{A}} x,\ a+1)$ possibilities of completing $\mathcal{A}$ to a partition of n. As already observed in [8], $\rho(n,m)$ is nondecreasing in n for $n \geq m$, is 0 for $1 \leq m < n$, and 1 for $n = 0$. Thus we have

$$Q(n,a) \leq 3^{a/2} \rho(n, a+1).$$

From Theorem 1 of [8], we have

$$\rho(n, a+1) \leq \rho(n,a) \leq \frac{1}{2^{a-2}} q\left(n + \frac{a^2}{4}\right),$$

and from Lemma 3 of [8] (which is an easy consequence of (1.14)),

$$q\left(n + \frac{a^2}{4}\right) \sim q(n) \exp \frac{\pi a^2}{8\sqrt{3}\sqrt{n}}$$

and Theorem 5 is proved. For $a \geq 0.64\sqrt{n}$, the quantity in the exponent in (1.15) is positive, and thus the trivial bound $Q(n,a) \leq q(n)$ is better.

Now consider the case when a is of the order of magnitude $\sqrt{n}$. In [4], Theorem 2.18 claims that if a is odd and $a \sim \sqrt{n}$, then we have for n large enough

$$\log R(n,a) \geq 2.0138\sqrt{n}. \tag{1.16}$$

This result can be extended to the case when a is odd, $a \sim \lambda\sqrt{n}$, and we obtain

$$\log R(n,a) \geq \varphi(\lambda)\sqrt{n} \tag{1.17}$$

for some function φ.

Our guess is that, when a is odd, such a result is best possible. But when a is even, we have no precise conjecture. J. Dixmier has proved (cf. [6]) that for $\epsilon > 0$ there exists $\delta < 1$ such that, for n large enough,

$$\epsilon\sqrt{n} \leq a \leq n - \epsilon\sqrt{n} \Rightarrow R(n,a) \leq (p(n))^{\delta} \tag{1.18}$$

for all n. The proof is short, and starts with the results of [7].

In the same way, it can be deduced from Theorem 5 above that for $\epsilon > 0$, there exists $\delta < 1$ such that, for n large enough,

$$\epsilon\sqrt{n} \leq a \leq n - \epsilon\sqrt{n} \Rightarrow Q(n,a) \leq (q(n))^{\delta}. \tag{1.19}$$

The aim of [6] is to give a fairly good estimation of $R(2n,n)$, and to study $R(n,a)$ for $\lambda_2 n \leq a \leq n/2$, where λ_2 is a fixed constant.

The proofs of Theorems 1, 2 and 3 are based on results from additive number theory (cf. [11] and [12]). In §2, we shall give some estimates involving partitions. In §3, we shall prove some lemmas on additive properties of dense sequences, from which, in §4, the proof of our three theorems will follow.

All these proofs are effective, but the constants are rather large and we did not attempt to optimize them.

A table of $R(n,a)$ for $n \leq 40$ was given in [7]. Here in the appendix, we give a table of $Q(n,a)$ for $n \leq 40$. It has been computed by M. DeLéglise, and we are very pleased to thank him. For a fixed a, first he determines, by a backtracking programming method, all the subsets $\mathcal{A}$ of $\{1,2,\dots,a-1\}$ having no subsum equal to a. Then for all $\mathcal{A}$ such that $S(\mathcal{A}) = \sum_{x\in\mathcal{A}} x$ is smaller than n,

$$Q(n,a) = \sum_{\mathcal{A}} \rho(n - S(\mathcal{A}), a+1).$$

As can be seen in [8], $\rho(n,m)$ is easy to calculate.

We thank J. Dixmier for many helpful remarks, and for an improvement of Lemma 11 below.

Notations: $p(n,m)$ will denote the number of unrestricted partitions of n into parts $\leq m$ (or into atmost m parts); here m is not necessarily an integer.

$$\mathbb{N} \quad \text{is the set of positive integers} \quad \{1,2,...\}.$$

$$\mathbb{N}_M = \{1,2,\dots,M\}.$$

If $\mathcal{A}$ is a finite set of not necessarily distinct integers, then $|\mathcal{A}|$ denotes cardinality of $\mathcal{A}$, $\mathcal{A}'$ the set of distinct elements of $\mathcal{A}$, $S(\mathcal{A}) = \sum_{a\in\mathcal{A}} a$,

$$P(\mathcal{A}) = \left\{ \sum_{a\in\mathcal{A}} \epsilon_a a; \quad \epsilon_a = 0 \quad \text{or} \quad 1, \sum_{a\in\mathcal{A}} \epsilon_a \neq 0 \right\}$$

the set of the nonzero subsums of $\mathcal{A}$,

$$\mathcal{L}(\mathcal{A},d) = \{i;\ 1 \leq i \leq d, \text{ there exist at least 2 elements of } \mathcal{A} \text{ which are } \equiv i \pmod d\}$$

and

$$L(\mathcal{A},d) = |\mathcal{L}(\mathcal{A},d)|.$$

2. Partition Lemmas

Lemma 1. *Let $s(m)$ be the smallest integer which does not divide m. Then for all $m \geq 2$, we have*

$$s(m) < \frac{3}{\log 2} \log m < 4.5 \log m. \tag{2.1}$$

Proof: First, if m is odd, $s(m) = 2$ and (2.1) holds. So we may suppose that m is even, and $s(m) \geq 3$. Let $\psi(x)$ denote the Chebychef function:

$$\psi(x) = \sum_{p^k \leq x} \log p.$$

It follows from Chebychef's results that for all integers $n \geq 2$,

$$\psi(n)/n \geq (\log 2)/2.$$

Then

$$\log m \geq \psi(s(m) - 1) \geq \frac{\log 2}{2}(s(m) - 1)$$

which implies

$$s(m) \leq 1 + \frac{2 \log m}{\log 2} < \frac{3}{\log 2} \log m.$$

It can be shown similarly that

$$m \geq m_0(\epsilon) \Longrightarrow s(m) < (1 + \epsilon) \log m. \tag{2.2}$$

Lemma 2. *Let α be a real number satisfying $0 < \alpha \leq 1.05$. For $m \leq \alpha\sqrt{n}$ we have*

$$p(n, m) < \exp\left(\left(2\alpha \log \frac{3.6}{\alpha}\right) \sqrt{n}\right). \tag{2.3}$$

This inequality can be used to obtain upper bounds for $\rho(n, m)$ and $r(n, m)$ since

$$\rho(n, m) \leq r(n, m) \leq p(n, [n/m]). \tag{2.4}$$

Proof: From the classical inequality

$$m!\, p(n, m) \leq \binom{n + \frac{m(m+1)}{2} - 1}{m - 1},$$

it has been proved in [3] that for all $\alpha > 0$, and $m \leq \alpha\sqrt{n}$,

$$p(n, m) \leq \exp\left(\left(\alpha^3/2 + 2\alpha(1 - \log \alpha)\right) \sqrt{n}\right).$$

Observing that $\alpha \leq 1.05$ implies $\alpha^2/4 + 1 < \log(3.6)$, (2.3) follows easily.

For $\alpha \geq 1.06$, the obvious inequality $p(n, m) \leq p(n)$ and (1.14) give a better upper bound.

As $p(n, m)$ is also the number of partitions of n with at most m parts, (2.4) can be proved easily.

Lemma 3. *Let $Y(n,t,m)$ denote the number of partitions of n into unequal parts such that at most t parts not exceeding m may occur. We have*

$$Y(n,t,m) \le p(n, t + n/m). \tag{2.5}$$

Proof: A partition counted in $Y(n,t,m)$ has at most t parts $\le m$, and n/m parts $> m$. The right–hand side of (2.5) is certainly greater than the number of partitions of n with at most $t + n/m$ unequal parts.

Lemma 4. *Let $Z(n,t,m)$ denote the number of unrestricted partitions of n such that at most t distinct parts not exceeding m may occur. For $1 \le t \le m \le n$ we have*

$$Z(n,t,m) \le 6tn^2 \binom{m}{min(t,[m/2])} p(n,t)p(n,n/m). \tag{2.6}$$

proof: For $\mathcal{A} \subset \{1,2,\ldots,m\}$, $\mathcal{A} = \{a_1 < a_2 < \cdots < a_s\}$, let $P(n,\mathcal{A},m)$ denote the number of partitions of n with all the parts not exceeding m in $\mathcal{A}$, i.e., $P(n,\mathcal{A},m)$ denotes the number of solutions of

$$a_1x_1 + \cdots + a_sx_s + \sum_{i=1}^{n-m} (m+i)x_{s+i} = n, \quad (x_i \ge 0, 1 \le i \le n). \tag{2.7}$$

Then we have

$$P(n,\mathcal{A},m) \le \sum_{k=0}^{n} P(k,\{1,2,\ldots,s\},m)$$

since replacing a_j by j in (2.7), we have

$$x_1 + 2x_2 + \cdots + sx_s + \sum_{i=1}^{n-m} (m+i)x_{s+i} = k \tag{2.8}$$

for some $k \le n$. It follows that

$$Z(n,t,m) \le \sum_{s=0}^{t} \binom{m}{s} \sum_{k=0}^{n} P(k,\{1,\ldots,s\},m) \tag{2.9}$$

since $\mathcal{A}$ with $|\mathcal{A}| = s$ can be selected in $\binom{m}{s}$ ways from $\{1,\ldots,m\}$. Hence

$$Z(n,t,m) \le (t+1)\binom{m}{min(t,[m/2])} \sum_{k=0}^{n} P(k,\{1,\ldots,t\},m). \tag{2.10}$$

Now, counting the partitions according to the sum j of the parts not exceeding t, we obtain

$$P(k,\{1,\dots,t\},m) = \sum_{j=0}^{k} p(j,t)r(k-j,m+1)$$
$$\leq (k+1)p(k,t)r(k,m+1)$$

since it is easy to see that $p(n,m)$ and $r(n,m)$ are not decreasing in n. Then (2.10) yields (2.6) observing that $t+1 \leq 2t$,

$$\sum_{k=0}^{n}(k+1) \leq 3n^2,$$

and using (2.4).

Lemma 5. *Given integers $M \geq 2$, $D \geq 2$, let $V(n,M,D)$ denote the number of partitions of $n \geq M$ into distinct parts:*

$$n = n_1 + \cdots + n_t \quad (n_1 < \cdots < n_t)$$

with the set $\mathcal{N} = \{n_1,\dots,n_t\}$ of parts of n having the following property: there exists an integer d,

$$2 \leq d \leq D, \tag{2.11}$$

and integers $i_1,\dots,i_{[d/2]}$ satisfying

$$1 \leq i_1 < \cdots < i_{[d/2]} \leq d, \tag{2.12}$$

such that, if $\mathcal{N}_1 = \{n_\ell : n_\ell \in \mathcal{N},\ d \leq n_\ell \leq M,\ n_\ell \equiv i_j \mod d \text{ for some } j\}$, the cardinality of the set $\mathcal{N}_2 = \{n_\ell : n_\ell \in \mathcal{N}\backslash\mathcal{N}_1,\ n_\ell \leq M\}$ satisfies

$$|\mathcal{N}_2| \leq 2D, \tag{2.13}$$

then

$$V(n,M,D) \leq n^{5D} q([n/2]) p(n, n/M). \tag{2.14}$$

Proof: Let $\mathcal{N}_3 = \mathcal{N}\backslash(\mathcal{N}_1 \cup \mathcal{N}_2) = \{n_\ell : n_\ell \in \mathcal{N}, n_\ell > M\}$.

Let us fix $d, i_1, i_2, \dots, i_{[d/2]}$ in (2.11), (2.12) and (2.13). By the defintion of $\mathcal{N}_1$, every element m of $\mathcal{N}_1$ can be written in the form

$$m = i_{j(m)} + \ell(m)d$$

where

$$1 \leq j(m) \leq [d/2] \quad \text{and} \quad 1 \leq \ell(m). \tag{2.15}$$

To every $m \in \mathcal{N}_1$ we assign the integer

$$m^* = j(m) + (\ell(m) - 1)[d/2],$$

and write $\mathcal{N}_1^* = \{m^* : m \in \mathcal{N}_1\}$. Clearly, (2.15) implies that to distinct elements of $\mathcal{N}_1$, distinct elements of $\mathcal{N}_1^*$ are assigned. Furthermore, we have

$$\begin{aligned} m^* &= j(m) + (\ell(m) - 1)[d/2] \le [d/2] + (\ell(m) - 1)[d/2] \\ &= \ell(m)[d/2] \le \ell(m)d/2 < m/2 \end{aligned}$$

whence

$$S(\mathcal{N}_1^*) = \sum_{m^* \in \mathcal{N}_1^*} m^* < \frac{1}{2} \sum_{m \in \mathcal{N}_1} m = \frac{1}{2} S(\mathcal{N}_1).$$

Thus writing $S(\mathcal{N}_1) = u$, the elements of $\mathcal{N}_1^*$ form partition of an integer $v < u/2$ into distinct parts, so that for fixed $d, i_1, \ldots, i_{[d/2]}$ and u, $\mathcal{N}_1$ can be selected in at most

$$\sum_{v=0}^{[u/2]} q(v) \le ([n/2] + 1) q([n/2]) \le n q([n/2])$$

ways.

Furthermore, the elements of $\mathcal{N}_2$ are selected from $\{1, 2, \ldots, M\}$ and, by (2.13), their number is at most $2D$, so that $\mathcal{N}_2$ can be chosen in at most $M^{2D} \le n^{2D}$ ways.

Finally, if $d, i_1, \ldots, i_{[d/2]}$ and u are fixed, then

$$\begin{aligned} S(\mathcal{N}_3) &= (S(\mathcal{N}_1) + S(\mathcal{N}_2) + S(\mathcal{N}_3)) - S(\mathcal{N}_1) - S(\mathcal{N}_2) \\ &= n - u - S(\mathcal{N}_2) \le n - u \end{aligned}$$

so that

$$\sum_{n_i \in \mathcal{N}_3} n_i = z$$

is a partition of $z (\le n - u)$ into parts $\ge M + 1$. Thus $\mathcal{N}_3$ can be chosen in at most

$$\sum_{z=0}^{n-u} \rho(z, M+1) \le (n+1) \rho(n, M) \le n^2 \rho(n, M)$$

ways (note that $\rho(n, m)$ is non-decreasing function of n for $n \ge m$).

Collecting the results above, we obtain that for fixed $d, i_1, \ldots, i_{[d/2]}$, the partition $\mathcal{N}$ can be chosen in at most

$$n^{2D+3} q([n/2]) \rho(n, M)$$

ways. Furthermore, for fixed d, the numbers $i_1, i_2, \ldots, i_{[d/2]}$ in (2.12) can be chosen in at most 2^d ways, and summation over the d's in (2.11) gives

$$\sum_{d \le D} 2^d < 2^{D+1} \le n^{D+1}$$

whence, by (2.4), the result follows.

Lemma 6. *With the notation of Lemma 5 but considering unrestricted partitions $n = n_1 + \cdots + n_t$ ($n_1 \le \cdots \le n_t$) of $n \ge M$, so that now parts in $\mathcal{N}_1$ and $\mathcal{N}_2$ have to be counted according to multiplicity, suppose here that*

$$|\mathcal{N}_2'| \le 2D. \tag{2.16}$$

Then the number $W(n, M, D)$ of unrestricted partitions of n that corresponds to $V(n, M, D)$ in Lemma 5 satisfies

$$W(n, M, D) \le n^{7D} p([n/2]) p(n, n/M).$$

Proof: Again first fix $d, i_1, i_2, \ldots, i_{[d/2]}$, and define $\mathcal{N}_3$, m^*, $\mathcal{N}_1^*$ in the same way as in the proof of Lemma 5. The same argument shows that, writing $S(\mathcal{N}_1) = u$, $\mathcal{N}_1$ can be chosen in at most

$$\sum_{v=0}^{[u/2]} p(v) \le np([n/2])$$

ways.

Furthermore, the elements of $\mathcal{N}_2$ are selected from $\{1, 2, \ldots, M\}$ and, by (2.16), the number of distinct elements of $\mathcal{N}_2$ is at most $2D$, so that they can be chosen in at most $M^{2D} \le n^{2D}$ ways; and if we have selected the distinct elements of $\mathcal{N}_2$, then the multiplicity of each of them can be chosen in at most n ways, so that the multiplicities of the distinct elements of $\mathcal{N}_2$ can be chosen altogether in at most n^{2D} ways. Thus $\mathcal{N}_2$ can be chosen in at most $n^{2D} \cdot n^{2D} = n^{4D}$ ways.

Finally, the same argument as in Section 2 shows that $\mathcal{N}_3$ can be chosen in at most

$$\sum_{z=0}^{n-u} r(z, M+1) \le n^2 r(n, M)$$

ways.

Collecting the results above and using also (2.4), we obtain that for fixed $d, i_1, \ldots, i_{[d/2]}$, the partition $\mathcal{N}$ can be chosen in at most

$$n^{4D+3} p([n/2]) p(n, n/M)$$

ways. Finally, as in Lemma 5, $d, i_1, \ldots, i_{[d/2]}$ can be chosen in at most n^{D+1} ways whence the result follows.

3. Additive Lemmas

First we need the following well known fact (see, e.g. [12], Lemma 3).

Lemma 7. *If $d \in \mathbb{N}$ and $n_1, n_2, \dots, n_d$ are integers, then there is a sum of the form $n_{i_1} + \dots + n_{i_t} (1 \le i_i < \dots < i_t \le d)$ such that $d | (n_{i_1} + \dots + n_{i_t})$.*

The next lemma is variant of Lemma 4 in [12].

Lemma 8. *If $N \in \mathbb{N}$, $d \in \mathbb{N}$, and $\mathcal{B}$ is a finite set of not necessarily distinct positive integers such that*

$$\text{the elements of } \mathcal{B} \text{ do not exceed } N, \tag{3.1}$$

then for every integer n such that $0 \le n \le \frac{1}{d}S(\mathcal{B}) - N$ there is a number x_n in the set $\{n+1, n+2, \dots, n+N\}$ such that $dx_n \in \mathcal{P}(\mathcal{B})$.

Proof: It suffices to show the existence of integers $y_1, y_2, \dots, y_t$ such that $0 < y_1 \le N$, $0 < y_i - y_{i-1} \le N$ (for $i = 2, 3, \dots, t$), $\frac{1}{d}S(\mathcal{B}) - N < y_t$ and $dy_i \in \mathcal{P}(\mathcal{B})$ for $i = 1, 2, \dots, t$. Afterwards we shall define x_n by

$$x_n = \begin{cases} y_1 & \text{for } 0 \le n < y_1, \\ y_i & \text{for } y_{i-1} \le n < y_i \; (2 \le i \le t-1), \\ y_t & \text{for } y_{t-1} \le n \le \frac{1}{d}S(\mathcal{B}) - N. \end{cases}$$

We are going to define these integers y_i by recursion.

We may suppose that $S(\mathcal{B}) \ge dN$ which, by (3.1), implies $|\mathcal{B}| \ge d$. Let $\mathcal{B}_1 \subset \mathcal{B}$, $|\mathcal{B}_1| = d$. Then by Lemma 7, there is a (non-empty) subset $\mathcal{B}_1^*$ of $\mathcal{B}_1$ such that $d | S(\mathcal{B}_1^*)$; write $S(\mathcal{B}_1^*)/d = y_1$. Then $dy_i \in \mathcal{P}(\mathcal{B}_1^*) \subset \mathcal{P}(\mathcal{B}_1)$, $0 < y_1$, and

$$dy_1 = S(\mathcal{B}_1^*) = \sum_{b \in \mathcal{B}_1^*} b \le \sum_{b \in \mathcal{B}_1^*} N = N|\mathcal{B}_1^*| \le N|\mathcal{B}_1| = Nd$$

so that $y_1 \le N$.

Assume now that $y_1, y_2, \dots, y_{i-1}$ have been defined and

$$y_{i-1} \le \frac{1}{d}S(\mathcal{B}) - N. \tag{3.2}$$

By the definition of y_{i-1}, there is a subset $\mathcal{B}_{i-1}^* \subset \mathcal{B}$ such that $S(\mathcal{B}_{i-1}^*) = dy_{i-1}$. Then by (3.2) we have

$$\begin{aligned} S(\mathcal{B} \backslash \mathcal{B}_{i-1}^*) &= S(\mathcal{B}) - S(\mathcal{B}_{i-1}^*) = S(\mathcal{B}) - dy_{i-1} \\ &\ge S(\mathcal{B}) - (S(\mathcal{B}) - Nd) = Nd. \end{aligned} \tag{3.3}$$

(3.1) implies that

$$S(\mathcal{B}\backslash\mathcal{B}^*_{i-1}) = \sum_{b\in(\mathcal{B}\backslash\mathcal{B}^*_{i-1})} b \leq \sum_{b\in(\mathcal{B}\backslash\mathcal{B}^*_{i-1})} N = N|\mathcal{B}\backslash\mathcal{B}^*_{i-1}|.$$

It follows from (3.3) and (3.4) that

$$|\mathcal{B}\backslash\mathcal{B}^*_{i-1}| > d.$$

Thus there is a subset $\mathcal{B}_i$ of $\mathcal{B} - \mathcal{B}^*_{i-1}$ with $|\mathcal{B}_i| = d$. By Lemma 7, there is a (non-empty) subset $\mathcal{B}'_i$ of $\mathcal{B}_i$ such that $d|S(\mathcal{B}'_i)$; let $y_i = y_{i-1} + d^{-1}S(\mathcal{B}'_i)$. Then we have

$$\begin{aligned} y_{i-1} < y_i &= y_{i-1} + d^{-1}S(\mathcal{B}'_i) = y_{i-1} + d^{-1}\sum_{b\in\mathcal{B}'_i} b \\ &\leq y_{i-1} + d^{-1}\sum_{b\in\mathcal{B}'_i} N = y_{i-1} + Nd^{-1}|\mathcal{B}'_i| \leq y_{i-1} + Nd^{-1}|\mathcal{B}_i| \\ &= y_{i-1} + N \end{aligned}$$

and

$$\begin{aligned} dy_i &= dy_{i-1} + S(\mathcal{B}'_i) = S(\mathcal{B}^*_{i-1}) + S(\mathcal{B}'_i) \\ &= \sum_{b\in\mathcal{B}^*_{i-1}} b + \sum_{b\in\mathcal{B}'_i} b \in \mathcal{P}(\mathcal{B}) \end{aligned}$$

which completes the proof of the lemma.

Lemma 9. *Let* $N \in \mathbb{N}$,

$$N > 2500, \tag{3.5}$$

$A \subset \mathbb{N}_N$ *and*

$$|A| > 100(N\log N)^{1/2}. \tag{3.6}$$

Then there exist integers d, y, z *such that*

$$1 \leq d < 10^4\frac{N}{|A|}, \tag{3.7}$$

$$z > \frac{|A|^2}{7\cdot 10^4}, \tag{3.8}$$

$$1 \leq y < 7\cdot 10^4\frac{N}{|A|^2}z \tag{3.9}$$

and

$$\{yd, (y+1), ..., zd\} \subset P(A). \tag{3.10}$$

Proof: This is Theorem 4 in [**12**].

Lemma 10. *Let* $N \in \mathbb{N}$, $N \le 2500$, $m \in \mathbb{N}$,

$$7Ns(m) \le m \le 10^3 \frac{N^2}{s(m)^s}, \tag{3.11}$$

$$\mathcal{A} \subset \mathbb{N}_N \tag{3.12}$$

and

$$|\mathcal{A}| \ge 10^4 \frac{N}{s(m)}. \tag{3.13}$$

Then we have

$$m \in \mathcal{P}(\mathcal{A}). \tag{3.14}$$

Remarks: This lemma is non-trivial only when $s(m) > 10^4$, which implies that m must be a multiple of all integers up to 10^4, and N must be much greater than 2500.

It is easy to see that, from Lemma 1, (3.11) holds for every m and N such that $N > 2500$ and

$$63\,N \log N < m < 12\,N^2(\log N)^{-2}.$$

A slightly weaker version of this lemma follows from the results of Alon, Freiman, Lipkin [1], [2], [10].

Proof: It follows from (3.13) and Lemma 1 that

$$|\mathcal{A}| \ge 10^4 \frac{N}{s(m)} \ge \frac{10^4 N}{4 \cdot 5 \log m}. \tag{3.15}$$

Now by (3.11), $\log m \le \log(10^3 N^2) \le 3 \log N$, and thus (3.15) yields

$$|\mathcal{A}| \ge \frac{10^4}{13.5} \frac{N}{\log N} \ge \frac{8600}{13.5} \sqrt{N \log N},$$

because $\frac{N}{\log N} \ge 0.86\sqrt{N \log N}$ for all $N > 1$. So (3.6) holds and thus we may apply Lemma 9. We obtain that there exist integers d, y, z, satisfying (3.7), (3.8), (3.9) and (3.10). It follows from (3.7) and (3.13) that

$$d < 10^4 \frac{N}{|\mathcal{A}|} \le 10^4 \frac{N}{10^4 N/s(m)} = s(m)$$

so that

$$d | m. \tag{3.16}$$

Furthermore, by (3.10) we have $zd \in \mathcal{P}(\mathcal{A})$ whence

$$zd \leq S(\mathcal{A}) = \sum_{a\in\mathcal{A}} a \leq \sum_{a\in\mathcal{A}} N = N|\mathcal{A}|. \tag{3.17}$$

It follows from (3.9), (3.17), (3.13) and (3.11) that

$$yd < 7.10^4 \frac{N}{|\mathcal{A}|^2} zd \leq 7.10^4 \frac{N^2}{|\mathcal{A}|} \leq 7Ns(m) \leq m. \tag{3.18}$$

Now, (3.8), (3.13) and (3.11) imply

$$zd \geq z > \frac{|\mathcal{A}|^2}{7.10^4} \geq 10^3 \frac{N^2}{s(m)^2} \geq m, \tag{3.19}$$

and (3.14) follows from (3.10), (3.16), (3.18) and (3.19). This completes the proof of the lemma.

Lemma 11. *Assume that* $N \in \mathbb{N}$,

$$N > 10^{10}, \tag{3.20}$$

δ *is a real number with*

$$0 < \delta \leq \frac{1}{2}, \tag{3.21}$$

$\mathcal{A}$ *is a finite set of (not necessarily distinct) integers not exceeding* N,

$$|\mathcal{A}'| > 10^3(\delta^{-1}N)^{3/4} \tag{3.22}$$

and there is an integer m *with*

$$2 \cdot 10^7 \delta^{-2} N^2 |\mathcal{A}'|^{-1} < m < (1-\delta)S(\mathcal{A}) \quad \text{and} \quad m \notin \mathcal{P}(\mathcal{A}). \tag{3.23}$$

Then there is an integer d *with*

$$d < 11 \cdot 10^4 \delta^{-1} N |\mathcal{A}'|^{-1} \tag{3.24}$$

and

$$L(\mathcal{A}, d) \leq d/2. \tag{3.25}$$

Proof: In a first step we shall prove Lemma 11 when $\delta = 1/2$. In this step (3.21) should be read $\delta = 1/2$.

Let $D = 11 \cdot 10^4 \delta^{-1} N |\mathcal{A}'|^{-1}$. To every $d \leq D$, $i \in \mathcal{L}(\mathcal{A}, d)$, we assign two numbers $a(d,i) \in \mathcal{A}$, $a'(d,i) \in \mathcal{A}$, $a(d,i) \equiv a'(d,i) \equiv i \pmod d$ so that either $a(d,i) \neq a'(d,i)$, or $a(d,i) = a'(d,i)$ and $a(d,i)$ occurs with multiplicity at least 2 in $\mathcal{A}$. Let

$$\mathcal{A}_0 = \cup_{d \leq D} \cup_{i \in \mathcal{L}(\mathcal{A},d)} \{a(d,i), a'(d,i)\}.$$

Then clearly we have

$$|\mathcal{A}_0| \leq 2 \sum_{d \leq D} d \leq 2D^2 < 3 \cdot 10^{10} \delta^{-2} N^2 |\mathcal{A}'|^{-2} \tag{3.26}$$

(where in $|\mathcal{A}_0|$ we count the elements of $\mathcal{A}_0$ with multiplicity). Furthermore we have

$$S(\mathcal{A}) \geq S(\mathcal{A}') = \sum_{a \in \mathcal{A}} a \geq \sum_{a \leq |\mathcal{A}'|} a > \frac{1}{2} |\mathcal{A}'|^2. \tag{3.27}$$

It follows from (3.22), (3.26) and (3.27) that

$$\begin{aligned} S(\mathcal{A}_0) &= \sum_{a \in \mathcal{A}_0} a \leq \sum_{a \in \mathcal{A}_0} N \leq |\mathcal{A}_0| N \\ &< 3 \cdot 10^{10} \delta^{-2} N^3 |\mathcal{A}'|^{-2} < 3 \cdot 10^{10} \delta^{-2} N^3 |\mathcal{A}'|^{-2} \cdot 2S(\mathcal{A}) |\mathcal{A}'|^{-2} \\ &= 6 \cdot 10^{-2} \delta (10^{12} (\delta^{-1} N)^3 |\mathcal{A}'|^{-4}) S(\mathcal{A}) < 10^{-1} \delta S(\mathcal{A}). \end{aligned} \tag{3.28}$$

Let us write $(\mathcal{A} \backslash \mathcal{A}_0)' = \{a_1, a_2, \ldots, a_t\}$ where $a_1 < a_2 < \cdots < a_t$. (Here and in what follows, $\mathcal{A} \backslash \mathcal{A}_0$ is defined so that the multiplicity of a in $\mathcal{A} \backslash \mathcal{A}_0$ is the difference of the multiplicities of a in $\mathcal{A}$ and $\mathcal{A}_0$, respectively.) By (3.20), (3.21), (3.22), (3.26) and (3.28) we have

$$\begin{aligned} t &= |(\mathcal{A} \backslash \mathcal{A}_0)'| \geq |\mathcal{A}'| - |\mathcal{A}_0'| \\ &> |\mathcal{A}'| - 3 \cdot 10^{10} \delta^{-2} N^2 |\mathcal{A}'|^{-2} = |\mathcal{A}'| (1 - 3 \cdot 10^{10} \delta^{-2} N^2 |\mathcal{A}'|^{-3}) \\ &> |\mathcal{A}'| (1 - 3 \cdot 10^{10} \delta^{-2} N^2 10^{-9} (\delta^{-1} N)^{-9/4}) \\ &= |\mathcal{A}'| (1 - 3(\delta N^{-1})^{1/4}) > \frac{10}{11} |\mathcal{A}'| \end{aligned} \tag{3.29}$$

and

$$S(\mathcal{A} \backslash \mathcal{A}_0) = S(\mathcal{A}) - S(\mathcal{A}_0) > S(\mathcal{A}) - 10^{-1} \delta S(\mathcal{A}) = (1 - 10^{-1} \delta) S(\mathcal{A}). \tag{3.30}$$

Let $u = [\frac{\delta}{8} t]$. It follows easily from (3.20), (3.21), (3.22) and (3.29) that

$$\frac{\delta}{10} t < u \leq \frac{\delta}{8} t < \frac{t}{2}. \tag{3.31}$$

Write $\mathcal{A}_1 = \{a_1, a_2, \ldots, a_n\}$, $\mathcal{A}_2 = (\mathcal{A} \backslash \mathcal{A}_0) \backslash \mathcal{A}_1$ so that

$$\mathcal{A} \backslash \mathcal{A}_0 = \mathcal{A}_1 \cup \mathcal{A}_2 \tag{3.32}$$

(in the sense that the multiplicity of a in $\mathcal{A}\backslash\mathcal{A}_0$ is the sum of the multiplicities of a in $\mathcal{A}_1$ and $\mathcal{A}_2$). Clearly,

$$\begin{aligned} S(\mathcal{A}) &= \sum_{a\in\mathcal{A}} a \geq \sum_{a\in(\mathcal{A}\backslash\mathcal{A}_0)'} a = \sum_{i=1}^{t} a_i \\ &\geq \sum_{j=0}^{[t/u]-1}\left(\sum_{i=1}^{u} a_{ju+i}\right) \geq \sum_{j=0}^{[t/u]-1}\left(\sum_{i=1}^{u} a_i\right) = \left[\frac{t}{u}\right] S(\mathcal{A}_1). \end{aligned} \tag{3.33}$$

It follows from (3.29), (3.30), (3.31) and (3.33) that

$$|\mathcal{A}_1| = u > \frac{\delta}{10}t > \frac{\delta}{11}|\mathcal{A}'| \tag{3.34}$$

and

$$S(\mathcal{A}_1) \leq \frac{S(\mathcal{A})}{[t/u]} < \frac{2uS(\mathcal{A})}{t} < \frac{\delta}{4}S(\mathcal{A}). \tag{3.35}$$

By (3.21), (3.22) and (3.34) we have

$$|\mathcal{A}_1| > \frac{\delta}{11}|\mathcal{A}'| > \frac{\delta}{11}\cdot 10^3(\delta^{-1}N)^{3/4} > 90\delta^{3/4}N^{3/4} > 75N^{3/4}$$

so that (3.6) in Lemma 9 holds with $\mathcal{A}_1$ in place of $\mathcal{A}$. (Note that also (3.5) holds by (3.20).) Thus by Lemma 9, there exist integers d, y, z satisfying (3.7), (3.8), (3.9) and (3.10) (with $\mathcal{A}_1$ in place of $\mathcal{A}$) so that, in view of (3.34), we have

$$1 \leq d < 10^4\frac{N}{|\mathcal{A}_1|} < 11\cdot 10^4\delta^{-1}N|\mathcal{A}'|^{-1}, \tag{3.36}$$

$$z > \frac{|\mathcal{A}_1|^2}{7\cdot 10^4} > 10^{-7}\delta^2|\mathcal{A}'|^2, \tag{3.37}$$

$$y < 7\cdot 10^4\frac{N}{|\mathcal{A}_1|^2} < 10^7\delta^{-2}N|\mathcal{A}'|^{-2}z \tag{3.38}$$

and

$$\{yd, (y+1)d, \ldots, zd\} \subset \mathcal{P}(\mathcal{A}_1). \tag{3.39}$$

This integer d satisfies (3.24) by (3.36). It remains to show that if there is an m satisfying (3.23), then this implies (3.25). To show this, we start out from the indirect assumption

$$L(\mathcal{A}, d) > d/2. \tag{3.40}$$

First we shall show that

$$\nu \in \mathbb{N}, y \le \nu < (1-\delta)S(\mathcal{A})/d \quad \text{imply} \quad \nu d \in \mathcal{P}(\mathcal{A}_1 \cup \mathcal{A}_2). \tag{3.41}$$

It follows from (3.20), (3.21), (3.22), (3.37) and (3.38) that

$$\begin{aligned} z - y &> z(1 - 10^7\delta^{-2}N|\mathcal{A}'|^{-2}) \\ &> 10^{-7}(\delta|\mathcal{A}'|)^2(1 - 10^7\delta^{-2}N(10^3(\delta^{-1}N)^{3/4})^{-2}) \\ &> 10^{-7}(10^3\delta^{1/4}N^{3/4})^2(1 - 10(\delta N)^{-1/2}) \\ &> 7 \cdot 10^{-2}N^{3/2}(1 - 10(2N)^{-1/2}) > 3 \cdot 10^{-2}N^{3/2} > N \end{aligned}$$

so that (3.39) implies

$$\{yd, (y+1)d, \dots, (y+N-1)d\} \subset \mathcal{P}(\mathcal{A}_1). \tag{3.42}$$

Thus (3.41) holds for $y \le \nu < y + N$. Assume now that

$$y + N \le \nu < (1-\delta)S(\mathcal{A})/d. \tag{3.43}$$

Let us write $n = \nu - y - N$. Then by (3.43) we have

$$0 \le n. \tag{3.44}$$

Furthermore, it follows from (3.30), (3.32) and (3.35) that

$$\begin{aligned} S(\mathcal{A}_2) &= S(\mathcal{A}\backslash\mathcal{A}_0) - S(\mathcal{A}_1) > (1 - 10^{-1}\delta)S(\mathcal{A}) - \delta S(\mathcal{A})/4 \\ &= \left(1 - \frac{7}{20}\delta\right) S(\mathcal{A}) > (1-\delta)S(\mathcal{A}). \end{aligned} \tag{3.45}$$

(3.43) and (3.45) imply

$$n = \nu - y - N < \nu - N < (1-\delta)S(\mathcal{A})/d - N < S(\mathcal{A}_2)/d - N. \tag{3.46}$$

By (3.44) and (3.46), Lemma 8 can be applied with $\mathcal{A}_2$ in place of $\mathcal{B}$. We obtain that there is a number $x_n \in \mathbb{N}$ such that

$$n + 1 \le x_n \le n + N \tag{3.47}$$

and

$$dx_n \in \mathcal{P}(\mathcal{A}_2). \tag{3.48}$$

(3.47) can be rewritten in the equivalent form

$$0 \le n + N - x_n \le N - 1. \tag{3.49}$$

Furthermore, we have

$$\nu - x_n = (n + y + N) - x_n = y + (n + N - x_n). \tag{3.50}$$

By (3.49) and (3.50), $(\nu - x_n)d$ belongs to the arithmetic progression $\{yd, (y+1)d, \dots, (y+N-1)d\}$ and thus by (3.42) we have

$$(\nu - x_n)d \in \mathcal{P}(\mathcal{A}_1). \tag{3.51}$$

It follows from (3.48) and (3.51) that

$$\nu d = dx_n + (\nu - x_n)d \in \mathcal{P}(\mathcal{A}_1 \cup \mathcal{A}_2)$$

which proves (3.41).

Assume now that m satisfies (3.23). Let

$$\mathcal{L}_m(\mathcal{A}, d) = \{m - i : i \in \mathcal{L}(\mathcal{A}, d)\}.$$

The the elements of both $\mathcal{L}(\mathcal{A}, d)$ and $\mathcal{L}_m(\mathcal{A}, d)$ are pairwise incongruent modulo d and, in view of the indirect assumption (3.40), the total number of them is

$$|\mathcal{L}(\mathcal{A}, d)| + |\mathcal{L}_m(\mathcal{A}, d)| = 2L((\mathcal{A}, d) > d.$$

Thus by the box principle, there is a number i_1 in $\mathcal{L}(\mathcal{A}, d)$ which is congruent to a number $m - i_2$ in $\mathcal{L}_m(\mathcal{A}, d)$ modulo d:

$$i_1 \equiv m - i_2 \pmod{d}$$

whence

$$m - i_1 - i_2 \equiv 0 \pmod{d}. \tag{3.52}$$

Let us write $a_m = a(d, i_1)$ and

$$a'_m = \begin{cases} a(d, i_2) & \text{for } i_1 \neq i_2 \\ a'(d, i_1) = a'(d, i_2) & \text{for } i_1 = i_2. \end{cases}$$

Then if follows from (3.52) that

$$d | m - a_m - a'_m \tag{3.53}$$

and from the definition of $\mathcal{A}_0$ that $a_m \in \mathcal{A}_0$, $a'_m \in \mathcal{A}_0$, and thus

$$a_m + a'_m \in \mathcal{P}(\mathcal{A}_0). \tag{3.54}$$

Furthermore, in view of (3.21) and (3.23) we have

$$m - a_m - a'_m < m < (1-\delta)S(\mathcal{A}) \tag{3.55}$$

and

$$\begin{aligned} m - a_m - a'_m &\geq 2\cdot 10^7\delta^{-2}N^2|\mathcal{A}'|^{-1} - N - N \\ &= 2N(10^7\delta^{-2}N|\mathcal{A}'|^{-1} - 1) \\ &\geq 2N(10^7\delta^{-2}N|\mathcal{A}'|^{-1} - \delta^{-2}(N|\mathcal{A}'|^{-1})) \\ &> 2N\cdot 5\cdot 10^6\delta^{-2}N|\mathcal{A}'|^{-1} = 10^7\delta^{-2}N^2|\mathcal{A}'|^{-1}. \end{aligned} \tag{3.56}$$

It follows from (3.38), (3.39) and (3.56) that

$$\begin{aligned} yd &< 10^7\delta^{-2}N|\mathcal{A}'|^{-2}zd = 10^7\delta^{-2}N|\mathcal{A}'|^{-2}(zd) \\ &< 10^7\delta^{-2}N|\mathcal{A}'|^{-2}(|\mathcal{A}_1|N) < 10^7\delta^{-2}N|\mathcal{A}'|^{-2}|\mathcal{A}'|N \\ &= 10^7\delta^{-2}N^2|\mathcal{A}'|^{-1} < m - a_m - a'_m. \end{aligned} \tag{3.57}$$

By (3.53), $(m - a_m - a'_m)/d = \nu$ is an integer. It follows from (3.41), (3.55) and (3.57) that

$$\nu d = m - a_m - a'_m \in \mathcal{P}(\mathcal{A}_1 \cup \mathcal{A}_2).$$

Thus

$$m = \nu d + (a_m + a'_m)$$

where $\nu d \in \mathcal{P}(\mathcal{A}_1 \cup \mathcal{A}_2)$ and, in view of (3.54), $a_m - a'_m \in \mathcal{P}(\mathcal{A}_0)$. This implies

$$m \in \mathcal{P}(\mathcal{A}_1 \cup \mathcal{A}_2 \cup \mathcal{A}_0) = P(\mathcal{A})$$

which contradicts (3.23) and this completes the proof of the lemma, when in (3.21) it is assumed $\delta = 1/2$.

In a second step, we have to prove Lemma 11, with $0 < \delta \leq 1/2$. Let us refer to the particular case of Lemma 11 with $\delta = \delta_0 \stackrel{\text{def}}{=} 1/2$ as Lemma 11_o. So, Lemma 11_o has been proved. We assert that Lemma 11 is an easy consequence of Lemma 11_o.

Since $\delta \leq \delta_0$, one has:

$$|\mathcal{A}'| > 10^3(\delta_0^{-1}N)^{3/4}.$$

If $m \leq \frac{1}{2}S(\mathcal{A})$, one has:

$$2\cdot 10^7\delta_0^{-2}N^2|\mathcal{A}'|^{-1} < m \leq (1-\delta_0)S(\mathcal{A}).$$

If $m > \frac{1}{2}S(\mathcal{A})$ replace m by $m' = S(\mathcal{A}) - m$ (use the symmetry of $\mathcal{P}(\mathcal{A})$). Then $m' \le (1-\delta_0)S(\mathcal{A})$. Since $m \le (1-\delta)S(\mathcal{A})$, one has (see proof of (3.27))

$$m' \ge \delta S(\mathcal{A}) > \frac{1}{2}\delta|\mathcal{A}'|^2 > \frac{1}{2}\delta 10^6 \delta^{-3/2} N^{3/2} = \frac{10^6}{2}\delta^{-1/2}N^{3/2}.$$

Now, by (3.20), (3.21), and (3.22),

$$\frac{(1/2)10^6\delta^{-1/2}N^{3/2}}{2\cdot 10^7 \delta_0^{-2} N^2 |\mathcal{A}'|^{-1}} = \frac{1}{10}\delta^{-1/2}N^{-1/2}|\mathcal{A}'| > 10^2\delta^{-5/4}N^{1/4}$$
$$\ge 10^2 \delta_0^{-5/4} N^{1/4} \ge 75000 > 1$$

whence Lemma 11_o can be applied with either m or m'. We get d with

$$d < 11 \cdot 10^4 \delta_0^{-1/2} N |\mathcal{A}'|^{-1} < 11 \cdot 10^4 \delta^{-1} N |\mathcal{A}'|^{-1}$$

and $L(\mathcal{A}, d) \le d/2$, and Lemma 11 is completely proved.

4. Proofs of Theorems 1, 2, and 3

To every (unequal or unrestricted) partition

$$n = n_1 + n_2 + \cdots + n_t$$

of n which does not represent a, we assign the set $\mathcal{N} = \{n_1, n_2, \ldots, n_t\}$ (so that in case of unrestricted partitions the parts are taken with multiplicity). For an M which will be defined later, let $\mathcal{N}_0$ denote the set of the parts not exceeding M, and let $\mathcal{N}_0'$ denote the set of the distinct parts not exceeding M (so that, in case of unequal partitions, we have $\mathcal{N}_0 = \mathcal{N}_0' = \mathcal{N} \cap \{1, 2, \ldots, M\}$).

To prove Theorem 1, we shall choose

$$M = [10^{-3}(an)^{1/3}]. \tag{4.1}$$

We have to distinguish two cases.

CASE 1: Assume that

$$|\mathcal{N}_0'| > 8 \cdot 10^7 M^2 a^{-1} \stackrel{\text{def}}{=} B. \tag{4.2}$$

We are going to show that, if n is large enough, then in this case, Lemma 11 can be applied with M, N_0, N_0', $1/2$ and a in place of N, $\mathcal{A}$, $\mathcal{A}'$, δ, and m, respectively.

In fact, (3.20) follows from (1.1) and (4.1) for n large enough, and (3.21) holds trivially. Furthermore, it follows from (1.1), (4.1) and (4.2) that for large n we have

$$\begin{aligned}|\mathcal{N}_0'| > 8 \cdot 10^7 M^2 a^{-1} &= (4 \cdot 10^4 M^{5/4} a^{-1})(2 \cdot 10^3 M^{3/4}) \\ &> 3 \cdot 10^4 (10^{-3}(an)^{1/3})^{5/4} a^{-1} 10^3 (\delta^{-1} M)^{3/4} \\ &> 3a^{-7/12} n^{5/12} 10^3 (\delta^{-1} M)^{3/4} \\ &\geq 3 \cdot 10^3 (\delta^{-1} M)^{3/4}\end{aligned}$$

and thus (3.22) is verfied. The left–hand side inequality of (3.23) follows immediately from (4.2), and by (1.1), (4.1) and (4.2), we have for large n

$$\begin{aligned}(1-\delta)S(\mathcal{N}_0) = \frac{1}{2}S(\mathcal{N}_0) \geq \frac{1}{2}S(\mathcal{N}_0') \geq \frac{1}{2}\sum_{i=1}^{|\mathcal{N}_0'|} i > \frac{|\mathcal{N}_0'|^2}{4} \\ > 16 \cdot 10^{14} M^4 a^{-2} > 10 \cdot 10^{14} (10^{-3}(an)^{1/3})^4 a^{-2} \\ = 10^3 a^{-5/3} n^{4/3} a \\ > 10^3 (n^{5/7})^{-5/3} n^{4/3} a = 10^3 n^{1/7} a > a.\end{aligned}$$

Thus all the assumptions in Lemma 11 hold so that the lemma can be applied. We deduce that there is an integer d with

$$d < 11 \cdot 10^4 \cdot 2 \cdot M \cdot |\mathcal{N}_0'|^{-1} \tag{4.3}$$

and

$$L(\mathcal{N}_0, d) \leq d/2.$$

It follows from (4.1), (4.2) and (4.3) that, if we set

$$D = 4a^{2/3} n^{-1/3}, \tag{4.4}$$

then

$$d < 22 \cdot 10^4 M (8 \cdot 10^7 M^2 a^{-1})^{-1} < 3 \cdot 10^{-3} M^{-1} a < D. \tag{4.5}$$

Now let $\{i_1, \ldots, i_{[d/2]}\}$ be any set containing $\mathcal{L}(\mathcal{N}_0, d)$ and such that $1 \leq i_1 < \cdots < i_{[d/2]} \leq d$. As in Lemma 5 or 6, we can define $\mathcal{N}_1$ and $\mathcal{N}_2$, and it follows from the definition of $\mathcal{L}(\mathcal{N}_0, d)$ that

$$|\mathcal{N}_2| \leq (d-1) + d - [d/2] \leq 2d \leq 2D.$$

Therefore the number of partitions of n which do not represent a and satisfy (4.2) is smaller than $V(n, M, D)$ or $W(n, M, D)$.

CASE 2: Assume that

$$|\mathcal{N}_0'| \leq 8 \cdot 10^7 M^2 a^{-1} \stackrel{\text{def}}{=} B. \tag{4.6}$$

With the notation of Lemma 3 or 4, the total number of partitions of n is certainly smaller than $Y(n,B,M)$ or $Z(n,B,M)$.

So, we have proved that

$$Q(n,a) \leq V(n,M,D) + Y(n,B,M) \tag{4.7}$$

and

$$R(n,a) \leq W(n,M,D) + Z(n,B,M). \tag{4.8}$$

By Lemma 3 and Lemma 5, (4.7) yields

$$Q(n,a) \leq p(n, B + n/M) + n^{5D} q([n/2]) p(n, n/M). \tag{4.9}$$

Now, by (4.1) we have

$$n/M \leq (3/2) 10^3 a^{-1/3} n^{1/6} \sqrt{n},$$

$$20 a^{-1/3} n^{2/3} \leq B \leq 80 a^{-1/3} n^{2/3}, \tag{4.10}$$

and if we set $\alpha = 2 \cdot 10^3 a^{-1/3} n^{1/6}$, we have

$$n/M + B \leq \alpha \sqrt{n}.$$

By (1.1), we have $\alpha \leq 1$, so that we may apply Lemma 2. We obtain

$$\begin{aligned} Q(n,a) &\leq 2n^{5D} q([n/2]) p(n, \alpha\sqrt{n}) \\ &\leq q([n/2]) exp \left(5D \log n + \log 2 + \left(2\alpha \log \frac{3.6}{\alpha} \right) \sqrt{n} \right). \end{aligned}$$

But, from (4.4) and (1.1) we have $D = O(n^{1/7})$, and

$$\alpha \geq 2 \cdot 10^3 n^{-1/14},$$

and thus, for n large enough, (1.2) is proved.

It remains to deduce (1.3) from (4.8). We are going to apply Lemma 4. First observe that by (4.1) and (1.1) we have

$$B = 8 \cdot 10^7 M a^{-1} M \leq 8 \cdot 10^4 a^{-2/3} n^{1/3} M \leq 8 \cdot 10^{-2} M \leq M/2,$$

and thus

$$Z(n,B,M) \leq 6B \binom{M}{B} n^2 p(n,B) p(n, n/M). \tag{4.11}$$

By (1.1) and (4.10), one has

$$B \le 8 \cdot 10^{-5} n^{1/2}, \tag{4.12}$$

so that Lemma 2 gives

$$p(n, B) \le \exp(2 \cdot 10^{-3}\sqrt{n}). \tag{4.13}$$

Now, using Stirling's formula, (4.1) and (4.10), we have

$$\binom{M}{B} \le \frac{M^B}{B!} \le \left(\frac{Me}{B}\right)^B \le \left(\frac{10^{-5}e(an)^{1/3}}{20a^{-1/3}n^{2/3}}\right)^B \le (a^{2/3}n^{-1/3})^B$$
$$\le \exp(80n^{2/3}a^{-1/3}\log(a^2/3n^{-1/2})).$$

But, the above quantity is a decreasing function of a for $a \ge e^3\sqrt{n}$, so that by (1.1),

$$\binom{M}{B} \le \exp(8 \cdot 10^{-5}\sqrt{n}\log(10^{12})) \le \exp(3 \cdot 10^{-3}\sqrt{n}). \tag{4.14}$$

Therefore, for n large enough, (1.14), (4.11), (4.12), (4.13) and (4.14) give

$$Z(n, B, M) \le p([n/2])p(n, n/M),$$

and by Lemma 6, (4.8) gives

$$R(n, a) \le 2\, n^{7D} p([n/2])p(n, n/M).$$

The end of the proof of (1.3) is similar to the end of the proof of (1.2).
To prove Theorem 2, we shall choose

$$M = [n^{15/28}] \tag{4.15}$$

and

$$\delta = 10^4 \cdot n^{-1/28}. \tag{4.16}$$

Again we start to prove (1.5) and (1.6) simultaneously. Define $\mathcal{N}$, $\mathcal{N}_0$, $\mathcal{N}_0'$ in the same way as in the proof of Theorem 1, and also write $\mathcal{N}^* = \mathcal{N} - \mathcal{N}_0$ (so that $\mathcal{N}^*$ is the set of the parts greater than M). We have to distinguish three cases.

CASE 1: Assume that

$$(1-\delta)S(\mathcal{N}_0) \le n/2 \tag{4.17}$$

whence, for large n,

$$S(\mathcal{N}_0) \le \frac{n}{2}\frac{1}{1-\delta} \le (\frac{1}{2}+\delta)n.$$

If we fix $S(\mathcal{N}_0) = k$, then in the case of unequal partitions, $\mathcal{N}_0$ can be chosen in at most $q(k)$ ways, while $\mathcal{N}^* = \mathcal{N} - \mathcal{N}_0$ can be chosen in at most $\rho(n-k, M+1)$ ways (since $S(\mathcal{N}^*) = S(\mathcal{N}) - S(\mathcal{N}_0) = n-k$, and the elements of $\mathcal{N}^*$ are greater than M). Therefore the total number of unequal partitions with property (4.17) is at most

$$T \stackrel{\text{def}}{=} \sum_{k<(\frac{1}{2}+\delta)n} q(k)\rho(n-k, M+1). \tag{4.18}$$

Similarly the total number of unrestricted partitions with property (4.17) is at most

$$U \stackrel{\text{def}}{=} \sum_{k<(\frac{1}{2}+\delta)n} p(k)r(n-k, M+1). \tag{4.19}$$

We have

$$\rho(n, M) \le r(n, M) \le p(n, n/M), \tag{4.20}$$

and by Lemma 2,

$$p(n, n/M) < \exp((3n^{-1/28}\log 3.6n^{1/28})\sqrt{n}) \tag{4.21}$$

Now, from (1.14) and for n large enough, we have:

$$\begin{aligned} q([(\frac{1}{2}+\delta)n]) &< \exp\left(\frac{\pi}{\sqrt{3}}\sqrt{\left(\frac{1}{2}+\delta\right)n}\right) < \exp\left(\frac{\pi}{\sqrt{3}}\sqrt{\frac{n}{2}}(1+\delta)\right) \\ &< nq([n/2])\exp\left(\frac{\pi}{\sqrt{6}}\delta\sqrt{n}\right). \end{aligned} \tag{4.22}$$

From (4.18), (4.20), (4.21) and (4.22), we have

$$T < nq([(\frac{1}{2}+\delta)n])\rho(n, M) < q([n/2])\exp(n^{1/2-1/29}) \tag{4.23}$$

for n large enough.

Similarly, from (4.19) we obtain

$$U < p([n/2])\exp(n^{1/2-1/23}). \tag{4.24}$$

CASE 2: Assume that

$$\frac{n}{2} < (1-\delta)S(\mathcal{N}_0) \tag{4.25}$$

and

$$|\mathcal{N}_0'| > B \stackrel{\text{def}}{=} 10^3(\delta^{-1}M)^{3/4} > \frac{1}{2}n^{3/7}. \tag{4.26}$$

We are going to show that, if n is large enough, then Lemma 11 can be applied with M, $\mathcal{N}_0$, $\mathcal{N}_0'$ and a in place of N, $\mathcal{A}$, $\mathcal{A}'$, and m, respectively, and with $\delta = 10^4 n^{-1/28}$.

In fact, (3.20) and (3.21) hold trivially, while (3.22) holds by (4.26). Furthermore by (1.4), (4.15), (4.16) and (4.25), we have for large n

$$2 \cdot 10^7 \delta^{-2} M^2 |\mathcal{N}_0'|^{-1} < 2 \cdot 10^7 \cdot 10^{-8} n^{2/28+30/28-3/7} < n^{5/7} < a. \tag{4.27}$$

The assumption (3.23) follows from (4.25) and (4.27), and thus Lemma 11 can be applied. We obtain that there is an integer d with

$$d < 11 \cdot 10^4 \delta^{-1} M |\mathcal{N}_0'|^{-1} \tag{4.28}$$

and

$$L(\mathcal{N}_0, d) \leq d/2.$$

It follows from (4.15), (4.16), (4.26) and (4.28) that

$$d < 11 \cdot 10^4 \delta^{-1} M \cdot 2n^{-3/7} < 22 n^{1/7} \stackrel{\text{def}}{=} D. \tag{4.29}$$

Therefore, as in the proof of Theorem 1, the number of partitions of n which do not represent a and satisfy (4.25) and (4.26), is smaller than $V(n, M, D)$ or $W(n, M, D)$.

CASE 3: Assume that

$$|\mathcal{N}_0'| \leq B = n^{3/7}. \tag{4.30}$$

As in Case 2 in the proof of Theorem 1, the number of partitions of n is bounded above by $Y(n, B, M)$ or $Z(n, B, M)$.

So we have proved that, under the assumptions of Theorem 2, we have

$$Q(n, a) \leq T + V(n, M, D) + Y(n, B, M) \tag{4.31}$$

and

$$R(n, a) \leq U + W(n, M, D) + Z(n, B, M). \tag{4.32}$$

We have

$$B + n/M < B + 2n^{13/28} < 3n^{13/28},$$

and by Lemmas 2 and 3,

$$Y(n,B,M) \le p(n, B+n/M) < \exp(n^{1/2-1/29}) \tag{4.33}$$

for n large enough.

Moreover, by Lemma 5, (4.21) and (4.29),

$$V(n,M,D) < q([n/2]) \exp(n^{1/2-1/29}), \tag{4.34}$$

and by (4.23), (4.31), (4.33) and (4.34), (1.5) holds.

Similarly, from Lemma 4, using

$$\binom{M}{B} \le n^B$$

and since $p(n,B) \le p(n,n/M)$ by $B < n/M$, we have

$$Z(n,B,M) < \exp(n^{1/2-1/29}). \tag{4.35}$$

From Lemma 6, we have

$$W(n,M,D) < p([n/2]) \exp(n^{1/2-1/29}), \tag{4.36}$$

(4.24), (4.32), (4.35) and (4.36) yield (1.6).

To prove Theorem 3, we choose

$$M = [10^{-2}\sqrt{n\, s(a)}]. \tag{4.37}$$

To a partition of n which does not represent a, we associate $\mathcal{N}$, $\mathcal{N}_0$, $\mathcal{N}_0'$ in the same way as in the proofs of Theorems 1 and 2. We apply Lemma 10 with a, M, $\mathcal{N}_0'$ in place of m, N, $\mathcal{A}$, respectively. By $n \ge (2500)^2$ and $s(a) \ge 40000$, (4.37) yields $M \ge 2500$. It is easily seen that (1.7) implies (3.11), and we conclude that

$$|\mathcal{N}_0'| < 10^4 M/s(a) \le 10^2\sqrt{n/s(a)} \stackrel{\text{def}}{=} t. \tag{4.38}$$

So, with the notation of Lemmas 3 and 4, we have

$$Q(n,a) \le Y(n,t,M) \tag{4.39}$$

and

$$R(n,a) \le Z(n,t,M) \tag{4.40}$$

As $M \geq 2500$, from (4.37) we deduce that $n/M < 101\sqrt{n/s(a)}$. By Lemma 3, (4.39) gives

$$Q(n,a) \leq p(n, t + n/M) \leq p(n, 201\sqrt{n/s(a)})),$$

and Lemma 2 yields (1.8).

Now, by Lemma 4 and (4.40), we have

$$R(n,a) \leq 6tn^2 \binom{M}{t} p(n,t)p(n,n/M) \tag{4.41}$$

since

$$t/M < 2 \cdot 10^4/s(a) \leq 1/2. \tag{4.42}$$

From (4.38) and Lemma 1, we have

$$\begin{aligned} n &= 10^{-4}t^2 s(a) < (4.5) \cdot 10^{-4}t^2 \log a < (4.5)10^{-4}t^2 \log n \\ &< t^2 \log(n^{1/3}) < t^2 n^{1/3}, \end{aligned}$$

whence $n < t^3$ and

$$6tn^2 < 6t^7 = \exp(\log 6 + 7\log t) < \exp(\log 6 + 7(t-1)) < e^{7t}.$$

By Stirling's formula, (4.37) and (4.38), we have

$$6tn^2 \binom{M}{t} < e^{7t} \left(\frac{Me}{t}\right)^t = \left(e^8 M t^{-1}\right)^t \leq (e^8 10^{-4} s(a))^t < s(a)^t$$

and (4.41) and Lemma 2 give (1.9).

TABLE OF Q(n,a)

n	$q(n)$	$a=1$	2	3	4	5	6	7	8	9	10
1	1	0									
2	1	1									
3	2	1									
4	2	1	2								
5	3	2	2								
6	4	2	2	3							
7	5	3	3	3							
8	6	3	4	3	5						
9	8	5	5	4	5						
10	10	5	6	5	6	7					
11	12	7	7	7	7	7					
12	15	8	9	8	8	8	11				
13	18	10	11	10	10	10	10				
14	22	12	13	11	13	11	12	15			
15	27	15	16	14	15	13	15	15			
16	32	17	19	16	19	16	17	16	23		
17	38	21	22	20	21	20	20	20	20		
18	46	25	27	23	26	23	23	23	25	30	
19	54	29	32	28	29	28	28	27	28	28	
20	64	35	37	32	35	32	34	31	34	31	43
21	76	41	44	38	41	38	38	35	38	37	38

n	$q(n)$	$a = 1$ 11	2 12	3 13	4 14	5 15	6 16	7 17	8 18	9 19	10 20
22	89	48 57	52	44	48	43	46	42	45	42	45
23	104	56 51	60	52	56	50	52	51	50	49	50
24	122	66 57	70 79	60	66	58	62	57	57	55	59
25	142	76 67	82 67	70	75	68	70	67	69	65	67
26	165	89 74	95 78	81 102	88	77	81	76	81	73	77
27	192	103 88	110 88	94 90	101	91	93	89	91	81	88
28	222	119 96	127 99	108 97	116 138	104	107	101	106	97	99
29	256	137 110	146 113	124 114	134 114	119	123	116	119	114	114
30	296	159 126	169 133	143 127	154 133	137 174	140	131	139	127	126
31	340	181 144	194 145	164 145	176 147	157 149	161	150	156	147	150
32	390	209 160	221 166	188 161	202 166	177 162	184 232	170	180	164	173
33	448	239 177	254 187	214 185	231 188	204 188	209 191	196	201	189	194
34	512	273 210	291 212	245 203	262 213	232 205	239 215	219 192	229	211	221
35	585	312 241	331 239	279 233	300 240	262 232	271 239	251 242	258	241	247
36	668	356 267	377 260	318 260	340 267	299 259	307 271	284 265	294 375	272	281
37	760	404 305	429 308	360 293	386 299	340 293	348 298	321 302	333 303	306	314
38	864	460 337	487 350	409 327	438 334	383 322	394 336	363 329	373 341	341 471	357
39	982	522 385	553 387	463 356	496 376	434 364	445 375	412 375	424 376	387 386	399
40	1113	591 427	626 439	525 417	560 420	491 402	501 423	460 410	476 420	433 415	451 602

REFERENCES

[1] N. Alon, Subset sums, J. Number Theory **27** (1987), 196-205.

[2] N. Alon and G. Freiman, On sums of subsets of a set of integers, Combinatorica (to appear).

[3] J. Dixmier and J. L. Nicolas, Partitions without small parts, Number Theory, Coll. Math. Soc. J. Bolyai (to appear).

[4] J. Dixmier and J. L. Nicolas, Partitions sans petits sommants II. To be published.

[5] J. Dixmier, Sur les sous sommes d'une partition, Bull. Soc. Math. Fance. Mémoire no. 35, 1988.

[6] J. Dixmier, Sur l'évaluation de $R(n, a)$. Submitted to Bull. Soc. Math. Belgique.

[7] P. Erdös, J. L. Nicolas and A. Sárközy, On the number of partitions of n without a given subsum (I), Discrete Math **74** (1989), 155-166.

[8] P. Erdös, J. L. Nicolas and M. Szalay, Partitions into parts which are unequal and large. Proceedings of "Journeés Arithmétiques" of Ulm, Springer Verlag Lecture Notes, to appear.

[9] G. H. Hardy and S. Ramanujan, Asymptotic formulae in combinatory analysis, Proc. London Math. Soc. **(2) 17** (1918), 75-115. (Also in Collected Papers of S. Ramanujan pp. 276-309. Cambridge University Press 1927, reprinted by Chelsea, New-York 1962).

[10] E. Lipkin, On representation of r- powers by subset sums, Acta Arithmetica (to appear).

[11] A. Sárközy, Finite addition theorems I, J. Number Theory **32** (1989), 114-130.

[12] ———, Finite addition theorems II, J. Number Theory (to appear).

J. L. Nicolas
Département de Mathématiques
Université Claude Bernard (Lyon 1)
F-69622 Villeurbanne Cedex
France

P. Erdös and A. Sárközy
Mathematical Institute of
the Hungarian Academy of Sciences
Reáltanoda u. 13-15, Pf. 127
H-1364 Budapest
Hungary

On Gaps between Squarefree Numbers

MICHAEL FILASETA* AND OGNIAN TRIFONOV

Dedicated to Paul Bateman

1. Introduction.

A squarefree number is a positive integer not divisible by the square of an integer > 1. We investigate here the problem of finding small $h = h(x)$ such that for x sufficiently large, there is a squarefree number in the interval $(x, x+h]$. This problem was originally investigated by Fogels [3]; he showed that for every $\epsilon > 0$, $h = x^{2/5+\epsilon}$ is admissible. Later Roth [9] reported elementary arguments of Davenport and Estermann showing respectively that one can take $h \gg x^{1/3}$ and $h \gg x^{1/3}(\log x)^{-2/3}$ for sufficiently large choices of the implied constants. Roth then gave an elementary proof that $h = x^{1/4+\epsilon}$ is admissible, and by applying a result of van der Corput, he showed that one can take $h \gg x^{3/13}(\log x)^{4/13}$. Nair [6] later noted that the elementary proof could be modified to omit the ϵ in the exponent to get that $h \gg x^{1/4}$ is admissible, and more recently the first author [1] showed that one could obtain the result $h \gg x^{3/13}$ by elementary means. Using further exponential sum techniques, Richert [8], Rankin [7], Schmidt [10], and Graham and Kolesnik [4] obtained the improvements $h \gg x^{2/9}\log x$, $h = x^{\theta+\epsilon}$ where $\theta = 0.221982...$, $\theta = 109556/494419 = 0.221585...$, and $\theta = 1057/4785 = 0.2208986...$, respectively. The authors investigated the problem further. They independently were able to show by using only elementary methods that $h \gg x^{2/9}$ is admissible. Using exponential sum techniques, the second author [12] (also see [13]) in addition obtained that one may take θ above to be $17/77 = 0.220779...$, and the first author [2] obtained that the value $\theta = 47/217 = 0.216589...$ is admissible. The purpose of this paper is to make the following improvement:

---

*Research was supported in part by the NSF under grant number DMS-8903123.

Theorem. *There is a constant $c > 0$ such that for x sufficiently large, the interval $(x, x + cx^{3/14}]$ contains a squarefree number.*

We note that $3/14 = 0.2142857....$ Furthermore, we will show that one can obtain $h \gg x^\theta$ with $\theta = 8/37 = 0.216216...$ by using only elementary techniques. Indeed, the advancements here are at the elementary level and even obtaining the theorem will rely on adding the use of only the exponent pair $(1/2, 1/2)$. The proof of the theorem given here is based on combining the previous independent elementary approaches of the authors which implied that $h \gg x^{2/9}$ is admissible.

2. Preliminaries

Throughout this paper, we will make use of the following notation:

c is a sufficiently large constant.
x is a sufficiently large real number (i.e., $x \geq x_0$ for some $x_0 = x_0(c)$).
a, a', b, b', d, and u denote positive integers.
p denotes a prime.
θ satisfies $1/5 < \theta \leq 1/4$. In section 6, we will restrict our attention to $\theta \leq 2/9$.
$c_1, c_2, \ldots$ denote positive constants. All such constants and implied constants in the asymptotic notation of Vinogradov do not depend on c.
$h = cx^\theta$.
ϕ is a number $> \theta$. More specifically, $x^\phi > x^\theta\sqrt{\log x}$.
α is a real number.
u_1, v_1, u_2, v_2, and β are positive real numbers.

Let S denote the number of integers in $(x, x+h]$ which are not squarefree. Since $[(x+h)/p^2] - [x/p^2]$ denotes the number of integers in $(x, x+h]$ which are divisible by p^2, we get that

$$S \leq \sum_p \left(\left[\frac{x+h}{p^2}\right] - \left[\frac{x}{p^2}\right] \right).$$

Now, since x is sufficiently large, we get that if $p > 2\sqrt{x}$, then $p^2 > x + h > x$. Hence, if $p > 2\sqrt{x}$, then $[(x+h)/p^2] = [x/p^2] = 0$. Therefore,

$$S \leq S_1 + S_2,$$

where

$$S_1 = \sum_{p \leq x^\theta\sqrt{\log x}} \left(\left[\frac{x+h}{p^2}\right] - \left[\frac{x}{p^2}\right] \right)$$

and

$$S_2 = \sum_{x^\theta \sqrt{\log x} < p \leq 2\sqrt{x}} \left(\left[\frac{x+h}{p^2} \right] - \left[\frac{x}{p^2} \right] \right).$$

We will estimate S_1 and S_2 separately. Note that

$$\begin{aligned} S_1 &= \sum_{p \leq x^\theta \sqrt{\log x}} \left(\left[\frac{x+h}{p^2} \right] - \left[\frac{x}{p^2} \right] \right) \\ &\leq \sum_{p \leq x^\theta \sqrt{\log x}} \left(\frac{h}{p^2} + 1 \right) \\ &< h \sum_{n=2}^{\infty} \frac{1}{n^2} + \pi(x^\theta \sqrt{\log x}) \\ &= h(\frac{\pi^2}{6} - 1) + \pi(x^\theta \sqrt{\log x}). \end{aligned}$$

By the prime number theorem or a Chebyshev estimate, we get that

$$S_1 \leq \frac{2}{3} h.$$

Therefore, in order to prove the Theorem, it suffices to show that $S_2 \ll c^\sigma x^\theta$, where $\sigma < 1$.

Now,

$$\begin{aligned} S_2 &= \sum_{x^\theta \sqrt{\log x} < p \leq 2\sqrt{x}} \left(\left[\frac{x+h}{p^2} \right] - \left[\frac{x}{p^2} \right] \right) \\ &\leq \sum_{x^\theta \sqrt{\log x} < d \leq 2\sqrt{x}} M_d, \end{aligned}$$

where

$$M_d = \left[\frac{x+h}{d^2} \right] - \left[\frac{x}{d^2} \right].$$

Define

$$S(t_1, t_2) = \{ u \in (t_1, t_2] : \exists \text{ an integer } m \text{ such that } mu^2 \in (x, x+h] \}.$$

Suppose that $d \in (x^\theta \sqrt{\log x}, 2\sqrt{x}]$ and that for some integer m, $md^2 \in (x, x+h]$. Since x is sufficiently large, $d^2 > h$ so that there is at most one multiple of d^2 in $(x, x+h]$. We get that

$$M_d = 1 \iff d \in S(x^\theta \sqrt{\log x}, 2\sqrt{x})$$

and

$$M_d = 0 \iff d \notin S(x^\theta \sqrt{\log x}, 2\sqrt{x}).$$

Therefore, $S_2 \leq |S(x^\theta \sqrt{\log x}, 2\sqrt{x})|$. To get bounds for $|S(x^\theta \sqrt{\log x}, 2\sqrt{x})|$ we will use the following lemma.

Lemma 1. *If*

$$|S(x^{\phi}, 2x^{\phi})| \ll x^{\alpha-\beta\phi} \quad \text{for } u_1 \le \phi \le v_1, \tag{1}$$

then

$$|S(x^{u_1}, 2x^{v_1})| \ll_{\beta} x^{\alpha-\beta u_1}. \tag{2}$$

If

$$|S(x^{\phi}, 2x^{\phi})| \ll x^{\gamma+\delta\phi} \quad \text{for } u_2 \le \phi \le v_2, \tag{3}$$

then

$$|S(x^{u_2}, 2x^{v_2})| \ll_{\delta} x^{\gamma+\delta v_2}. \tag{4}$$

A proof of the above lemma is fairly simple and can be found in [1]. The lemma as it applies here is essentially contained in Roth's paper [9]. We shall not elaborate on the proof here. Note that this lemma converts our problem from estimating $|S(x^{\theta}\sqrt{\log x}, 2\sqrt{x})|$ to estimating $|S(x^{\phi}, 2x^{\phi})|$.

3. The Halberstam-Roth Method

In this section, we describe an elementary method which was developed by Halberstam and Roth [5] in studying the more general problem of gaps between k-free numbers (see also [1]). For the current problem, the method was first described by Roth [9] to obtain his elementary result mentioned in the introduction. We modify the method as in Nair [6].

Suppose that u and $u + a \in S(x^{\phi}, 2x^{\phi})$. Then there are integers m_1 and m_2 such that $m_1 u^2 \in (x, x + h]$ and $m_2(u + a)^2 \in (x, x + h]$. Observe that since u and $u + a \in (x^{\phi}, 2x^{\phi}]$,

$$m_1 = \frac{x}{u^2} + O\left(\frac{h}{u^2}\right) = \frac{x}{u^2} + O\left(cx^{\theta-2\phi}\right) \tag{5}$$

and

$$m_2 = \frac{x}{(u+a)^2} + O\left(cx^{\theta-2\phi}\right). \tag{6}$$

Next, we construct polynomials $P = P(u, a)$ and $Q = Q(u, a)$ in $\mathbb{Z}[u, a]$ which are homogeneous and of degree 1 such that if a is sufficiently small, then $m_1 P - m_2 Q = 0$. The idea is to make $m_1 P - m_2 Q$ small, and then to see what restrictions are necessary to place on a to obtain that $m_1 P - m_2 Q = 0$. Note that since $m_1 P - m_2 Q$ is an integer, to obtain $m_1 P - m_2 Q = 0$, it suffices to show that $|m_1 P - m_2 Q| < 1$.

Since u and $u + a \in S(x^{\phi}, 2x^{\phi})$, we get from (5) and (6) that

$$m_1 P - m_2 Q = \frac{x}{u^2} P - \frac{x}{(u+a)^2} Q + O\left(cx^{\theta-\phi}\right) \tag{7}$$

$$= \frac{x}{u^2(u+a)^2}\left((u+a)^2 P - u^2 Q\right) + O\left(cx^{\theta-\phi}\right).$$

We view the second of the three expressions above as a difference for x/u^2 modified by the appearance of the polynomials P and Q. To make this modified difference small, (7) indicates that we should make $(u+a)^2P-u^2Q$ small. Note that restricting P and Q to being polynomials of degree 1 led to the error term $O\left(cx^{\theta-\phi}\right)$ above which (since x is sufficiently large) is small in absolute value and, in particular, $< 1/4$. It is easy to see that making $(u+a)^2P - u^2Q$ small is equivalent to finding a good approximation to $(u+a)^2/u^2$ as a rational function Q/P. By considering the continued fraction convergents of

$$\frac{(u+a)^2}{u^2} = 1 + \cfrac{1}{(2u-a)/(4a) + \cfrac{1}{(8u+4a)/a}},$$

we easily arrive at the choices $P = 2u - a$ and $Q = 2u + 3a$. This approach emphasizes that we wish to make $(u+a)^2P - u^2Q$ small, but there are other approaches to obtaining the polynomials P and Q (cf. [1], [5]). We get from (7) that

$$m_1P - m_2Q = \frac{-a^3x}{u^2(u+a)^2} + O\left(cx^{\theta-\phi}\right). \tag{8}$$

Hence,

$$|m_1P - m_2Q| < a^3x^{1-4\phi} + \frac{1}{4}.$$

Thus, if $a \le x^{(4\phi-1)/3}/2$, then $m_1P - m_2Q = 0$.

Now, we will prove that if $I \subseteq (x^\phi, 2x^\phi]$ with $|I| \le x^{(4\phi-1)/3}/4$, then $|S(x^\phi, 2x^\phi) \cap I| \le 2$. Consider such an I. We may suppose that $|S(x^\phi, 2x^\phi) \cap I| \ge 2$. Fix u and $u+a$ to be the minimal elements in $S(x^\phi, 2x^\phi) \cap I$. Let m_1 and m_2 be integers such that m_1u^2 and $m_2(u+a)^2 \in (x, x+h]$. Assume that there is a positive integer b such that $u+a+b \in S(x^\phi, 2x^\phi)$, and let m_3 be the integer such that $m_3(u+a+b)^2 \in (x, x+h]$. Thus, by the above,

$$m_1P(u,a) - m_2Q(u,a) = 0, \quad m_1P(u,a+b) - m_3Q(u,a+b) = 0,$$

and

$$m_2P(u+a,b) - m_3Q(u+a,b) = 0.$$

In other words,

$$m_1(2u-a) - m_2(2u+3a) = 0, \tag{9}$$

$$m_1(2u-a-b) - m_3(2u+3a+3b) = 0, \tag{10}$$

and

$$m_2(2u+2a-b) - m_3(2u+2a+3b) = 0. \tag{11}$$

Combining (10) and (11) gives that

$$m_1(2u-a-b)(2u+2a+3b)-m_2(2u+3a+3b)(2u+2a-b)=0.$$

Now, from (9), we get that

$$m_1((2u+3a)(2u-a-b)(2u+2a+3b)-(2u-a)(2u+3a+3b)(2u+2a-b))$$
$$=-12ab(a+b)m_1=0.$$

Clearly, this last equation is impossible. Hence, $|S(x^\phi,2x^\phi)\cap I|\le 2$. By dividing $(x^\phi,2x^\phi]$ into $[4x^{(1-\phi)/3}]+1$ intervals of size $\le x^{(4\phi-1)/3}/4$, we get that

$$|S(x^\phi,2x^\phi)|\ll x^{(1-\phi)/3} \qquad \text{for} \quad x^\theta\sqrt{\log x}<x^\phi\le 2\sqrt{x}. \tag{12}$$

It follows from Lemma 1 that $|S(x^\theta\sqrt{\log x},2\sqrt{x})|\ll x^{(1-\theta)/3}$, which upon taking $\theta=1/4$ is sufficient to show that $h=cx^{1/4}$ is admissible. (Actually, one may take $u_1=\theta+\log\log x/(2\log x)$ in Lemma 1 and obtain the slightly stronger result $h=cx^{1/4}(\log x)^{-1/8}$. The exponent on $\log x$ can be decreased further by extending the range on the summation in the definition of S_1 in section 2.)

Before leaving this section, we note that we have established that in intervals of length $\le x^{(4\phi-1)/3}/4$, there are ≤ 2 elements of $S(x^\phi,2x^\phi)$.

4. Divided Differences

This section is mainly based on work by the second author in [12]. The basic idea is to replace the use of modified differences in the previous section with the use of divided differences. Let I be an interval in $(x^\phi,2x^\phi]$ with $|I|\le x^{(5\phi-1)/6}/3$. We will prove that

$$|I\cap S(x^\phi,2x^\phi)|\le 7. \tag{13}$$

Assume (13) does not hold. Then there exists non-consecutive elements $u, u+a, u+a+b$, and $u+a+b+b'$ of $I\cap S(x^\phi,2x^\phi)$. Recall that in the previous section, we showed that in intervals of length $\le x^{(4\phi-1)/3}/4$, there are ≤ 2 elements of $S(x^\phi,2x^\phi)$. Thus, each of a,b, and b' must be $> x^{(4\phi-1)/3}/4$. Let $M=\max(a,b,b')$. Let m_1,m_2,m_3, and m_4 be integers such that $m_1u^2, m_2(u+a)^2, m_3(u+a+b)^2$, and $m_4(u+a+b+b')^2\in(x,x+h]$. Let T be the integer defined by

$$T=bb'(b+b')m_1-b'(a+b)(a+b+b')m_2+a(b+b')(a+b+b')m_3-ab(a+b)m_4.$$

Then using (5), (6), and the corresponding expressions for m_3 and m_4, we obtain that

$$T=\frac{x}{u^2}bb'(b+b')-\frac{x}{(u+a)^2}b'(a+b)(a+b+b')+\frac{x}{(u+a+b)^2}a(b+b')(a+b+b')$$

$$-\frac{x}{(u+a+b+b')^2}ab(a+b)+O\left(cM^3x^{\theta-2\phi}\right).$$

The above expression represents a divided difference of $\frac{x}{u^2}$. Note that $M \le x^\phi < u$ so that by the theory of divided differences or by a direct computation, one gets that

$$T = 4abb'(a+b)(b+b')(a+b+b')\frac{x}{u^5}\left(1+O(Mx^{-\phi})\right)+O\left(cM^3x^{\theta-2\phi}\right). \quad (14)$$

Since $\phi > \theta$ and x is sufficiently large, the error term $O(cM^3x^{\theta-2\phi})$ has absolute value $< M^3x^{-\phi}/1024$. Also, since $M \le |I| \le x^{(5\phi-1)/6}/3$, $Mx^{-\phi} \le x^{(-\phi-1)/6}/3$ so that the absolute value of the expression $O(Mx^{-\phi})$ in (14) is $< 1/2$. Recalling that $\min(a,b,b') > x^{(4\phi-1)/3}/4$ and $\max(a,b,b') = M$, we obtain easily that

$$abb'(a+b)(b+b')(a+b+b') \ge \frac{1}{64}x^{(4\phi-1)}M^3.$$

Now, since $u \le 2x^\phi$, we get from (14) that

$$T > \frac{1}{16}x^{(4\phi-1)}M^3\frac{x}{u^5}\left(1-\frac{1}{2}\right) - M^3\frac{x^{-\phi}}{1024} \ge 0. \quad (15)$$

On the other hand, since $M > x^{(4\phi-1)/3}/4, \phi > \theta$, and x is sufficiently large, it is easy to check that the term in (14) corresponding to $O(cM^3x^{\theta-2\phi})$ is $< M^6x^{1-5\phi}$. Clearly, $abb'(a+b)(b+b')(a+b+b') \le 12M^6$. Hence, since $u > x^\phi$, we get from (14) that

$$\begin{aligned} T &< 48M^6\frac{x}{u^5}(3/2) + M^6x^{1-5\phi} \\ &< 73M^6x^{1-5\phi}. \end{aligned}$$

Now, since $M \le |I| \le x^{(5\phi-1)/6}/3$, we get that $T < 73/3^6 < 1$. But T is an integer, and from (15), $T > 0$; thus, we obtain a contradiction. Hence, (13) must hold.

Dividing the interval $(x^\phi, 2x^\phi]$ into $\left[3x^{(1+\phi)/6}\right]+1$ subintervals of length $\le x^{(5\phi-1)/6}/3$ and using (13), we are led to

$$|S(x^\phi, 2x^\phi)| \ll x^{(1+\phi)/6} \qquad \text{for} \quad x^\theta\sqrt{\log x} < x^\phi \le 2\sqrt{x}. \quad (16)$$

Note that (16) is better than (12) precisely when $\phi < 1/3$. In fact, we get from Lemma 1 and (16) that

$$|S(x^\theta\sqrt{\log x}, x^{1/3})| \ll x^{(1+(1/3))/6} = x^{2/9},$$

and we get from Lemma 1 and (12) that

$$|S(x^{1/3}, 2\sqrt{x})| \ll x^{(1-(1/3))/3} = x^{2/9}.$$

These imply that one can take $h = cx^{2/9}$ for the gap problem.

Before leaving this section, we note that (12) implies that in intervals of length $\le x^{(5\phi-1)/6}/3$, there are ≤ 7 elements of $S(x^\phi, 2x^\phi)$.

5. Further Differences

In this section, we will follow the work of the first author in [1] and [2], modifying it with the results of the second author described in the previous section. First, we fix ϕ and define

$$T(a) = \{u : u \text{ and } u + a \text{ are consecutive elements in } S(x^{\phi}, 2x^{\phi})\},$$

and

$$t(a) = |T(a)|.$$

Note that

$$|S(x^{\phi}, 2x^{\phi})| \leq 1 + \sum_{a=1}^{\infty} t(a). \tag{17}$$

Recall that in section 3, we proved that in intervals of length $\leq x^{(4\phi-1)/3}/4$, there are ≤ 2 elements of $S(x^{\phi}, 2x^{\phi})$. Also, we established in the previous section that in intervals of length $\leq x^{(5\phi-1)/6}/3$, there are ≤ 7 elements of $S(x^{\phi}, 2x^{\phi})$. Let $R = \max\{x^{(4\phi-1)/3}/8, x^{(5\phi-1)/6}/21\}$. Thus, of every 8 consecutive elements in $S(x^{\phi}, 2x^{\phi})$, there exist 2 consecutive elements of distance at least R from one another. In other words, we get that

$$\sum_{a \leq R} t(a) \leq 6 + 6 \sum_{a > R} t(a).$$

Hence, from (17), we get that

$$|S(x^{\phi}, 2x^{\phi})| \leq 7 + 7 \sum_{a > R} t(a). \tag{18}$$

Next, we estimate the right-hand side of (18). Fix $B > 0$ to be specified later. We break up the sum on the right into 2 sums, considering $R < a \leq B$ and $a > B$ separately. In the latter case, we use that

$$x^{\phi} \geq \sum_{a=1}^{\infty} a t(a) \geq \sum_{a > B} a t(a) \geq B \sum_{a > B} t(a)$$

so that

$$\sum_{a > B} t(a) \leq \frac{x^{\phi}}{B}. \tag{19}$$

Now, we estimate $\sum_{R < a \leq B} t(a)$. Fix a, and let r be a positive integer. Partition $(x^{\phi}, 2x^{\phi}]$ into r disjoint subintervals, say $J_1, J_2, \ldots, J_r$. Let J denote such a subinterval and define

$$T_J(a) = \{u : u \text{ and } u + a \text{ are consecutive elements in } J \cap S(x^{\phi}, 2x^{\phi})\}$$

and

$$t_J(a) = |T_J(a)|.$$

Note that for $A > 0$,

$$\sum_{A<a\leq 2A} t(a) \leq r + \sum_{A<a\leq 2A} (t_{J_1}(a) + t_{J_2}(a) + \cdots + t_{J_r}(a)). \tag{20}$$

We will complete our estimation of $\sum_{a>R} t(a)$ in three steps. First, we will find an upper bound for $t_J(a)$. Second, we will show that for many values of a we actually have that $t_J(a) = 0$. And third, we will choose B and combine our estimates, using sums of the form given in (20), to obtain the results we want.

Fix a. We now establish that

$$t_J(a) \ll |J| a^{1/3} x^{(1-5\phi)/3} + 1. \tag{21}$$

It suffices to show that if $I \subseteq (x^\phi, 2x^\phi]$ and $|I| \leq a^{-1/3} x^{(5\phi-1)/3}/5$, then $|I \cap T(a)| \leq 2$. Fix such an interval I. Suppose there exist u and $u+b \in I \cap T(a)$ (otherwise, $|I \cap T(a)| \leq 1 \leq 2$). Let m_1, m_2, m_3, and m_4 be integers such that $m_1 u^2, m_2(u+a)^2, m_3(u+b)^2$, and $m_4(u+a+b)^2 \in (x, x+h]$. Define

$$T' = m_1(2u+a-2b) - m_2(2u+a-2b) - m_3(2u+a+4b) + m_4(2u+a+4b).$$

Note that T' is an integer. Also, using (5), (6), and the corresponding expressions for m_3 and m_4, we may view T' as a second difference for x/u^2 modified by the presence of the polynomials $2u+a-2b$ and $2u+a+4b$. Now, we get that

$$\begin{aligned} T' &= (m_1 - m_2)(2u+a-2b) - (m_3 - m_4)(2u+a+4b) \qquad (22) \\ &= \left(\frac{x}{u^2} - \frac{x}{(u+a)^2}\right)(2u+a-2b) \\ &\quad - \left(\frac{x}{(u+b)^2} - \frac{x}{(u+a+b)^2}\right)(2u+a+4b) + O\left(cx^{\theta-\phi}\right) \\ &= \frac{a(2u+a)x}{u^2(u+a)^2}(2u+a-2b) \\ &\quad - \frac{a(2u+a+2b)x}{(u+b)^2(u+a+b)^2}(2u+a+4b) + O\left(cx^{\theta-\phi}\right). \end{aligned}$$

When simplifying this last quantity, we are led to the expression

$$(2u+a)(2u+a-2b)(u+b)^2(u+a+b)^2-(2u+a+2b)(2u+a+4b)u^2(u+a)^2.$$

A direct computation shows that we may expand this expression to a sum of 82 terms of the form $\pm a^i b^j u^k$ where $i+j+k=6$ and $k \le 3$. Since u and $u+b \in T(a)$, we get that $a \le b < u$. Hence,

$$|(2u+a)(2u+a-2b)(u+b)^2(u+a+b)^2-(2u+a+2b)(2u+a+4b)u^2(u+a)^2|$$
$$\le 82b^3u^3.$$

Now, using that $u, u+a, u+b$, and $u+a+b \in (x^\phi, 2x^\phi]$, we get from (22) that

$$|T'| \le \frac{82ab^3u^3x}{u^2(u+a)^2(u+b)^2(u+a+b)^2} + O\left(cx^{\theta-\phi}\right)$$
$$\le 82ab^3x^{1-5\phi} + O\left(cx^{\theta-\phi}\right).$$

Since $\phi > \theta$ and x is sufficiently large, the term $O\left(cx^{\theta-\phi}\right)$ is $< 1/4$. Since u and $u+b \in I$ and $|I| \le a^{-1/3}x^{(5\phi-1)/3}/5$, we get that $b \le a^{-1/3}x^{(5\phi-1)/3}/5$. Hence, $|T'| \le (82/5^3)+(1/4) < 1$. Since T' is an integer, we now get that $T' = 0$.

Now, fix u and $u+b'$ as the minimal elements in $I \cap T(a)$, and assume that there is a $b > b'$ such that $u+b \in I \cap T(a)$. Let m'_3 and m'_4 be integers such that $m'_3(u+b')^2$ and $m'_4(u+a+b')^2 \in (x, x+h]$, and let m_1, m_2, m_3, and m_4 be as before. Then we get from the above that

$$m_1(2u+a-2b')-m_2(2u+a-2b')-m'_3(2u+a+4b')+m'_4(2u+a+4b')=0$$

and

$$m_1(2u+a-2b)-m_2(2u+a-2b)-m_3(2u+a+4b)+m_4(2u+a+4b)=0.$$

Also, we get by considering the elements $u+b'$ and $u+b'+(b-b')$ in $I \cap T(a)$ that

$$m'_3\left(2(u+b')+a-2(b-b')\right)-m'_4\left(2(u+b')+a-2(b-b')\right)$$
$$-m_3\left(2(u+b')+a+4(b-b')\right)+m_4\left(2(u+b')+a+4(b-b')\right)$$
$$= m'_3(2u+a-2b+4b')-m'_4(2u+a-2b+4b')$$
$$-m_3(2u+a+4b-2b')+m_4(2u+a+4b-2b')=0.$$

A simple computation produces from the first two of these three equations that

$$m'_3(2u+a-2b)(2u+a+4b') - m'_4(2u+a-2b)(2u+a+4b')$$

$$-m_3(2u+a-2b')(2u+a+4b) + m_4(2u+a-2b')(2u+a+4b) = 0.$$

And now using the third equation to eliminate the variables m'_3 and m'_4, we obtain that

$$(m_3 - m_4)\big((2u+a-2b')(2u+a+4b)(2u+a-2b+4b')$$

$$-(2u+a-2b)(2u+a+4b')(2u+a+4b-2b')\big) = 0,$$

which simplifies to

$$48(m_3 - m_4)bb'(b-b') = 0. \tag{23}$$

Now, this last equation is possible only if $m_3 - m_4 = 0$. On the other hand,

$$m_3 - m_4 = \frac{x}{(u+b)^2} - \frac{x}{(u+a+b)^2} + O\left(cx^{\theta-2\phi}\right)$$

$$= \frac{a(2u+a+2b)x}{(u+b)^2(u+a+b)^2} + O\left(cx^{\theta-2\phi}\right).$$

The main term in this last expression is $\gg x^{1-3\phi}$. Since we are only interested in θ satisfying $\theta \le 1/4$ and we are only interested in ϕ satisfying $x^\phi < 2\sqrt{x}$, we easily get that $1-3\phi > \theta - 2\phi$. Thus, the main term above dominates the error term, and we get that $m_3 - m_4 \ne 0$. This contradicts (23) and, therefore, establishes that $|I \cap T(a)| \le 2$ and that (21) holds.

Fix $A \ge R$. Let c_1 be a large positive constant to be specified momentarily. Take $r = [A/(c_1 c)] + 1$ where $[\,]$ denotes the greatest integer function. Thus, we can choose each $J \in \{J_1, \ldots, J_r\}$ with $|J| \le c_1 c A^{-1} x^\phi$. We will prove that the number of $a \in (A, 2A]$ for which $t_J(a) \ge 1$ is $\ll cA^3 x^{(1-4\phi)/3}$. In other words, for certain values of A, $t_J(a) = 0$ for many $a \in (A, 2A]$.

Suppose that $u \in T_J(a)$ and $u' \in T_J(a+a')$ where a and $a+a' \in (A, 2A]$. In particular, $A < a \le 2A$ and $1 \le a' < A$. By (8), we get that there are integers m and m' such that

$$\frac{a^3 x}{u^2(u+a)^2} = m + O\left(cx^{\theta-\phi}\right)$$

and

$$\frac{(a+a')^3 x}{u'^2(u'+a+a')^2} = m' + O\left(cx^{\theta-\phi}\right).$$

By subtracting the first of these two equations from the second, we are led to the identity (cf. [2] for further details)

$$\frac{(3a'a^2+3a'^2a+a'^3)x}{u^4}=m'-m+O\left(A^3|J|x^{1-5\phi}\right)+O\left(cx^{\theta-\phi}\right), \quad (24)$$

where the implied constants are absolute and, in particular, do not depend on c_1. Denote the left-hand side of (24) by M, and note that since $x^\phi < u \le 2x^\phi$ and $a' < A < a \le 2A$, we get that $3/16 \le M/(a'A^2x^{1-4\phi}) \le 19$.

We now consider $A \ge R \ge x^{(4\phi-1)/3}/8$. We also restrict our attention to

$$\frac{1}{4}<\phi\le 1-3\theta. \quad (25)$$

In particular, since x is sufficiently large, $A/(c_1c) \ge 1$ so that $r \le 2A/(c_1c)$. Note that since $\phi \le 1-3\theta$, we get that $(4\phi-1)/3 \ge (3\phi+\theta-1)/2$. Hence, for any constant $c_2 > 64$, either $a' < c_2c_1c$ or $a' \ge c_2c_1c > c_1cA^{-2}x^{3\phi+\theta-1}$. We prove that in fact if c_1 and c_2 are sufficiently large, then either

$$a' < c_2c_1c \quad \text{or} \quad a' > A^{-2}x^{4\phi-1}/60. \quad (26)$$

Assume otherwise so that $c_2c_1c \le a' \le A^{-2}x^{4\phi-1}/60$. Then, as we have just shown, $a' > c_1cA^{-2}x^{3\phi+\theta-1}$ so that

$$a'A^2x^{1-4\phi} > c_1cx^{\theta-\phi}.$$

Also, $|J| \le c_1cA^{-1}x^\phi$ implies that

$$a'A^2x^{1-4\phi} \ge c_2c_1cA^2x^{1-4\phi} \ge c_2A^3|J|x^{1-5\phi}.$$

Finally, note that

$$a'A^2x^{1-4\phi} \le \frac{1}{60}.$$

Now, $3/16 \le M/(a'A^2x^{1-4\phi}) \le 19$ implies that $M < 1/3$. Also, we get that if c_1 and c_2 are sufficiently large, then by the above the error terms in (24) have absolute value $< M/3$ implying that $0 < M/3 < m'-m < 5M/3 < 5/9$, contradicting that $m'-m$ is an integer. Hence, fixing c_1 and c_2 sufficiently large, we get that (26) holds.

We now get that in subintervals of $(A, 2A]$ of length $\le A^{-2}x^{4\phi-1}/60$ there are $< c_2c_1c+1$ choices of a for which there is a $u \in T_J(a)$. Hence, $t_J(a) \ge 1$ for

$$\ll \left(A^3x^{1-4\phi}+1\right)\left(c_2c_1c+1\right) \ll cA^3x^{1-4\phi}$$

choices of $a \in (A, 2A]$. This implies that for $A \geq R$ and ϕ satisfying (25), we get now from (20) and (21) that

$$\sum_{A<a\leq 2A} t(a) \ll r + \sum_{A<a\leq 2A} \left(\sum_{k=1}^{r} t_{J_k}(a)\right) \tag{27}$$

$$\ll r + r\left(\big(\max_{1\leq k\leq r}\{|J_k|\}A^{1/3}x^{(1-5\phi)/3} + 1\big)(cA^3x^{1-4\phi})\right)$$

$$\ll r + r\left(\big(cA^{-1}x^{\phi}A^{1/3}x^{(1-5\phi)/3} + 1\big)(cA^3x^{1-4\phi})\right)$$

$$\ll \frac{1}{c}A + (\frac{1}{c}A)\left(cA^3x^{1-4\phi} + c^2A^{7/3}x^{(4-14\phi)/3}\right)$$

$$\ll A + A^4x^{1-4\phi} + cA^{10/3}x^{(4-14\phi)/3}.$$

Now, fix

$$B = c^{-3/10}x^{(17\phi-4)/13}.$$

By considering subintervals $(A, 2A]$ of $(R, B]$ which are pairwise disjoint with the exception of possibly one pair, we then get from (27) by summing that

$$\sum_{R<a\leq B} t(a) \ll x^{(17\phi-4)/13} + x^{(16\phi-3)/13} + x^{(4-4\phi)/13} \tag{28}$$

provided (25) holds. Although (28) will be sufficient for our main results, we will be able to improve on our estimates for $|S(x^{\phi}, 2x^{\phi})|$ for certain values of ϕ by replacing $B = c^{-3/10}x^{(17\phi-4)/13}$ above with $B = c^{-3/10}x^{(5\phi-1)/5}$. From (27), we get that

$$\sum_{R<a\leq B} t(a) \ll x^{(5\phi-1)/5} + x^{1/5}\log x + x^{(2-4\phi)/3} \tag{29}$$

provided (25) holds. Note that from (19) and (28), we get that since $\phi < 1$

$$\sum_{a>R} t(a) \ll x^{(16\phi-3)/13} + c^{3/10}x^{(4-4\phi)/13}.$$

Now, using (18) and recalling (25), we deduce that

$$|S(x^{\phi}, 2x^{\phi})| \ll x^{(16\phi-3)/13} + c^{3/10}x^{(4-4\phi)/13} \quad \text{for} \quad \frac{1}{4} < \phi \leq 1 - 3\theta. \tag{30}$$

For $\theta = 8/37$, we get from Lemma 1 that

$$|S(x^{11/37}, x^{13/37})| \ll c^{3/10} x^{8/37}.$$

Also, from (12), we get that

$$|S(x^{13/37}, 2\sqrt{x})| \ll x^{(1-\frac{13}{37})/3} = x^{8/37},$$

and from (16), we get that

$$|S(x^{\theta}\sqrt{\log x}, x^{11/37})| \ll x^{(1+\frac{11}{37})/6} = x^{8/37}.$$

Thus, for c a sufficiently large constant, the above combine to complete an elementary proof that for x sufficiently large, the interval $(x, x + cx^{8/37}]$ contains a squarefree number.

Before proceeding, we comment on how one can make use of (29) instead of (28) above. Using (29) and the above argument gives that

$$|S(x^{\phi}, 2x^{\phi})| \ll x^{(5\phi-1)/5} + x^{1/5}\log x + x^{(2-4\phi)/3} \quad \text{for} \quad \frac{1}{4} < \phi \le 1 - 3\theta.$$

In particular, this implies from Lemma 1 that whenever $\theta > 1/5$

$$\begin{aligned} |S(x^{7/20}, x^{1-3\theta})| &\ll x^{(5(1-3\theta)-1)/5} + x^{1/5}\log x \\ &\ll x^{(4-15\theta)/5} + x^{1/5}\log x \ll x^{\theta}. \end{aligned}$$

Also, from (12), we get that in general

$$|S(x^{1-3\theta}, 2\sqrt{x})| \ll x^{\theta}.$$

Thus, we get that

$$|S(x^{7/20}, 2\sqrt{x})| \ll x^{\theta} \quad \text{for every } \theta > 1/5. \tag{31}$$

In particular, to establish a gap result for $\theta > 1/5$, it suffices to estimate $|S(x^{\phi}, 2x^{\phi})|$ for $\phi < 7/20$.

6. The Use of Exponential Sums

We continue to follow the work of the first author in [2] making appropriate modifications to take advantage of the work of the second author in [12]. The strategy in this section is essentially the same as in the previous section. We will again make use of (18), (19), and (21). For $A \geq R$ and $1/4 < \phi \leq 1 - 3\theta$, we showed that $t_J(a) \geq 1$ for $\ll cA^3x^{1-4\phi}$ choices of $a \in (A, 2A]$. In this section, we will improve on that estimate for θ satisfying $1/5 < \theta \leq 2/9$ and for ϕ satisfying $6/23 < \phi \leq 6/19$. We will do this by making use of two further lemmas. The first lemma is fairly simple and can be found in [2]. The second lemma is well known (cf. [11]) and corresponds to the exponent pair (1/2,1/2). Their proofs are omitted here.

Lemma 2. *Let $E = [0, \gamma] \cup [1 - \gamma, 1)$ where $0 < \gamma < 1/2$. Let $f : \Re \mapsto \Re$ be any function. Let S be a set of positive integers. Then for any positive integer $K \leq 1/(4\gamma)$, we get that*

$$|\{s \in S : \{f(s)\} \in E\}| \leq \frac{\pi^2}{2(K+1)} \sum_{1 \leq j \leq K} \left| \sum_{s \in S} e(jf(s)) \right| + \frac{\pi^2}{4(K+1)} \sum_{s \in S} 1,$$

where $\{f(s)\}$ denotes the fractional part of $f(s)$ and $e(jf(s)) = e^{2\pi i j f(s)}$.

Lemma 3. *For fixed positive real numbers A, ϕ, j, w, and x, with $A \geq 1$ and $w \in (x^\phi, 2x^\phi]$, one has that*

$$\sum_{A < a \leq 2A} e\left(j\frac{a^3x}{w^4}\right) \ll A(jAx^{1-4\phi})^{1/2} + (jAx^{1-4\phi})^{-1/2}.$$

Fix $A \geq R$, and let ϕ be such that $6/23 < \phi \leq 6/19$. From (8) we get that if u and $u + a \in S(x^\phi, 2x^\phi)$, then there is an integer m such that

$$\frac{a^3x}{u^2(u+a)^2} = m + O\left(cx^{\theta-\phi}\right). \tag{32}$$

Take $r = \left[A^3x^{1-3\phi-\theta}/c\right] + 1$ so that we can choose each $J \in \{J_1, ..., J_r\}$ with

$$|J| \leq cA^{-3}x^{4\phi+\theta-1}.$$

Fix $w \in J$, and suppose now that $a \in (A, 2A]$ and $u \in T_J(a)$. It is easy to deduce from (32) (cf. [2]) that

$$\begin{aligned} \frac{a^3x}{w^4} &= m + O\left(cx^{\theta-\phi}\right) + O\left(|J|A^3x^{1-5\phi}\right) \\ &= m + O\left(cx^{\theta-\phi}\right). \end{aligned}$$

It now follows that there is a constant c_3 such that

$$\left\{\frac{a^3x}{w^4}\right\} \in [0,\gamma] \cup [1-\gamma, 1),$$

where $\gamma = c_3 c x^{\theta-\phi}$. The above holds for each $a \in (A, 2A]$ with $t_J(a) \geq 1$. Thus, we are in a position now to use the above lemmas to estimate the number of $a \in (A, 2A]$ for which $t_J(a) \geq 1$. Indeed, if we denote this quantity by $Q(A)$, we get that for any positive intger $K \leq x^{\phi-\theta}/(4c_3c)$,

$$Q(A) \leq \frac{\pi^2}{2(K+1)} \sum_{1\leq j\leq K} \left| \sum_{A<a\leq 2A} e\left(j\frac{a^3x}{w^4}\right)\right| + \frac{\pi^2}{4(K+1)} \sum_{A<a\leq 2A} 1 \tag{33}$$

$$\ll \frac{1}{K} \sum_{1\leq j\leq K} \left(A\left(jAx^{1-4\phi}\right)^{1/2} + \left(jAx^{1-4\phi}\right)^{-1/2}\right) + K^{-1}A$$

$$\ll \frac{1}{K} \sum_{1\leq j\leq K} \left(A^{3/2}x^{(1-4\phi)/2}j^{1/2} + A^{-1/2}x^{(4\phi-1)/2}j^{-1/2}\right) + K^{-1}A$$

$$\ll A^{3/2}x^{(1-4\phi)/2}K^{1/2} + A^{-1/2}x^{(4\phi-1)/2}K^{-1/2} + K^{-1}A.$$

Our choice for K will depend on A. Let

$$B = x^{(9\phi-2)/8},$$

and set

$$B' = \max\{R, x^{(8\phi-2)/5}\}.$$

We take

$$K = \begin{cases} [A^{-2}x^{4\phi-1}] + 1 & \text{for} \quad R \leq A < B' \\ [A^{-1/3}x^{(4\phi-1)/3}] + 1 & \text{for} \quad B' \leq A \leq B. \end{cases}$$

We do not concern ourselves with the case that $B' \geq x^{(9\phi-2)/8}$ since then $\sum_{B'<a\leq B} t(a)$ is vacuously 0 and our final estimates for this sum will easily hold. Straight forward calculations verify that for $1/5 < \theta \leq 2/9$ and $\theta < \phi \leq 6/19$, $K \leq x^{\phi-\theta}/(4c_3c)$ so that (33) holds. Note that if there is an $A \in [R, B')$, then $B' = x^{(8\phi-2)/5}$. Thus, for $R \leq A < B'$, we get that since $\phi > 6/23$,

$$A^{-2}x^{4\phi-1} > x^{(4\phi-1)/5} \geq 1.$$

Also, for $A \leq B$, we get that since $\phi > 6/23$,

$$A^{-1/3}x^{(4\phi-1)/3} \geq x^{(23\phi-6)/24} \geq 1.$$

Hence, for $R \le A < B'$,

$$A^{-2}x^{4\phi-1} < K \le 2A^{-2}x^{4\phi-1},$$

and for $B' \le A \le B$,

$$A^{-1/3}x^{(4\phi-1)/3} < K \le 2A^{-1/3}x^{(4\phi-1)/3}.$$

Thus, from (33), we get that

$$Q(A) \ll A^{1/2} + A^3x^{1-4\phi} \ll A^{1/2} \quad \text{for } R \le A < B',$$

and

$$Q(A) \ll A^{4/3}x^{(1-4\phi)/3} + A^{-1/3}x^{(4\phi-1)/3} \ll A^{4/3}x^{(1-4\phi)/3} \qquad \text{for } B' \le A \le B.$$

Now, we recall that $r = [A^3x^{1-3\phi-\theta}/c] + 1$ and $\max_{1\le k\le r}\{|J_k|\} \le cA^{-3}x^{4\phi+\theta-1}$. We may consider only $c \ge 1$. Now, since $R \ge x^{(4\phi-1)/3}/8$,

$$A^3x^{1-3\phi-\theta} \ge R^3x^{1-3\phi-\theta} \ge x^{\phi-\theta}/8^3 \ge 1$$

so that $r \le 2A^3x^{1-3\phi-\theta}$. Using this estimate for r with $R \le A < B'$, we get as in (27) that

$$\sum_{A<a\le 2A} t(a) \ll r + r\left(\max_{1\le k\le r}\{|J_k|\}A^{1/3}x^{(1-5\phi)/3} + 1\right)Q(A)$$
$$\ll A^{7/2}x^{1-3\phi-\theta} + A^{5/6}x^{(1-2\phi)/3}.$$

And with $B' \le A \le B$, we get that

$$\sum_{A<a\le 2A} t(a) \ll A^3x^{1-3\phi-\theta} + A^{13/3}x^{(4-13\phi-3\theta)/3} + A^{5/3}x^{(2-6\phi)/3}.$$

We now restrict our attention to $\theta \ge 4/19$ and recall that $\phi \le 6/19$. By dividing the intervals $(R, B']$ and $(B', B]$ into intervals of the form $(A, 2A]$ and summing, we now obtain that

$$\sum_{R<a\le B'} t(a) \ll B'^{7/2}x^{1-3\phi-\theta} + B'^{5/6}x^{(1-2\phi)/3}$$
$$\ll \left(x^{(8\phi-2)/5}\right)^{7/2}x^{1-3\phi-\theta} + \left(x^{(8\phi-2)/5}\right)^{5/6}x^{(1-2\phi)/3}$$
$$\ll x^{(13\phi-5\theta-2)/5} + x^{2\phi/3} \ll x^{2\phi/3},$$

and

$$\sum_{B'<a\leq B} t(a) \ll B^3 x^{1-3\phi-\theta} + B^{13/3} x^{(4-13\phi-3\theta)/3} + B^{5/3} x^{(2-6\phi)/3}$$

$$\ll x^{(3\phi-8\theta+2)/8} + x^{(13\phi-24\theta+6)/24} + x^{(2-\phi)/8}$$

$$\ll x^{(2-\phi)/8}.$$

The above now imply that

$$\sum_{R<a\leq B} t(a) \ll x^{(2-\phi)/8}.$$

On the other hand, we get from (19) that

$$\sum_{a>B} t(a) \leq x^{(2-\phi)/8}.$$

Assuming that $\theta \geq 4/19$ and combining the above with (18) now gives that

$$|S(x^\phi, 2x^\phi)| \ll x^{(2-\phi)/8} \quad \text{for} \quad 6/23 < \phi \leq 6/19. \tag{34}$$

We take $\theta = 3/14 > 4/19$. Since $2/7 > 6/23$, we now obtain from Lemma 1 and (34) that

$$|S(x^{2/7}, x^{6/19})| \ll x^{3/14}.$$

Using Lemma 1 with (16) and (30), we have that

$$|S(x^\theta \sqrt{\log x}, x^{2/7})| \ll x^{3/14}$$

and

$$|S(x^{6/19}, x^{7/20})| \ll c^{3/10} x^{4/19} \ll x^{3/14},$$

respectively. Combining these estimates with (31) now implies the theorem stated in the introduction.

REFERENCES

[1] M. Filaseta, An elementary approach to short intervals results for k-free numbers, J. Number Theory **30** (1988), 208–225.

[2] M. Filaseta, Short interval results for squarefree numbers, J. Number Theory (to appear).

[3] E. Fogels, On the average values of arithmetic functions, Proc. Cambridge Philos. Soc. **37** (1941), 358–372.

[4] S. W. Graham and G. Kolesnik, On the difference between consecutive squarefree integers, Acta Arith. **49** (1988), 435–447.

[5] H. Halberstam and K. F. Roth, On the gaps between consecutive k-free integers, J. London Math. Soc. (2) **26** (1951), 268–273.

[6] M. Nair, Power free values of polynomials II, Proc. London Math. Soc. (3) **38** (1979), 353–368.

[7] R. A. Rankin, Van der Corput's method and the theory of exponent pairs, Quart. J. Math. Oxford Ser. (2) **6** (1955), 147–153.

[8] H. E. Richert, On the difference between consecutive squarefree numbers, J. London Math. Soc. (2) **29** (1954), 16–20.

[9] K. F. Roth, On the gaps between squarefree numbers, J. London Math. Soc. (2) **26** (1951), 263–268.

[10] P. G. Schmidt, *Abschätzungen bei unsymmetrischen Gitterpunktproblemen*, Dissertation zur Erlangung des Doktorgrades der Mathematisch-Naturwissenschaftlichen Fakultät der Georg–August–Universität zu Göttingen, 1964.

[11] E. C. Titchmarsh, *The Theory of the Riemann Zeta–Function*, Second edition, revised by D. R. Heath–Brown, Oxford Univ. Press, Oxford, 1986, p. 104.

[12] O. Trifonov, On the squarefree problem, C. R. Acad. Bulgare Sci. **41** (1988), 37–40.

[13] O. Trifonov, On the squarefree problem II, Mathematica Balcanika (to appear).

Michael Filaseta
Mathematics Department
University of South Carolina
Columbia, SC 29208

Ognian Trifonov
Institute of Mathematics
Bulgarian Academy of Sciences
1090 Sofia
Bulgaria

Some Arithmetical Semigroups

A. S. FRAENKEL, H. PORTA, AND K. B. STOLARSKY

Dedicated to Paul T. Bateman on the occasion of his retirement

1. Introduction

The peculiar multiplication rule

$$n \times m = (3/2)nm$$

is associative but does not always produce an integer for integers n and m. If we try to force the result to be an integer by truncation, *i. e.*, by

$$n \times m = \lfloor (3/2)nm \rfloor$$

we no longer have an associative multiplication. For example $(3 \times 5) \times 7 = 231$, while $3 \times (5 \times 7) = 234$. It would seem exceptional to find associativity in operations whose definition involves truncation. Our object is to study several such exceptional operations.

Our investigation involves properties of sequences of the form $\lfloor n\alpha \rfloor$ where α is real and $n = 1, 2, 3, \ldots$ These are known as Beatty sequences. They arise, for example, in the study of continued fractions, ergodic theory, Wythoff's game, and quasicrystallography. An extensive bibliography on them can be put together from **[10]**, **[15]**, **[17]**, and the references therein; see also **[1]**, **[4]** (Volume 1, pp. 62, 76-77), **[5]**, **[6]**, **[7]**, **[8]**, **[11]**, **[12]**, **[13]**, and **[14]**. It is in this framework that we place the present paper. In section 5 we announce a result stemming from this study that relates to a paper of P. T. Bateman and A. L. Duquette (**[2]**).

Research by A. S. Fraenkel partially supported by BSF Grant 85-00368, Israel
Research by H. Porta partially supported by CONICET, Argentina
Research by K. B. Stolarsky partially supported by ONR Grant N00014-85-K-0368

2. Wythoff pairs and the KLM formula

Let $\phi = (1+\sqrt{5})/2$ denote the golden mean and for any real number x denote by $\lfloor x \rfloor$ its integer part and $\{x\}$ its fractional part. Thus $x = \lfloor x \rfloor + \{x\}$, $\lfloor x \rfloor$ is an integer, and $0 \le \{x\} < 1$. For n an integer, set $a(n) = \lfloor n\phi \rfloor$, $b(n) = \lfloor n\phi^2 \rfloor (= a(n)+n)$ and $c(n) = \lfloor n\phi^{-1} \rfloor (= a(n)-n)$. The vectors $W(n) = (a(n), b(n))$ are called *Wythoff pairs* ([8], [14]). The sets $A = \{a(n); n \ge 1\}$ and $B = \{b(n); n \ge 1\}$ form a partition of the set Z^+ of positive integers by Beatty's "$\alpha^{-1} + \beta^{-1} = 1$" theorem (see [3]), since $\phi^{-1} + \phi^{-2} = 1$. There is an equivalent way to see this property: a positive integer m is in A or in B according to whether $\{m\phi\} > \phi^{-2}$ or $\{m\phi\} < \phi^{-2}$ (notice that $\{m\phi\} \ne \phi^{-2}$ for $m \ge 1$). Carlitz et al. (see [5]) have proved that all words $d_1 d_2 \ldots d_k(n)$, where $d_i = a$ or $d_i = b$ for each $i = 1, 2, \ldots k$, can be calculated as maps $Z^+ \to Z^+$ in "linearized" form: $d_1 d_2 \ldots d_k(n) = Ka(n) + Ln + M$ for appropriate integers of K, L, M. Here we use "word" to indicate a finite composition of a's and b's, in any order, as a map form Z to Z. Not all choices of K, L, M represent words: $K = 1, L = 0, M = 1$ fails to give a word. The linearized forms for arbitrary integers K, L, M are not always closed under composition, but the following result is of considerable help in calculating such compositions.

Theorem 1 (the KLM formula). *For any integers K, L, M, n we have*

$$a(Ka(n) + Ln + M) = Kb(n) + La(n) + \lfloor M\phi + (L\phi - K)\{n\phi\}/\phi \rfloor$$

Proof: Denote the right hand side of the formula by R. Then

$$\begin{aligned} R &= Ka(n) + La(n) + Kn + \lfloor M\phi + (L\phi - K)(n\phi - a(n))/\phi \rfloor \\ &= Ka(n) + \lfloor M\phi + Ln\phi + Ka(n)(\phi - 1) \rfloor \\ &= \lfloor (M + Ln + Ka(n))\phi \rfloor = a(Ka(n) + Ln + M). \end{aligned}$$

3. Operations on integers

We introduce two maps: $w : Z \to Z[\phi]$, and $z : Z \to Z[\phi]$ as follows:

$$w(n) = b(n) - a(n)\phi$$

$$z(n) = -a(n) + n\phi.$$

We have $z(n) = \{n\phi\}$ by definition of $a(n)$ and $w(n) = b(n) - a(n)\phi = \{n\phi\}/\phi$. The irrationality of ϕ implies that both w and z are one to one.

Theorem 2. *The operation on integers* $n \dagger m = nm - na(m) - a(n)m$ *satisfies* $z(n \dagger m) = z(n)z(m)$. *In particular* $n \dagger m$ *is associative.*

Proof: $z(n)z(m) = (n\phi - a(n))(m\phi - a(m)) = k + l\phi$ where $k = nm + a(n)a(m)$ and $l = n \dagger m$. Since $k + l\phi = z(n)z(m)$ satisfies $0 \le k + l\phi < 1$ we conclude that $z(n)z(m) = k + l\phi = \{k + l\phi\} = \{l\phi\} = z(l)$.

Corollary 1. *The set of real numbers* $\{n\phi\}$, $n \in Z$ *is closed under ordinary multiplication.*

Proof: $z(n \dagger m) = z(n)z(m)$ means $\{(n \dagger m)\phi\} = \{n\phi\}\{m\phi\}$.

Corollary 2. $a(n \dagger m) = -nm - a(n)a(m)$.

Proof: $0 = \lfloor z(l) \rfloor = \lfloor k + l\phi \rfloor = k + \lfloor l\phi \rfloor = nm + a(n)a(m) + a(l)$.

We can interpret Corollary 1 as saying that the equation $\{p\phi\} = \{n\phi\}\{m\phi\}$ in the integer p has the solution $p = n \dagger m$. Here is a related equation:

$$\frac{\{p\phi\}}{\phi} = \frac{\{n\phi\}}{\phi} \cdot \frac{\{m\phi\}}{\phi}.$$

The solution of this equation is $p = n \dagger 1 \dagger m$. In fact, this follows from $z(n \dagger 1 \dagger m) = z(n)z(1)z(m) = z(n)\{\phi\}z(m)$ and from $\phi^{-1} = \phi - 1 = \{\phi\}$. More generally:

Theorem 3. *For any* $s = 1, 2, \ldots$ *the Diophantine equation*

$$\frac{\{p\phi\}}{\phi^s} = \frac{\{n\phi\}}{\phi^s} \cdot \frac{\{m\phi\}}{\phi^s}$$

has the solution

$$p = n \dagger ((-1)^{s+1} F_s) \dagger m$$

where F_s *is the* s^{th} *Fibonacci number defined by* $F_1 = F_2 = 1$, *and* $F_n = F_{n-1} + F_{n-2}$ *for* $n \ge 3$.

Proof: Since $\phi^{-1} = \{\phi\}$, the formula amounts to

$$\{p\phi\} = \{n\phi\}\{m\phi\}\{\phi\}^s.$$

The result follows from the identity $z((-1)^{s+1}F_s) = \{\phi\}^s$.

This suggests that we define an operation

$$n \odot_s m = n \dagger ((-1)^{s+1} F_s) \dagger m$$

for each $s \in Z^+$. This operation will necessarily be associative. In fact it is possible to say a little more:

Theorem 4. *The operations* $n \odot_s m$ *are associative and the maps* $z_s(n) = \{n\phi\}\phi^{-s}$ *satisfy* $z_s(n \odot_s m) = z_s(n) z_s(m)$. *Further, the operation* $n \odot_s m$ *has the following explicit formula*

$$n \odot_s m = (-1)^{s+1}\big(F_{s-2}nm + F_{s-1}na(m) + F_{s-1}a(n)m + F_s a(n)a(m)\big).$$

Proof: Only the explicit formula needs proof. We use the identity

$$a((-1)^{l+1}F_l) = (-1)^{l+1}F_{l+1}.$$

Then

$$\begin{aligned} k \dagger ((-1)^{l+1}F_l) &= (-1)^{l+1}kF_l - k(-1)^{l+1}F_{l+1} - a(k)((-1)^{l+1}F_l) \\ &= (-1)^{l+1}(k(F_l - F - l - 1) - a(k)F_l) \\ &= (-1)^l(kF_{l-1} + a(k)F_l). \end{aligned}$$

Now, using $a(n \dagger m) = -nm - a(n)a(m)$ (Corollary 2):

$$\begin{aligned} n \odot_s m &= n \dagger ((-1)^{s+1}F_s) \dagger m \\ &= (-1)^l(n \dagger mF_{s-1} + a(n \dagger m)F_s) \\ &= (-1)^l((nm - a(n)m - na(m))F_{s-1} + (-nm - a(n)a(m))F_s) \\ &= (-1)^l(nmF_{s-2} + a(n)mF_{s-1} + na(m)F_{s-1} + a(n)a(m))F_s) \end{aligned}$$

as claimed.

The special case of $n \odot_{-1} m$ is denoted by $n \star m$ in [12], and has the formula $n \odot_{-1} m = n \star m = nm + a(n)a(m)$.

We present now a slightly more general case. Suppose g_k, $k = 1, 2, \ldots$ is a sequence of integers satisfying $g_{k+2} = g_k + g_{k+1}$. This means that

$$\begin{aligned} g_k - (g_1 - g_0\phi^{-1})\phi^k + (g_0\phi - g_1)\phi^{-k} \\ = g_0(\phi^{-k+1} - \phi^{k-1}) + g_1(\phi^k - \phi^{-k}). \end{aligned} \qquad (i)$$

Define next

$$A_m^k = g_k m + g_{k+1}a(m),$$
$$\xi_k = g_{k+1} - g_k\phi.$$

Notice that $-\xi_0$ is the coefficient of ϕ^{-k} in (i).

Lemma 1.

$$\xi_j = \left(\frac{-1}{\phi}\right)^{j-i}\xi_i \qquad (ii)$$

$$A_m^k\phi - A_m^{k+1} = \xi_{k+1}\{m\phi\}. \qquad (iii)$$

Proof:

$$\begin{aligned}\xi_n\phi &= g_{n+1}\phi - g_n\phi - g_n\\ &= (g_{n+1} - g_n)\phi - g_n\\ &= -(g_n - g_{n-1}\phi) = -\xi_{n-1}\end{aligned}$$

and therefore by induction $\xi_n\phi^l = (-1)^l\xi_{n-l}$, which is equivalent to (ii). Also:

$$\begin{aligned}A_m^k\phi - A_m^{k+1} &= (g_k m + g_{k+1}a(m))\phi - (g_{k+1}m + g_{k+2}a(m))\\ &= (g_k\phi - g_{k+1})m + (g_{k+1}\phi - g_{k+2})a(m)\\ &= -\xi_k m - \xi_{k+1}a(m)\\ &= \xi_{k+1}(m\phi - a(m)) = \xi_{k+1}\{m\phi\},\end{aligned}$$

and this proves (iii).

Remark: Taking $g_0 = 0$, $g_1 = 1$, $g_2 = 1$, and $k = 0$ in formula (iii) we get an alternate proof of $b(m) - a(m)\phi = \{m\phi\}/\phi$.

Proposition 1. *If* $0 \le \xi_{k+1} < \phi$, *then*

$$a(A_m^k a(n) + A_m^k n) = A_m^{k+2}a(n) + A_m^{k+1}n \tag{iv}$$

Proof: We have by the KLM formula:

$$a(A_m^{k+1}a(n) + A_m^k n) = A_m^{k+1}b(n) + A_m^k a(n) + l$$

where

$$l = \lfloor (A_m^k\phi - A_m^{k+1})\{n\phi\}/\phi \rfloor.$$

By formula (iii),

$$l = \lfloor \xi_{k+1}\{m\phi\}\{n\phi\}/\phi \rfloor$$

and therefore $l = 0$. Also

$$\begin{aligned}A_m^{k+1}b(n) + A_m^k a(n) &= (A_m^{k+1} + A_m^k)a(n) + A_m^{k+1}n\\ &= A_m^{k+2}a(n) + A_m^{k+1}n,\end{aligned}$$

and (iv) follows.

Theorem 5. *If* $0 \le \xi_{k+1} < 1$, *then the operation*

$$m \times_k n = g_k mn + g_{k+1}(ma(n) + a(m)n) + g_{k+2}a(m)a(n)$$

is associative and

$$\begin{aligned}l \times_k m \times_k n = (g_k^2 + g_{k+1}^2)lmn\\ + (g_k g_{k+1} + g_{k+1}g_{k+2})(lma(n) + la(m)n + a(l)mn)\\ + (g_{k+1}^2 + g_{k+2}^2)(la(m)a(n) + a(l)ma(n) + a(l)a(m)n)\\ + (g_{k+1}g_{k+2} + g_{k+2}g_{k+3})a(l)a(m)a(n)\end{aligned} \tag{v}$$

Proof: Observe that

$$m \times_k n = A_m^{k+1} a(n) + A_m^k n$$

and therefore

$$l \times_k (m \times_k n) = A_l^{k+1} a(A_m^{k+1} a(n) + A_m^k n) + A_m^k (A_m^{k+1} a(n) + A_m^k n)$$

Apply now Lemma 1 to obtain

$$l \times_k (m \times_k n) = A_l^{k+1} A_m^{k+2} a(n) + A_l^{k+1} A_m^{k+1} n + A_l^k A_m^{k+1} a(n) + A_l^k A_m^k n$$

Expanding according to the definition of A_i^j we get that $l \times_k (m \times_k n)$ is equal to the right hand side in formula (v). This in turn proves that $\times_k$ is associative by symmetry.

An alternate way to write (v) is the following: for $\epsilon = 0, 1$ let us set $a^\epsilon(n) = n$ if $\epsilon = 0$ and $a^\epsilon(n) = a(n)$ if $\epsilon = 1$. Then:

$$l \times_k m \times_k n = \sum (g_k g_{k+\alpha+\beta+\gamma} + g_{k+1} g_{k+1+\alpha+\beta+\gamma}) a^\alpha(l) a^\beta(m) a^\gamma(n)$$

where α, β, γ range over all choices of 0, 1 (cf. **[11]**).

Example 1: Take for g_k the Fibonacci numbers, i.e., $g_1 = 1,\ g_2 = 1$. Hence $\xi_0 = 1$ and $\xi_{k+1} = (-1/\phi)^{-(k+1)}$. Therefore the condition in Proposition 1 is satisfied for $k = -1, 1, 3, 5, \ldots$ The corresponding associative operations are:

$$m \times_{-1} n = mn + a(m)a(n)$$
$$m \times_1 n = mn + ma(n) + a(m)n + 2a(m)a(n)$$
$$m \times_3 n = 2mn + 3ma(n) + 3a(m)n + 5a(m)a(n)$$

and so on.

Example 2: Take for g_k the Lucas numbers, i.e., $g_1 = 1,\ g_2 = 3$. Then $\xi_0 = 1 - 2\phi$, $\xi_{k+1} = (-1/\phi)^{-(k+1)}(1 - 2\phi)$ and the condition of Proposition 1 holds for $k = 0, 2, 4, \ldots$ The associated operations are

$$m \times_0 n = 2mn + ma(n) + a(m)n + 3a(m)a(n)$$
$$m \times_2 n = 3mn + 4ma(n) + 4a(m)n + 7a(m)a(n)$$
$$m \times_4 n = 7mn + 11ma(n) + 11a(m)n + 18a(m)a(n)$$

and so on. Writing the triple product as

$$l \times_k m \times_k n = P_k lmn + Q_k(lma(n) + \ldots) + R_k(la(m)a(n) + \ldots) + S_k a(l)a(m)a(n)$$

one can prove easily that $P_k = Q_k = R_k = S_k \pmod 5$. This congruence holds also for the numbers generated by $g_1 = 1,\ g_2 = 8$ but fails for the Fibonacci numbers and for all the choices $g_1 = 1,\ 4 \leq g_2 \leq 7$.

4. Operations on pairs of integers

We consider now operations on pairs of integers. By the interpretation of Wythoff pairs $W(n)$ as the real numbers $w(n)$ or $z(n)$ we have associative operations on the Wythoff pairs themselves. In particular the product induced by $\star$ coincides with the matrix product of the associated matrices

$$W(n) \leftrightarrow M(n) = \begin{pmatrix} b(n) & a(n) \\ a(n) & n \end{pmatrix}.$$

This can also be viewed as a two dimensional representation of $(Z,\star)$ defined by

$$\rho : n \rightarrow M(n) = \begin{pmatrix} b(n) & a(n) \\ a(n) & n \end{pmatrix}.$$

As a real representation ρ is totally reducible with invariant subspaces generated by the vectors

$$\begin{pmatrix} 1 \\ -\phi \end{pmatrix}, \quad \begin{pmatrix} \phi \\ 1 \end{pmatrix}.$$

In other words, ρ is the sum $z \oplus z'$ where z is the function defined above and $z'(n) = b(n)\phi + a(n)$. These functions can be alternatively characterized as the bounded and unbounded one–dimensional representations of $(Z,\star)$.

Similar operations can be defined on all pairs of integers.

Theorem 6. *Let*

$$t_{ij}(n) = (j - l)a(n) + (c(j) - c(l))n + j.$$

Then the operation $\otimes$ on $Z \times Z$ defined by

$$(j,l) \otimes (n,m) = (t_{jl}(n), t_{jl}(m)).$$

is noncommutative but associative.

The natural setting for this theorem involves the integers extended with an additional element θ. Details appear in **[14]**. In particular the following proposition can be proved with the help of the KLM formula.

Proposition 2.

$$(a(n),\theta) \otimes (a(m),\theta) = (a(n \star m),\theta)$$

$$(0,n) \otimes (0,m) = (0, n \dagger m)$$

Theorem 7. *Define $n \oplus m = -n \dagger m + n + m$. Then*

$$(j,\theta) \otimes (k,\theta) = (j \oplus k, \theta).$$

In particular $\oplus$ *is associative.*

The key to much of this is to identify the Wythoff pairs with 3×1 column vectors with components $a(n)$, $b(n)$, and 1, and to determine all affine transformations (3×3 matrices with final row $(0,0,1)$) that map the set of all these 3×1 vectors back into itself.

Note that

$$(j,j) \otimes (l,l) = (j,j)$$

so that multiplication of diagonal elements is not commutative. Also the operation

$$n \diamond m = a(n)a(m) + na(m) + a(n)m$$

is *not* associative; in fact $(1 \diamond 2) \diamond 3 \neq 1 \diamond (2 \diamond 3)$. We introduce it because of the following relations between $\star$, $\dagger$ and $\diamond$.

Proposition 3.

1) $a(n \dagger m) = -n \star m$

2) $a(n \star m) = n \diamond m$

3) $n \star m = n \diamond m + n \dagger m$.

It is also of interest to note the following identites, vaguely reminiscent of the scalar triple product of vector analysis, a product that also involves a nonassociative operation.

Proposition 4.

1) $(n \star m) \diamond p = n \diamond (m \star p)$

2) $(n \dagger m) \diamond p = n \diamond (m \dagger p)$.

The associative operations also satisfy a relation of this sort, namely:

Proposition 5.

$$(k \dagger m) \star n = k \star (m \dagger n) = k \dagger (m \star n) = -a(n \dagger m \dagger k).$$

All of the above are verified by straightforward calculations.

5. PV Analogues

For an introduction to PV theory see [16]. The function field analogue of PV theory was first developed by P. T. Bateman and A. L. Duquette (see [2]). The p-adic analogue goes back to Chabauty (see [6]); for more references see [9].

The PV theory is nicely motivated by the question "under what circumstances can $\{a_n\}$ tend (or tend rapidly) to 0 modulo 1 if

$$\begin{cases} a_1 = x \\ a_{n+1} = \lambda a_n, \end{cases}$$

where $\lambda > 1$?". We shall ask a sort of "naive p-adic question" consonant with the subject of the paper. If $p^{e(n)}$ is the highest power of the prime p dividing $\lfloor a_n \rfloor$, can $e(n) \to \infty$ as $n \to \infty$? In other words, can $\nu_p(\lfloor a_n \rfloor) \to 0$ where ν_p is the usual p-adic valuation? What if we replace the second recurrence by $a_{n+1} = \lfloor \lambda a_n \rfloor$ or by something similar with more terms and possibly other multipliers?

Theorem 8. *There is a prime p and real numbers α, b, c all greater than 1, with α badly approximable and b/p, c/p not integers, such that $\nu_p(a_n) \to 0$ where a_n is defined inductively by*

$$\begin{cases} a_1 = 1 \\ a_{n+1} = b a_n + c\lfloor \alpha a_n \rfloor. \end{cases}$$

The theorem is phrased so as to suggest various more general problems. we in fact can show that the highest power of 5 dividing a_n, where

$$\begin{cases} a_1 = 1 \\ a_{n+1} = 7a_n + 11\lfloor \phi a_n \rfloor, \end{cases}$$

and ϕ is the golden mean, tends weakly monotonically to ∞ with n. For α chosen at random we conjecture that the existence of such a prime has probability zero.

6. Further extensions

A. Fraenkel has generalized many of the results of this paper, and of [12], [13], and [14], to algebraic integers of the form

$$\alpha = \frac{2 - d + \sqrt{d^2+4}}{2},$$

where d is any positive integer (the case $d = 1$ gives $\alpha = \phi$). In particular, he established the following generalized KLM-formula:

Proposition 6. *Let K, L, M be integers such that $d|K$ and $d|M$. For every integer n we have*

$$a\left(\frac{K}{d}a(n) + Ln + \frac{M}{d}\right) = \left(\frac{K}{d}(2-d) + L\right)a(n) + Kn + \left\lfloor \left(L - \frac{K}{\alpha}\right)\{n\alpha\} + \frac{M}{d}\alpha \right\rfloor.$$

REFERENCES

[1] P. Arnoux, Some remarks about Fibonacci multiplication, in preparation.

[2] P. T. Bateman and A. L. Duquette, The analogue of the Pisot–Vijayaraghavan numbers in fields of formal power series, Illinois J. Math., 6(1962 594–606.

[3] S. Beatty, Problem 3177, Amer. Math. Monthly, 33(1926), 159; 34(1927) 159.

[4] E. R. Berlekamp, J. H. Conway, and R. K. Guy, "Winning Ways", Academi Press, London, 1982.

[5] L. Carlitz, R Scoville and V. E. Hoggatt, "Fibonacci Representations", *Th Fibonacci Quarterly*, 10 (1972) pp 1–28.

[6] G. Chabauty, Sur la répartition modulo 1 des certaines suites p–adiques C. R. Acad. Sci. Paris, 231(1950), 465–466.

[7] I. G. Connell, A generalization of Wythoff's game, Canadian Math. Bull. 2(1959), 181–190.

[8] H. S. M. Coxeter, The golden section, phyllotaxis and Wythoff's game Scripta Mathematica 19(1953), 135–143.

[9] A. Decomps-Guilloux, Généralization des nombres de Salem aux adèles Acta Arith. 16(1969/70), 265–314.

[10] A. S. Fraenkel, M. Mushkin, and U. Tassa, Determination of $[n\theta]$ by it sequence of differences, Canadian Math. Bull., 21(1978), 441–446.

[11] D. Knuth, The Fibonacci multiplication, Appl. Math. Lett., 1(1988) 57–60.

[12] H. Porta and K. B. Stolarsky, The edge of a golden semigroup, to appea in Proc. 1987 János Bolyai Math. Soc. Conf. Number Theory.

[13] H. Porta and K. B. Stolarsky, A number system related to iterated map whose ultimately periodic set is $\mathbf{Q}(\sqrt{5})$, preprint.

[14] H. Porta and K. B. Stolarsky, Wythoff pairs as semigroup invariants, t appear in Advances in Mathematics.

[15] H. Porta and K. B. Stolarsky, Half–silvered mirrors and Wythoff's game to appear in Canadian Math. Bull.

[16] R. Salem, "Algebraic Numbers and Fourier Analysis", Heath, 1963.

[17] K. B. Stolarsky, Beatty sequences, continued fractions, and certain shif operators, Canadian Math. Bull., 19(1976), 472–482.

A. S. Fraenkel
Dept. of Applied Mathematics
The Weizman Institute of Science
Rehovot, 76100, Israel

H. Porta and K. B. Stolarsky
Department of Mathematics
University of Illinois
Urbana, IL 61801

Norms in Arithmetic Progressions

J. B. FRIEDLANDER AND H. IWANIEC

To Paul Bateman, with friendship and respect

The study of the distribution of various number theoretic functions in arithmetic progressions has long been a topic of concern and has in recent years received new impetus from outside sources such as exponential sums over varieties [**D, B, H**] and from those occurring in the theory of automorphic forms [**D-I**]. Particular examples are squarefree numbers [**HB1**], primes [**Fo-I, B-F-I**] and divisor functions [**Fo, F-I1, F-I2, HB2**].

The case of the divisor functions serves as a good model for the general study. Let $\tau_k(n)$ be the number of representations of n as the product of k positive integers. The generating Dirichlet series of $\tau_k(n)$ is

$$\zeta_k(s) = \sum_{n\geq 1} \tau_k(n) n^{-s} = \zeta(s)^k$$

and this offers one way to attack the problem by twisting with Dirichlet characters and using the theory of Dirichlet polynomials [**Hu, F-I2**]. This technique works for other arithmetic sequences (b_n) whose generating Dirichlet series factors in a similar way. Thus let F be a cyclic extension of $\mathbf{Q}$ of degree k, discriminant Δ and let $b_k(n)$ be the number of representations of n as a norm. In this case the generating Dirichlet series is the Dedekind zeta function of F which is known to factor as a product of k Dirichlet L-functions. The results of [**F-I2**] carry over to give corresponding statements. An example is

Theorem 1. *As $x \to \infty$ we have*

$$\sum_{\substack{n\leq x \\ n\equiv a\,(q)}} b_5(n) \sim \frac{c_F}{\phi(q)} \prod_{\wp | q} \left(1 - \frac{1}{N\wp}\right) x$$

for $a\Delta$ and q coprime, uniformly in $q < x^{9/20-\epsilon}$.

The other statements of [**F-I2**] extend in the analogous fashion; in fact in certain cases for $k \geq 7$ these statements may be strengthened using recent work of Burgess [**Bu**]. If the field extension is not cyclic, the lack of a complete factorization seems to be a serious obstacle to getting corresponding results for the general case. An analogous situation occurs for the symmetric powers of L-functions attached to modular forms [**S**].

For τ_2 and τ_3 better results follow by the employment of Fourier analysis and estimates for Kloosterman sums. Thus independently Selberg and Hooley (unpublished) obtained, using the Weil bound for Kloosterman sums,

$$\sum_{\substack{n<x \\ n\equiv a\,(q)}} \tau_2(n) \sim \frac{c_q}{\phi(q)} x \log x$$

uniformly for $(a,q)=1$ and $q < x^{2/3-\epsilon}$. Fouvry [**Fo**] was able to obtain this for the additional range $x^{2/3+\epsilon} < q < x^{1-\epsilon}$, for almost all q, by employing estimates from the theory of automorphic forms.

For τ_3 the authors [**F-I1**] used estimates for exponential sums in three variables (deduced from Deligne's work [**D, F-I1** Appendix]) to obtain

$$\sum_{\substack{n<x \\ n\equiv a\,(q)}} \tau_3(n) \sim \frac{c_q}{\phi(q)} x (\log x)^2$$

in the range $(a,q)=1$, $q < x^{\frac{1}{2}+\frac{1}{230}-\epsilon}$. Further improvements have been given in [**HB2**]. Here we consider sums of the type

$$\sum\sum\sum_{\substack{n_1n_2n_3\leq x \\ n_1n_2n_3\equiv a\,(q)}} 1,$$

count integers $n_1 \leq x/n_2n_3$, $n_1 \equiv a\overline{n_2n_3}\,(q)$ by Fourier analysis, and in so doing we are led to the consideration of incomplete Kloosterman type sums. Two theorems [**F-I1**] were proved to estimate the latter sums. In each case the method may be extended to sums weighted by an arbitrary additive character, and this allows the treatment of sums of the type

$$\sum\sum\sum_{\substack{n_1n_2n_3\leq x \\ n_1n_2n_3\equiv a\,(q)}} e(\alpha n_1 + \beta n_2 + \gamma n_3),$$

where α, β, γ are fixed rational numbers. By way of Gauss sums we return to sums with multiplicative characters

$$\sum\sum\sum_{\substack{n_1n_2n_3\leq x \\ n_1n_2n_3\equiv a\,(q)}} \psi_1(n_1)\psi_2(n_2)\psi_3(n_3).$$

Combining this with the factorization of the cyclic cubic field

$$\zeta_F(s) = L(s,\chi_0)L(s,\chi)L(s,\chi^2)$$

with $\psi_1 = \chi_0$, $\psi_2 = \chi$, $\psi_3 = \chi^2$, we prove the following result.

Theorem 2. *As $x \to \infty$ we have*

$$\sum_{\substack{n \le x \\ n \equiv a\,(q)}} b_3(n) \sim \frac{c_F}{\phi(q)} \prod_{\wp | q} \left(1 - \frac{1}{N\wp}\right) x$$

for $a\Delta$ and q coprime, q squarefree, uniformly in $q < x^{\frac{1}{2}+\frac{1}{230}-\epsilon}$.

Presumably this holds also for q not squarefree and can be sharpened as in [**HB2**]. Analogously one may get

$$\sum_{\substack{n \le x \\ n \equiv a\,(q)}} b_2(n) \sim \frac{c_F}{\phi(q)} \prod_{\wp | q} \left(1 - \frac{1}{N\wp}\right) x$$

for all $q < x^{2/3-\epsilon}$ and for almost all q in the range $x^{2/3+\epsilon} < q < x^{1-\epsilon}$. Also one should be able to evaluate the contribution of ideals in a fixed class, using ideas from Linnik's ergodic method. Note that the restriction of the summation to one class means we are counting values of a quadratic form in an arithmetic progression. It would be interesting to do the same thing in the cubic case.

REFERENCES

[B] E. Bombieri, On exponential sums in finite fields, II, Invent. Math. **47** (1978), 29–39.

[B-F-I] E. Bombieri, J. Friedlander and H. Iwaniec, Primes in arithmetic progressions to large moduli III, J. Amer. Math. Soc. **2** (1989), 215–224.

[Bu] D. A. Burgess, Estimation of character sums modulo a power of a prime, Proc. London Math. Soc. (3) **52** (1986), 215–235.

[D] P. Deligne, La conjecture de Weil I, Publ. Math. IHES **43** (1974), 273–307.

[D-I] J.-M. Deshouillers and H. Iwaniec, Kloosterman sums and Fourier coefficients of cusp forms, Invent. Math. **70** (1982), 219–288.

[F-I1] J. Friedlander and H. Iwaniec, Incomplete Kloosterman sums and a divisor problem, Ann. Math. **121** (1985), 319–344; Appendix by B. J. Birch and E. Bombieri, 345–350.

[F-I2] J. Friedlander and H. Iwaniec, The divisor problem for arithmetic progressions, Acta Arith. **45** (1985), 273–277.

[Fo] E. Fouvry, Sur le problème des diviseurs de Titchmarsh, J. Reine Angew. Math. **357** (1985), 51–76.

[Fo-I] E. Fouvry and H. Iwaniec, Primes in arithmetic progressions, Acta Arith. **42** (1983), 197–218.

[H] C. Hooley, On exponential sums and certain of their applications, in *Journées Arithmétiques 1980, J. V. Armitage (ed.)*, Cambridge, 1982, pp. 92–122.

[HB1] D. R. Heath-Brown, The least square-free number in an arithmetic progression, J. Reine Angew. Math. **332** (1982), 204–220.

[HB2] D. R. Heath-Brown, The divisor function $d_3(n)$ in arithmetic progressions, Acta Arith. **47** (1986), 29–56.

[Hu] M. Huxley, Large values of Dirichlet polynomials, III, Acta Arith. **26** (1975), 435–444.

[S] F. Shahidi, On the Ramanujan conjecture and finiteness of poles for certain L-functions, Ann. Math. **127** (1988), 547–584.

J. B. Friedlander
Department of Mathematics
University of Toronto
Toronto, Ontario M5S 1A1
Canada

H. Iwaniec
Department of Mathematics
Rutgers University
New Brunswick, NJ 08903

Lower Bounds for Least Quadratic Non-Residues

S. W. GRAHAM AND C. J. RINGROSE

Dedicated to Paul Bateman

1. Introduction

Let p be a prime, and let n_p denote the least positive integer n such that n is a quadratic non-residue mod p. In 1949, Fridlender [F] and Salié [Sa] independently showed that $n_p = \Omega(\log p)$; in other words, there are infinitely many primes p such that $n_p \geq c \log p$ for some absolute constant c. In 1971, Montgomery showed that if the Generalized Riemann Hypothesis is true, then

$$n_p = \Omega(\log p \log\log p) . \tag{1.1}$$

In this paper, we give an unconditional sharpening of the Fridlender-Salié result.

Theorem 1. $n_p = \Omega(\log p \log\log\log p)$.

Ron Graham has pointed out that our result has an application to the quasi-random graphs. Define the Paley graph Q_p as follows. Let p be a prime with $p \equiv 1 \bmod 4$. The vertices of Q_p are the integers $1, 2, \cdots, p$, and $\{i, j\}$ is an edge if and only if

$$\left(\frac{i-j}{p}\right) = 1 .$$

Q_p is a quasi-random graph in the sense of Chung, Graham, and Wilson [CGW]. However, Q_p deviates from a random graph in the following way. The expected size of the largest clique in a random graph is

$(1 + o(1))2\log p/\log 2$. Our results show that for infinitely many p, Q_p has a clique of size $\gg \log p \log\log\log p$. (Actually, this requires a slight modification of our results; see the comments at the end of the paper.)

To explain the ideas behind our proof, we first reproduce Montgomery's conditional argument. Let y be a real number to be chosen later, and set

$$\mathbf{P}_y = \{\, p : p \text{ is prime and } \left(\frac{p_1}{p}\right) = 1 \text{ for all } p_1 \le y \,\} . \tag{1.2}$$

Then

$$\sum_{\substack{p\in \mathbf{P}_y \\ x<p\le 2x}} \log p = 2^{-\pi(y)} \sum_{x<p\le 2x} \prod_{p_1\le y} \left(1 + \left(\frac{p_1}{p}\right)\right) \tag{1.3}$$

$$= 2^{-\pi(y)} \sum_{m|P_y} \sum_{x<p\le 2x} \log p \, \chi_m(p) .$$

Here, $P_y = \prod_{p_1\le y} p_1$ and χ_m is the character determined by quadratic reciprocity so that

$$\chi_m(p) = \prod_{p_1|m} \left(\frac{p_1}{p}\right) . \tag{1.4}$$

Note that χ_m is primitive and that its modulus is either m or $4m$. If the Generalized Riemann Hypothesis is true, then

$$\sum_{x<p\le 2x} \chi_m(p)\log p = E(m)x + O(x^{1/2}\log^2 P + x^{1/2}\log^2 x) ,$$

where $E(m) = 1$ if $m = 1$ and $E(m) = 0$ otherwise. Therefore

$$\sum_{\substack{x<p\le 2x \\ p\in \mathbf{P}_y}} \log p = 2^{-\pi(y)}x + O(x^{1/2}y^2 + x^{1/2}\log^2 x)$$

This is positive if $y \le c\log x \log\log x$ for some sufficiently small positive constant c, and (1.1) follows.

To prove Theorem 1, we need an unconditional bound for the sum on the right-hand side of (1.3). (For technical reasons, we use a weighted version of this sum.) One way of obtaining bounds for character sums is to use zero-density estimates and zero-free regions for Dirichlet L-functions. The classical theorem on zero-free regions ([**D**], Chapter 14) states that if χ is a Dirichlet character mod q, then there is a positive constant c_1 such that $L(s, \chi)$ has at most one zero in the region

$$\sigma \geq 1 - \frac{c_1}{\log q(|t|+1)} .$$

This result does not suffice for our purposes. However, in his thesis [**R**], Ringrose showed that for characters of certain moduli, the above zero-free region may be widened. We shall prove a variant of this result, which we state as

Theorem 2. *Let χ_m be as defined in (1.4), and define*

$$\mathbf{L}(s, P_y) = \prod_{m|P_y} L(s, \chi_m) .$$

Then there is a constant C_1 such that $\mathbf{L}(s, P_y)$ has at most one zero in the region

$$\sigma > 1 - \frac{C_1(\log\log P_y)^{1/2}}{\log P_y} , |t| \leq \log P_y .$$

The exceptional zero, if it exists, is real.

For our purposes, we can avoid the exceptional zeros entirely. We do this by employing an argument of Maier ([**Ma**], Lemma 1.)

Theorem 3. *Let C_1 be as in Theorem 2. There is a sequence of values $\{y_\nu\}_{\nu=0}^{\infty}$ such that $y_\nu \to \infty$ and whenever $y = y_\nu$, $\mathbf{L}(s, P_y)$ has no zeros in the region*

$$\sigma > 1 - \frac{C_1(\log\log P_y)^{1/2}}{2\log P_y} , \quad |t| \leq \log P_y .$$

We also need the following result on the density of zeros of L functions.

Theorem 4. *Let $N_1(\alpha)$ denote the number of zeros of $\mathbf{L}(s, P_y)$ in the rectangle*

$$\alpha \le \sigma \le 1 \ , \ |t| \le \log P_y \, .$$

Let $k_0 = [\sqrt{\log\log P_y}]$ and

$$\eta_1 = \frac{k_0}{2(2^{k_0} - 2)} \, .$$

Then there is a constant C_2 such that

$$N_1(\alpha) \ll \sqrt{\log\log P}\exp(C_2(1-\alpha)\log P(\log\log P)^{-1/2}) \ \text{ if } \alpha \ge 1 - \eta_1$$

and

$$N_1(\alpha) \ll \exp(C_2(1-\alpha)\log P(\log\frac{1}{1-\alpha})^{-1}) \ \text{ if } \alpha < 1 - \eta_1 \, .$$

An important feature of Theorem 4 is that it is sharp near $\sigma = 1$. Another important feature is that the bound for $N_1(\alpha)$ is smaller than $P^{c(1-\alpha)}$ for any c. The key tool for proving Theorems 2 and 4 is the following character sum estimate.

Theorem 5. *Suppose that $q = 2^{\alpha} r$, where $0 \le \alpha \le 3$ and r is an odd square-free integer, and that χ is a non-principal character* mod q. *Let p denote the largest prime factor of q. Suppose also that k is a non-negative integer, and let $K = 2^k$. Finally, assume that $N \le M$. Then*

$$\sum_{M < n \le M+N} \chi(n) \ll M^{1 - \frac{k+3}{8K-2}} p^{\frac{k^2+3k+4}{32K-8}} q^{\frac{1}{8K-2}} (d(q))^{\frac{3k^2+11k+8}{16K-4}} (\log q)^{\frac{k+3}{8K-2}} \sigma_{-1}(q) \, .$$

A classical result of van der Corput states that

$$\sum_{M < n \le M+N} n^{-it} \ll M^{1-\frac{k+3}{8K-2}} t^{\frac{1}{8K-2}} + tM^{-1}.$$

Thus Theorem 5 is a q-analogue of van der Corput's result. Heath-Brown [**He**] gave a qT-analog of van der Corput's result in the case $k = 0$. In other words, he gave an estimate for $\sum_{M < n \le M+N} \chi(n) n^{it}$ that is non-trivial in terms of t and reduces to Theorem 5 when $k = 0$ and $t = 0$. For our purposes, we need a q-analog for arbitrary values of k, but we do not need qT-analogs.

Theorems similar to Theorems 2 and 5 were given by Ringrose in his thesis [**R**]. However, since that material is not widely available, we shall give the complete proofs here. We note that in fact, Ringrose proved the following result about zero free regions. He showed that there exists a positive constant C_3 with the following property. Suppose that q is a square-free integer whose largest prime factor is p. Then the function $\prod_{\chi \bmod q} L(s, \chi)$ has at most one zero in the region $1 - \psi(q,T) \le \sigma \le 1$, $|t| \le T(q)$, where $\psi(q,T) =$

$$C_3 \min \left\{ \frac{\log\log q}{\log q}, \frac{1}{(\log q \log d(q))^{1/2}}, \frac{1}{(\log q \log p)^{1/2}}, \frac{\log(\log q/(\log(T(q)+2)))}{\log q} \right\}.$$

The proof of Theorem 4 is modeled on Jutila's proof of Linnik's Density Theorem [**J**].

Notation: We use the variables $C_1, C_2, \cdots$ to denote unspecified positive constants that occur in the statements of our main theorems. The meanings of $C_1, C_2, \cdots$ will stay fixed throughout the paper. The variables $c_1, c_2, \cdots$ are used for other, less important, unspecified positive constants. The numbering of these constants will begin anew in each section.

We use $d(q)$ to denote the number of divisors of q, and we use $\sigma_a(q)$ to denote $\sum_{d|q} d^a$.

Acknowledgements. We thank Ron Graham and Hugh Montgomery for suggesting this problem and for helpful discussions. We also thank Adolf Hildebrand for pointing out Maier's trick with exceptional zeros. Finally, we thank Roger Heath-Brown for facilitating our collaboration.

2. Preliminaries

We will consider the weighted sum

$$\sum_{\substack{p \in \mathbf{P}_y \\ x^{1/2} < p \leq x^2}} (\log p)\,(e^{-p/2x} - e^{-p/x})$$

For technical reasons, we assume that $y < x^{1/2}$. To prove the theorem, it suffices to show that there are arbitrarily large x for which the above sum is positive when $y = c \log x \log\log\log x$.

We may rewrite the above sum as

$$\begin{aligned} &= 2^{-\pi(y)} \sum_{x^{1/2} < p \leq x^2} (\log p)\,(e^{-p/2x} - e^{-p/x}) \prod_{p_1 \leq y} (1 + (\frac{p_1}{p})) && (2.1) \\ &= 2^{-\pi(y)} \sum_{m \mid P_y} \sum_{x^{1/2} < p \leq x^2} (\log p)\,(e^{-p/2x} - e^{-p/x}) \chi_m(p) \ . \end{aligned}$$

Now let

$$S(m) = \sum_{x^{1/2} < p \leq x^2} (\log p)\,(e^{-p/2x} - e^{-p/x}) \chi_m(p) \ . \tag{2.2}$$

Then

$$S(m) = \frac{-1}{2\pi i} \int_{2+i\infty}^{2-i\infty} \frac{L'}{L}(s\,,\chi_m) x^s K(x,s)\, ds \ + \ O(x^{1/2} \log x)\,, \tag{2.3}$$

where

$$K(x,s) = ((2x)^s - x^s)\Gamma(s) \ .$$

We pull the line of integration to Re $s = -3/4$. $K(s)$ has no poles in the strip $-3/4 \le \text{Re } s \le 2$ since the pole of $\Gamma(s)$ at $s = 0$ is canceled by the zero of $((2x)^s - x^s)$. Thus the only poles of the integrand in (2.3) occur at the zeros $\rho = \beta + i\gamma$ of $L(s, \chi_m)$, and

$$S(m) = E(m)x - \sum_{\rho} ((2x)^{\rho} - x^{\rho})\Gamma(\rho) + O(x^{1/2}\log x) \ .$$

Here, $E(m) = 1$ if $m = 1$ and 0 otherwise, and the sum is over all zeros with $0 \le \beta < 1$. Now the number of zeros up to height T is $\ll T\log PT$ and $K(\sigma + it) \ll x^{\sigma}|t|e^{-\pi t/2}$. Therefore, the contribution of the zeros with $|\gamma| \ge \log P$ is $\ll 1$. The zeros with $\beta \le 1 - \varepsilon$ and $|\gamma| \le \log P$ contribute

$$\ll N(1 - \varepsilon, \log P; \chi_m)x^{1-\varepsilon} \ll x^{1-\varepsilon}\log^2 P \ .$$

Combining these observations with equations (2.1) through (2.3) yields

$$\sum_{\substack{p \in \mathbf{P}_y \\ p > x^{1/2}}} (\log p)(e^{-p/2x} - e^{-p/x}) \tag{2.4}$$
$$= 2^{-\pi(y)}x + O(x^{1-\varepsilon}\log^2 P + 2^{-\pi(y)}T) \ ,$$

where

$$T = \sum_{m|P_y} \sum_{\substack{\rho \\ \beta \ge 1-\varepsilon, |\gamma| \le \log P}} x^{\beta} .$$

Let $N_1(\alpha)$ be as defined in the statement of Theorem 3. Then

$$T = -\int_{1-\varepsilon}^{1} x^{\alpha}\, dN_1(\alpha) = x^{1-\varepsilon}N_1(1 - \varepsilon) + (x\log x)I \tag{2.5}$$

where

$$I = \int_{1-\varepsilon}^{1} x^{\alpha - 1} N_1(\alpha)\, d\alpha . \tag{2.6}$$

We will return to the evaluation of I in Section 9.

3. A q-analog of the A-process

The method of exponent pairs is a method for bounding sums of the form $\sum_{M < n \le M+N} e(f(n))$. The method is based on two processes, which have come to be known as A and B. (See [T], Section 5.20 or [I], Section 2.3 for more details.) The B-process is essentially the Poisson summation formula, and the A-process is an application of Cauchy's inequality.

In this section, we develop a q-analog of the A-process, and we use it to study sums of the form $\sum_{M < n \le M+N} \chi(n)$, where χ is a primitive character mod q. This involves the use of several divisors of q, whose product also divides q. We then decide how large we would like these divisors to be in order to obtain the optimal estimate.

Throughout this section, we will be making the following assumptions:

$q = 2^{\alpha} r$, $0 \le \alpha \le 3$, and r is an odd square-free integer. (3.1a)

χ is a primitive non-principal character mod q. (3.1b)

$q_0, q_1, \cdots q_k$ are divisors of q. (3.1c)

$$\prod_{i=0}^{k} q_i | q . \tag{3.1d}$$

$$N \le M . \tag{3.1e}$$

It will also be convenient to set

$$\xi(n) = \begin{cases} \chi(n) & \text{if } M + 1 \le n \le M + N , \\ 0 & \text{otherwise.} \end{cases} \tag{3.1f}$$

and to write

$$S = \sum_{n} \xi(n) .$$

Let H be a positive integer. Then

$$S = \frac{1}{H} \sum_{h=1}^{H} \sum_{m} \xi(m + hq_0) = \frac{1}{H} \sum_{m} \sum_{h=1}^{H} \xi(m + hq_0) \; .$$

By Cauchy's inequality,

$$\begin{aligned} |S|^2 &\le \frac{1}{H^2} \sum_{m=M-Hq_0+1}^{M+N-q_0} 1 \; \sum_{m} \Big| \sum_{h=1}^{H} \xi(m + hq_0) \Big|^2 \\ &\le \frac{N + Hq_0}{H^2} \sum_{h=1}^{H} \sum_{j=1}^{H} \sum_{m} \xi(m + hq_0) \overline{\xi(m + jq_0)} \\ &= \frac{N + Hq_0}{H} \sum_{|h| \le H} \Big(1 - \frac{|h|}{H}\Big) \sum_{m} \xi(m) \overline{\xi(m + hq_0)} \\ &= \frac{N + Hq_0}{H} \sum_{|h| \le H} \Big(1 - \frac{|h|}{H}\Big) S_0(h) \, , \end{aligned}$$

say. If $H \le N/q_0$ then

$$|S|^2 \le \frac{2N}{H} \sum_{|h| < H} |S_0(h)| \; .$$

Up to now, we have assumed that H is a positive integer. However, if we assume only that $H > 0$ and let $H' = [H] + 1$, then we may write

$$|S|^2 \le \frac{2N}{H'} \sum_{|h| < H'} |S_0(h)| \le \frac{2N}{H} \sum_{|h| \le H} |S_0(h)| \; .$$

Since $|S_0(0)| \le N$ and $S_0(h) = \overline{S_0(-h)}$, we obtain

$$|S|^2 \le \frac{2N^2}{H} + \frac{4N}{H} \sum_{1 \le h \le H} |S_0(h)| \; . \tag{3.2}$$

If we set $H = H_0 = N/q_0$ then we get

$$|S|^2 \le 2Nq_0 + 4q_0 \sum_{1 \le h \le H_0} |S_0(h)| .$$

We now iterate the above process. Upon squaring (3.2) and applying Cauchy's inequality, we obtain

$$|S|^4 \le \frac{8N^4}{H_0^2} + \frac{32N^2}{H_0^2} H_0 \sum_{1 \le h_0 \le H_0} |S_0(h_0)|^2 .$$

Applying (3.2) to the inner sum, we obtain

$$|S|^4 \le \frac{8N^4}{H_0^2} + \frac{32N^2}{H_0} \sum_{1 \le h_0 \le H_0} \left\{ \frac{2N^2}{H_1} + \frac{4N}{H_1} \sum_{1 \le h_1 \le H_1} |S_1(h_0, h_1)| \right\}$$

$$= \frac{8N^4}{H_0^2} + \frac{64N^4}{H_1} + \frac{128N^3}{H_0 H_1} \sum_{1 \le h_0 \le H_0} \sum_{1 \le h_1 \le H_1} |S_1(h_0, h_1)|$$

where

$$S_1(h_0, h_1) = \sum_m \xi(m) \overline{\xi(m + h_0 q_0) \xi(m + h_1 q_1)} \xi(m + h_0 q_0 + h_1 q_1) .$$

If we set $H_0 = Nq_0^{-1}$ and $H_1 = Nq_1^{-1}$, then we may write the above as

$$|S|^4 \le 8^3 \{ \max(N^2 q_0^2, N^3 q_1) \} + \\ + \frac{N^3}{H_0 H_1} \sum_{1 \le h_0 \le H_0} \sum_{1 \le h_1 \le H_1} |S_1(h_0, h_1)| \} . \tag{3.3}$$

By continuing this argument with a straightforward induction, we can prove

Lemma 3.1 *Assume hypotheses* (3.1a) *through* (3.1f). *Let* $k \ge 0$ *be an integer, and* $K = 2^k$. *Let* $H_i = N/q_i$ *for* $i = 1, \cdots, k$. *Then*

$$|\sum_n \xi(n)|^{2K} \le 8^{2K-1} \{ \max_{0 \le j \le k} (N^{2K - K/J} q_j^{K/J}) + \\ + \frac{N^{2K-1}}{H_0 \cdots H_k} \sum_{h_0=1}^{H_0} \cdots \sum_{h_k=1}^{H_k} |S_k(\mathbf{h})| \} ,$$

where $J = 2^j$ *and* $\mathbf{h} = (h_0, h_1, \cdots, h_k)$. *Here, the sum* $S_k(\mathbf{h})$ *is defined as*

$$S_k(\mathbf{h}) = \sum_m f_j(m)\overline{g_j(m)},$$

where

$$f_j(m) = \prod_{V \in T_j} \xi(m + \sum_{i \in V} h_i q_i),$$

$$g_j(m) = \prod_{V \in U_j} \xi(m + \sum_{i \in V} h_i q_i),$$

$$T_j = \{V \subseteq \{0, 1, \cdots, j\} : |V| \not\equiv j \bmod 2\},$$

and

$$U_j = \{V \subseteq \{0, 1, \cdots, j\} : |V| \equiv j \bmod 2\}.$$

Observe that

$$S_k(\mathbf{h}) = \sum_{n \in I} \chi\left(\frac{f_k(n)}{g_k(n)}\right),$$

where $I = I(\mathbf{h})$ is a subinterval of $(M+1, M+N]$. Furthermore,

$$\begin{aligned}
\sum_{n \in I} \chi\left(\frac{f_k(n)}{g_k(n)}\right) &= \sum_{r=1}^{q} \sum_{\substack{n \in I \\ n \equiv r \bmod q}} \chi\left(\frac{f_k(r)}{g_k(r)}\right) \\
&= \frac{1}{q} \sum_{r=1}^{q} \sum_{n \in I} \chi\left(\frac{f_k(r)}{g_k(r)}\right) \sum_{s=1}^{q} e\left(\frac{s(r-n)}{q}\right) \\
&= \frac{1}{q} \sum_{s=1}^{q} \sum_{r=1}^{q} \chi\left(\frac{f_k(r)}{g_k(r)}\right) e\left(\frac{sr}{q}\right) \sum_{n \in I} e\left(\frac{-sn}{q}\right)
\end{aligned}$$

$$= \frac{1}{q} \sum_{s=1}^{q} \sum_{n \in I} e\left(\frac{-sn}{q}\right) \sum_{r=1}^{q} \chi\left(\frac{f_k(r)}{g_k(r)}\right) e\left(\frac{sr}{q}\right)$$

Now

$$\sum_{n \in I} e\left(\frac{-sn}{q}\right) \ll \begin{cases} N & \text{if } s = 0 , \\ \dfrac{q}{s} & \text{if } 1 \le s \le q/2 , \\ \dfrac{q}{q-s} & \text{if } q/2 < s < q . \end{cases}$$

Thus

$$S_k(\mathbf{h}) \ll N q^{-1} \, |S(q\,;\chi\,, f_k\,, g_k\,, 0)| + \sum_{0 < |s| \le q/2} |s|^{-1} \, |S(q\,;\chi\,, f_k\,, g_k\,, s)| \, , \tag{3.4}$$

where

$$S(q\,;\chi\,, f_k\,, g_k\,, s) = \sum_{r=1}^{q} \chi\left(\frac{f_k(r)}{g_k(r)}\right) e\left(\frac{sr}{q}\right) .$$

In the next section, we shall prove a series of lemmas devoted to analyzing the last sum.

4. The sum $S(q\,;\chi\,, f_k\,, g_k\,, s)$

Lemma 4.1. *Assume that χ is a primitive non-principal character* mod q. *Suppose that $(u,v) = 1$ and $q = uv$. Define $\bar{u}$ and $\bar{v}$ by the relations*

$$u\bar{u} \equiv 1 \bmod v \quad \text{and} \quad v\bar{v} \equiv 1 \bmod u \, .$$

Then there are primitive characters χ_u and χ_v (modulo u and v respectively) such that

$$S(q;\chi,f_k,g_k,s) = S(u;\chi_u,f_k,g_k,s\overline{v})\,S(v;\chi_v,f_k,g_k,s\overline{u})\ .$$

Proof. Define

$$\chi_u(n) = \chi(n + u\overline{u}(1-n))\ ,\ \chi_v(n) = \chi(n + v\overline{v}(1-n))\ .$$

Then χ_u, χ_v are primitive characters modulo u and v respectively, and $\chi_u(n)\chi_v(n) = \chi(n)$. Therefore

$$\begin{aligned} S(q;\chi,f_k,g_k,s) &= \sum_{r=1}^{q}\chi\Big(\frac{f_k(r)}{g_k(r)}\Big)e\Big(\frac{sr}{q}\Big) \\ &= \sum_{l=1}^{u}\sum_{m=1}^{v}\chi\Big(\frac{f_k(vl+um)}{g_k(vl+um)}\Big)e\Big(\frac{s(vl+um)}{q}\Big) \\ &= \sum_{l=1}^{u}\chi_u\Big(\frac{f_k(vl)}{g_k(vl)}\Big)e\Big(\frac{sl}{u}\Big)\ \sum_{m=1}^{v}\chi_v\Big(\frac{f_k(um)}{g_k(um)}\Big)e\Big(\frac{sl}{v}\Big) \\ &= S(u;\chi_u,f_k,g_k,s\overline{v})\,S(v;\chi_v,f_k,g_k,s\overline{u}) \end{aligned}$$

as required.

In the next two lemmas, we consider sums of the form $S(p;\chi,f_k,g_k,s)$, where p is a prime. For brevity, we denote this sum by S.

Lemma 4.2. *Let p be a prime, and suppose that $p|h_iq_i$ for some i.*

(1) *If $p|s$ then $|S| \le p$.*

(2) *If $p \nmid s$ then $|S| \le 2^k$.*

Proof. In this case, $f_k(r) = g_k(r) \bmod p$, so

$$S = \sum_{\substack{r=1 \\ p\nmid g_k(r)}}^{p} e\Big(\frac{rs}{p}\Big)\ .$$

If $p \nmid s$ then

$$|S| = \Big|\sum_{r=1}^{p}e\Big(\frac{rs}{p}\Big) - \sum_{\substack{r=1 \\ p|g_k(r)}}^{p} e\Big(\frac{rs}{p}\Big)\Big| \le k \le 2^k$$

since g_k has degree k. If $p|s$, then we can do no better than the trivial estimate $|S| \le p$.

Lemma 4.3. *Suppose* $p \nmid \prod_{i=1}^{k} h_i q_i$. *Then* $|S| \le 2^{k+1} p^{1/2}$.

Proof. We divide the proof into two cases, according as to whether or not $p|s$.

First, assume that $p \nmid s$. In this case, we use the following estimate due to Weil [W]. Let $R(m)$ be any rational function, ψ any non-trivial additive character mod p, d the total number of zeros and poles of R, and suppose that m runs over all values mod p for which $R(m)$ is defined. Then

$$|\sum_m \chi(R(m))\,\psi(m)| \le dp^{1/2} .$$

(This is the unnumbered displayed formula at the bottom of p. 206 of [W].) In our case, $\psi(m) = e(\frac{sm}{p})$ and $R(m) = f_k(m)/g_k(m)$. We have $d = 2^{k+1}$ since f_k and g_k each have 2^k zeros. Consequently

$$|S| \le 2^{k+1} p^{1/2} .$$

Now consider the case $p|s$. In this case

$$S = \sum_{m=1}^{p} \chi(\frac{f_k(m)}{g_k(m)}) .$$

Let the order of χ be h. Since the group of multiplicative characters mod p is isomorphic to $\mathbf{Z}_p^*$, we know that $h|(p-1)$. We will now prove that f_k/g_k can not be identically an h–th power in $\mathbf{Z}_p(x)$, and then apply a result from Schmidt [Sc] to estimate $|S|$.

Suppose that $f_k/g_k \equiv F^d \bmod p$ for some F in $\mathbf{Z}_p(x)$ and some integer $d > 1$. Let ρ be a p-th root of unity, and consider

$$G(\rho) = \prod_{i=0}^{k} (1 - \rho^{h_i q_i}) . \tag{4.1}$$

On expanding (4.1), we see that

$$G(\rho) = dG_0(\rho) \tag{4.2}$$

for some $G_0 \in \mathbf{Z}[x]$, by the assumption on f_k/g_k. We consider both sides of (4.2) as elements of $\mathbf{Z}[\rho]$, the ring of integers in $\mathbf{K} = \mathbf{Q}(\rho)$, and take norms to obtain

$$N_{\mathbf{K}}(\prod_{i=0}^{k} (1 - \rho^{h_i q_i})) = d^{p-1} N_{\mathbf{K}}(G_0(\rho)) . \tag{4.3}$$

Now

$$N_{\mathbf{K}}(1 - \rho^{h_i q_i}) = \prod_{j=1}^{p-1} (1 - \rho^{h_i q_i j}) .$$

Since $(h_i q_i , p) = 1$, we see that

$$N_{\mathbf{K}}(1 - \rho^{h_i q_i}) = \prod_{j=1}^{p-1} (1 - \rho^{j}) = p .$$

It follows from (4.3) that $p^k = d^{p-1} N_{\mathbf{K}}(G_0(\rho))$, and therefore that d is a non-negative power of p. Consequently, we can not have $d = h$, which is what we wanted to show. Notice that $\chi(f_k(m)/g_k(m)) = \chi(f_k(m) g_k(m)^{p-2})$; therefore, $f_k(m) g_k(m)^{p-2}$ is not an h-th power in $\mathbf{Z}_p[x]$. The desired estimate now follows from Theorem 2C′ in Chapter II of [Sc].

Lemma 4.4. *Assume hypotheses (3.1a) through (3.1e). Define* $Q_k = \prod_{i=1}^{k} h_i q_i$. *Then*

$$|S(q ; \chi , f_k , g_k , s)| \le 4\, d(q)^{k+1} \left(\frac{q}{(q , Q_k)}\right)^{1/2} (q , Q_k , |s|) .$$

Proof. We note that if $u = 2^\alpha$ and χ_u is a character mod u, then

$$|S(u\,;\chi_u, f_k, g_k, s)| \le 2^{\alpha-1}$$

by the trivial estimate. We then combine this with Lemmas 4.1 through 4.3 to get the desired result.

Lemma 4.5. *Assume hypotheses* (3.1a) *through* (3.1e). *Suppose* $k \ge 0$ *is an integer and* $K = 2^k$. *Let* p *denote the largest prime factor of* q, *and write*

$$S = \sum_{M < n \le M+N} \chi(n)\,.$$

There is an absolute constant c *such that*

$$|S|^{4K} \ll c^{4K}\left(AA_0^{2K} + BA_0^{-2K+1} + CA_0^{2K-1}\right)\,, \tag{4.4}$$

where A_0 *is any real number with* $A_0 \ge 1$,

$$A = M^{2K}\,,\; B = M^{6K-k-4}p^{k+1}q\,d(q)^{2k+4}(\log q)^2\,,$$

and

$$C = M^{2K+k+2}q^{-1}d(q)^{4k+4}\,.$$

The constant implied by $\ll$ *in (4.4) is independent of* k.

Proof. This proof is a combination of Lemmas 3.1 and 4.4, so we use the notation of those lemmas. Define $R_j = \prod_{i=1}^{j} q_i$ and $Q_j = \prod_{i=1}^{j} h_i q_i$. Then $R_j | Q_j$ and $(q, Q_k, |s|) \le (q, |s|)$, so that

$$\sum_{h_k=1}^{H_k} \sum_{0<|s|\le q/2} |s|^{-1}|S(q\,;\chi, f_k, g_k, s)|$$

$$\ll d(q)^{k+1}q^{1/2}R_k^{-1/2}H_k \sum_{0<|s|\le q/2} |s|^{-1}(q, |s|)$$

$$\ll d(q)^{k+2} q^{1/2} R_k^{-1/2} H_k \log q \ . \tag{4.5}$$

Let $S_j = \prod_{i=0}^{j} h_i$; then

$$(q\ , S_j) \le (q\ , S_{j-1})\,(q\ , h_j)\ .$$

On applying Lemma 4.4 together with the bound

$$\sum_{h_j=1}^{H_j} (q\ , h_j)^{1/2} \le d(q) H_j\ ,$$

we find that

$$\begin{aligned} Nq^{-1} \sum_{h_k=1}^{H_k} &|S(q\,;\chi\ ,f_k\ ,g_k\ ,0)| \\ &\ll Nq^{-1} d(q)^{k+1} q^{1/2} \sum_{h_k=1}^{H_k} (q\ , Q_k)^{1/2} \\ &\ll N\Big(\frac{R_k}{q}\Big)^{1/2} d(q)^{k+2} H_k (q\ , S_{k-1})^{1/2} \ . \end{aligned} \tag{4.6}$$

Substituting (4.5) and (4.6) into (3.4), we see that

$$\begin{aligned} S^{2K} \ll c^{2K} &\max_{0 \le j \le k} \{M^{2K - K/J} q_j^{K/J}\} + \\ &c^{2K} M^{2K-1} d(q)^{k+2} (\log q)\, q^{1/2} R_k^{-1/2} + \\ &c^{2K} M^{2K} d(q)^{2k+2} q^{-1/2} R_k^{1/2} \ . \end{aligned} \tag{4.7}$$

In order to calculate the optimal estimate for S from (4.7), we must determine how closely we can approximate to the ideal values for $q_0\ , \cdots\ , q_k$. Suppose that we want a divisor of q to be approximately equal to A, where $1 \le A \le qp$. (Recall that p is the largest prime divisor of q.) We form a chain of divisors $\{d_n\}$ by setting $d_n = \prod_{i=1}^{n} p_i$, where $p_1\ , p_2\ , \cdots$ are the prime divisors of q in some order. Let d_s be the first of these with $d_s \geq A/p$. If $s \geq 2$ then $d_{s-1} < A/p$, so $d_s \leq d_{s-1} p < A$. If $s = 1$ but $d_s > A$ then we take 1 for the approximation, as in this case $A/p \leq 1 \leq A$. Otherwise, we take d_s. In any case, we have produced a divisor d with

$$A/p \leq d \leq A \,.$$

Similarly, if $A_0, \cdots, A_k$ satisfy $A_j \geq 1$ and $\prod_j A_j \leq qp$, then we can find divisors $q_0, \cdots, q_k$ of q, whose product also divides q, with q_j lying in the interval $[A_j/p, A_j]$ (we construct q_j from the original list of primes less those already used in constructing $q_0, \cdots, q_{j-1}$). Choosing q_j as above, we may modify (4.7) to read

$$\begin{aligned} S^{2K} \ll \max_{0 \leq j \leq k} c^{2K} \; &\{M^{2K - K/J} A_j^{K/J}\} \\ &+ c^{2K} M^{2K-1} q^{1/2} d(q)^{k+2} (\log q) \{\prod_j (A_j p^{-1})\}^{-1/2} \\ &+ c^{2K} M^{2K} q^{-1/2} d(q)^{2k+2} (\prod_j A_j)^{1/2} \,. \end{aligned} \tag{4.8}$$

This estimate is trivial if $\prod A_j > q$, so the condition $\prod A_j \leq qp$ is unnecessary. To complete the proof, we choose A_j so that

$$M^{2K - K/J} A_j^{K/J} = M^K A_0^K \,;$$

in other words, we take

$$A_j = A_0^J M^{-J+1} \,.$$

The desired estimate now follows by plugging these choices of A_j into (4.8).

5. Proof of Theorem 5

To analyze the estimate given in Lemma 4.5, we shall use the following two simple lemmas.

Lemma 5.1. *If $X_1, \cdots, X_k$ are positive numbers and $a_1, \cdots, a_k$ are non-negative numbers such that $a_1 + \cdots + a_k = 1$, then*

$$\min(X_1, \cdots, X_k) \le X_1^{a_1} \cdots X_k^{a_k} \le \max(X_1, \cdots, X_k).$$

The proof is obvious. We shall often refer to this as the *convexity principle*.

Our next lemma generalizes the following well known principle. Suppose we have an estimate of the form

$$U \ll AH^a + BH^{-b},$$

where $A, B, a,$ and b are positive constants and H is at our disposal. We obviously want to take H so as to minimize the right hand side. By choosing H to satisfy $AH^a = BH^{-b}$, we get

$$U \ll (A^b B^a)^{1/(a+b)},$$

and this is best possible apart from the value of the implied constant. To generalize this principle, we prove

Lemma 5.2. *Suppose that*

$$L(H) = \sum_{i=1}^{m} A_i H^{a_i} + \sum_{j=1}^{n} B_j H^{-b_j},$$

where $A_i, B_j, a_i,$ and b_j are positive. Assume that $H_1 \le H_2$. Then there is some H with $H_1 \le H \le H_2$ and

$$L(H) \ll \sum_{i=1}^{m} \sum_{j=1}^{n} (A_i^{b_j} B_j^{a_i})^{1/(a_i+b_j)} + \sum_{i=1}^{m} A_i H_1^{a_i} + \sum_{j=1}^{n} B_j H_2^{-b_j}.$$

The implied constants depend only on m and n.

Proof. Define

$$L_+(H) = \max(A_1 H^{a_1}, \cdots, A_m H^{a_m})$$

and

$$L_-(H) = \max(B_1 H^{-b_1}, \cdots, B_n H^{-b_n}) .$$

Now $L \le mL_+ + nL_-$, so it suffices to bound L_+ and L_-. We observe that L_+ is a strictly increasing continuous function, $L_+(0) = 0$, and $L_+(\infty) = \infty$. Similarly L_- is a strictly decreasing continuous function, $L_-(0) = \infty$, and $L_-(\infty) = 0$. Therefore there is a unique H_0 such that $L_+(H_0) = L_-(H_0)$. We distinguish three cases: (a) $H_1 \le H_0 \le H_2$, (b) $H_0 < H_1$, (c) $H_0 > H_2$.

If $H_1 \le H_0 \le H_2$ then there is some i and some j such that

$$L_+(H_0) = A_i H_0^{a_i} = L_-(H_0) = B_j H_0^{-b_j} .$$

Consequently

$$L_+(H_0) = L_-(H_0) = (A_i^{b_j} B_j^{a_i})^{1/(a_i+b_j)} .$$

If $H_0 < H_1$ then

$$L_-(H_1) < L_+(H_1) \le \sum_{i=1}^{m} A_i H_1^{a_i} .$$

If $H_0 > H_2$ then

$$L_+(H_2) < L_-(H_2) \le \sum_{j=1}^{n} B_j H_2^{-b_j} .$$

The above lemma was first stated and proved by Srinivasan [Sr], although van der Corput [vdC] gave the special case $H_0 = 0$, $H_1 = \infty$ some 40 years previously.

Lemma 5.3. *Assume the hypotheses of Lemma 4.5. Then*

$$S \ll M^{1 - \frac{k+3}{8K-2}} p^{\frac{k+1}{8K-2}} q^{\frac{1}{8K-2}} (d(q))^{\frac{2k+4}{8K-2}} (\log q)^{\frac{2}{8K-2}} + \tag{5.1}$$
$$M^{1 - \frac{1}{4K}} p^{\frac{k+1}{8K}} (d(q))^{\frac{3k+4}{4K}} (\log q)^{\frac{1}{4K}} .$$

Proof. We apply Lemma 5.2 with $H_1 = 1$ and $H_2 = \infty$ to the estimate in Lemma 4.5. In this way, we get

$$S \ll M^{1-\frac{k+3}{8K-2}} p^{\frac{k+1}{8K-2}} q^{\frac{1}{8K-2}} (d(q))^{\frac{2k+4}{8K-2}} (\log q)^{\frac{2}{8K-2}} +$$
$$M^{1-\frac{1}{4K}} p^{\frac{k+1}{8K}} (d(q))^{\frac{3k+4}{4K}} (\log q)^{\frac{1}{4K}} +$$
$$M^{\frac{1}{2}} + M^{\frac{1}{2}+\frac{k+2}{4K}} q^{\frac{-1}{4K}} (d(q))^{\frac{4k+4}{4K}} = G_1 + G_2 + G_3 + G_4 ,$$

say. Now G_3 is obviously dominated by G_1, so it may be omitted. By the Polya-Vinogradov estimate and the trivial estimate,

$$S \ll \max(G_1 , G_2 , \min(G_4 , M , q^{1/2}\log q)) .$$

When $k \geq 2$, we note that

$$\min(G_4 , M) \leq G_4^{3/4} M^{1/4} \leq G_2 .$$

When $k = 1$,

$$G_4 = M^{7/8} q^{-1/8} d(q) \leq M^{7/8} q^{7/8} \leq G_2$$

since $d(q) \leq q$. When $k = 0$,

$$\min(G_4 , q^{1/2}\log q) \leq G_4^{1/2}(q^{1/2}\log q)^{1/2} \leq G_1 .$$

Thus G_4 may also be eliminated.

Lemma 5.4. *Assume the hypotheses of Lemma 4.5. Then*

$$S \ll M^{1-\frac{k+3}{8K-2}} p^{\frac{k^2+3k+4}{32K-8}} q^{\frac{1}{8K-2}} (d(q))^{\frac{3k^2+11k+8}{16K-4}} (\log q)^{\frac{k+3}{8K-2}} . \tag{5.2}$$

Proof. Let the right-hand side of the proposed theorem be denoted by E_k, and let F_k denote the second term on the right-hand side of (5.1). Then (5.1) may be written $S \ll E_k + F_k$. To prove the theorem in the case $k = 0$, we use (5.1) and the Polya-Vinogradov inequality to get

$$S \ll E_0 + \min(F_0, q^{\frac{1}{2}} \log q) \ll E_0 + F_0^{\frac{2}{3}} (q^{\frac{1}{2}} \log q)^{\frac{1}{3}} \ll E_0 .$$

We complete the proof by induction. Suppose (5.2) is true with k replaced by $k-1$. This together with (5.1) yields

$$S \ll \max(E_k, \min(F_k, E_{k-1})) \le \max(E_k, F_k^{\frac{4K}{8K-2}} E_{k-1}^{\frac{4K-2}{8K-2}}) \le E_k$$

and this completes the proof of Lemma 5.4.

Lemma 5.4 proves Theorem 5 when χ is primitive. To complete the proof of Theorem 5, suppose that χ is a non-principal character mod q that is induced by the primitive character χ_1 mod q_1. On writing $q = q_1 q_2$, we observe that

$$\sum_{M < n \le M+N} \chi(n) = \sum_{M < n \le M+N} \chi_1(n) \sum_{\substack{d|n \\ d|q}} \mu(d) = \sum_{d|q_2} \mu(d) \chi_1(d) \sum_{\frac{M}{d} < n \le M + \frac{N}{d}} \chi_1(n) .$$

By Lemma 5.4, the above is

$$\ll M^{1 - \frac{k+3}{8K-2}} p^{\frac{k^2+3k+4}{32K-8}} q_1^{\frac{1}{8K-2}} (d(q_1))^{\frac{3k^2+11k+8}{16K-4}} (\log q)^{\frac{k+3}{8K-2}} \sigma_{-\alpha}(q_2) ,$$

where $\alpha = 1 - (k+3)/(8K-2)$. By Hölder's Inequality,

$$\sigma_{-\alpha}(q_2) \le \Big(\sum_{d|q_2} 1\Big)^{1-\alpha} \Big(\sum_{d|q_2} d^{-1}\Big)^{\alpha} = (d(q_2))^{1-\alpha} (\sigma_{-1}(q_2))^{\alpha} .$$

Since

$$1-\alpha = \frac{k+3}{8K-2} \le \frac{3k^2+11k+8}{16K-4} ,$$

we have

$$d(q_1)^{\frac{3k^2+11k+8}{16K-4}} d(q_2)^{\frac{k+3}{8K-2}} \le d(q)^{\frac{3k^2+11k+8}{16K-4}} ,$$

and the desired result follows.

6. Zero-free regions: The proofs of Theorem 2 and 3

We begin this section by proving an upper bound for L-functions associated with the characters considered in Theorem 5.

Lemma 6.1. *Assume the hypothesis of Theorem 5, and let*

$$\sigma_k = 1 - \frac{k+3}{8K-2} .$$

Let $s = \sigma + it$ and $\tau = |t| + 1$. If $\sigma \ge \sigma_k$ then

$$L(s, \chi) \ll \left[q^{\frac{1}{8K-2}} p^{\frac{k^2+3k+4}{32K-8}} (d(q))^{\frac{3k^2+11k+8}{16K-4}} \tau \right]^{\frac{1-\sigma}{1-\sigma_k}} (\log q\tau)^3 .$$

Proof: If $\sigma > 1$, then

$$L(s, \chi) = \sum_{n=1}^{\infty} \chi(n) n^{-s} = \int_{1^-}^{\infty} u^{-\sigma} \, dS(u,t) ,$$

where $S(u,t) = \sum_{n \le u} \chi(n) n^{-it}$. Integrating by parts yields

$$L(s, \chi) = \sigma \int_1^{\infty} u^{-\sigma-1} S(u,t) \, du . \tag{6.1}$$

Now

$$S(u,t) = S(u,0)\,u^{-it} + it\int_1^u w^{-it-1}\,S(w,0)\,dw\,. \tag{6.2}$$

By the Polya-Vinogradov estimate,

$$S(u,t) \ll q^{1/2}\tau\log q\,(1+\log u)\,. \tag{6.3}$$

The integral in (6.1) therefore converges for $\sigma > 0$, and so we have an analytic continuation of $L(s\,,\chi)$ to the half-plane $\sigma > 0$.

We will prove the lemma by combining (6.1) with appropriate estimates for $S(u,t)$. Assume that $\sigma \ge \sigma_k$, and let $M = q\tau^2$. By (6.3)

$$\int_M^\infty u^{-\sigma-1}\,S(u,t)\,du \ll q^{1/2}\tau\,(\log q)\int_M^\infty u^{-\sigma_k-1}\log u\,du \ll (\log q\,\tau)^2\,.$$

since $\sigma_k \ge 1/2$.

Now consider $S(u,t)$ for $u \le M$. We have the trivial estimate $|S(u,t)| \le u$. We may also use Theorem 5 in (6.2) to obtain

$$S(u,t) \ll u^{1-\frac{k+3}{8K-2}}\,B\,\tau\,(\log q\,\tau)^2\,, \tag{6.4}$$

where, for notational convenience, we have written

$$B = q^{\frac{1}{8K-2}}\,p^{\frac{k^2+3k+4}{32K-8}}\,(d(q))^{\frac{3k^2+11k+8}{16K-4}}\,.$$

Let α be a real number with $0 \le \alpha \le 1$. By (6.4), the trivial estimate, and convexity, we see that

$$S(u,t) \ll (u^{1-\frac{k+3}{8K-2}}\,B\,\tau\,(\log q\,\tau)^2)^{\alpha}\,u^{1-\alpha}\,.$$

We choose α so that the exponent on u is σ; i.e.

$$\alpha = \frac{1-\sigma}{1-\sigma_k}\,.$$

This gives the bound

$$S(u,t) \ll (B\tau)^{\frac{1-\sigma}{1-\sigma_k}} u^{\sigma} (\log q\tau)^2 .$$

Using this in (6.1) finishes the proof.

Lemma 6.2. *There exist effectively computable constants c_1, c_2 with the following property. Suppose that $\phi = \phi(P_y)$ and $\theta = \theta(P_y)$ satisfy $0 < \theta \le 1$, $\phi \ge 1$, and $\phi\theta^{-1} \le c_1 e^{\phi}$. Suppose also that for every $m > 1$ such that $m | P_y$, $|L(s, \chi_m)| \le e^{\phi}$ in the region $1 - \theta \le \sigma \le 3$, $|t| \le 2T(P_y) + 1$, where $\log(T(P_y) + 2) \le \phi$. Then the function $\prod_{m|P_y} L(s, \chi_m)$ has at most one zero in the region*

$$1 - c_2 \frac{\theta}{\phi} \le \sigma \le 2 \, , \, |t| \le T(P_y) .$$

This is a q-analog of a classical theorem of Landau ([**T**], Theorem 3.10). It can be proved by combining the arguments of Landau with the arguments used to prove the usual zero-free regions for L-functions. The proof is routine but rather lengthy, and we shall omit it here.

Now we are ready to prove Theorem 2. We apply Lemma 6.2 with $T = \log P_y$. From Lemma 6.1, we see that we may take

$$\theta = \frac{k+3}{8K-2} \quad \text{and} \quad \phi = \frac{\log P_y}{8K-2} + \frac{3k^2 + 11k + 8}{16K-4} \log d(P) + 4 \log\log P_y .$$

Now $\log P_y \asymp y$ and $\log d(P) \asymp \pi(y) \asymp y/\log y \asymp \log P_y / \log\log P_y$. (We use $A \asymp B$ to denote $A \ll B \ll A$.) Therefore

$$\frac{\theta}{\phi} \asymp \min \left[\frac{k}{\log P_y} \, , \, \frac{\log\log P_y}{k \log P_y} \, , \, \frac{k}{K \log\log P_y} \right] .$$

Taking $k = [\sqrt{\log\log P_y}]$ gives

$$\frac{\theta}{\phi} \approx \frac{(\log\log P_y)^{1/2}}{\log P_y} ,$$

and the result follows.

Next, we prove Theorem 3. Suppose, for example, that for some v, $L(s, P_v)$ has an exceptional zero β_0. Then

$$\beta_0 > 1 - \frac{C_1(\log\log P_v)^{1/2}}{\log P_v} .$$

Now if we choose y such that

$$1 - \frac{C_1(\log\log P_y)^{1/2}}{\log P_y} < \beta_0 \leq 1 - \frac{C_1(\log\log P_y)^{1/2}}{2\log P_y}$$

then we see that $\mathbf{L}(s, P_y)$ has no zeros with

$$\beta > 1 - \frac{C_1(\log\log P_y)^{1/2}}{2\log P_y} , \quad |\gamma| \leq \log P_y . \tag{6.5}$$

It follows that we can find a sequence of values $\{y_\nu\}_{\nu=0}^{\infty}$ such that $y_\nu \to \infty$ and $\mathbf{L}(s, P_{y_\nu})$ has no zeros satisfying (6.5).

7. Preliminaries for the Zero-Density Theorem

Our proof of Theorem 4 is along the same lines as Jutila's proof of the Linnik zero-density estimates [J], but we employ some of the modifications introduced by Graham [G2]. We use what has become known as "Halász's method." This involves showing that a zero of an L-function forces a certain Dirichlet polynomial to be large. We then show that this cannot be happen too often.

One method of producing large values of Dirichlet polynomials is to use the mollifier

$$M^*(s, \chi) = \sum_{n \leq X} \mu(n)\chi(n)n^{-s} .$$

This is approximately $(L(s,\chi))^{-1}$, so if s is a zero of $L(s,\chi)$, then $1 - LM^*(s,\chi)$ is large. (See Montgomery's book [**Mo**], Chapter 12, for more details.)

The use of M^* leads to results that are, for our purposes, too weak in the neighborhood of $\sigma = 1$. According, we introduce the mollifier

$$M(s,\chi) = \sum_{\substack{d \le R \\ e \le Qz}} \theta_d \lambda_e \chi([d,e])[d,e]^{-s} \tag{7.1}$$

where R, Q, z are real parameters to be specified later. The coefficients λ_d are defined by

$$\lambda_d = \begin{cases} \mu(d) & \text{if} \quad 1 \le d \le z\,, \\ \mu(d)\dfrac{\log(Qz/d)}{\log Q} & \text{if} \quad z < d \le Qz\,, \\ 0 & \text{if} \quad d > Qz\,. \end{cases}$$

We will need a result of Graham [G1], which states that

$$\sum_{z < n \le u} \Big[\sum_{d|n} \lambda_d\Big]^2 \le \frac{u}{\log Q} \cdot \left(1 + O\left(\frac{1}{\log Q}\right)\right).$$

Using this and partial summation, we see that if $1/2 < \alpha < 1$ then

$$\sum_{z < n \le u} \Big[\sum_{d|n} \lambda_d\Big]^2 n^{1-2\alpha} \le \frac{u^{2-2\alpha}}{(2-2\alpha)\log Q} \cdot \left(1 + O\left(\frac{1}{\log Q}\right)\right). \tag{7.2}$$

The coefficients θ_d are defined by

$$\theta_d = \frac{\mu(d)d}{\phi(d)} \sum_{\substack{r \le R/d \\ (r,d)=1}} \frac{\mu^2(r)}{\phi(r)} \left[\sum_{r \le R} \frac{\mu^2(r)}{\phi(r)}\right]^{-1}$$

if $d \le R$ and $\theta_d = 0$ if $d > R$. These coefficients occur in Selberg's upper bound sieve (see **[H-R]**, Chapter 3, for example). For later reference, we quote two known results on θ_d. It is well known (**[H-R]**, equation (3.1.8)) that

$$|\theta_d| \le 1 .$$

Moreover, Hooley (**[H]**, Section 3) has shown that

$$\sum_{\substack{d,e \\ (de,q)=1}} \theta_d \theta_e [d,e]^{-1} \le \frac{q}{\phi(q)} \frac{1}{\log R} . \tag{7.3}$$

For later use, we define

$$G(s, \chi) = \sum_{d, e \le R} \theta_d \theta_e \chi([d,e])[d,e]^{-s} . \tag{7.4}$$

We close this section with three lemmas that we will use in the proof of Theorem 3.

Lemma 7.1 *Let M and G be as defined in equations (7.1) and (7.4) respectively. Let $s = \sigma + it$. If $\sigma \le 1$ then*

$$M(s, \chi) \ll (RQz)^{(1-\sigma)} (\log RQz)^3 \text{ and } G(s, \chi) \ll R^{2(1-\sigma)} (\log R)^3 .$$

If $\sigma > 1$ then

$$M(s, \chi) \ll (\log RQz)^3 \text{ and } G(s, \chi) \ll (\log R)^3 .$$

Proof. Consider $M(s, \chi)$. If $\sigma \ge 1$, then

$$|M(s, \chi)| \le \sum_{\substack{d \le R \\ e \le Qz}} [d,e]^{-1} = \sum_{\substack{d \le R \\ e \le Qz}} \frac{(d,e)}{de} = \sum_{\substack{d \le R \\ e \le Qz}} \frac{1}{de} \sum_{\substack{r|d \\ r|e}} \phi(r)$$

$$= \sum_{r \le Qz} \frac{\phi(r)}{r^2} \sum_{d \le R/r} \frac{1}{d} \sum_{e \le Qz/r} \frac{1}{e} \ll \sum_{r \le Qz} \frac{(\log R)(\log Qz)}{r}$$

$$\ll (\log Qz)^2 \log R \ll (\log RQz)^3 .$$

If $\sigma < 1$ then

$$|M(s\,,\chi)| \le \sum_{\substack{d \le R \\ e \le Qz}} [d,e]^{-\sigma} \le (RQz)^{1-\sigma} \sum_{\substack{d \le R \\ e \le Qz}} [d,e]^{-1} \ll (RQz)^{1-\sigma} (\log RQz)^3 .$$

The proof for G is similar.

Our next lemma recasts Lemma 6.1 in a form more suitable for the next section.

Lemma 7.2 *Suppose $P = P_y$, and that χ is a non-principal character* mod q, *where $q | P$. Let $k \ge 3$ be an integer, and $K = 2^k$. Define $D = yd(P)$. Let σ_k, s, and τ be as in Lemma 6.1. If $\sigma \ge \sigma_k$ then*

$$L(s\,,\chi) \ll (P^{\frac{1}{K-2}} D^{\frac{2k^2}{K-2}} \tau)^{\frac{1-\sigma}{1-\sigma_k}} (\log P\tau)^3 .$$

Proof. This follows by replacing k with $k-3$ in the conclusion of Lemma 6.1.

Lemma 7.3 *Let χ be a primitive character* mod q. *Let $s = \sigma + it$ be a complex number with $1 < \sigma \le 2$, and set $\tau = |t| + 1$. Then*

$$\sum_{\substack{\rho \\ L(\rho\,,\chi)=0}} \operatorname{Re} \frac{1}{s-\rho} \ll \frac{1}{\sigma-1} + \log q\tau .$$

Proof. From equations (17) and (18) of Chapter 12 of Davenport's book, we see that

$$\operatorname{Re} \frac{L'}{L}(s,\chi) = -\frac{1}{2}\log\frac{q}{\pi} - \frac{1}{2}\operatorname{Re} \frac{\Gamma'}{\Gamma}(\frac{1}{2}s + \frac{1}{2}\kappa) + \sum_{\substack{\rho \\ L(\rho\,,\chi)=0}} \frac{1}{s-\rho} ,$$

where $\kappa = (1 - \chi(-1))/2$. Now

$$\left|\frac{L'}{L}(s,\chi)\right| \leq \sum_{n=1}^{\infty} \Lambda(n) n^{-\sigma} \ll (\sigma - 1)^{-1} ,$$

and

$$\mathrm{Re}\, \frac{\Gamma'}{\Gamma}(\frac{1}{2}s + \frac{1}{2}\kappa) \ll \log \tau ,$$

and the lemma follows.

8. Proof of Theorem 4

Recall from the statement of Theorem 4 that $N_1(\alpha)$ is the number of zeros of $\prod_{m|P} L(s, \chi_m)$ in the rectangle

$$\alpha \leq \sigma \leq 1 \ , \ |t| \leq \log P .$$

(For brevity, we shall write P in place of P_y throughout this section.)

We begin by making some simplifying assumptions. Since the number of zeros of $L(s, \chi_m)$ with imaginary part $\leq \log P$ is $\ll (\log P)^2$, we have the trivial estimate

$$N_1(\alpha) \ll 2^{\pi(y)} (\log P)^2 \ll \exp(\frac{c_0 \log P}{\log\log P}) . \tag{8.1}$$

This is better than the estimate we are trying to prove when

$$\alpha \geq 1 - \frac{c_1}{\log\log P} ,$$

which we henceforth assume. Now $\zeta(s)$ has no zeros in the range

$$\sigma \geq 1 - \frac{c_1}{\log\log P} \ , \ |t| \leq \log P .$$

(This follows, for example, from Theorem 5.17 of Titchmarsh [T].) Thus we may assume that all zeros being counted in $N_1(\alpha)$ are zeros of $L(s, \chi_m)$ with χ_m non-principal. We may also assume that P is larger than any fixed absolute constant, for otherwise, the desired result follows from (8.1).

Let $k \geq 3$ be a positive integer to be chosen later, and let $K = 2^k$. Let j be the real number defined by

$$\alpha = 1 - \frac{j}{K-2} \ .$$

We will eventually choose k so that

$$j \leq k/2 \ . \tag{8.2}$$

In the previous section, we introduced the parameters R, Q, and z. Now we introduce yet another parameter, x, and we define

$$R = Q = P^{\frac{1}{k}} \ , z = P^{\frac{b}{k}} \ , x = P^{\frac{c}{k}} \ ,$$

where b and c will be chosen later. We also define

$$Z - z/Q \ \text{ and } \ X - Qx \ .$$

Lemma 8.1 *Suppose* χ *is a non-principal character* mod q *and that* $q|P$. *Suppose* $\rho = \beta + i\gamma$ *is a zero of* $L(s, \chi)$ *with*

$$\alpha \leq \beta < 1 \ , \ |\gamma| \leq \log P \ .$$

If

$$k \leq \sqrt{\log\log P} \tag{8.3}$$

and if

$$14 + 2b \le c \tag{8.4}$$

then

$$\Big| \sum_{z<n\le X} \Big(\sum_{d|n} \lambda_d\Big)\Big(\sum_{e|n} \theta_e\Big)\chi(n) n^{-\rho} e^{-n/x} \Big| = 1 + O(P^{-1/(K-2)}) .$$

Proof. Using a well known Mellin transform, we see that

$$\begin{aligned} 1 + \sum_{z<n} \Big(\sum_{d|n} \lambda_d\Big)\Big(\sum_{e|n} \theta_e\Big)\chi(n) n^{-\rho} e^{-n/x} \\ = \frac{1}{2\pi i}\int_{2+i\infty}^{2-i\infty} L(s+\rho,\chi) M(s+\rho,\chi) x^s \Gamma(s)\, ds \end{aligned} \tag{8.5}$$

Now let $\sigma_k = 1 - k/(K-2)$, and move the contour to $\sigma = \sigma_k - \beta$. The vertical integral contributes

$$\begin{aligned} &\ll \int_0^\infty (PD^{2k^2} R^k Q^k z^k)^{\frac{1}{K-2}} x^{\sigma_k - \beta} (\log P\tau)^6 |\Gamma(\sigma_k - \beta + it)| (t+1)\, dt \\ &\ll (PD^{2k^2} R^k Q^k z^k x^{j-k})^{\frac{1}{K-2}} (\log P)^6 K . \end{aligned} \tag{8.6}$$

Our hypothesis $k \le \sqrt{\log\log P}$ means that $K \ll \log P$. Moreover, for any $c_2 \ge \log 2$,

$$d(P) = 2^{\pi(y)} \le \exp\Big(\frac{c_2 \log P}{\log\log P}\Big) .$$

Therefore $D^{2k^2} \le P^2$, and (8.6) is

$$\ll (P^3 R^k Q z^k x^{j-k})^{\frac{1}{K-2}} (\log P)^7 .$$

Using (8.2),(8.3), and (8.4), we see that the above is

$$\ll P^{-2/(K-2)} (\log P)^7 \ll P^{-1/(K-2)} .$$

For the tail of the series in (8.5), we note that

$$\sum_{n>X} (\sum_{d|n} \lambda_d)(\sum_{e|n} \theta_e)\chi(n) n^{-\rho} e^{-n/x} \ll \sum_{n>X} e^{-n/x} \ll e^{-X/x} \ll P^{-1/(K-2)} .$$

Lemma 8.1 shows that a zero of an L-function forces a certain Dirichlet polynomial to be unusually large. In the next two lemmas, we will show that this cannot happen to frequently. We begin by defining

$$J(s) = \frac{Z^s - (ZQ)^s - X^s + (XQ)^s}{s^2 \log Q} .$$

and

$$K(n) = \frac{1}{2\pi i}\int_{2+i\infty}^{2-i\infty} J(s) n^{-s}\, ds .$$

Using residue calculus, it is easy to show that

$$\frac{1}{2\pi i}\int_{2+i\infty}^{2-i\infty} \frac{w^s}{s^2}\, ds = \begin{cases} \log w & \text{if } w \ge 1 \\ 0 & \text{if } 0 < w < 1 . \end{cases}$$

Consequently, $K(n) = 1$ if $z = ZQ < n \le X$, $K(n) = 0$ if $n > XQ$ or $n < Z$, and $K(n) \ge 0$ for all n. Next, we define

$$B(s, \chi) = \sum_{n=1}^{\infty} K(n)\chi(n)(\sum_{d|n}\theta_d)^2 n^{-1-s} .$$

Lemma 8.2 *Suppose* $0 \le \sigma \le 2(1-\alpha)$ *and* $|t| \le \log P$. *If* $k \ge 4$ *and*

$$b \ge 6 \tag{8.7}$$

then

$$B(s, \chi) = E(\chi)\frac{\phi(q)}{q} G(1, \chi) J(-s) + O(P^{-1/(K-2)}) ,$$

where $E(\chi) = 1$ *if* χ *is principal and* 0 *otherwise.*

Proof. Note that

$$B(s, \chi) = \frac{1}{2\pi i} \int_{2+i\infty}^{2-i\infty} L(1+s+w, \chi)\, G(1+s+w, \chi) J(w)\, dw .$$

Move the contour to Re $w = \sigma_k - 1 - \sigma$. At $w = -s$ there is a residue of

$$E(\chi) \frac{\phi(q)}{q} G(1, \chi) J(-s) .$$

The integral along Re $w = \sigma_k - 1 - \sigma$ contributes

$$\ll \int_0^P (PD^{2k^2} R^{2k} Z^{-k})^{1/(K-2)} (\log P)^6 \frac{dt}{t+1}$$
$$+ \int_P^\infty (P^{\frac{1}{6}} D^{\frac{2}{3}} tR)^{2k/(K-2)} t^{-2} (\log t)^6\, dt .$$

Here, we have used Lemma 7.2 for the first integral and Lemma 6.1 with $k = 0$ for the second integral. The first integral contributes

$$\ll (PD^{2k^2} R^{2k} Z^{-k})^{\frac{1}{K-2}} (\log P)^7 \ll P^{-1/(K-2)}$$

Using the hypothesis $k \geq 4$, we see that the second integral contributes $\ll P^{-1/7} (\log P)^7 \ll P^{-1/(K-2)}$.

Now suppose that $\rho = \beta + i\gamma$ is a zero that we wish to count; i.e. $L(\rho, \chi_m) = 0$ for some $m | P$,

$$\alpha \leq \beta \leq 1, \text{ and } |\gamma| \leq \log P . \tag{8.8}$$

For each $m | P$, let

$$S(\chi_m) = \{\rho - \alpha : L(\rho, \chi_m) = 0 \text{ and } \rho \text{ satisfies (8.8)} \} ,$$

and let S be the set of all relevant ordered pairs of the form $(\rho - \alpha, \chi_m)$. In other words,

$$S = \bigcup_{m | P} \{(s, \chi_m) : s \in S(\chi_m)\} .$$

Lemma 8.3 *Let $a(n)$ be a sequence of complex numbers. Then*

$$\sum_{(s,\chi)\in S} \Big| \sum_{z<n\le X} a(n)\chi(n)\Big(\sum_{d|n}\theta_d\Big) n^{-s-1/2}\Big|^2 \le L \sum_{z\le n\le Y} |a(n)|^2 , \tag{8.9}$$

where

$$L \ll k(1-\alpha)\log Q + |S| P^{-1/(K-2)} .$$

Proof. By duality, proving (8.9) is equivalent to proving

$$\sum_{z<n\le X} \Big| \sum_{(s,\chi)\in S} b(s,\chi)\chi(n)\Big(\sum_{d|n}\theta_d\Big) n^{-s-1/2}\Big|^2 \le L \sum_{(s,\chi)\in S} |b(s,\chi)|^2 \tag{8.10}$$

for any set of complex numbers $b(s,\chi)$. Now the left-hand side of (8.10) is

$$\le \sum_n K(n) \Big| \sum_{(s,\chi)\in S} b(s,\chi)\chi(n)\Big(\sum_{d|n}\theta_d\Big) n^{-s-1/2}\Big|^2$$
$$= \sum_{(s,\chi)\in S} \sum_{(s',\chi')\in S} b(s,\chi)\, \overline{b(s',\overline{\chi}')} B(s+\overline{s}', \chi\overline{\chi}') .$$

Using Lemma 8.1 and (7.3), we see that the above is

$$\le \max_{\chi} \frac{1}{\log R} \sum_{s\in S(\chi)} \sum_{s'\in S(\chi)} |J(-s-\overline{s}')\, b(s,\chi)\, \overline{b(s',\chi)}|$$
$$+ \; O\Big(\sum_{(s,\chi)\in S} \sum_{(s',\chi')\in S} |b(s,\chi)|\,|b(s',\chi')|\, P^{-1/(K-2)}\Big) . \tag{8.11}$$

By Cauchy's inequality,

$$2|b(s,\chi)|\,|b(s',\chi')| \le |b(s,\chi)|^2 + |b(s',\chi')|^2 ,$$

and thus the error term in (8.11) is

$$\ll |S|\, P^{-1/(K-2)} \sum_{(s,\chi)\in S} |b(s,\chi)|^2 .$$

Now we need to bound

$$\max_{\chi} \max_{s' \in S(\chi)} \sum_{s \in S(\chi)} |J(-s - \overline{s'})| \tag{8.12}$$

First of all, we see from the definition of J that if Re $w \le 0$ then

$$|J(w)| \le \frac{4}{|w|^2 \log Q} .$$

We also have

$$J(w) = \frac{1}{\log Q} \int_0^{\log Q} \int_{\log Z}^{\log X} e^{-wu - wv}\, du\, dv ,$$

and therefore $|J(w)| \le \log(X/Z)$. Consequently, the inner sum in (8.12) is

$$\ll \log Q \sum_{s \in S(\chi)} \min\left\{\frac{1}{|s + \overline{s'}|^2 (\log Q)^2}, 1\right\}.$$

Now write $s = \rho - \alpha = \beta - \alpha + i\gamma$ and $s' = \rho' - \alpha = \beta' - \alpha + i\gamma'$, and let

$$s_0 = 1 + (\log Q)^{-1} + i\gamma' .$$

We claim that

$$\min\left\{\frac{1}{|s + \overline{s'}|^2 (\log Q)^2}, 1\right\} \ll \frac{(\sigma_0 - \beta)^2}{|s_0 - \rho|^2} . \tag{8.13}$$

For if $|\gamma - \gamma'| \le (\log Q)^{-1}$, then

$$1 \le \frac{(\sigma_0 - \beta)^2 + (\log Q)^{-2}}{(\sigma_0 - \beta)^2 + (\gamma - \gamma')^2} \ll \frac{|\sigma_0 - \beta|^2}{|s_0 - \rho|^2}$$

since $\sigma_0 - \beta \ge \sigma_0 - 1 = (\log Q)^{-1}$. On the other hand, if $|\gamma - \gamma'| \ge (\log Q)^{-1}$ then

$$\frac{|s_0-\rho|^2}{|s+\bar{s}'|^2(\log Q)^2} \le \frac{(\sigma_0-\beta)^2+(\gamma-\gamma')^2}{\{(\beta-\beta'+2\alpha)^2+(\gamma-\gamma')^2\}(\log Q)^2}$$
$$\le \frac{(\sigma_0-\beta)^2+(\gamma-\gamma')^2}{(\gamma-\gamma')^2(\log Q)^2} \le (\sigma_0-\beta)^2+(\log Q)^{-2} \ll (\sigma_0-\beta)^2 .$$

These last two inequalities show that (8.13) holds in any case.

Now

$$\sum_\rho \frac{|\sigma_0-\beta|^2}{|s_0-\rho|^2} \le (\sigma_0-\alpha)\sum_\rho \operatorname{Re}\frac{1}{s_0-\rho} \ll (1-\alpha)\log P$$

by Lemma 7.3. Thus we may take

$$L \ll \frac{(1-\alpha)\log P\log Q}{\log R} + |S|P^{\frac{-1}{K-2}} \ll k(1-\alpha)\log Q + |S|P^{-1/(K-2)}$$

in (8.9) and (8.10).

Now we are ready to prove Theorem 3. We take $b=6$ and $c=26$, so that (8.4) and (8.7) are satisfied. By Lemmas 8.1 and 8.3,

$$|S|\,(1+O(P^{-1/(K-2)})) \le$$
$$\sum_{(s,\chi)\in S}\Big|\sum_{z<n\le X}\Big(\sum_{d|n}\lambda_d\Big)\Big(\sum_{e|n}\theta_e\Big)\chi(n)n^{-\rho}e^{-n/x}\Big|^2 \ll$$
$$\{|S|P^{-1/(K-2)}+k(1-\alpha)\log Q\}\sum_{z<n\le X}\Big(\sum_{d|n}\lambda_d\Big)^2 n^{1-2\alpha} .$$

Using (7.2), we see that

$$N_1(\alpha) = |S| \ll kX^{2-2\alpha} \ll kP^{54(1-\alpha)/k} . \tag{8.14}$$

To complete the proof, we need to choose k and to verify that (8.2) holds. Suppose first that $\alpha \ge 1-\eta_1$. In this case, we take $k=k_0=[\sqrt{\log\log P}]$. Since

$$\alpha \ge 1-\frac{k}{2(K-2)} ,$$

(8.2) is true. The theorem follows from (8.14).

Now suppose that $\alpha < 1 - \eta_1$. We choose k to be the unique integer that satisfies

$$1 - \frac{k}{2(K-2)} \le \alpha < 1 - \frac{k+1}{2(2K-2)} .$$

Equation (8.2) is clearly satisfied. Moreover,

$$k \gg \log \frac{1}{1-\alpha} ,$$

so

$$N_1(\alpha) \ll \sqrt{\log\log P}\exp(c_3(1-\alpha)\log P(\log \frac{1}{1-\alpha})^{-1}) .$$

The $\sqrt{\log\log P}$ factor may be absorbed into the exponent at the cost of increasing c_3.

9. Proof of Theorem 1

At this point, we resume the analysis we started in Section 2. We need to bound the expression T that occurs in (2.4), and this in turn can be reduced to bounding the expression

$$I = \int_{1-\varepsilon}^{1} x^{\alpha - 1} N_1(\alpha)\, d\alpha .$$

Assume that y one of the values described in Theorem 3. We again write P in place of P_y, and we set

$$\eta = \frac{C_1(\log\log P)^{1/2}}{2 \log P} .$$

Let η_1 be as in Theorem 4, and let η_2 be a parameter to be chosen later.

Then

$$I = \int_{\eta}^{\varepsilon} x^{-\beta} N_1(1-\beta)\, d\beta = \int_{\eta}^{\eta_1} + \int_{\eta_1}^{\eta_2} + \int_{\eta_2}^{\varepsilon} .$$

Let us denote the three integrals in the last line as I_1 , I_2 , and I_3.

From the bounds for $N_1(\alpha)$ given in Theorem 4 and from the trivial bound (8.1), we have

$$I_1 \ll \int_{\eta}^{\eta_1} \exp(-\beta \log x + \frac{c_1 \beta \log P}{\sqrt{\log\log P}})\, d\beta,$$
$$I_2 \ll \int_{\eta_1}^{\eta_2} \exp(-\beta \log x + \frac{c_1 \beta \log P}{\log(1/\beta)})\, d\beta ,$$

and

$$I_3 \ll \int_{\eta_2}^{\varepsilon} \exp(-\beta \log x + \frac{c_1 \log P}{\log\log P})\, d\beta ,$$

where $c_1 = \max(C_2, c_0)$. (C_2 is the constant occurring in Theorem 4 and c_0 is the constant occurring in (8.1).) Motivated by the bounds for I_2 and I_3, we choose η_2 so that

$$\frac{\eta_2}{\log(1/\eta_2)} \approx \frac{1}{\log\log P} .$$

A choice that accomplishes this is $\eta_2 = (\log\log\log P)/\log\log P$. To make I_2 sufficiently small, we need to have

$$-\log x + \frac{c_1 \log P}{\log(1/\eta_2)} \le -\frac{1}{2} \log x ,$$

i.e.

$$\frac{2c_1 \log P}{\log x} \le \log\log\log P - \log\log\log\log P .$$

This will be satisfied if we choose x so that $\log P = c_2 \log x \log\log\log x$ for some c_2 sufficiently small. With these choices of x and η_2, we have

$$I_\nu = o\left(\frac{1}{\log x}\right)$$

for $\nu = 1, 2$, and 3. Thus the sum in (2.1) is $\gg x 2^{-\pi(y)}$ when $y = c \log x \log\log\log x$. Thus there is some prime p with $x^{1/2} \le p < x^2$ and $n_p > y$; i.e. $n_p > c \log p \log\log\log p$.

For the application to Paley graphs mentioned in Section 1, we need to show that there are infinitely many primes $p \equiv 1 \bmod 4$ for which $n_p \gg \log\log\log p$. To do this, we modify the sum in (2.1) to

$$2^{-\pi(y)-1} \sum_{x^{1/2} < p \le x^2} (\log p)(1 + \chi_4(p))\,(e^{-p/2x} - e^{-p/x}) \prod_{p_1 \le y} \left(1 + \left(\frac{p_1}{p}\right)\right),$$

where χ_4 is the non-principal character mod 4. This modified sum may be treated in a manner very similar to our treatment of (2.1); we leave the details to the reader.

Added in Proof: Andrew Odlyzko (private communication) has pointed out that the way we stated the results about random graphs leaves open the possibility that they might be consistent with this result on Paley graphs. That is, the expected value is $(1 + o(1)) 2 \log p / \log 2$, but the variance could be so large that cliques of size $\log p \log\log\log p$ could happen infinitely often. However, as Odlyzko points out, the distribution of maximal clique numbers is very sharply peaked, and this does not happen.

REFERENCES

[CGW] F.R.K. Chung, R.L. Graham, and R.M. Wilson. Quasi-random graphs, preprint.

[C] J. G. van der Corput, Verschärfung der Abschätzungen beim Teilerproblem, Math. Annalen **89** (1922), 39-65.

[D] H. Davenport, *Multiplicative Number Theory* (2nd edition, revised by H.L. Montgomery), GTM 74, Springer-Verlag, Berlin- Heidelberg-New York, 1980.

[F] V.R. Fridlender, On the least n-th power non-residue, Dokl. Akad. Nauk SSSR **66** (1949), 351-352.

[G1] S. W. Graham, An asymptotic formula related to the Selberg sieve, J. Number Theory **10** (1978), 83-94.

[G2] S. W. Graham, On Linnik's constant, Acta Arith. **39** (1980), 163-179.

[H-R] H. Halberstam and H.-E. Richert, *Sieve Methods*, Academic Press, London, 1974.

[He] D. R. Heath-Brown, Hybrid bounds for Dirichlet L-functions, Invent. Math. **47** (1978), 148-170.

[Ho] C. Hooley, On the Brun-Titchmarsh theorem, J. Reine Angew. Math. **255** (1972), 60-79.

[I] A. Ivić, *The Riemann zeta-function*, Wiley-Interscience, New York, 1985.

[J] M. Jutila, On Linnik's Constant, Math. Scand. **41** (1977), 45-62.

[Ma] H. Maier, Chains of large gaps between consecutive primes, Adv. in Math. **39** (3) (1981), 257-269.

[Mo] H.L. Montgomery, *Topics in Multiplicative Number Theory*, Lecture Notes in Math. 227, Springer-Verlag, New York, 1971.

[R] C. Ringrose, The q-analogue of van der Corput's method, Thesis, University of Oxford, Oxford, 1985.

[Sa] H. Salié, Über den kleinsten positiven quadratischen Nichtrest nach einer Primzahl, Math. Nachr. **3** (1949), 7-8.

[Sc] W. Schmidt, *Equations over finite fields: an elementary approach*, Lecture Notes in Math. 536, Springer-Verlag, New York, 1976.

[Sr] B. R. Srinivasan, The lattice point problem of many-dimensional hyperboloids II, Acta Arith. **8** (1963) 173-204.

[T] E. C. Titchmarsh, *The theory of the Riemann zeta-function*, (2nd edition, revised by D.R. Heath-Brown) Clarendon Press, Oxford, 1986.

[W] A. Weil, On some exponential sums, Proc. Nat. Acad. Sci. U.S.A., **34** (1948) 204-207.

C. J. Ringrose
3 The Laurels
Gledhow Lane
Leeds LS8 1PD
England

S. W. Graham
Department of Mathematics
Michigan Technological University
Houghton, MI 49931

Some Conjectures in Analytic Number Theory And their Connection With Fermat's Last Theorem

ANDREW GRANVILLE

Dedicated to P. Bateman on his retirement

1. Introduction

The first case of Fermat's Last Theorem is the assertion:

For all odd primes p there are no integer solutions x, y, z to

$$x^p + y^p + z^p = 0 \quad \text{with} \quad p \nmid xyz. \tag*{$(1)_p$}$$

In 1823 Sophie Germain showed that if $2p + 1$ is also prime then $(1)_p$ has no solutions and this has been generalized as follows (see [15]):

Lemma 1. *For any fixed positive integer m, with $m \equiv 2$ or $4 \pmod 6$, define N_m to be the product, over all pairs α, β of mth roots of unity, of $(1 + \alpha + \beta)$. If p and $q = mp+1$ are both primes, where p does not divide m and q does not divide N_m, then $(1)_p$ has no solutions in integers x, y and z.*

By finding prime pairs of the form p, $q = mp+1$ one hopes to be able to use Lemma 1 to establish the first case of Fermat's Last Theorem. Unfortunately it is not presently known how to formulate a 'reasonable' conjecture in analytic number theory that would achieve this goal; however, in this paper, we examine what exactly a number of quite different conjectures of analytic number theory actually imply about the set of primes p for which there is an integer solution x, y, z of $(1)_p$.

This paper may be seen as a continuation of [8] where we investigated the consequences (for Fermat's Last Theorem) of a variety of conjectures from algebraic, combinatorial and transcendental number theory.

2. Statement of Results

In order to exploit Lemma 1 it is obviously necessary to obtain information about primes q in the arithmetic progression $1(\mathrm{mod}\,p)$, for which 3 does not divide $q-1$. A famous result of Linnik [13] implies that there exists a constant $L>0$ such that the least such prime q is $<p^L$. A recent result of Bombieri, Friedlander and Iwaniec [3] implies that we may take $L=2$ for almost all primes p. However we actually need to use stronger estimates than these. We start by assuming a conjecture that has been formulated by each of Heath-Brown [12], McCurley [14] and Wagstaff [17] independently.

Conjecture 1. *There exists a constant* $c_1 > 0$ *such that, for any given integer* d*, the least prime in the arithmetic progression* $a(\mathrm{mod}\ d)$ *is less than* $c_1\phi(d)\log^2 d$ *whenever* $(a, d) = 1$.

From this we will deduce

Theorem 1. *If Conjecture 1 is true then*

$$\#\{\textit{primes } p \leq x\text{: } (1)_p \textit{ has solutions}\} \ll \log^7 x.$$

In a recent paper Adleman and Heath-Brown [1] showed what effect three conjectures in analytic number theory have on $(1)_p$. The third of these conjectures was proved by Fouvry [6] and allowed them to state that $(1)_p$ has no solutions for $\gg x^{2/3}$ prime exponents $p \leq x$. Michael Filaseta has noted that their results imply that there exist arbitrarily large values of x for which this can be improved to $\gg x/\log x$ prime exponents $p \leq x$ (we give his proof in Section 5). We now state a new conjecture, which is a modification of the one that Fouvry proved. Define, as usual, $\pi(x;d,a)$ to be the number of primes $\leq x$ that are $\equiv a(\mathrm{mod}\,d)$, and let

$$\pi^*(x;d) = \pi(x;d,1) - \pi(x;3d,1)\,.$$
$$(= \#\{\text{primes } q \leq x\text{: } q \equiv 1(\mathrm{mod}\,d),\ q \not\equiv 1(\mathrm{mod}\,3d)\})$$

Conjecture 2. *There exists* θ, $2/3 < \theta < 1$, *such that*

$$\sum_{x^\theta < p < 2x^\theta} \pi^*(x;p) \gg x/\log^2 x.$$

Of the three approaches presented in this paper, perhaps this one has the greatest chance of success (in the sense that we have real hope of Conjecture 2 being proved in the forseeable future). We will show

Theorem 2. *If Conjecture 2 is true then* $(1)_p$ *does not have solutions for* $\gg \pi(x)$ *primes* $p \le x$.

A minor modification of the proof of Theorem 2 leads to a new and shorter proof of the results of Adleman and Heath-Brown, and of Filaseta (see section 5).

As early as 1904, Dickson [4] had conjectured that, with certain obvious restrictions, an arbitrary set of linear polynomials will simultaneously take on prime values infinitely often. Hardy and Littlewood conjectured asymptotic formulae for how often this happens for various sets of polynomials in [11]. These conjectures were extended to arbitrary sets of polynomials by Schinzel and Sierpinski [16] and then modified to obtain greater accuracy by Bateman and Horn [2]. An explicit form of these conjectures restricted to certain linear polynomials is given here:

Conjecture 3. *Suppose that* $m_1, m_2, ..., m_k$ *are given positive integers and let* $N(x; m_1, m_2, ..., m_k)$ *be the number of primes* p, $x < p \le 2x$, *for which* $m_1p+1, m_2p+1, ..., m_kp+1$ *are also prime. Then*

$$N(x; m_1, m_2, ..., m_k) = C(m_1, ..., m_k)\frac{x}{(\log x)^{k+1}}\{1 + o(1)\}, \qquad (2)$$

where $C(m_1, ..., m_k) = \prod_{p \text{ prime}} \frac{(1-w_m(p)/p)}{(1-1/p)^{k+1}}$ *and* $w_m(p)$ *is the number of distinct solutions* $y(\text{mod } p)$ *of* $y(ym_1+1)(ym_2+1) \dots (ym_k+1) \equiv 0 \pmod p$.

Just as one should view Conjecture 3 as a generalization of Dirichlet's Theorem (for primes in arithmetic progressions) from one to many linear polynomials, so one should view the next conjecture as a generalization of a weak form of the Siegel-Walfisz Theorem from one to many linear polynomials.

Conjecture 3^u. *For any fixed integer* k *and positive real* d, *the error term* $o(1)$ *in (2) depends only on* k *and* d *whenever each* $m_i \le d \log x$.

A consequence of our Proposition 2 is that Conjecture 3^u implies

Conjecture 3^*. *For any given* $\varepsilon > 0$, *there exists a constant* $c(\varepsilon) > 0$ *such that if* x *is sufficiently large then there are less than* $\varepsilon\pi(x)$ *primes* $p \le x$ *with* $mp+1$ *composite for every* $m \le c(\varepsilon)\log x$ *and not divisible by 3.*

In Section 6 we will deduce from Conjecture 3^* and Lemmas 1 and 2 that

$$\#\{\text{primes } p \le x\text{: } (1)_p \text{ has solutions}\} = o(\pi(x)) .$$

Thus we will have proved

Theorem 3. *If Conjecture* 3^u *is true then* $(1)_p$ *has no solutions for almost all primes* p; *that is* $\#\{\text{primes } p \le x\text{: } (1)_p \text{ has solutions}\} = o(\pi(x))$.

In [9] we saw how the methods used in proving Sophie Germain's Theorem could be applied to studying any Diophantine Equation. For the rest of this section suppose that $f(X_1, \ldots, X_n) \in \mathbf{Z}[X_1, \ldots, X_n]$ is a given homogenous polynomial. For a given prime p we investigate whether there are integer solutions $x_1, x_2, \ldots, x_n$ to

$$f(x_1^p, x_2^p, \ldots, x_n^p) = 0 . \qquad (3)_p$$

In [9] we proved a rather technical analogue to Lemma 1:

Lemma 1'. *For any given homogenous polynomial f in n variables, there is a finite (computable) set of positive integers β such that if m is a positive, even integer, not divisible by any element of β, then there exists a non-zero integer $N_m(=N_m(f))$ such that if p and $q = mp+1$ are both primes, and q does not divide N_m then $(3)_p$ has no 'non-trivial' integer solutions. Moreover there are $\ll_f m^{n-1}$ primes q that divide N_m.*

It is clear that Lemma 1' is useless if 1 or 2 are in the set β (for then all positive even integers m are divisible by an element of β!). We call f "admissible" if neither 1 nor 2 are elements of β (it is easily shown that there are relatively few inadmissible polynomials f).

Now, as any integer ≥ 3 is divisible by some element of $Q := \{4\} \cup \{\text{the odd primes}\}$, we can certainly replace β in Lemma 1' by a finite subset $\beta(f)$ of Q, whenever f is admissible. Then, by the methods used to prove Theorems 1, 2 and 3 (and by the methods of [1]) we are able to give various results on $(3)_p$.

Theorem 1 generalized. *If f is an admissible polynomial and Conjecture 1 is true then*

$$\#\{\textit{primes } p \leq x : (3)_p \textit{ has non-trivial solutions}\} \ll \log^{2n+1} x .$$

For any odd prime p, $p \notin \beta(f)$, define

$$\begin{aligned} \pi_\beta(x;p) &= \#\{\text{primes } q \leq x : p \mid q-1 \text{ but } b \nmid q-1 \text{ for all } b \in \beta\} \\ &= \sum_{d \mid \prod_{b \in \beta} b} \mu(d)\, \pi(x; 2dp, 1) . \end{aligned}$$

Conjecture 2'. *For a given finite subset β of Q and positive integer $n \geq 3$, there exists θ, $1-1/n < \theta < 1$, for which*

$$\sum_{x^\theta < p < 2x^\theta} \pi_\beta(x;p) \gg x/\log^2 x .$$

Theorem 2 generalized. *If f is an admissible polynomial and Conjecture 2' is true then $(3)_p$ does not have non-trivial solutions for $\gg \pi(x)$ primes $\leq x$.*

Theorem 3 generalized. *If f is an admissible polynomial and Conjecture 3^u is true then $(3)_p$ has no non-trivial integer solutions for almost all primes p; that is*

$$\#\{\text{primes } p \le x : (3)_p \text{ has non-trivial solutions}\} = o(\pi(x)).$$

In a similar fashion we may use Lemma 1' to apply the ideas of Adleman and Heath-Brown [1] and of Filaseta, to equation $(3)_p$. The conjectures of [1] (given below as Conjecture 5) can be generalized as follows:

Conjecture 4. *For a given finite subset β of Q and integer $n \ge 3$, there exists θ, $1 - 1/n < \theta < 1$ for which*

(a) $$\sum_{\substack{2<p<x^\theta \\ p \notin \beta(f)}} \left| \pi_\beta(x;p) - \rho(\beta)\frac{\pi(x)}{(p-1)} \right| \ll x/\log^3 x$$

where $\rho(\beta) = \prod_{b\in\beta}\{1 - 1/\phi(b)\}$; and

(b) $$\sum_{x^\theta<p\le x} \pi_\beta(x;p) \gg x/\log x.$$

Evidently the Elliott-Halberstam conjecture implies (a) which itself implies (b). Moreover, as in [1], we can show

Theorem 4. *If f is an admissible polynomial and Conjecture 4(a) is true then*

$$\#\{\text{primes } p \le x : (3)_p \text{ has non-trivial solutions}\} \ll x/\log^2 x.$$

Let T be the set of primes for which $(3)_p$ has no solutions, and let

$$\pi_T(x) = \sum_{p\in T, p\le x} 1.$$

Theorem 5. *If f is an admissible polynomial and Conjecture 4(b) is true then*

(i) $$\sum_{p\in T, p\le x} \frac{\log p}{p} \gg \log x$$

(ii) $\pi_T(x) \gg x^\theta$

(iii) *There are arbitrarily large values of x for which $\pi_T(x) \gg \pi(x)$.*

Theorems 5(i) and (ii) generalize results in [1] while Theorem 5(iii) generalizes Lemma 4(iii) (due to Filaseta) given below.

3. Exceptional Prime Pairs p,q

In order to be able to apply lemma 1 it is evidently necessary to estimate how many values of q divide N_m.

Lemma 2. *There exists a constant* $c_2 > 0$ *such that*

$$\#\{\text{prime pairs } p, q = mp+1:\ p|m \text{ or } q|N_m\} \ \le\ c_2 m^2,$$

for all positive integers $m \equiv 2$ *or* 4 *(mod 6).*

Proof: For each α and β, $|1+\alpha+\beta| \le 3$ and so $|N_m| \le 3^{m^2}$. Therefore there are $0(m^2)$ distinct primes q dividing N_m, and trivially $0(m)$ dividing m.

4. The Proof of Theorem 1

Proof: For a given prime p in the range $x < p \le 2x$, we know, by Conjecture 1, that there is a prime $q_p < 7c_1 x \log^2 x$ in the arithmetic progression a_p (mod 3p) where $a_p = p+1$ if $p \equiv 1$ (mod 3), $2p+1$ otherwise. So if $q_p = mp+1$ then $m \equiv 2$ or 4 (mod 6) and $m < 7c_1\log^2 x$. Therefore, by Lemmas 1 and 2 we have

$$\#\{\text{primes } p:\ x < p \le 2x \text{ and } (1)_p \text{ has solutions}\} \ \le \sum_{\substack{m<7c_1\log^2 x\\ m\equiv 2 \text{ or } 4 \ (\text{mod } 6)}} c_2 m^2$$

$$\ll \log^6 x.$$

Summing over the intervals $[2^{-i-1}x, 2^{-i}x]$ gives the result.

5. The Adleman-Heath-Brown approach

The Bombieri-Vinogradov Theorem states that for any $\varepsilon, A > 0$,

$$\sum_{q\le Q} \max_{(a,q)=1} \max_{y\le x} \left|\pi(y;\, q,a) - \frac{\pi(y)}{\phi(q)}\right| \ll_{A,\varepsilon} x/\log^A x$$

where $Q = x^{1/2-\varepsilon}$. Elliott and Halberstam [3] conjectured that this can be extended to $Q = x^{1-\varepsilon}$. This implies the case $\beta = \{3\}$, n=3 of Conjecture 4a), namely

Conjecture 5a. *There exists* θ, $2/3 < \theta < 1$, *such that*

$$\sum_{3<p\le x^\theta} \left|\pi^*(x;\, p) - \frac{\pi(x)}{2(p-1)}\right| \ll x/\log^3 x\,.$$

This, in turn, implies the case $\beta = \{3\}$, $n=3$ of Conjecture 4b):

Conjecture 5b. *There exists* θ, $2/3 < \theta < 1$, such that

$$\sum_{x^\theta < p \le x} \pi^*(x;p) \gg x/\log x .$$

Adleman and Heath-Brown [1] showed

Lemma 3. *If Conjecture 5a) is true then* $(1)_p$ *has solutions for* $\ll x/\log^2 x$ *prime exponents* $p \le x$.

Let T be the set of primes for which $(1)_p$ has no solutions. Adleman and Heath-Brown [1] also showed the first two parts of Lemma 4; the last part is due to Filaseta.

Lemma 4. *If Conjecture 5b) is true then*

(i) $$\sum_{x^\theta < p \le x, p \in T} \frac{\log p}{p} \gg \log x;$$

(ii) $\pi_T(x) \gg x^\theta$.

(iii) *There are arbitrarily large values of* x *for which* $\pi_T(x) \gg \pi(x)$.

Lemma 4(iii) follows immediately from Lemma 4(i) and

Lemma 5. *If* T *is a set of primes such that* $\sum_{p \le x, p \in T} \frac{\log p}{p} > c_3 \log x$ *for all* $x > x_0$ *(for some constant* $c_3 > 0$*) then there are arbitrarily large values of* x *for which* $\pi_T(x) > \frac{c_3}{2} \frac{x}{\log x}$.

Proof. Suppose not, so that $\pi_T(x) \le \frac{c_3}{2} \frac{x}{\log x}$ for all $x > x_1 (> x_0)$. Then, by forming a Riemann-Stieltjes integral, we have

$$\sum_{p \le x, p \in T} \frac{\log p}{p} = \int_{x_1}^{x} \frac{\log z}{z} \, d\pi_T(z) + O(1)$$

$$= \frac{\log x}{x} \pi_T(x) + \int_{x_1}^{x} \frac{(\log z - 1)}{z^2} \pi_T(z) dz + O(1)$$

$$\le \frac{c_3}{2} \int_{x_1}^{x} \frac{\log z}{z^2} \frac{z}{\log z} \, dz + O(1)$$

$$< \frac{c_3}{2} \log x + O(1) ,$$

giving a contradiction.

In 1985, Fouvry [6] showed that Conjecture 5b) holds for some $\theta > 0.6687$. From Lemma 4 one can immediately deduce a number of consequences for the set T.

Our approach here (that is, through Conjecture 2) evidently corresponds to assuming Conjecture 5b) on diadic intervals. This slight strengthening of the (already proved) Conjecture 5b) implies a significantly stronger result.

The next result not only implies Theorem 2 but also a different proof of Lemmas 4(ii) and (iii).

Proposition 1. *Suppose that* $c_4 > 0$ *and* $1 < \lambda < 3/2$ *are fixed constants. If, for given values of* z *and* x, *with* $x \le z^\lambda$, *we have*

$$\sum_{z<p\le 2z} \pi^*(x;\, p) \;\ge\; c_4 x/\log^2 x \tag{4}$$

then

$$\#\{z < p \le 2z\colon\ p \text{ prime and } p \in T\} \gg z/\log z$$

We see that Theorem 2 follows immediately by taking $\lambda = 1/\theta$ in Proposition 1. Moreover Conjecture 5b) implies that, for any given x, there are $\gg \log x$ values of z of the form 2^k, satisfying (4), in the range $x^\theta < z < x$. Therefore we see, from Proposition 1:

Corollary 1. *If Conjecture 5b) is true then (ii) and (iii) follow.*

Proof of Proposition 1: Let $y = x/z$ so that

$$S_1 := \sum_{m\le y,\ 3\nmid m} \#\{p \text{ prime: } z<p<2z,\ q = mp+1 \text{ is prime}, p\nmid m,\ q\nmid N_m\}$$

$$= \sum_{z<p\le 2z} \#\{m \le y\colon 3\nmid m \text{ and } mp+1 \text{ is prime}\} + O\Big(\sum_{m\le y} m^2\Big)$$

by Lemma 2

$$\ge \sum_{z<p\le 2z} \pi^*(x;p) + O(y^3) \gg x/\log^2 x$$

by (4), as $y^3 = o(x/\log^2 x)$. On the other hand, it is well known that if $r < s \le y$ then $N(z;r,s) \ll C(r,s)\, z/\log^3 z$ (see [10], Theorem 5.7). Therefore for $\beta = \{3\}$,

$$S_2 := \sum_{\substack{r<s\le y\\ 3\nmid rs}} N(z;\, r,\, s)$$

$$\ll F_{2,\beta}(y) z/\log^3 z$$

where $F_{2,\beta}(y) = \sum_{r<s\le y,\ 3\nmid rs} C(r,s)$. In Proposition 3 below we shall accurately estimate $F_{2,\beta}(y)$, but here a crude argument suffices:

By noting that, for $d = s-r > 0$,

$$C(r,s) \ll \frac{r}{\phi(r)} \frac{d}{\phi(d)} \frac{r+d}{\phi(r+d)},$$

we see that

$$F_{2,\beta}(y) \ll \sum_{d\le y} \frac{d}{\phi(d)} \sum_{r\le y} \frac{r}{\phi(r)} \frac{r+d}{\phi(r+d)}$$

$$\ll \sum_{d\le y} \frac{d}{\phi(d)} \left[\sum_{r\le y}\left[\frac{r}{\phi(r)}\right]^2\right]^{1/2} \left[\sum_{r\le y}\left[\frac{r+d}{\phi(r+d)}\right]^2\right]^{1/2}$$

by Cauchy's inequality,

$$\le \left[\sum_{n\le 2y}\left[\frac{n}{\phi(n)}\right]^2\right]^2$$

$$\ll y^2$$

from elementary considerations. Thus $S_2 \ll y^2 z/\log^3 z$, and so by Cauchy's inequality and Lemma 1, we have

$$\pi_T(2z) - \pi_T(z) \gg S_1^2/S_2 \gg z/\log z.$$

6. The Number of Small Primes in Arithmetic Progressions

In order to prove Theorem 3, we will use Conjecture 3^u to count, in a very precise way, the number of "small" primes in the arithmetic progression $1 \pmod p$. More precisely, for given subset β of Q and $d > 0$ we define, for each $g \ge 0$, $B(x, g)$ to be the number of primes p, $x < p \le 2x$, for which there are exactly g distinct integers $m_1, \ldots, m_g$, not divisible by any $b \in \beta$ and less than $d \log x$, such that each of $m_1p+1, m_2p+1, \ldots, m_gp+1$ is prime. We shall prove:

Proposition 2. *Suppose that Conjecture 3^u is true. Given any finite subset β of Q and $d > 0$ we have*

$$B(x, t) \sim \frac{e^{-\lambda}\lambda^t}{t!} \frac{x}{\log x} \qquad (as\ x \to \infty)$$

for any fixed non-negative integer t, where $\lambda = d\rho(\beta)$.

Assuming Proposition 2 we can now give the

Proof of Theorem 3: Fix $\varepsilon > 0$. By taking $\beta = \{3\}$, $t = 0$ and $d = -4\log\varepsilon$ $(= c(\varepsilon))$ in Proposition 2, we see that Conjecture 3^* follows from Conjecture $3''$.

Now by taking the integers $m \equiv 2$ or $4 \pmod 6$ with $m < d\log x$ in Lemma 1 we have

$$\#\{\text{primes } p\colon\ x < p \le 2x \text{ and } (1)_p \text{ has solutions}\}$$

$$\le \#\{\text{primes } p\colon\ x < p \le 2x \text{and there **does not** exist a prime } mp{+}1$$

$$\text{with } m < d\log x \text{ and } m \equiv 2 \text{ or } 4 (\text{mod } 6)\}$$

$$+ \sum_{m < d\log x,\ m\equiv 2 \text{ or } 4 (\text{mod } 6)} \#\{\text{primes } p\colon\ p \mid m \text{ or } q = mp{+}1 \mid N_m\}$$

$$\le \varepsilon\pi(x) + O(\log^3 x)$$

by Conjecture 3^* and Lemma 2,

$$\le 2\varepsilon\pi(x)$$

for all sufficiently large x. Summing over the intervals $[2^{-i-1}x, 2^{-i}x]$ gives the result.

The proof of Proposition 2 is very similar to that of Theorem 5 in [7] where we estimated, for any fixed $a \ne 0$, the number of integers n, $x < n \le 2x$, for which there are exactly g integers $m_1, \dots, m_g$, each less than $d\log x$, such that each of $m_1n{+}a, m_2n{+}a, \dots, m_gn{+}a$, is prime. In our proof we shall miss out some technical details that are identical to the proof of that result.

Now, for any fixed k,

$$\sum_{g\ge k} \binom{g}{k} B(x, g) = \sum_{\substack{1\le m_1<\dots<m_k<d\log x \\ b\nmid m_i \text{ for all } b\in\beta}}{}_{1} N(x;\, m_1, m_2, \dots, m_k)$$

$$= F_{k,\beta}(d\log x)\frac{x}{(\log x)^{k+1}}\ \{1 + o(1)\} \qquad (5)$$

by Conjecture $3''$, where $F_{k,\beta}(y) = \sum_1 C(m_1, \dots, m_k)$ and $\sum_1$ is the sum over sets of k positive integers $m_1 < m_2 < \cdots < m_k \le y$, none of which are divisible by any $b \in \beta$.

In Section 7 we will prove

Proposition 3. *For any fixed subset* β *of* Q, *integer* $k \ge 1$ *and real* $\varepsilon > 0$, *we have the estimate*

$$F_{k,\beta}(x) = \frac{1}{k!}\,(\rho(\beta)x)^k\,\{1 + O(x^{\varepsilon-1/2})\}\,. \tag{6}$$

The main idea of the proof of Proposition 2 is to use the combinatorial identity

$$B(x,t) = \sum_{k\ge t} A_k(x), \text{ where } A_k(x) = (-1)^{k-t}\begin{bmatrix}k\\t\end{bmatrix}\sum_{g\ge k}\begin{bmatrix}g\\k\end{bmatrix}B(x,g), \tag{7}$$

for each $t \ge 0$, together with the estimates (5) and (6). Unfortunately, as the o(1) in (2) depends on k, we cannot use the infinite sum in (7), **but** we are able to approximate $B(x, t)$ by $\sum_{k=t}^{n} A_k(x)$ for n large to prove the result.

Now, by (2) and (6) we have

$$A_k(x) = \frac{\lambda^t}{t!}\,\frac{(-\lambda)^{k-t}}{(k-t)!}\,\frac{x}{\log x}\,\{1 + o_k(1)\}. \tag{8}$$

Moreover, as $\left|\sum_{k=0}^{r}(-1)^{k+1}\begin{bmatrix}s\\k\end{bmatrix}\right| \le \begin{bmatrix}s\\r\end{bmatrix}$ for any integers $r, s \ge 1$, we have, for any fixed $n \ge t+1$,

$$\begin{aligned}\left|B(x,t) - \sum_{k=t}^{n} A_k(x)\right| &= \left|\sum_{g\ge n+1}\left[\sum_{k=0}^{n-t}(-1)^{k+1}\begin{bmatrix}g-t\\k\end{bmatrix}\right]\begin{bmatrix}g\\t\end{bmatrix}B(x,g)\right| \\ &\le \sum_{g\ge n+1}\begin{bmatrix}g-t\\n-t\end{bmatrix}\begin{bmatrix}g\\t\end{bmatrix}B(x,g) \\ &\le \begin{bmatrix}n\\t\end{bmatrix}\sum_{g\ge n}\begin{bmatrix}g\\n\end{bmatrix}B(x,g) \\ &\le \frac{\lambda^n}{t!}\,\frac{1}{(n-t)!}\,\frac{x}{\log x}\,\{1 + o(1)\}\end{aligned} \tag{9}$$

by (8). Define $s_n = \sum_{k\ge n}(-\lambda)^k/k!$ which tends to 0 as $n \to \infty$. Then

$$\left|B(x,t) - e^{-\lambda}\frac{\lambda^t}{t!}\,\frac{x}{\log x}\right| \le \left|B(x,t) - \sum_{k=t}^{n} A_k(x)\right|$$
$$+ \left|\sum_{k=t}^{n}\left\{A_k(x) - \frac{\lambda^t}{t!}\,\frac{(-\lambda)^{k-t}}{(k-t)!}\,\frac{x}{\log x}\right\}\right| + \frac{\lambda^t}{t!}\,\frac{x}{\log x}\left|e^{-\lambda} - \sum_{k=t}^{n}\frac{(-\lambda)^{k-t}}{(k-t)!}\right|$$

$$\leq \frac{\lambda^t}{t!}\frac{x}{\log x}\left\{\frac{\lambda^{n-t}}{(n-t)!} + o_n(1) + |s_{n-t+1}|\right\}$$

by (8) and (9),

7. Technical stuff: The Proof of Proposition 3

We evaluate $F_{k,\beta}(x)$ as $x \to \infty$, using essentially the same method as in the proof of Theorem 6 of [7]. In keeping control of the error term the details become extremely technical. We avoid these details here as they are very similar and refer the reader to [7].

Now $w_m(p)$ counts precisely the number of distinct residue classes (mod p) that contain an m_i $(i = 0, 1, \ldots, k)$ where $m_0 = 0$.

We define $\phi_k(d) = \prod_{p|d, p>k} (p-k)$ for each $d \geq 1$, and

$$c_5 = \prod_p \frac{\phi_{k+1}(p)/p}{(1-1/p)^{k+1}} .$$

It is easy to see that

$$C(m_1, \ldots, m_k) = c_5 \prod_{p|\theta(m)} \frac{p - w_m(p)}{\phi_{k+1}(p)} \tag{10}$$

where $\theta(m) = \left[\prod_{i=1}^{k} m_i\right]\left[\prod_{1\leq i<j\leq k} (m_j - m_i)\right]$, and so the product in (10) is finite. Therefore

$$F_{k,\beta}(x) = c_5 \sum\nolimits_1 \sum_{d|\theta(m)} \mu^2(d) \prod_{p|d}\left[\frac{p - \phi_{k+1}(p) - w_m(p)}{\phi_{k+1}(p)}\right]$$

$$= g_{k,\beta}\frac{x^k}{k!}\{1 + O_{k,\beta}(x^{\varepsilon-1/2})\}$$

after a considerable amount of rearrangement (exactly as in [7]) where

$$g_{k,\beta} = c_5 \sum_{d\geq 1} \frac{\mu^2(d)}{\phi_{k+1}(d)} \frac{1}{(ad)^k} \sum\nolimits_2 \prod_{p|d} [p - \phi_{k+1}(p) - w_m(p)] , \tag{11}$$

$a = \prod_{b\in\beta} b$ and $\sum_2$ is the sum over $1 \leq m_1, \ldots, m_k \leq ad$ with $b \nmid m_i$ for each i and $b \in \beta$.

Now, in order to evaluate the sum in (11) we need:

Lemma 6. *For each* $k \geq 1$ *we have*

(a) $$\lambda_k(p) = \sum_{0 \leq m_1, \ldots, m_k \leq p-1} w_m(p) = p^{k+1} - (p-1)^{k+1},$$

(b) $$\overline{\lambda}_k(p) = \sum_{1 \leq m_1, \ldots, m_k \leq p-1} w_m(p) = p(p-1)^k - (p-1)(p-2)^k, \text{ and}$$

(c) $$\sum_{1 \leq m_1, \ldots, m_k \leq 3} w_m(2) = 2.3^k - 1.$$

Proof: (a) Let $\lambda_{k,j}(p)$ denote the number of k-tuples $(m_1, \ldots, m_k)$, with $0 \leq m_1, \ldots, m_k \leq p-1$, for which there are **exactly** j non-zero residue classes (mod p) that contain an m_i. We shall prove our result by induction on k: For $k = 1$,

$$\lambda_1(p) = 2.\lambda_{1,1}(p) + 1.\lambda_{1,0}(p) = 2(p-1) + 1 = p^2 - (p-1)^2 .$$

Now, by using the identity

$$\lambda_{k+1,j}(p) = (j+1)\, \lambda_{k,j}(p) + (p-j)\, \lambda_{k,j-1}(p) , \tag{12}$$

we have

$$\lambda_{k+1}(p) = \sum_{j=0}^{k+1} (j+1)\, \lambda_{k+1,j}(p)$$

$$= \sum_{j=0}^{k} \big(p + (p-1)(j+1)\big)\, \lambda_{k,j}(p) ,$$

using (12),

$$= p^{k+1} + (p-1)\, \lambda_k(p) = p^{k+2} - (p-1)^{k+2} ,$$

by the induction hypothesis.

(b) It is easy to see that $\lambda_h(p) = \sum_{j=0}^{h} \binom{h}{j} \overline{\lambda}_j(p)$ and so,

$$\overline{\lambda}_k(p) = \sum_{h=0}^{k} \binom{k}{h} (-1)^{k-h}\, \lambda_h(p)$$

$$= \sum_{h=0}^{k} \binom{k}{h} (-1)^{k-h} \left(p^{h+1} - (p-1)^{h+1}\right)$$

by (a),

$$= p(p-1)^k - (p-1)(p-2)^k .$$

(c) As $w_m(2) = 2$ unless each m_i equals 2, the result is immediate.

Now, for a fixed value of d we have

$$\sum_{d\theta(m)} \Omega \prod_{p|d} \left[p - \phi_{k+1}(p) - w_m(p)\right] = \Pi_1 \, \Pi_2 \, \Pi_3 \, \Pi_4 \tag{13}$$

where

$$\Pi_1 = \prod_{\substack{p \nmid a \\ p|d}} \sum_{\substack{0 \le m_1, \dots, m_k \le p-1 \\ p|\theta(m)}} \left[p - \phi_{k+1}(p) - w_m(p)\right]$$

$$= \prod_{\substack{p \nmid a \\ p|d}} \left[p^k(p-\phi_{k+1}(p)) - \lambda_k(p)\right] = \prod_{\substack{p \nmid a \\ p|d}} \left[(p-1)^{k+1} - p^k \, \phi_{k+1}(p)\right]$$

by Lemma 6(a),

$$= \prod_{\substack{p|d \\ p \nmid a}} (-p^k \phi_{k+1}(p)) \left[1 - \frac{(p-1)}{\phi_{k+1}(p)} \left(\frac{p-1}{p}\right)^k\right];$$

$$\Pi_2 = \prod_{\substack{b \in \beta \\ (b,d)=1}} \sum_{1 \le m_1, \dots, m_k \le b-1} 1 = \prod_{\substack{b \in \beta \\ (b,d)=1}} (b-1)^k;$$

$$\Pi_3 = \prod_{\substack{p|(a,d) \\ p \ge 3}} \sum_{\substack{0 \le m_1, \dots, m_k \le p^2-1 \\ p \nmid m_1 \dots m_k}} \left[p - \phi_{k+1}(p) - w_m(p)\right]$$

$$= \prod_{\substack{p|(a,d) \\ p \ge 3}} p^k \left[(p-1)^k(p-\phi_{k+1}(p)) - \overline{\lambda}_k(p)\right]$$

$$= \prod_{\substack{p|(a,d) \\ p \ge 3}} p^k \left[(p-1)(p-2)^k - (p-1)^k \, \phi_{k+1}(p)\right]$$

by Lemma 6(b),

$$= \prod_{\substack{p|(d,a) \\ p \ge 3}} (-p^k(p-1)^k \phi_{k+1}(p)) \left[1 - \frac{(p-1)}{\phi_{k+1}(p)} \left(\frac{p-2}{p-1}\right)^k\right];$$

$$\Pi_4 = \sum_{\substack{0\le m_1,\dots,m_k\le 7\\ 4\nmid m_1,\dots,m_k}} \left[1 - w_m(p)\right] = 2^k[1-3^k],$$

by Lemma 6(c), if $4\in\beta$ and $2|d$; $\pi_4 = 1$ otherwise.

Therefore, by (11) and (13), and a little rearrangement, we have

$$g_{k,\beta} = c_5 \prod_{b\in\beta}\left(1-\frac{1}{b}\right)^k \sum_{d\ge 1}\mu(d)\cdot\prod_{p|d}\left[1-\frac{(p-1)}{\phi_{k+1}(p)}\left(1-\frac{1}{p-\varepsilon_p}\right)^k\right]\cdot r_d$$

where $r_d = 1 - 1/3^k$ if $4\in\beta$ and $2|d$, 1 otherwise, and $\varepsilon_p = 1$ if $p|a$, 0 otherwise,

$$= c_5 \prod_{b\in\beta}\left(1-\frac{1}{b}\right)^k\left[\prod_p \frac{(p-1)}{\phi_{k+1}(p)}\left(\frac{p-1}{p}\right)^k\right]\left[\prod_{\substack{p|a\\ p\ge 3}}\left(\frac{p(p-2)}{(p-1)^2}\right)^k\right] s_\beta$$

where $s_\beta = (2/3)^k$ if $4\in\beta$, 1 otherwise,

$$= \prod_{b\in\beta}\left(1-\frac{1}{\phi(\beta)}\right)^k = \rho(\beta)^k.$$

The result follows immediately.

Acknowledgements: I'd like to thank Michael Filaseta for allowing me to include his unpublished results (Lemma 4(iii) and Lemma 5) in this paper; and the referee for a number of thoughtful comments.

REFERENCES

[1] Adleman, L.M. and Heath-Brown, D.R., The first case of Fermat's last theorem, Invent. Math., **79** (1985) 409-416.

[2] Bateman, P.T. and Horn, R.A., A heuristic asymptotic formula concerning the distribution of prime numbers, Math. Comp., **16** (1962), 363-367.

[3] Bombieri, E., Friedlander, J.B. and Iwaniec, H., Primes in arithmetic progressions to large moduli, III, J. Amer. Math. Soc., **2** (1989) 215-224.

[4] Dickson, L.E., A new extension of Dirichlet's theorem on prime numbers, Messenger of Math., **33** (1904) 155-161.

[5] Elliott, P.D.T.A. and Halberstam, H., A conjecture in prime number theory, Symp. Math., **4** (1968-9), 59-72.

[6] Fouvry, E., Théorème de Brun-Titchmarsh, application au théorème de Fermat, Invent. Math., **79** (1985), 383-407.

[7] Granville, A., Least Primes in Arithmetic Progressions, in: J.-M. de Koninck and C. Levesque (eds.), Théorie des nombres (Proceedings of the International Number Theory Conference at Laval, Quebec, 1987), de Gruyter, New York 1989, pp. 306-321.

[8] Granville, A., Some conjectures related to Fermat's Last Theorem, to appear in the Proceedings of the First Conference of the Canadian Number Theory Association, 1988.

[9] Granville, A., Diophantine Equations with varying exponents, (Ph.D. Thesis, Queen's University), 1987.

[10] Halberstam, H. and Richert, H.-E., Sieve Methods, (Academic Press), 1974.

[11] Hardy, G.H. and Littlewood, J., Some problems of partitio numerantium III. On the expression of a number as a sum of primes, Acta Math., **44** (1923), 1-70.

[12] Heath-Brown, D.R., Almost primes in arithmetic progressions and short intervals, Math. Proc. Camb. Phil. Soc., **83** (1978) 357-375.

[13] Linnik, U.V., On the least prime in an arithmetic progression II, The Deuring-Heilbronn phenomenon, Rec. Math. [Math. Sb.] N.S. **15(57)** (1944) 347-368.

[14] McCurley, K.S., The least r-free number in an arithmetic progression, Trans. Amer. Math. Soc., **293** (1986) 467-475.

[15] Ribenboim, P., 13 Lectures on Fermat's Last Theorem, (Springer-Verlag, New York) 1979.

[16] Schinzel, A and Sierpinski, W., Sur certaines hypothèses concernant les nombres premiers, Acta Arith. **4** (1958), 185-208; erratum **5** (1959), 259.

[17] Wagstaff, S.S. Jr., Greatest of the least primes in arithmetic progresssions having a given modulus, Math. Comp., **33** (1979) 1073-1080.

Andrew Granville
School of Mathematics
Institute for Advanced Study
Princeton, NJ 08540

Modular Integrals and Their Mellin Transforms

MARVIN KNOPP

To my teacher and friend, Paul Bateman, on his seventieth birthday

I. Introduction

My purpose is to give a succinct and readable account of recent developments in the theory of modular integrals (with associated rational period functions) and the Mellin transforms of these. As I have already given one such exposition [**13**] which emphasizes the rational period functions on the modular group $\Gamma(1)$ at the expense of the modular integrals (and their Mellin transforms), the present article will redress the balance, dealing mainly with the latter and putting aside discussion of rational period functions *per se*, whenever possible. Unavoidably there is a good bit of overlap between the present note and [**13**], to which it should be regarded as supplementary. Proofs are - or are to be - given elsewhere [**4,10,11,12**]. All of the results presented here can be generalized to accommodate (reasonably) general multiplier systems, but for the sake of clarity we restrict attention to multiplier system identically one.

My interest in this subject began in the academic year 1956-57, when at the suggestion of my teacher, Paul Bateman, I studied the dissertation of Hurwitz [**10**] in preparation for thesis work in the area of modular forms. I was particularly struck by Hurwitz's investigation of the series

$$G_2(z) = {\sum_{m,n \in Z}}' \frac{1}{(mz+n)^2}, \tag{1.1}$$

the Eisenstein series of weight 2 connected with the full modular group

$$\Gamma(1) = \{\begin{pmatrix} a & b \\ c & d \end{pmatrix} | a,b,c,d, \in Z, ad-bc = 1\}. \tag{1.2}$$

Hurwitz demonstrates that, in contrast to the Eisenstein series of higher weight,

$$G_{2k}(z) = \sum_{m,n\epsilon Z}{}' (mz+n)^{-2k}, k \geq 2,$$

which are modular forms of weight $2k$ on $\Gamma(1), G_2$ is a kind of modular "quasi-form," of weight 2, satisfying the transformation equations

$$G_2(z+1) = G_2(z), z^{-2}G_2(-\frac{1}{z}) = G_2(z) - \frac{2\pi i}{z}, \tag{1.3}$$

for $z\epsilon H = \{z = x + iy | y > 0\}$. Since $S = \begin{pmatrix} 1 & 1 \\ 0 & 1 \end{pmatrix}$ and $T = \begin{pmatrix} 0 & -1 \\ 1 & 0 \end{pmatrix}$ generate $\Gamma(1)$,(1.3) implies that G_2 has "reasonable" behavior under any transformation in $\Gamma(1)$. Hurwitz's work makes it plain that the appearance of the "period-function" $(-2\pi i)/z$ in (1.3) is due to the conditional convergence of the series (1.1). (Functions with the functional equations (1.3) arise also as the logarithmic-derivatives of modular forms.)

Somewhat later, in my dissertation **[9]**, I encountered a similar phenomenon, but this time the functions in question had negative weight and polynomial "periods." Specifically, in **[9]** I construct functions F analytic in H such that

$$F(z+1) = F(z), z^{2k}F(-\frac{1}{z}) = F(z) + p(z), \tag{1.4}$$

where $k\epsilon Z^+$ and $p(z)$ is polynomial of degree at most $2k$. We now recognize such F as "Eichler integrals," but at the time Eichler's classic work **[1]** was not yet known to me. (Either it had just appeared or it was about to appear.)

In the summer of 1975, after many attempts to prove that (1.3) and (1.4) represent essentially all cases of functions having transformation formulas with rational period functions, under transformation by elements of $\Gamma(1)$, I proved the opposite, devising a method for the construction of an infinite class of rational period functions for $\Gamma(1)$, entirely new in the sense that they are not described in (1.3) or (1.4)**[11]**. Since then, much further progress has been made in constructing and characterizing rational period functions for $\Gamma(1)$, most of it described in **[13]**.

II. Modular integrals and the generalized Poincaré series

Suppose f is meromorphic in H and there satisfies

$$f(z+1) = f(z), z^{-2k}f(-\frac{1}{z}) = f(z) + q(z), \tag{2.1}$$

where $k\epsilon Z$ and $q(z)$ is a rational function. Then we call f a *modular integral* (MI) on $\Gamma(1)$ *of weight* $2k$ with *rational period function* (RPF) q. This definition can, of course, be generalized to odd integral weights, nontrivial multiplier systems and groups other than $\Gamma(1)$. (We shall have something to say later about RPF's on subgroups of $\Gamma(1)$.) Now, because $T^2 = (ST)^3 = I$ as linear fractional transformations (the defining relations of $\Gamma(1)$), it follows from (2.1) that

$$z^{-2k}q(-\frac{1}{z})+q(z) = 0, (z-1)^{-2k}q(\frac{-1}{z-1})+z^{-2k}q(\frac{z-1}{z})+q(z) = 0. \quad (2.2a)$$

Letting

$$F|\begin{pmatrix} a & b \\ c & d \end{pmatrix} = (cz+d)^{-2k}F(\frac{az+b}{cz+d}) \quad (2.3)$$

for $M = \begin{pmatrix} a & b \\ c & d \end{pmatrix} \epsilon\Gamma(1)$ and F defined on H, we can rewrite (2.2a) as

$$q|T + q = 0, q|(ST)^2 + q|ST + q = 0, \quad (2.2b)$$

a direct consequence of (2.1), as we have indicated.

In fact, even more is true; (2.2) is *equivalent* to (2.1) in the following sense. Suppose q is a rational function (or, less restrictively, simply holomorphic in H and of polynomial growth, both at ∞ and upon vertical approach to the real axis from within H) satisfying the relations (2.2). Then there exists f holomorphic in H such that (2.1) holds. The proof of this involves the *generalized Poincaré series* of Eichler ([2],[3]), an often useful device which is easy to describe yet apparently not widely known. For a discrete Γ acting on H, the collection $\{q_M \mid M\epsilon\Gamma\}$ is called a *cocycle in weight* $2k$ if

$$q_{M_1M_2} = q_{M_1}|M_2 + q_{M_2}, \text{ for } M_1, M_2\epsilon \quad \Gamma, \quad (2.4)$$

where $q_{M_1}|M_2$ is defined by (2.3). To obtain a cocycle, we simply need to assign a q_M to each M in a set of generators for $\Gamma(1)$ in such a way that the choice is consistent with the group relations among the generators. Then q_M for general M in Γ is defined by (2.4). That is, write M as a word in the generators and apply (2.4) repeatly. In the case $\Gamma = \Gamma(1)$, the modular group, we can choose S and T as generators, with the defining relations $T^2 = (ST)^3 = I$. In particular, we wish to construct a cocycle $\{q_M\}$ such that $q_S = 0, q_T = q$, consistent with (2.1); then the conditions (2.2) on q are precisely the conditions of consistency with the two group relations in $\Gamma(1)$.

Now, given a rational function q satisfying (2.2) define the cocycle $\{q_M \mid M\epsilon\Gamma(1)\}$ by application of (2.4) and form the *generalized Poincaré series*

$$H(z) = \sum_{\substack{c,d\epsilon Z\\(c,d)=1}} q_M(z)(cz+d)^{-2\rho}, \tag{2.5}$$

where $M = \begin{pmatrix} * & * \\ c & d \end{pmatrix} \epsilon\Gamma(1)$ and $\rho\epsilon Z^+$, chosen sufficiently large to guarantee absolute convergence of the series. Note that q_M depends only upon the lower row c, d of M as a consequence of $q_S = 0$. (There is a good deal of estimation required for the proof of absolute convergence; see [**14**,§II].) A function f satisfying (2.1) is then given by

$$f(z) = -H(z)/E_{2\rho}(z), \tag{2.6}$$

where $E_{2\rho}$ is the Eisenstein series,

$$E_{2\rho}(z) = \sum_{\substack{c,d\epsilon Z\\(c,d)=1}} (cz+d)^{-2\rho}.$$

($E_{2\rho}(z)$ is of weight 2ρ on $\Gamma(1)$; it is in fact virtually the same as $G_{2\rho}(z)$, defined above: $G_{2\rho}(z) = \zeta(2\rho)E_{2\rho}(z)$.) However, $f(z)$ defined by (2.6) falls short of what we require, since it may have poles at the zeros of $E_{2\rho}$. f can be modified to remove the poles. When $2k \geq 2$, we can accomplish this by application of a "Mittag-Leffler theorem" for automorphic forms; if $2k \leq 0$, elimination of the poles in H (but not at $i\infty$) is still possible by use of a technically more complicated procedure based upon results of Douglas Niebur [**17**]. (See [**14**,§III] for details.)

This construction is general in the sense that, with the weight fixed, any two MI's corresponding to the same RPF on $\Gamma(1)$ differ by a modular form of that weight. Thus, given the RPF q on $\Gamma(1)$ any corresponding MI is the sum of the generalized Poincaré series we have constructed and a modular form on $\Gamma(1)$. (Within the present context, modular forms on $\Gamma(1)$ are regarded as known, either as linear combinations of Poincaré series or in terms of the discriminant function-the cusp form of weight 12- and the Eisenstein series G_4, G_6.)

III. Modular integrals and the quadratic Eisenstein series

The initial aims of our study, clearly, should be the characterization of the RPF's on $\Gamma(1)$ within the class of all rational functions, and the determination of their corresponding MI's. There is good reason to believe that Choie and Zagier are very close to reaching the first of these two goals.

One might conclude that the second will be reached with it by means of the generalized Poincaré series. However, while the construction of these series resolves the existence question for MI's on $\Gamma(1)$ (given the RPF on $\Gamma(1)$), it does not shed much light upon the relationship between the MI and its RPF, or upon the structure of the Fourier coefficients of the integral. For this reason it is desirable to have an alternative to the generalized Poincaré series for the construction of modular integrals.

Here and in § IV we shall present two variants of an approach to the construction of MI's based upon Zagier's *quadratic Eisenstein series*,

$$f_{k,D}(z) = \sum (az^2 + bz + c)^{-k}. \tag{3.1}$$

Here k is a positive even integer and $D \epsilon Z^+$ is a discriminant: $D \equiv 0$ or $1 (\bmod 4)$; the summation is over all triples $a, b, c \epsilon Z$ such that $b^2 - 4ac = D$. $f_{k,D}$ is a cusp form on $\Gamma(1)$ of weight $2k \equiv 0 (\bmod 4)$. This follows easily from absolute convergence of the sum when $k \geq 4$; when $k = 2$ the proof is more subtle, requiring the introduction of a Hecke convergence factor [20,39-42] or careful handling of the order of summation.

It is elementary that $f_{k,D} = 0$ when k is odd, but L.A. Parson has observed that in this case *half* of the series for $f_{k,D}$ yields a nontrivial MI on $\Gamma(1)$. That is to say, if one defines

$$\varphi_{k,D}(z) = \sum_{a>0} (az^2 + bz + c)^{-k}, \tag{3.2}$$

with k odd and again subject to the summation condition $b^2 - 4ac = D$, then

$$\varphi_{1,D}(z+1) = \varphi_{k,D}(z), z^{-2k}\varphi_{k,D}(-1/z) = \varphi_{k,D}(z) + q_{k,D}(z). \tag{3.3}$$

Here,

$$q_{k,D}(z) = 2 \sum_{a<0<c} (az^2 + bz + c)^{-k}, \tag{3.4}$$

once again with $b^2 - 4ac = D$ in the summation. Of course, the condition $a < 0 < c$ guarantees that the sum is finite, hence a rational function. Again, (3.3) is a direct consequence of absolute convergence of the sum (3.2), provided $k \geq 3$.

The case $k = 1$ is more difficult, dependent upon the use of the Hecke convergence factor (as when $k = 2$). Specifically, when $k = 1$ we consider, in place of $\varphi_{k,D}$, the function

$$\varphi_{1,D}(z|s) = \sum_{a>0} (az^2 + bz + c)^{-1} |az^2 + bz + c|^{-s}, \tag{3.5}$$

defined initially for complex s of sufficiently large real part. The next step entails analytic continuation(in the variable s) of $\varphi_{1,D}(z|s)$ into an open half-plane containing the point $s = 0$. Finally, one studies $\varphi_{1,D}(z|0)$, where this is the analytic continuation evaluated at $s = 0$.

The major part of the work in the case $k = 1$ is in establishing the analytic continuation; as we learn from the work of Hecke [6], Maass [**16**] and Siegel [**18**], this can be done by calculating the Fourier coefficients of $\varphi_{1,D}(z|s)$, which, like $\varphi_{k,D}(z)$, is periodic in z, with period 1. This calculation, in turn, depends upon an application of Poisson summation. Zagier has carried it out for $f_{2,D}(z|s)$, consequently for the cusp form $f_{2,D}(z) = f_{2,D}(z|0)$[**20**,39-42]. The coefficients of $\varphi_{1,D}(z|s)$ will be similar.

To avoid technical difficulties we omit the case $k = 1$ and give the result of Zagier's calculation when applied to $\varphi_{k,D}$ in the case of absolute convergence. For $k \geq 3$ the method of [**20**, 43-45], under the simplifying assumption $D \neq a$ square, yields the following:

$$\varphi_{k,D}(z) = \sum_{n=1}^{\infty} c_n(D) e^{2\pi i n z}, \tag{3.6}$$

$$c_n(D) = \frac{2^{k+\frac{1}{2}}\pi^{k+2} n^{k-\frac{1}{2}}}{D^{\frac{k}{2}-\frac{1}{4}}(k-1)!} \sum_{a=1}^{\infty} a^{-\frac{1}{2}} S_a(n,D) J_{k-\frac{1}{2}}\left(\frac{\pi n\sqrt{D}}{a}\right), \tag{3.7}$$

where

$$S_a(n,D) = \sum_{\substack{b(\bmod 2a) \\ b^2 \equiv D(\bmod 4a)}} \exp(\pi i r b/a)$$

and $J_{k-1/2}$ is the usual Bessel function.

IV. More on modular integrals

A less direct modification of the quadratic Eisenstein series arises from Kohnen and Zagier's explicit calculation [**15**, §2] of the even period polynomials $r^+(f_{k,D})$, which result from $(2k-1)$- fold integration of $f_{k,D}$.

To define $r^+(f_{k,D})$, we introduce the period polynomial $r(f)$ associated to the cusp form f of weight $2k$ (and the element T of $\Gamma(1)$) by

$$r(f)(X) = \int_{0=T(i\infty)}^{i\infty} f(z)(X-z)^{2k-2} dz. \tag{4.1}$$

One can show, without much difficulty, that such $r(f)$ are RPF's of weight $2-2k$, and since $\Gamma(1)$ has the automorphism

$$\begin{pmatrix} a & b \\ c & d \end{pmatrix} \to \begin{pmatrix} a & -b \\ -c & d \end{pmatrix},$$

it follows that the polynomials $r^*(f)$ defined by $r^*(f)(X) = r(f)(-X)$ are again RPF's. The even and odd periods, $r^+(f)$ and $r^-(f)$ respectively, are then defined by

$$r^+(f) = \frac{1}{2}\{r(f) + r^*(f)\}, r^-(f) = \frac{1}{2}\{r(f) - r^*(f)\}. \tag{4.2}$$

Once again these are (polynomial) RPF's. By [**15**, Theorem 4], $r^+(f)$ can be expressed in the form

$$r^+(f)(z) = \alpha \sum (az^2 + bz + c)^{k-1} + \beta(z^{2k-2} - 1), \tag{4.3}$$

where α, β are constants and the summation conditions in (4.3) are

$$a, b, c \quad \epsilon \quad Z, b^2 - 4ac = D, a < 0 < c. \tag{4.4}$$

As before, the last condition in (4.4) guarantees that the sum in (4.3) is finite, hence that $r^+(f)$ is a polynomial of degree $\leq 2k - 2$.

Now the sums in (4.3) are RPF's, whether the (odd) exponent $k - 1$ is positive or negative. When k is even and $k \leq 0$ these sums are RPF's of the same general type as (3.4) and, in fact, if we put $D = 5$ and replace $k - 1$ by $-k$, now with k odd and > 0, they reduce to my first new examples of 1975 [**11**]. Motivated by these observations, I carried out a formal $(2k-1)$-fold integration, term-by-term, of the series (3.1) for $f_{k,D}(z)$. While the integrated series clearly diverges, if we simply replace k by $-k$ formally, we obtain the series

$$\psi_{k,D}(z) = \sum \frac{\log((z-\beta)/(z-\alpha))}{(az^2 + bz + c)^{k+1}}, \tag{4.5}$$

where $k+1$ is odd and > 0. Here $\alpha = \frac{-b+\sqrt{D}}{2a}, \beta = \frac{-b-\sqrt{D}}{2a}$ and the summation is again over all triples $a, b, c \epsilon Z$ with $b^2 - 4ac = D$ (as in 3.1)). When $k+1 \geq 3$, the series (4.5) converges absolutely; if $k+1 = 1$ the series again requires a convergence factor. Note that when $k+1$ is even, $\psi_{k,D} = 0$.

Study of the function $\psi_{k,D}$, for $k+1$ odd, shows that it is holomorphic in H, periodic in z with period 1 and that it satisfies the following transformation equation when subjected to the inversion T:

$$\begin{aligned} \psi_{k,D}|T - \psi_{k,D} &= \sum_{b^2-4ac=D} \frac{\log(\alpha/\beta)}{(az^2+bz+c)^{k+1}} \\ &\quad + 2\pi i \sum_{\substack{b^2-4ac=D \\ a<0<c}} (az^2+bz+c)^{-k-1} \\ &= r_o(z) + r_e(z), \end{aligned} \tag{4.6}$$

say, with summation again over $a, b, c \epsilon Z$. The condition $a < 0 < c$ implies that $r_e(z)$ is a finite sum, and thus a rational function. Since $k > 0$, it has poles at the points $\alpha, \beta \epsilon Q(\sqrt{D})$. Further consideration shows that r_e is an even function, while r_o is odd. However, since the real line is a natural boundary for r_o, it is not possible to conclude from this alone (as we did above for the polynomials $r^+(f), r^-(f)$ defined in (4.2)) that r_o and r_e satisfy (2.2). Notwithstanding this, these functions do in fact satisfy (2.2); for, as mentioned above, this has been verified directly for r_e, so it follows as well for r_o.

$\psi_{k,D}$, then, is a periodic function having as even period under T the RPF r_e, but $\psi_{k,D}$ fails to satisfy the condition (2.1) for a MI since its period $r_o + r_e$, under T, is not a rational function. Indeed, this is only one respect in which the MI $\varphi_{k,D}$ of § III is more satisfactory than $\psi_{k,D}$. Another is the relative complexity of $\psi_{k,D}$; as a consequence of this, though the Fourier coefficients of $\psi_{k,D}$ presumably can be calculated by a suitable modification of Zagier's method of [**20**, 44-45], their structure is undoubtedly more complicated than that of the $c_n(D)$, given in (3.7).

On the other hand, the $\psi_{k,D}$ and similar series certainly have a role in the theory of MI's and bear further study, particularly since both $\varphi_{k,D}$ and $\psi_{k,D}$ are of interest only for weights $2k \equiv 2(\bmod 4)$. Indeed, aside from the generalized Poincaré series, no principle for the construction of MI's is yet available in weights $2k \equiv 0 \bmod 4)$.

V. Mellin transforms of MI's on $\Gamma(1)$

As is well known, Hecke, following Riemann, discovered - by applying the Mellin transform and its inverse - the systematic relationship between modular forms, on the one hand, and Dirichlet series with a simple functional equation, on the other [5,7]. In [**12**, Theorems 3 and 4], I showed that the same kind of bilateral relationship obtains between MI's with RPF's having poles in Q only (thus, at 0 and ∞) and a larger class of Dirichlet series with *precisely the same* functional equation as for the Mellin transform of a modular form. It follows, as a consequence of this relationship, that when the RPF of a MI has poles outside of Q (i.e. in $Q(\sqrt{N})$, with $N \neq$ a square), the Mellin transform of the MI cannot satisfy this same simple functional equation.

This observation serves as the starting point of my recent joint work with Hawkins [**4**], which in fact establishes a more complex functional equation for the Mellin transform of a MI with RPF, whether or not the poles of the RPF lie in Q. Compared with that of Hecke, this functional equation contains an additional term which is a finite sum of beta functions, the number of terms depending upon the number of and orders of the poles of the RPF associated with the MI.

To make this specific, suppose that f is an *entire* MI on $\Gamma(1)$. That is to say: (i) f satisfies (2.1);(ii) f is holomorphic in H; (iii) f has a Fourier expansion of the form

$$f(z) = \sum_{n=0}^{\infty} a_n e^{2\pi i n z}. \tag{5.1}$$

It follows from these three conditions that $a_n = O(n^\gamma), n \to +\infty$, for some $\gamma > 0$, and this in turn guarantees the absolute convergence of the Dirichlet series $\sum_{n=1}^{\infty} a_n n^{-s}$ in the half-plane $Re s > 1+\gamma$. This series arises naturally from term-by-term integration when one forms the Mellin transform

$$\Phi_f(s) = \int_0^{\infty} \{f(iy) - a_o\} y^s \frac{dy}{y} = (2\pi)^{-s}\Gamma(s) \sum_{n=1}^{\infty} a_n n^{-s}, \tag{5.2}$$

of $f(z) - a_o$. Note that $\Phi_f(s)$, like the Dirichlet series, is holomorphic in $Re s > 1 + \gamma$.

The classic work of Hecke [**5**,7] shows that if f is an entire modular *form* (that is, if $q = 0$ in (2.1)), then $\Phi_f(s)$ has certain desirable properties, the most striking among them the functional equation

$$\Phi_f(2k - s) = (-1)^k \Phi_f(s). \tag{5.3}$$

In order to make sense of (5.3) he must show as well that Φ_f can be continued to a function meromorphic in the entire s-plane. Furthermore, there is a converse theorem, but we shall not state it explicity.

After showing in [**12**, Theorem 1] that any RPF on $\Gamma(1)$ with poles in Q has the form

$$q(z) = \sum \alpha_l z^{-l}, \quad -L \le l \le M, \tag{5.4}$$

I proved the following generalization of Hecke's celebrated correspondence (between entire modular forms and Dirichlet series having a functional equation).

Theorem 1. *[**12**, Theorem 3]. Suppose f is an entire MI on $\Gamma(1)$ with RPF of the form (5.4). Let Φ_f be the Mellin transform of $f(z) - a_o$, defined by (5.2). Then Φ_f can be continued analytically to a function meromorphic in the entire s- plane, with at worst simple poles at a finite number of integer values of s. Furthermore, Φ_f satisfies the functional equation (5.3).*

The striking aspect of this result is that RPF's of the form (5.4) have no effect upon the functional equation of Φ_f: the Mellin transform of an entire MI on $\Gamma(1)$ with RPF of the form (5.4) has *precisely the same functional equation* as does the Mellin transform of an entire modular form on $\Gamma(1)$. In [**12**, Theorem 3] the number and position of the possible simple poles of Φ_f are determined completely in terms of the weight $2k$. Here, as in Hecke's work, there is a straightforward converse.

Recently, Hawkins and I proved a generalization of Theorem 1 that removes the restriction (5.4) on the RPF q of the MI:

Theorem 2. *[4, Theorem 2]. Suppose f is an entire MI on $\Gamma(1)$. Then Φ_f can be continued analytically to a function meromorphic in the entire s - plane, with at worst simple poles at integer values of s. Furthermore, Φ_f has a functional equation*

$$\Phi_f(2k-s) = (-1)^k \Phi_f(s) + R_f(s), \tag{5.5}$$

where $R_f(s)$ is a finite complex linear combination, summed over integral j and r, of terms of the form

$$(\frac{-1}{\alpha_j})^r \{B(s, r-s) e^{-\pi \frac{s}{2}} \alpha_j^s - (-1)^k B(2k-s, r-2k+s) e^{-\pi i \frac{s}{2}} \alpha_j^{2k-s}\}.$$

Here, B is the beta function and the α_j are the poles of the RPF q in the set

$$P = \{Re z > 0, Im z \le 0\} \cup \{Re z = 0, 1 \le Im z < 0\}.$$

Remarks 1. P does not intersect H, in conformity with the fact that f is holomorphic in H. Note that q may have a pole at 0, but that leaves unaffected the formula for R_f, as is consistent with Theorem 1 and, indeed, follows from it.

2. To each pole α in P corresponds another pole $-\frac{1}{\alpha}$ of q, outside of P. This follows directly from the first relation $q|T = -q$ of (2.2b). The proof of Theorem 2 does not involve the second relation , $q|(ST)^2 + q|ST + q = 0$, of (2.2b), so that it actually holds in the far broader context of MI's on the subgroup Γ_θ, of index 3 in $\Gamma(1)$: $\Gamma_\theta = < S^2, T >$. (The sole relation in these two generators is $T^2 = I$; thus the RPF's for Γ_θ, subject only to the first of the two relations in (2.2b), form a much larger class than do the RPF's on $\Gamma(1)$.)

3. With j fixed (i.e., α_j a fixed pole of q) r is summed over the various powers of $(z - \alpha_j)^{-1}$ that occur in the principal part of q at α_j.

4. It is the explicit simple form of R_f that conveys the significance of the functional equation (5.5). Lacking such an expression, $R_f(s)$ would simply be another label for $\Phi_f(2k-s) - (-1)^k \Phi_f(s)$, and (5.5) a tautology, hardly a functional equation.

5. The ordinary Gaussian hypergeometric function ${}_2F_1$ figures prominently in the analytic continuation of Φ_f and, as well, in the derviation of the closed form of R_f. It surprised the authors that, despite the presence of several terms involving ${}_2F_1$ throughout most of the calculation of R_f, in

the (very) end they dropped out, leaving a remarkably simple formula for R_f.

6. The expression for the analytic continuation of Φ_f, involving ${}_2F_1$, is not given here. It shows that, while Φ_f is holomorphic in a right half-plane (as follows from the definition (5.2)), it has infinitely many poles - at integral values of s- in the corresponding left half-plane. This is reflected in our closed expression for R_f.

7. Theorem 2 has a converse, but unlike the converse to Theorem 1, its statement is not evident without a more explicit version of Theorem 2 than we care to give here. The same holds true for the converse to Theorem 4(§VI).

VI. Generalization to $\Gamma_0^*(N)$

We now turn to results analogous to Theorems 1 and 2 that emerge when $\Gamma_0^*(N) =< \Gamma_0(N), \omega(N) >, N \epsilon Z^+$, replaces the modular group $\Gamma(1)$. Here $\Gamma_0(N)$ is the congruence subgroup of $\Gamma(1)$ defined by the condition $N|c$ in the modular matrix $M = \begin{pmatrix} a & b \\ c & d \end{pmatrix} \epsilon \Gamma(1)$, and $\omega(N) = \begin{pmatrix} 0 & -\frac{1}{\sqrt{N}} \\ \sqrt{N} & 0. \end{pmatrix}$ For $N > 1, \Gamma_0^*(N)$ is not a subgroup of $\Gamma(1)$.

In **[19]** Weil develops an important generalization to $\Gamma_0^*(N)$ of the Hecke correspondence. His work goes beyond straightforward generalization, however : Weil obtains functional equations not only for the Dirichlet series attached directly to the modular form by the Mellin transform, but in addition for the infinite class of Dirichlet series arising from the Mellin transform through a "twisting" of the coefficients by Dirichlet characters of conductor relatively prime to N.

Weil's converse theorem is stronger than the expected one: it postulates functional equations only for the "twisted Mellin transforms" corresponding to a suitable infinite subclass of Dirichlet characters with conductor relatively prime to N, rather than for the entire class. Weil's functional equation for the twisted Mellin transform is new even for $N = 1$, in which case $\Gamma_0^*(N) = \Gamma_0(N) = \Gamma(1)$.

Theorems 3 and 4 of this section bear the same relationship to Theorems 1 and 2, respectively, as does Weil's work to Hecke's. That is, Theorem 3 (Theorem 4) extends Theorem 1 (Theorem 2) to $\Gamma_0^*(N)$. In both cases the single functional equation satisfied by Φ_f in Theorems 1 and 2 is replaced by an infinite class of functional equations, one for each twisted Mellin transform. Again, both have converses, but we omit discussion of these.

Suppose that in H the function f is holomorphic and satisfies the transformation formulae

$$f|V = f + q_V, V \epsilon \Gamma_0(N); f|\omega(N) = Cf + q_\omega, \tag{6.1}$$

where q_V, q_ω are rational functions and C is a complex number. Assume further that f has the expansion at ∞:

$$f(z) = \sum_{n=0}^{\infty} a_n e^{2\pi i n z}, z \epsilon H, \tag{6.2}$$

where

$$a_n = O(n^\gamma), n \to +\infty, \tag{6.3}$$

with fixed $\gamma > 0$. Then we call f an *entire modular integral on* $\Gamma_0^*(N)$ *of weight* $2k$.

Because $\omega(N)$ has order 2 as a linear fractional transformation, it follows that $C = \pm 1$. If $C = 1$ this definition simply carries over to $\Gamma_0^*(N)$ our earlier definition of a MI on $\Gamma(1)$ (that is, with multiplier system identically one).

The greater generality allowed here ($C = \pm 1$) arises from the circumstance that Weil's focus of interest in [19] is actually $\Gamma_0(N)$ rather than the extended group $\Gamma_0^*(N)$. But, since $\omega(N)$ is in the normalizer of $\Gamma_0^*(N)$, and since entire modular forms on $\Gamma_0(N)$, of fixed weight $2k$, constitute a finite dimensional Hilbert space (with respect to the Petersson inner product), this space has a basis consisting entirely of eigenfunctions of the operator $f \to f|\omega(N)$. Thus Weil may -and does-assume from the outset that

$$f|\omega(N) = Cf, \tag{6.4}$$

and this makes available to Weil Hecke's line of reasoning, which depends upon invariance of a $\Gamma(1)$-modular form with respect to the inversion $Tz = -\frac{1}{z}$. (For $N > 1$, T is not in $\Gamma_0(N)$, or in $\Gamma_0^*(N)$.) The appropriate analogue of Weil's condition (6.4) within the context of MI's on $\Gamma_0^*(N)$, is given in the second part of (6.1).

For $m \epsilon Z^+, \chi$ a Dirichlet character mod m and f given by (6.2), put

$$(a) \quad f_\chi(z) = \sum_{n=1}^{\infty} a_n \chi(n) e^{2\pi i n z};$$

$$(b) \quad \Phi_{f,\chi}(s) = \left(\frac{m}{2\pi}\right)^s \Gamma(s) \sum_{n=1}^{\infty} a_n \chi(n) n^{-s}.$$

Then,

$$\Phi_{f,\chi}(s) = m^s \int_0^{\infty} f_\chi(iy) y^s \frac{dy}{y} = \int_0^{\infty} f_\chi\left(\frac{iy}{m}\right) y^s \frac{dy}{y},$$

so that $\Phi_{f,\chi}(s)$ is the Mellin transform of $f_\chi(\frac{z}{m})$.

Theorem 3. *[10,Theorem 1]. Let f be an entire MI on $\Gamma_0^*(N)$ of weight $2k \le 0$, such that the RPF's q_V, q_ω of (6.1) are polynomials of degree $\le -2k$. Suppose χ is a primitive Dirichlet character modulo m with $(m, N) = 1$. Then Φ_f and $\Phi_{f,\chi}$ have analytic continuations to the entire s- plane with at most finitely many simple poles at nonpositive, integer values of s. Furthermore,*

$$(A) \quad N^{k-s}\Phi_f(2k-s) = C(-1)^k\Phi_f(s)$$

and

$$(B) \quad (Nm^2)^{k-s}\Phi_{f,\bar{\chi}}(2k-s) = C\frac{g(\bar{\chi})}{g(\chi)}\bar{\chi}(-N)(-1)^k\Phi_{f,\chi}(s),$$

where $g(\chi)$ is the Gaussian sum,

$$g(\chi) = \sum_{a(\bmod m)} \chi(a)e^{2\pi i\frac{a}{m}}.$$

Remarks 1. Here, as in Theorem 1, the special nature of the RPF's corresponding to the MI leads to functional equations for the Mellin transforms which are precisely the same as those that occur in **[19]** for entire modular *forms* . In contrast to Theorem 1, the RPF's in Theorem 3 are not allowed to have poles at $z = 0$. This reflects the true mathematical situation, not a defect in the method of proof.

2. The assumption that q_V, q_ω are polynomials implies both that $k \le 0$ and that their degrees are at most $-2k$. The hypotheses in Theorem 3 may be simplified accordingly. Assuming q_V, q_ω to be polynomials has the further consequence that the poles of Φ_f and $\Phi_{f,\chi}$ are restricted to $s \le 0$.

Since the weight is ≤ 0 in Theorem 3 and > 0 in **[19]** (there are no non-trivial entire modular forms of weight ≤ 0), Theorem 3 should be regarded as a supplement to Weil's theorem rather than a generalization. The following result, containing no assumption upon the RPF's q_V, q_ω of (6.1), generalizes both Weil's theorem and Theorem 3.

Theorem 4. *Let f be an arbitrary entire MI on $\Gamma_0^*(N)$ of weight $2k, k \epsilon Z$. Let χ be as in Theorem 3. Then we have :*

(i) $f_\chi|\omega(Nm^2) = C\frac{g(\chi)}{g(\bar{\chi})}\chi(-N)f_{\bar{\chi}} + Q_\chi$, *with Q_χ a rational function holomorphic in H.*

(ii) *Both Φ_f and $\Phi_{f,\chi}$ have continuations to the entire s- plane, with at worst simple poles (usually infinite in number) at integer values of* s.

Furthermore,

$$(A)\quad N^{k-s}\Phi_f(2k-s) = C(-1)^k\Phi_f(s) + R_f(s)$$

and

$$(B)\quad (Nm^2)^{k-s}\Phi_{f,\bar{\chi}}(2k-s) = C\frac{g(\bar{\chi})}{g(\chi)}\bar{\chi}(-N)(-1)^k\Phi_{f,\chi}(s) + R_{f,\chi}(s).$$

In (A) $R_f(s)$ is a finite complex linear combination, summed over integral j and r, of terms having the form

$$(\sqrt{N})^{r-s}\{\Gamma(r-s)\Gamma(s)(i\sqrt{N}\alpha_j)^{s-r} \\ -(-1)^k\Gamma(r-2k+s)\Gamma(2k-s)(i\sqrt{N}\alpha_j)^{2k-s-r}\},$$

where the α_j are those poles of RPF q_ω which lie in the set $\{Rez > 0, Imz \leq 0\} \cup \{Re = 0, \frac{1}{\sqrt{N}} \leq Imz < 0\}$. In (B) $R_{f,\chi}(s)$ is a complex linear combination, summed on integral j and r, of terms having the form

$$(\sqrt{N})^{r-s}m^{s-1}g(\bar{\chi})\{(\bar{\chi})(Nb)\Gamma(r-s)\Gamma(s)(i\sqrt{N}\alpha_j)^{s-r} \\ -(-1)^k\chi(-b)\Gamma(r-2k+s)\Gamma(2k-s)(i\sqrt{N}m\alpha_j)^{2k-s-r}\},$$

where the α_j now are those poles of Q_χ in the set

$$\{Rez > 0, Imz \leq 0\} \cup \{Rez = 0, -1/\sqrt{N}m \leq Imz < 0\},$$

and b is an integer such that $1 \leq b \leq m, (b,m) = 1$.

The remarks following Theorem 2 are applicable here as well, with the obvious exception of the third and fourth sentences of Remark 2. For these two sentences there is an analogue only for $N = 2$ and 3.

REFERENCES

[1] M. Eichler, Eine Verallgemeinerung der Abelschen Integrale, Math. Z. **67** (1957), 267-298.

[2] M. Eichler, Grenzkreisgruppen und kettenbruchartige Algorithmen, Acta Arith. **11** (1965), 169-180.

[3] M. Eichler, Lectures on modular correspondences, Tata Institute of Fundamental Research, Bombay 1955-56.

[4] J. Hawkins and M. Knopp, A Hecke correspondence theorem for automorphic integrals with rational period functions, preprint.

[5] E. Hecke, Lectures on Dirichlet series, modular functions and quadratic forms, Edwards Bros., Inc., Ann Arbor, 1938. (Revised and reissued, Vandenhoeck and Ruprecht, Göttingen, 1983, ed. B. Schoeneberg).

[6] E. Hecke, Theorie der Eisensteinschen Reihen höherer Stufe und ihre Anwendung auf Funktionentheorie und Arithmetik, Mathematische Werke, Vandenhoeck und Ruprecht, Göttingen (1959), 461-486.

[7] E. Hecke, Über die Bestimmung Dirichletscher Reihen durch ihre Funktionalgleichung, Math. Annalen **112** (1936), 664-699. Also, paper no. 33, pp. 591-626 in *Mathematische Werke* (ed. B. Schoeneberg), Vandenhoeck and Ruprecht, Götttingen, 1959.

[8] A. Hurwitz, Grundlagen einer independenten Theorie der elliptischen Modulfunctionen und Theorie der Multiplicatorgleichungen erster Stufe, Math. Annalen **18** (1881), 528-591.

[9] M. Knopp, Fourier series of automorphic forms of nonnegative dimension, Illinois J. Math. **5** (1961), 18-42.

[10] M. Knopp, Modular integrals on $\Gamma_0(N)$ and Dirichlet series with functional equations, in Number theory, Lecture Notes in Mathematics, Springer Verlag, New York, **1135** (1985), 211-224.

[11] M. Knopp, Rational period functions of the modular group, Duke Math. J. **45** (1978), 47-62.

[12] M. Knopp, Rational period functions of modular group II, Glasgow Math. J. **22** (1981), 185-197.

[13] M. Knopp, Recent developments in the theory of rational period functions, in Number theory seminar, Lecture Notes in Mathematics, Springer Verlag, New York, to appear.

[14] M. Knopp, Some new results on the Eichler cohomology of automorphic forms, Bull. Amer. Math. Soc. **80** (1974), 607-632.

[15] W. Kohnen and D. Zagier, Modular forms with rational periods, Chapter 9, in: R. Rankin (ed.), *Modular forms*, Halsted Press, New York, 1984, pp. 197-249.

[16] H. Maass, Konstruktion ganzer Modulformen halbzahliger Dimension mit θ- Multiplikatoren in einer und zwei Variablen, Abh. Math. Sem. Hansische Univ. **12** (1938), 133-162.

[17] D. Niebur, Construction of automorphic forms and integrals, Trans. A.M.S. **191** (1974), 373-385.

[18] C.L. Siegel, Die Funktionalgleichungen einiger Dirichletscher Reihen, Math. Z. **63** (1956), 363-373.

[19] A. Weil, Über die Bestimmung Dirichletscher Reihen durch Funktionalgleichungen, Math. Annalen **168** (1967), 149-156.

[20] D. Zagier, Modular forms associated to real quadratic fields, Invent. Math. **30** (1975), 1-46.

Marvin Knopp
Department of Mathematics
Bryn Mawr College
Bryn Mawr, PA 19010

Dept. of Mathematics
Temple University
Philadelphia, PA 19122

A Congruence for Generalized Frobenius Partitions with 3 Colors Modulo Powers of 3

LOUIS W. KOLITSCH

Dedicated to Paul Bateman

In 1919 Ramanujan conjectured congruences for certain classes of ordinary partitions modulo powers of 5 and 7 which were later proved by G.N. Watson. The corresponding congruences for colored generalized Frobenius partitions with 5 and 7 colors were recently derived by establishing a relationship between these partitions and ordinary partitions [5]. In this paper we prove similar congruences for colored generalized Frobenius partitions with 3 colors modulo powers of 3 using certain generating function identities and the modular equation of order 3.

We will show that $\overline{c\phi_3}(m) = c\phi_3(m) - p(m/3)$, the number of generalized Frobenius partitions of m with 3 colors whose order is three under cyclic permutation of the colors, is the coefficient of q^m in the generating function $\frac{9q(q^9;q^9)^3_\infty}{(q;q)^3_\infty(q^3;q^3)_\infty}$. Using this result we will then show that $\overline{c\phi_3}(m)$ is congruent to zero modulo large powers of 3 for certain values of m. Specifically, if λ_α is the reciprocal of 8 modulo 3^α, then

$$\overline{c\phi_3}(3^\alpha n + \lambda_\alpha) \equiv 0 \begin{cases} \bmod 3^{2\alpha+2} & \text{if } \alpha \text{ is even} \\ \bmod 3^{2\alpha+1} & \text{if } \alpha \text{ is odd} \end{cases}$$

The technique used in deriving the congruence will be analogous to those used by Atkin [2], Hirschhorn and Hunt [4], and Garvan [3].

Before we start on the proofs, let me remind the reader of the definition of a generalized Frobenius partition. As defined in [1] a generalized Frobenius partition is a two-lined array of nonnegative integers of the form $\begin{pmatrix} a_1, a_2, \cdots, a_r \\ b_1, b_2, \cdots, b_r \end{pmatrix}$ where the entries in each row are ordered from largest to smallest. The number being partitioned is $m = \sum_{i=1}^r (a_i + b_i + 1)$.

In our case we are considering a special class of generalized Frobenius partitions–generalized Frobenius partitions of m with 3 colors. These are all of the arrays of the above form where the entries are taken from three distinct copies of the nonnegative integers distinguished by three different colors and ordered first according to size and then by color.

We begin by proving

Theorem 1.

$$\sum_{m=0}^{\infty} \overline{c\phi_3}(m)q^m = \frac{9q(q^9;q^9)_\infty^3}{(q;q)_\infty^3(q^3;q^3)_\infty}.$$

The proof of this result is based on the following two lemmas and the fact that $\sum_{m=0}^{\infty} p\left(\frac{m}{3}\right) q^m = 1/(q^3;q^3)_\infty$ where $p\left(\frac{m}{3}\right) = 0$ if $m/3$ is not an integer.

Lemma 1.

$$\sum_{m=0}^{\infty} c\phi_3(m)q^m = \frac{1}{(q;q)^3}\left[1 + 6\sum_{i=0}^{\infty}\left(\frac{q^{3i+1}}{1-q^{3i+1}} - \frac{q^{3i+2}}{1-q^{3i+2}}\right)\right].$$

Lemma 2.

$$\frac{(q;q)_\infty^3}{(q^3;q^3)_\infty} = 1 + 6\sum_{i=0}^{\infty}\left(\frac{q^{9i+3}}{1-q^{9i+3}} - \frac{q^{9i+6}}{1-q^{9i+6}}\right) - \frac{3q(q^9;q^9)_\infty^3}{(q^3;q^3)_\infty}.$$

The first lemma is a result due to Andrews [1]. The second follows by applying Jacobi's triple product identity and the logarithmic derivative of Jacobi's triple product identity to $(q;q)_\infty^3 = \sum_{n=-\infty}^{\infty} n(-1)^n q^{\binom{n+1}{2}}$, another result due to Jacobi, after rewriting this series according to the congruence class of n modulo 3.

Combining these two lemmas we have

$$\begin{aligned}
\sum_{m=0}^{\infty} \overline{c\phi_3}(m)q^m &= \sum_{m=0}^{\infty}\left(c\phi_3(m) - p\left(\frac{m}{3}\right)\right)q^m \\
&= \frac{1}{(q;q)_\infty^3}\left[1 + 6\sum_{i=0}^{\infty}\left(\frac{q^{3i+1}}{1-q^{3i+1}} - \frac{q^{3i+2}}{1-q^{3i+2}}\right) - \frac{(q;q)_\infty^3}{(q^3;q^3)_\infty}\right] \\
&= \frac{1}{(q;q)_\infty^3}\left[6\sum_{i=0}^{\infty}\left(\frac{q^{3i+1}}{1-q^{3i+1}} - \frac{q^{3i+2}}{1-q^{3i+2}}\right.\right. \\
&\qquad \left.\left. - \frac{q^{9i+3}}{1-q^{9i+3}} + \frac{q^{9i+6}}{1-q^{9i+6}}\right) + \frac{3q(q^9;q^9)_\infty^3}{(q^3;q^3)_\infty}\right]
\end{aligned}$$

$$= \frac{1}{(q;q)_\infty^3}\left[6\sum_{i=0}^{\infty}\left(\frac{q^{3i+1}}{1-q^{9i+3}} - \frac{q^{6i+4}}{1-q^{9i+6}}\right) + \frac{3q(q^9;q^9)_\infty^3}{(q^3;q^3)_\infty}\right]$$

$$= \frac{1}{(q;q)_\infty^3}\left[6\sum_{i=-\infty}^{\infty}\frac{q^{3i+1}}{1-q^{9i+3}} + \frac{3q(q^9;q^9)_\infty^3}{(q^3;q^3)_\infty}\right]$$

$$= \frac{1}{(q;q)_\infty^3}\left(\frac{6q(q^9;q^9)_\infty^2}{(q^6;q^9)_\infty(q^3;q^9)_\infty} + \frac{3q(q^9;q^9)_\infty^3}{(q^3;q^3)_\infty}\right)$$

(which follows from Ramanujan's ${}_1\psi_1$-identity). Our theorem follows immediately from this.

Defining $\xi = \frac{E^3}{qE_9^3}$, $T = \frac{q^3E_9^3}{E_3^3}$, and $S = \frac{E_3^4}{E_9^4} + 9\frac{q^3E_3E_{27}^3}{E_9^4}$, our second main theorem is

Theorem 2. *If $\alpha \geq 1$, then*

$$\sum_{n=0}^{\infty}\overline{c\phi_3}(3^\alpha n + \lambda_\alpha)q^n =$$

$$\begin{cases} \frac{1}{qE_3}\sum_{j=1}^{\infty} 9x_{\alpha,j}T^{-4j}q^{9j}\xi^{-4j-1}(\xi+9) & \text{if } \alpha \text{ is even,} \\ \frac{T}{q^3E_3}\sum_{j=1}^{\infty} 9x_{\alpha,j}T^{-4j}q^{9j}\xi^{-4j}(\xi+9) & \text{if } \alpha \text{ is odd,} \end{cases}$$

where $E(q) = (q;q)_\infty$ and $E_i = E(q^i)$ with $E_1 = E$,

$$x_1 = (3,0,0,\cdots),$$

$$x_{\alpha+1} = \begin{cases} x_\alpha A & \text{if } \alpha \text{ is even,} \\ x_\alpha B & \text{if } \alpha \text{ is odd,} \end{cases}$$

$$A = (a_{ij})_{i,j\geq 1}, \quad a_{ij} = 9m_{4i+1,i+j} + m_{4i,i+j},$$

$$B = (b_{ij})_{i,j\geq 1}, \quad b_{ij} = m_{4i-1,i+j} + 9m_{4i,i+j},$$

$$m_{1,1} = 3, \quad m_{2,1} = 1, \quad m_{i,1} = 0 \quad \text{for } i \geq 3,$$

$$m_{1,2} = 0, \quad m_{2,2} = 3^4, \quad m_{3,2} = 2\cdot 3^3,$$

$$m_{1,3} = 0, \quad m_{2,3} = 0, \quad m_{3,3} = 3^7,$$

$$m_{i,j} = 0 \quad \text{for } i \leq 3 \text{ and } j \geq 4,$$

$$m_{i,j} = m_{i-3,j-1} + 9m_{i-2,j-1} + 27m_{i-1,j-1} \quad \text{for } i \geq 4,\ j \geq 2,$$

λ_α *is the reciprocal of 8 modulo 3^α.*

The proof of Theorem 2 relies on the following four lemmas.

Lemma 3. $\xi^3 + 9\xi^2 + 27\xi = q^9T^{-4}$.

Lemma 4.

$$H_2(\xi^{-i}) = \frac{S}{q}\sum_{j=1}^{\infty} m_{i,j}T^{4j}q^{-9j},$$

where H_k is the operator which acts on a power series of q and picks out those terms in which the power of q is congruent to k modulo 3.

Lemma 5.

$$H_2\left(9\xi^{-4i-1}+\xi^{-4i}\right)=\frac{S}{q}\sum_{j=1}^{\infty}a_{ij}T^{4i+4j}q^{-9i-9j},$$

$$H_2\left(\xi^{-4i+1}+9\xi^{-4i}\right)=\frac{S}{q}\sum_{j=1}^{\infty}b_{ij}T^{4i+4j}q^{-9i-9j}.$$

Lemma 6. *If λ_α represents the reciprocal of 8 modulo 3^α, then*

$$\lambda_\alpha=\begin{cases}\frac{1}{8}(7\cdot 3^\alpha+1) & \text{if } \alpha \text{ is even,}\\ \frac{1}{8}(5\cdot 3^\alpha+1) & \text{if } \alpha \text{ is odd,}\end{cases}$$

and

$$\lambda_{\alpha+1}=\begin{cases}3^\alpha+\lambda_\alpha & \text{if } \alpha \text{ is even,}\\ 2\cdot 3^\alpha+\lambda_\alpha & \text{if } \alpha \text{ is odd.}\end{cases}$$

Lemma 3 is the modular equation of order three. Lemma 4 follows by observing that $\xi^{-i}=T^4q^{-9}(\xi^{3-i}+9\xi^{2-i}+27\xi^{1-i})$ from Lemma 3 and $\xi=\frac{E_3^4}{qE_9^4}+\frac{9q^2E_{27}^3E_3}{E_9^4}-3$ from Lemma 2 and the proof of Theorem 1. Lemma 5 follows immediately from Lemma 4 after observing that $m_{i,j}=0$ if $j\le i/3$. The first part of Lemma 6 is easily verified and the second follows by an easy induction argument.

To prove Theorem 2 we proceed by induction on α. It is easily verified that

$$\sum_{n=0}^{\infty}\overline{c\phi_3}(3n+2)q^n=\frac{27q^6T^{-3}}{E_3}(\xi^{-3}+9\xi^{-4})$$

which is consistent with the statement of the theorem for $\alpha=1$.

Now suppose α is odd and

$$\sum_{n=0}^{\infty}\overline{c\phi_3}(3^\alpha n+\lambda_\alpha)q^n=\frac{T}{q^3E_3}\sum_{i=1}^{\infty}9x_{\alpha,i}T^{-4i}q^{9i}\xi^{-4i}(\xi+9).$$

Picking out the powers of q congruent to 2 modulo 3 we have

$$\sum_{n=0}^{\infty}\overline{c\phi_3}(3^\alpha(3n+2)+\lambda_\alpha)q^{3n+2}=\frac{T}{q^3E_3}\sum_{i=1}^{\infty}9x_{\alpha,i}T^{-4i}q^{9i}H_2(\xi^{-4i+1}+9\xi^{-4i}).$$

Using Lemma 5 and Lemma 6 it follows that

$$\sum_{n=0}^{\infty}\overline{c\phi_3}(3^{\alpha+1}n+\lambda_{\alpha+1})q^{3n+2}=\frac{TS}{q^4E_3}\sum_{j=1}^{\infty}9x_{\alpha+1,j}T^{4j}q^{9j}.$$

Noting that $T(q^{1/3}) = \frac{qE_9^3}{E^3} = T^{-1}\xi^{-1}q^3$ and $S(q^{1/3}) = \frac{E^4}{E_3^4}(1+9\xi^{-1})$, we have

$$\sum_{n=0}^{\infty} \overline{c\phi_3}(3^{\alpha+1}n+\lambda_{\alpha+1})q^n = \frac{1}{qE_3}\sum_{j=1}^{\infty} 9x_{\alpha+1,j}T^{-4j}q^{9j}\xi^{-4j-1}(\xi+9).$$

To complete the proof, suppose α is even and

$$\sum_{n=0}^{\infty} \overline{c\phi_3}(3^{\alpha}n+\lambda_{\alpha})q^n = \frac{1}{qE_3}\sum_{j=1}^{\infty} 9x_{\alpha,j}T^{-4j}q^{9j}\xi^{-4j-1}(\xi+9).$$

Picking out the powers of q congruent to 1 modulo 3 we have

$$\sum_{n=0}^{\infty} \overline{c\phi_3}(3^{\alpha}(3n+1)+\lambda_{\alpha})q^{3n+1} = \frac{1}{qE_3}\sum_{i=1}^{\infty} 9x_{\alpha,i}T^{-4i}q^{9i}H_2(9\xi^{-4i-1}+\xi^{-4i}).$$

Using Lemma 5 and Lemma 6 it follows that

$$\sum_{n=0}^{\infty} \overline{c\phi_3}(3^{\alpha+1}n+\lambda_{\alpha+1})q^{3n+1} = \frac{S}{q^2E_3}\sum_{j=1}^{\infty} 9x_{\alpha+1,j}T^{4j}q^{-9j}.$$

Hence

$$\sum_{n=0}^{\infty} \overline{c\phi_3}(3^{\alpha+1}n+\lambda_{\alpha+1})q^n = \frac{T}{q^3E_3}\sum_{j=1}^{\infty} 9x_{\alpha+1,j}T^{-4j}q^{9j}\xi^{-4j}(\xi+9).$$

Turning our attention to the divisibility of $\overline{c\phi_3}(3^{\alpha}n+\lambda_{\alpha})$ by powers of 3, let $\nu(m)$ be the highest power of 3 dividing m. We have the following three Lemmas.

Lemma 7. $\nu(m_{i,j}) \geq \left[\frac{9j-3i-3}{2}\right]$.

Lemma 8. $\nu(a_{ij}) \geq \left[\frac{9j-3i-3}{2}\right]$ and $\nu(b_{ij}) \geq \left[\frac{9j-3i}{2}\right]$.

Lemma 9.

$$\nu(x_{1,1}) = 1,$$
$$\nu(x_{2\beta,j}) \geq 4\beta + \left[\frac{9j-9}{2}\right],$$
$$\nu(x_{2\beta+1,j}) \geq 4\beta + 1 + \left[\frac{9j-8}{2}\right].$$

To prove Lemma 7 we define $v_{i,j} = [(9j-3i-3)/2]$ and note that $\nu(m_{i,j}) \geq v_{i,j}$ for $i \leq 3$ and for $i > 3$ and $j = 1$. For $i \geq 4$ and $j \geq 2$, we have

$$\begin{aligned}\nu(m_{i,j}) &\geq \min\{\nu(m_{i-3,j-1}),\ 2+\nu(m_{i-2,j-1}),\ 3+\nu(m_{i-1,j-1})\}\\ &\geq \min\{v_{i-3,j-1},\ 2+v_{i-2,j-1},\ 3+v_{i-1,j-1}\}\\ &= \left[\frac{9j-3i-3}{2}\right] = v_{i,j}.\end{aligned}$$

Lemma 8 follows immediately from the fact that

$$\begin{aligned}\nu(a_{ij}) &= \nu(9m_{4i+1,i+j} + m_{4i,i+j})\\ &\geq \min\{2+v_{4i+1,i+j},\ v_{4i,i+j}\} = \left[\frac{9j-3i-3}{2}\right].\end{aligned}$$

Similarly $\nu(b_{ij}) \geq [(9j-3i)/2]$. Lemma 9 follows easily using induction and noting that

$$\nu(x_{2\beta,j}) \geq \min_{i\geq 1}\{x_{2\beta-1,i}\ b_{ij}\} \quad\text{and}\quad \nu(x_{2\beta+1,j}) \geq \min_{i\geq 1}\{x_{2\beta,i}\ a_{ij}\}.$$

Combining these three lemmas with Theorem 2 we immediately have the following congruence.

Theorem 3.

$$\overline{c\phi_3}(3^\alpha n + \lambda_\alpha) \equiv 0 \begin{cases} \bmod 3^{2\alpha+2} & \textit{if } \alpha \textit{ is even},\\ \bmod 3^{2\alpha+1} & \textit{if } \alpha \textit{ is odd}.\end{cases}$$

REFERENCES

[1] George E. Andrews, "Generalized Frobenius Partitions", Memoirs of the American Mathematical Society, Volume 301, Providence, RI, May 1984.

[2] A.O.L. Atkin, Ramanujan Congruences for $p_{-k}(n)$, J. London Math. Soc. **(2) 39** (1935), 142–149.

[3] F.G. Garvan, A Simple Proof of Watson's Partition Congruences for Powers of 7, J. Austral. Math. Soc. (Series A) **36** (1984), 316–334.

[4] M.D. Hirschhorn and D.C.Hunt, A Simple Proof of the Ramanujan Conjecture for Powers of 5, J. Reine Angew. Math. **326** (1981), 1–17.

[5] Louis W. Kolitsch, A Relationship between Certain Colored Generalized Frobenius Partitions and Ordinary Partitions, Journal of Number Theory **33** (1989), 220–223.

Louis Kolitsch
Department of Mathematics
and Computer Science
The University of Tennessee at Martin
Martin, TN 38238

The Coefficients of Cyclotomic Polynomials

HELMUT MAIER

Dedicated to Professor Paul Bateman

1. Introduction and Statement of Results

Let $\Phi_n(z) = \sum_{m=0}^{\phi(n)} a(m,n)z^m$ be the nth cyclotomic polynomial. Let

$$A(n) = \max_{0 \leq m \leq \phi(n)} |a(m,n)| \tag{1.1}$$

and let

$$S(n) = \sum_{0 \leq m \leq \phi(n)} |a(m,n)|.$$

The coefficients $a(m,n)$, and especially $A(n)$ and $S(n)$, have been the subject of numerous investigations (see [1] and the references given there). All these investigations concern a very thin set of integers n. Here we deal with properties that hold on a set of integers of asymptotic density 1. We say that a property holds for almost all integers if it holds on a sequence of asymptotic density 1. P. Erdös conjectured that, for any constant $c > 0$, $A(n) \geq c$ holds for almost all n. We shall prove a theorem that implies Erdös' conjecture.

Theorem. *Let $\epsilon(n)$ be a function defined for all positive integers such that $\lim_{n\to\infty} \epsilon(n) = 0$. Then $S(n) \geq n^{1+\epsilon(n)}$ for almost all n.*

The author wishes to thank Professor C. Pomerance for valuable advice concerning the presentation of the paper.

2. Outline of proof and basic lemmas

We start with a well-known identity.

Lemma 1. *For complex z we have*

$$\log|\Phi_n(z)| = \sum_{d|n} \mu(n/d)\log|1-z^d|, \tag{2.1}$$

wherever these expressions are defined.

Lemma 2. *If m is odd and i an arbitrary positive integer, then $\Phi_{2^i m}(z) = \Phi_m(-z^{2^{i-1}})$. Further, if m is odd and $p_1, p_2, \dots, p_k$ are the distinct primes dividing m, then $\Phi_m(z) = \Phi_{p_1 p_2 \cdots p_k}(z^{m/(p_1 p_2 \cdots p_k)})$.*

Proof: See [**1**].

Let $\psi(x)$ be any function with $\lim_{x\to\infty}\psi(x) = \infty$. For all but $o(x)$ integers $n \le x$ we have that the squarefree kernel $q(n)$ satisfies $q(n)\psi(n) \ge n$. To each squarefree integer m there belong thus $\ll \psi(m)$ integers n with $q(n) = m$, $n < m\psi(m)$. Since $\psi(x)$ is arbitrary and by Lemma 2

$$S(n) \ge \max_{|z|=1} |\Phi_n(z)| = \max_{|z|=1} |\Phi_{q(n)}(z)|,$$

we see that it suffices to prove that $\max_{|z|=1}|\Phi_n(z)| \ge n^{1+\epsilon(n)}$ holds for almost all squarefree integers n.

For $z = e(\alpha)$ Lemma 1 takes the form

$$\log|\Phi_n(e(\alpha))| = \sum_{d|n} \mu(n/d)\log|1-e(\alpha d)|. \tag{2.2}$$

(Here and in the sequel we write $e^{2\pi i\beta} = e(\beta)$.) The only divisors $d|n$ that give large contributions to the sum in (2.2) are those for which $\mu(n/d) < 0$ and $|1-e(\alpha d)|$ are small. The latter happens if $\|\alpha d\|$ is small. (Here and in the sequel $\|\beta\|$ denotes the distance of β to the nearest integer.) Any too simple construction of large values is however prevented by a certain cancellation effect which can be described as follows:

If we write $\alpha d = m + \rho$ with an integer m and $\rho \in (-1/2, 1/2]$, then we have by Taylor's formula $e(\alpha d) = 1 + 2\pi i\rho + O(\rho^2)$ and thus

$$\log|1-e(\alpha d)| = \log|\rho| + O(1). \tag{2.3}$$

(The constants implied in the O- and $\ll$-symbols are absolute unless indicated otherwise.)

Assume now $t|(n/d)$ with $\omega(t) > 1$, where ω denotes the number of prime factors, and $|\rho t| < 1$. Then

$$\sum_{s|t} \mu(n/ds) \log|1 - e(\alpha ds)|$$
$$= \sum_{s|t} \mu(n/ds)(\log|\rho| + \log s) + O(2^{\omega(t)}) = O(2^{\omega(t)}). \tag{2.4}$$

Thus the large term $\mu(n/d)\log|1-e(\alpha d)|$ is cancelled by contributions from other divisors.

A way to prevent this cancellation is to construct a triple (α, d, t) with $\mu(n/d) < 0$, $t|(n/d)$, such that $\alpha d = m+\rho$ with small $|\rho|$ and $|\rho s| < 1$ for $s|t$ with $s \le s_0$, but $|\rho s| > 1$ for $s|t$ with $s > s_0$. For $s > s_0$ the approximation by Taylor's formula is no longer valid, and (2.4) is to be replaced by

$$\sum_{s|t} \mu(n/ds) \log|1 - e(\alpha ds)|$$
$$= \sum_{\substack{s|t \\ s\le s_0}} \mu(n/ds)(\log|\rho| + \log s) + \sum_{\substack{s|t \\ s>s_0}} \mu(n/ds)(\log|1 - e(\alpha ds)|)$$
$$= \Sigma_1 + \Sigma_2, \quad \text{say.} \tag{2.5}$$

By an appropriate choice of α and d we try to achieve that the 'incomplete convolution' Σ_1 is large compared to the sum Σ_2. In fact, we will construct an α and an entire system of sums of the form (2.5) with this property.

The successful realisation of this plan depends on the solution of two kinds of problems.

(i) We need an appropriate set of divisors $D_0, \dots, D_k$. The existence of such a set depends on a certain configuration of the prime factors of n. In Lemma 3 it will be shown that almost all integers n have this special configuration.

(ii) We must find an $\alpha \in [0, 1)$ that satisfies a system of inequalities of the form

$$\Delta_j^{(1)} \le \|\alpha D_j\| \le \Delta_j^{(2)} \quad (0 \le j \le k).$$

We further have to ensure that α does not satisfy any unwanted inequalities of this kind. This will be established in Lemma 4.

We now formulate Lemmas 3 and 4; the next two sections will be devoted to their proofs.

Before stating Lemma 3 we need to introduce some definitions. For a squarefree integer $n > 1$ let

$$n = p_1(n)p_2(n)\dots p_{\omega(n)}(n), \qquad p_1(n) > p_2(n) > \dots > p_{\omega(n)}(n)$$

be the prime factorization of n. We fix a sufficiently small constant $\epsilon_0 > 0$ once and for all (e. g., $\epsilon_0 = 10^{-3}$ will suffice). We say that the prime factor $p_k(n)$ is *special* if

$$p_k(n)^{\epsilon_0/2} < p_{k+1}(n) < p_k(n)^{\epsilon_0}, \quad p_{k+1}(n) \le p_{k+7}(n)^{1+\epsilon_0},$$
$$\prod_{j\ge k+8} p_j(n) \le p_{k+7}(n)^{\epsilon_0}. \tag{2.6}$$

We set

$$\begin{aligned} \mathcal{S}_k(x) = \{n \le x : \mu^2(n) = 1,\ &\log p_k(n) \ge e^{-4k} \log x, \\ &p_k(n) \text{ is special, } \omega(n) \le 1.1 \log\log x, \\ &\prod_{j\ge k+8} p_j(n) \ge \exp((\log\log x)^6)\} \end{aligned} \tag{2.7}$$

Lemma 3. *Let $\psi(x) \to \infty$ as $x \to \infty$. There are at most $o(x)$ squarefree integers $n \le x$ that do not belong to $\mathcal{S}_k(x)$ for some even $k \le \psi(x)$.*

Let $n \in \mathcal{S}_k(x)$. We define

$$D_0 = D_0(n) = \prod_{1\le i\le k+7} p_i(n),$$
$$D_j = D_j(n) = \prod_{\substack{1\le i\le k+7 \\ i\ne j}} p_i(n) \quad (1 \le j \le k),$$
$$E = E(n) = \prod_{k+8\le j\le \omega(n)} p_j(n).$$

We further set

$$\begin{aligned} U &= \exp((\log\log x)^6), \\ V &= V(n) = p_1(n)\dots p_{k+6}(n) p_{k+7}(n)^{-3}, \\ \eta &= \exp(-(\log\log x)^3). \end{aligned} \tag{2.8}$$

We define $\mathcal{P}(n)$, a set of certain divisors of n, by

$$\mathcal{P}(n) = \{E(n)D_j(n) : 1 \le j \le k\} \cup \{dE(n) : d \mid \prod_{k+1\le j\le k+7} p_j,\ \omega(d) < 3\}. \tag{2.9}$$

For $n \in \mathcal{S}_k(x)$ we set

$$\mathcal{M}(n) = \{\alpha \in [1/5, 4/5] : \text{(2.10i), (2.10ii), (2.10iii)}\},$$

where

$$(U/3)p_j^{-1} \le \|ED_j\alpha\| \le 2Up_j^{-1} \ (1 \le j \le k), \tag{2.10i}$$
$$(1/3)p_{k+7}^{-4}U \le \|E\alpha\| \le 2p_{k+7}^{-4}U, \tag{2.10ii}$$
$$\|d_0\alpha\| \ge \eta^2 \text{ for all } d_0|n,\ d_0 \notin \mathcal{P}(n). \tag{2.10iii}$$

We can now state Lemma 4.

Lemma 4. *Let x be sufficiently large and $k \le (\log\log x)^{1/2}$. For all $n \in \mathcal{S}_k(x)$ except those from a set of cardinality $\le x\exp(-(\log\log x)^2)$ the set $\mathcal{M}(n)$ is non-empty.*

For the proof of our Theorem we shall apply (2.5) for some $\alpha \in \mathcal{M}(n)$. The condition that k be even in Lemma 3 is needed to ensure that the Möbius function carries the right sign.

3. The appearance of special prime factors in almost all integers

In this section we shall prove Lemma 3.

Let

$$\mathcal{E}_k(x) = \{n \le x : \mu^2(n) = 1, \text{ there is no even } j \text{ with } p_j(n) \text{ special } \\ e^{-4k}\log x \le \log p_j(n) \le e^{-2k}\log x\}. \tag{3.1}$$

The proof is based on two auxiliary lemmas.

Lemma 5. *Let $k \le (\log\log x)^{1/2}$. Then $|\mathcal{E}_k(x)| \ll k^{-1/2}x$.*

Proof: To prove this result we use an idea from [5]. We consider, for $l < 2k$, the relative frequency of the exceptional set of integers that have no special prime factor $p_j(n)$ with

$$e^{-4k}\log x \le \log p_j(n) \le e^{-4k+l}\log x.$$

We shall show that if n has no exceptional prime factor $p_j(n)$ in this range then the conditional probability that n still does not have such a prime factor in the larger range

$$e^{-4k}\log x < \log p_j(n) \le \exp(-4k + l + [3|\log\epsilon_0|])\log x$$

is not too close to one. This will prove that the exceptional set is shrinking as l increases and lead to the desired estimate for $|\mathcal{E}_k(x)|$.

For positive integers k and l with $l < 2k$ let

$$\mathcal{E}(k,l,x) = \{n \le x : \text{ there is no even } j \text{ with } p_j(n) \text{ special}, \\ e^{-4k}\log x < \log p_j(n) < e^{-4k+l}\log x\}.$$

Let $\mathcal{F}(k,l,x)$ be the set of those n in $\mathcal{E}(k,l,x)$ of the form $n = aq_7q_6\dots q_1b$ with primes q_j such that

$$q_7 < q_6 < \cdots q_1 < q_7^{1+\epsilon_0},$$
$$\epsilon_0{}^{-1}e^{-4k+l}\log x \le \log q_7 \le 2\epsilon_0{}^{-1}e^{-4k+l}\log x,$$
$$q_1^{1/\epsilon_0} \le p_-(b) \le q_1^{2/\epsilon_0}.$$

Obviously all $n \in \mathcal{F}(k,l,x)$ have the special prime factor $p_-(b)$ with $\log p_-(b) \le \exp(-4k + l + [3|\log\epsilon_0|])\log x$. Thus, $n \in \mathcal{F}(k,l,x)$ implies that $n \notin \mathcal{E}(k, l + [3|\log\epsilon_0|], x)$. Hence

$$\mathcal{E}(k, l + [3|\log\epsilon_0|], x) \le \mathcal{E}(k,l,x) - \mathcal{F}(k,l,x). \tag{3.2}$$

Let $C = k^{1/2}$. We shall show that $|\mathcal{E}(k,l,x)| \ge c_4 C^{-1} x$ implies that

$$|\mathcal{F}(k,l,x)| \gg C^{-1}|\mathcal{E}(k,l,x)| \tag{3.3}$$

and thus

$$|\mathcal{E}(k, l + [3|\log\epsilon_0|], x)| \le (1 - c_5 C^{-1})|\mathcal{E}(k,l,x)|, \tag{3.4}$$

where c_4 and c_5 are suitable positive constants. Iteration of (3.4) gives Lemma 5 if we observe the well-known fact that

$$|\{n \le x : \omega(n) \ge 1.1\log\log x\}| \ll \frac{x}{(\log x)^{c_6}}$$

for some appropriate c_6.

Assuming that $x \ge q_1^{\epsilon_0^{-3}}$, $k \ge k_0$ (with k_0 sufficiently large), $l \le 2k$, we have

$$|\mathcal{F}(k,l,x)| \gg \sum_{\substack{a \in \mathcal{E}(k,l,x) \\ \log a \le e^{-4k+l}\log x}} \quad \sum_{\substack{(q_1,\dots,q_7):\, q_j \text{ prime } (1\le j\le 7) \\ \epsilon_0{}^{-1}e^{-4k+l}\log x \le \log q_7 \le 2\epsilon_0{}^{-1}e^{-4k+l}\log x \\ q_7 < q_6 < \cdots < q_1 < q_7^{1+\epsilon_0}}} \quad \sum_{\substack{b \le x/(aq_1\dots q_7) \\ q_1^{1/\epsilon_0} \le p_-(b) \le q_1^{2/\epsilon_0} \\ \mu(b)=1}} 1. \tag{3.5}$$

By standard results from sieve theory the inner sum is $\gg x/(aq_1 \dots q_7 \log q_1)$ (see [4]). Also, by the prime number theorem we have

$$\sum_{(q_1,\dots,q_7)}{}' \frac{1}{q_1 \dots q_7 \log q_1} \gg \frac{e^{4k-l}}{\log x},$$

where the range of summation in $\sum'$ is the same as in (3.5). Thus

$$|\mathcal{F}(k,l,x)| \gg \frac{x}{\log x} e^{4k-l} \sum_{\substack{a \in \mathcal{E}(k,l,x) \\ \log a \le e^{-4k+l}\log x}} \frac{1}{a}. \tag{3.6}$$

This lower bound for $|\mathcal{F}(k,l,x)|$ we compare with an upper bound for $|\mathcal{E}(k,l,x)|$. We write $a = a(n) = \prod_{j\geq r} p_j(n)$ with r being the minimal odd index such that

$$\log\Big(\prod_{j\geq r} p_j\Big) \leq e^{-4k+l}\log x.$$

We then use the decomposition $n = aq(n/aq)$, where $q = p_{r-1}$, and obtain

$$|\mathcal{E}(k,l,x)| \ll \sum_{\substack{a:\,\log a\leq e^{-4k+l}\log x\\ a\in\mathcal{E}(k,l,x)}} \sum_{\substack{q\text{ prime}\\ p_+(a)\leq q}} \sum_{\substack{n\leq x:\, n\equiv 0 \bmod aq\\ \log p_-(n/aq)>\max(\log q,\, e^{-4k+l}\log x-\log(aq))\\ \mu(n/aq)=1}} 1$$

$$= \sum\nolimits^{(1)} + \sum\nolimits^{(2)}, \tag{3.7}$$

where in $\sum^{(1)}$ we sum over all a with $\log p_+(a) > C^{-1}e^{-4k+l}\log x$, and in $\sum^{(2)}$ over all a with $\log p_+(a) \leq C^{-1}e^{-4k+l}\log x$, where $C = k^{-1/2}$.

For the estimate of $\sum^{(2)}$ we use the well-known results on integers free of large prime factors. With the notation

$$\psi(w,y) = \sum_{n\leq w:\, P_+(n)\leq y} 1$$

we have

$$\psi(w,y) \ll w\exp\left(-\frac{1}{2}\frac{\log w}{\log y}\log\frac{\log w}{\log y}\right)$$

as $w\to\infty$ and $y \geq \exp((\log w)^{5/8+\epsilon})$ (see [**2**]). We obtain

$$\sum\nolimits^{(2)} \ll \sum_{\substack{a:\,\log a\leq e^{-4k+l}\log x\\ \log p_+(a)\leq C^{-1}e^{-4k+l}\log x}} \sum_{\substack{q\text{ prime}\\ q\geq p_+(a)}} \sum_{\substack{n\leq x:\, n\equiv 0 \bmod aq\\ \log p_-(n/aq)>\max(\log q,\, e^{-4k+l}\log x-\log(aq))}} 1$$

$$\ll \sum_{\substack{a:\,\log a\leq e^{-4k+l}\log x\\ \log p_+(a)\leq C^{-1}e^{-4k+l}\log x}} \sum_{\substack{q\text{ prime}\\ q\geq p_+(a)}} \min\left(\frac{x}{aq(e^{-4k+l}\log x-\log(aq))}, \frac{x}{aq\log q}\right)$$

$$\ll \sum_{\substack{a:\,\frac{1}{10}e^{-4k+l}\log x\le\log a\le e^{-4k+l}\log x\\ \log p_+(a)\le C^{-1}e^{-4k+l}\log x}} \frac{x}{a\log p_+(a)}$$

$$+ \sum_{\substack{a:\,\log a\le\frac{1}{10}e^{-4k+l}\log x\\ \log p_+(a)\le C^{-1}e^{-4k+l}\log x}} \frac{x}{ae^{-4k+l}\log x}\left(\sum_{\substack{q\text{ prime}\\ \log p_+(a)\le\log q<\frac{1}{2}e^{-4k+l}\log x}} \frac{1}{q}\right).$$

By dividing both sums into partial sums extending over intervals of the form

$$2^{-g-1}e^{-4k+l}\log x \le \log a \le 2^{-g}e^{-4k+l}\log x,$$
$$2^{-h-1}C^{-1}e^{-4k+l}\log x \le \log p_+(a) \le 2^{-h}C^{-1}e^{-4k+l}\log x,$$

we obtain

$$\sum\nolimits^{(2)} \ll C^{-1}x. \tag{3.8}$$

We thus obtain from (3.7)

$$|\mathcal{E}(k,l,x)| \ll \sum_{\substack{a:\,\log a\le e^{-4k+l}\log x\\ a\in\mathcal{E}(k,l,x)}} \sum_{\substack{n\le x,\, n\equiv 0 \bmod a\\ \log p_-(n/a)>C^{-1}e^{-4k+l}\log x}} 1 + C^{-1}x$$
$$\ll \frac{Cx}{\log x}e^{4k-l}\sum_{\substack{a\in\mathcal{E}(k,l,x)\\ \log a\le e^{-4k+l}\log x}}\frac{1}{a} + C^{-1}x. \tag{3.9}$$

A comparison of (3.6) and (3.9) shows that if $|\mathcal{E}(k,l,x)| \ge c_4C^{-1}x$ for an appropriate $c_4 > 0$ then (3.3) holds. Lemma 5 is thus proved.

Lemma 6. *Let $j \le (\log\log x)^{1/2}$. For all $n \le x$ except those of a set of cardinality $\le xe^{-c_5 j}$ for some appropriate fixed $c_5 > 0$ we have $\log p_j(n) \ge e^{-2j}\log n$.*

Proof: This lemma is easily proved by the Turán-Kubilius inequality (see [3]).

Lemma 3 now immediately follows from Lemmas 5 and 6.

4. The construction of the set $\mathcal{M}(n)$

In this section we shall prove Lemma 4. The basic idea behind the proof is as follows. We shall construct the set $\mathcal{M}(n)$ defined by (2.10i)–(2.10iii) as a union of certain intervals of the form $S_l = [(l-1)/n, l/n]$. To this end we replace the inequalities in (2.10) by congruence conditions for l. We are then faced with the task to count those l-values. It is easy to get a lower bound for the l-values satisfying the congruence-conditions corresponding to the inequalities (2.10i) and (2.10ii). This will simply be achieved by an application of the Chinese Remainder Theorem. It is harder to eliminate the l-values that satisfy any one of the congruence-conditions corresponding to one of the inequalities $||d_0\alpha|| < \eta^2$ for $d_0|n$, $d_0 \notin \mathcal{P}(n)$ that have been forbidden in (2.10iii). This will be done mainly by a complicated counting argument that involves averaging over $n \in \mathcal{S}_k(x)$.

Proof of Lemma 4: Let $\mathcal{C}_k(n)$ be the set of all l with $(1/4)n \le l \le (3/4)n$ satisfying

$$\begin{aligned} &l \equiv m_j \bmod p_j \text{ for some } m_j && (4.1\text{i})\\ &\qquad \text{with } U/2 \le |m_j| \le U \ (1 \le j \le k), \\ &l \equiv m_{k+1} \bmod D_0 \text{ for some } m_{k+1} && (4.1\text{ii})\\ &\qquad \text{with } UV/2 \le |m_{k+1}| \le UV. \end{aligned}$$

One easily sees that the congruences (4.1) for l imply that each $\alpha \in S_l$ satisfies the inequalities (2.10i) and (2.10ii). For a fixed $(k+1)$-tuplet $(m_1, \ldots, m_k, m_{k+1})$ the system (4.1i), (4.1ii) is solvable if and only if

$$m_{k+1} \equiv m_j \bmod p_j \quad (1 \le j \le k).$$

We call such $(k+1)$-tuplets *admissible*. There are $U^k(1+O(k/U))$ possible k-tuplets $(m_1, \ldots, m_k)$ satisfying (4.1i). For each such k-tuplet there are $UV(p_1 \cdots p_k)^{-1} + O(1) = Up_{k+1} \cdots p_{k+6}p_{k+7}^{-3} + O(1)$ choices for m_{k+1} that lead to admissible $(k+1)$-tuplets by the Chinese Remainder Theorem, and there are $n/2D_0 + O(1) = E/2 + O(1)$ choices for l for each admissible $(k+1)$-tuplet. Thus

$$|\mathcal{C}_k(n)| = \frac{1}{2} p_{k+1} \cdots p_{k+6} p_{k+7}^{-3} E U^{k+1}(1 + O(k/U)). \tag{4.2}$$

To ensure (2.10iii) we try to remove from $\mathcal{C}_k(n)$ all l that satisfy at least one of the congruences $l \equiv m \bmod n/d_0$ for some m with $|m| \le \eta^2 n/d_0$ and $d_0 \notin \mathcal{P}(n)$. In a first step we ensure the existence of an α which besides (2.10i) and (2.10ii) also satisfies (2.10iii) for all divisors $d_0 = D_j E t_j^{-1}$ with $t_j \le (2\eta)^{-1}$ and $d_0 = E t_{k+1}^{-1}$ with $t_{k+1} \le (2\eta)^{-1}$. For this purpose we remove from $\mathcal{C}_k(n)$ all l that satisfy at least one of the congruences

$$\begin{aligned} &l \equiv m_j \bmod p_j t_j \text{ for some pair } (m_j, t_j) && (4.1\text{iii})\\ &\qquad \text{with } |m_j| \le \eta p_j t_j,\ t_j | n,\ t_j \le (1/2)\eta^{-1} \ (1 \le j \le k) \\ &l \equiv m_{k+1} \bmod D_0 t_{k+1} \text{ for some pair } (m_{k+1}, t_{k+1}) && (4.1\text{iv})\\ &\qquad \text{with } |m_{k+1}| \le \eta D_0 t_{k+1},\ t_{k+1} | n,\ t_{k+1} \le (1/2)\eta^{-1}. \end{aligned}$$

(We can afford to replace η^2 in (2.10) by the larger number η.) The set of the remaining l-values we denote by $\mathcal{H}_k(n)$.

The conditions (4.1iii) and (4.1iv) may be written as

$$l = u_j t_j p_j + m_j \quad (1 \le j \le k)$$

and

$$l = u_{k+1} t_{k+1} D_0 + m_{k+1},$$

where the coefficients u_j are uniquely determined for fixed l_j, p_j and t_j because of the inequalities $|m_j| \le \eta p_j t_j$ $(1 \le j \le k)$ and $|m_{k+1}| \le \eta D_0 t_{k+1}$, and since $t_j \le (1/2)\eta^{-1}$. Therefore the system (4.1i), (4.1iii) for some fixed j with $1 \le j \le k$ reduces to

$$\begin{aligned} &l \equiv m_j \bmod p_j t_j \text{ for some } m_j \\ &\quad \text{with } U/2 \le |m_j| \le U \ (1 \le j \le k), \end{aligned} \tag{4.3i}$$

whereas the system (4.1ii), (4.1iv) reduces to

$$\begin{aligned} &l \equiv m_{k+1} \bmod D_0 t_{k+1} \text{for some } m_{k+1} \\ &\quad \text{with } UV/2 < |m_{k+1}| \le UV. \end{aligned} \tag{4.3ii}$$

Thus we can construct $\mathcal{H}_k(n)$ by removing from $\mathcal{C}_k(n)$ all l-values that satisfy one of the congruences (4.3i) and (4.3ii) for some *prime* divisor $t_j < (1/2)\eta^{-1}$.

Let $(t_1, \dots, t_k, t_{k+1})$ be a $(k+1)$-tuplet of divisors $t_j | E$, $1 \le j \le k+1$. We define $\mathcal{C}_k(n; t_1, \dots, t_{k+1})$ as the set of l with $(1/4)n \le l \le (3/4)n$ satisfying (4.3i) and (4.3ii). Let $q_1, q_2, \dots, q_s$ be the prime divisors of n that are $\le (1/2)\eta^{-1}$. We write $\mathcal{C}_{k,j,r}(n)$ for $\mathcal{C}_k(n; 1, \dots, q_r, \dots, 1)$, where the j-th entry is q_r and the other entries are 1. Then

$$\mathcal{H}_k(n) = \mathcal{C}_k(n) - \bigcup_{\substack{1 \le r \le s \\ 1 \le j \le k+1}} \mathcal{C}_{k,j,r}(n). \tag{4.4}$$

Also

$$\mathcal{C}_k(n; t_1, \dots, t_{k+1}) = \bigcap_{\substack{1 \le j \le k+1 \\ 1 \le l \le r(j)}} \mathcal{C}_k(n; 1, \dots, q_j^{(l)}, \dots, 1), \tag{4.5}$$

where

$$t_j = \prod_{1 \le l \le r(j)} q_j^{(l)} \quad (1 \le j \le k+1)$$

are the prime factorizations of t_j. From (4.4) and (4.5) it follows that $|\mathcal{H}_k(n)|$ can be determined by the inclusion-exclusion principle: we have

$$|\mathcal{H}_k(n)| = \sum_{(t_1,\dots,t_{k+1})}{}' (-1)^{\omega(t_1)+\cdots+\omega(t_{k+1})}|\mathcal{C}_k(n;t_1,\dots,t_{k+1})|, \qquad (4.6)$$

where the sum is extended over all $(k+1)$-tuplets of divisors $t_j|n$, all of whose prime divisors are $\leq (1/2)\eta^{-1}$.

We now determine $|\mathcal{C}_k(n;t_1,\dots,t_{k+1})|$. Let

$$[t_1,\dots,t_{k+1}] = \prod_{1\leq r\leq L} q_r,$$

where $[\dots]$ denotes the least common multiple. For a fixed L-tuplet $(a_1,a_2,\dots,a_L)$ with $0\leq a_r < q_r$ $(1\leq r\leq L)$ we count the number of $(k+1)$-tuplets $(m_1,\dots,m_k,m_{k+1})$ with $U/2\leq |m_j|\leq U$ $(1\leq j\leq k)$ and $(1/2)UV\leq |m_{k+1}|\leq UV$, such that $m_j\equiv a_r \bmod q_r$ for $q_r|t_j$ $(1\leq j\leq k)$, $m_{k+1}\equiv a_r \bmod q_r$ for $q_r|t_{k+1}$ and $m_{k+1}\equiv m_j \bmod p_j$ for $1\leq j\leq k$. For each m_j $(1\leq j\leq k)$ we have $Ut_j^{-1}+O(1)$ possibilities, whereas for m_{k+1} there are $UV(p_1\dots p_k)^{-1}t_k^{-1}+O(1)$ possibilities. Thus the total number of $(k+1)$-tuplets belonging to $(a_1,\dots,a_L)$ is

$$U^{k+1}V\prod_{j=1}^{k+1} t_j^{-1}(1+O(kU^{-1/2})).$$

The congruences (4.3i) and (4.3ii) determine l uniquely modulo $D_0[t_1,\dots,t_{k+1}]$. The number of l-values for a fixed $(k+1)$-tuplet $(m_1,\dots,m_{k+1})$ thus is $(1/2)E([t_1,\dots,t_{k+1}])^{-1}$. The number of all L-tuplets $(a_1,\dots,a_L)$ is $[t_1,\dots,t_{k+1}]$. Thus the total number of integers l satisfying the system (4.3i), (4.3ii) is

$$\frac{1}{2}EU^{k+1}p_{k+1}\cdots p_{k+6}p_{k+7}^{-3}\Bigl(\prod_{1\leq j\leq k+1} t_j\Bigr)^{-1}(1+O(kU^{-1/2})).$$

The inclusion-exclusion principle (4.6) now gives

$$|\mathcal{H}_k(n)| = |\mathcal{C}_k(n)|\Bigl\{\prod_{\substack{p\leq(1/2)\eta^{-1}\\ p|n}}(1-p^{-1})^{k+1} + O\Bigl(kU^{-1/2}\sum_{(t_1,\dots,t_{k+1})}(t_1\dots t_{k+1})^{-1}\Bigr)\Bigr\}. \qquad (4.7)$$

We observe that

$$\sum_{(t_1,\ldots,t_{k+1})} (t_1 \ldots t_{k+1})^{-1} = \prod_{p|n} (1+p^{-1})^{k+1} \ll (\log x)^{2k}$$

and obtain

$$|\mathcal{H}_k(n)| = |\mathcal{C}_k(n)| \prod_{\substack{p \le (1/2)\eta^{-1} \\ p|n}} (1-p^{-1})^{k+1}(1+O(U^{-1/3})). \tag{4.8}$$

The other divisors of the form $d_0 = D_j E t_j^{-1}$, $d_0 = E t_{k+1}^{-1}$ with $t_j | E$, $t_j > \eta^{-1}/2$ can be treated by a simple subtraction. The number of $k+1$-tuplets $(t_1, \ldots, t_{k+1})$ for which any $t_j > \eta^{-1}/2$ is $\ll (\log x)^{k+1}$, whereas the number of $l \in \mathcal{E}_k(n)$ satisfying the system (4.1iii), (4.1iv) for such a $k+1$-tuplet is $\ll \eta |\mathcal{C}_k(n)| = \exp(-(\log\log x)^3)|\mathcal{C}_k(n)|$.

Let $\mathcal{J}_k(n)$ be the set of all $l \in \mathcal{C}_k(n)$ that satisfy (4.1i), (4.1ii), but none of the congruences

$$l \equiv m_j \bmod p_j t_j \text{ for some pair } (m_j, t_j) \text{ with}$$
$$|m_j| \le \eta p_j t_j, \quad t_j | E \quad (1 \le j \le k) \tag{4.9i}$$

or

$$l \equiv m_{k+1} \bmod D_0 t_{k+1} \text{ for some pair } (m_{k+1}, t_{k+1}) \text{ with}$$
$$|m_{k+1}| \le \eta d_0 t_{k+1}, \quad t_{k+1} | E. \tag{4.9ii}$$

We obtain from (4.8) and the argument for $t_j > \eta/2$ that

$$|\mathcal{J}_k(n)| \ge \frac{1}{2}|\mathcal{C}_k(n)| \prod_{\substack{p \le (2\eta)^{-1} \\ p|n}} (1-p^{-1})^{k+1}. \tag{4.10}$$

We now treat the other divisors. For $d_0 | n$, where $n \in S_k(x)$, let $M(n, d_0)$ be the number of l with $(1/4)n \le l \le (3/4)n$ and

$$l \equiv m_j \bmod p_j \text{ with } U/2 \le |m_j| \le U \ (1 \le j \le k), \tag{4.11i}$$

$$l \equiv m^{(2)} \bmod D_0 \text{ with } UV/2 \le |m^{(2)}| \le UV, \tag{4.11ii}$$

$$l \equiv m^{(3)} \bmod n/d_0 \text{ with } |m^{(3)}| \le \eta n/d_0. \tag{4.11iii}$$

Our aim is to show that for most n, $M(n, d_0)$ is small for all $d_0 \notin \mathcal{P}(n)$ except for divisors $d_0 = nt^{-1}$ with $t \le (1/2)\eta^{-1}$. These divisors must be treated in a different manner.

Let $n \in \mathcal{S}_k(x)$, $d_0|n$, where in the usual notation $n = p_1 \dots p_{k+7}E$. We write $n/d_0 = F_1F_2F_3$, where

$$F_1|p_1 \dots p_k, \ F_2|p_{k+1} \dots p_{k+7}, \ F_3|E. \tag{4.12}$$

Let

$$F_1 = p_{j_1} \dots p_{j_t}, \quad F_2 = p_{j_{t+1}} \dots p_{j_s}. \tag{4.13}$$

We call $P(n, d_0) = (j_1, \dots, j_s) = \tau$ the divisor pattern for the pair (n, d_0) and write $\tau \cap [1, k] = (j_1, \dots, j_t)$.

We distinguish two cases according to whether $F_1F_2 < UV$ or $F_1F_2 \geq UV$.

We first deal with the case when $F_1F_2 < UV$. We decompose $M(n, d_0) = M_1(n, d_0) + M_2(n, d_0)$, where $M_1(n, d_0)$ is the number of l that satisfy (4.11i)–(4.11iii) with $m^{(3)} \neq m_{j_r}$ for $1 \leq r \leq t$, and $M_2(n, d_0)$ is the number of such l where $m^{(3)} = m_{j_r}$ for some r with $1 \leq r \leq t$. For a fixed s-tuplet $\tau = (j_1, \dots, j_s)$ with $1 \leq j_1 < \dots j_s \leq k + 7$ we collect the contributions from all $d_0|n$ for which $P(n, d_0) = \tau$. We write

$$N_1(n, \tau) = \sum_{\substack{d_0|n,\, F_1F_2<UV \\ P(n,d_0)=\tau}} M(n, d_0) \tag{4.14}$$

and have the decomposition

$$N_1(n, \tau) = N_{1,1}(n, \tau) + N_{1,2}(n, \tau),$$

where

$$N_{1,i}(n, \tau) = \sum_{\substack{d_0|n,\, F_1F_2<UV \\ P(n,d_0)=\tau}} M_i(n, d_0) \ (i = 1, 2).$$

For some positive constant C we also introduce

$$N_1(n, C, \tau) = \sum_{\substack{d_0|n,\, C<F_3\leq 2C, \\ F_1F_2<UV,\, P(n,d_0)=\tau}} M(n, d_0) \tag{4.15}$$

with the decomposition

$$N_1(n, C, \tau) = N_{1,1}(n, C, \tau) + N_{1,2}(n, C, \tau), \tag{4.16}$$

where $N_{1,i}(n, C, \tau)$ denotes the above sum with $M(n, d_0)$ replaced by $M_i(n, d_0)$.

We subdivide the set $\mathcal{S}_k(x)$ into $\ll (2\log x)^{k+8}$ subsets of the form

$$\mathcal{S}_k(x; A_1, A_2, \dots, A_{k+8}) = \{n \in \mathcal{S}_k(x) : n = p_1 \cdots p_{k+7}E(n) \text{ with } A_j \le p_j \le 2A_j \ (1 \le j \le k+7),\ A_{k+8} \le E \le 2A_{k+8}\}.$$

We will assume that

$$s \neq 0 \text{ or } C > \eta^{-1}. \tag{4.17}$$

The remaining case corresponds to divisors

$$d_0 = nt_0^{-1} \text{ with } t_0 \ll \eta^{-1}, \tag{4.18}$$

which will be treated separately later.

To estimate $\sum_{n\in\mathcal{S}_k(s;A_1,\dots,A_{k+8})} N_{1,i}(n,C,\tau)$ we interchange the order of summation. For a fixed $(k+2)$-tuplet $(m_1,\dots,m_k,m^{(2)},m^{(3)})$ we collect all the n-values which satisfy congruences of the form (4.11) with these given values of m_j, $m^{(2)}$, $m^{(3)}$. Let

$$\begin{aligned} \{g_1,\dots,g_u\} &= [1,k] - \{j_1,\dots,j_t\}, \\ \{h_1,\dots,h_v\} &= \{h : 1 \le h \le k+7;\ h \notin \tau\}. \end{aligned} \tag{4.19}$$

The congruences (4.11i) and (4.11iii) are only compatible if $m_{j_r} \equiv m^{(3)} \bmod p_{j_r}$ for $1 \le r \le t$. We reformulate this as $p_{j_r} | m^{(3)} - m_{j_r}$ (in $\sum^{(3)}$ below). The congruences (4.11i) and (4.11ii) are only compatible if $m^{(2)} \equiv m_j \bmod p_j$ for $1 \le j \le k$. We formulate these conditions separately as $m^{(2)} \equiv m_{j_r} \bmod p_{j_r}$ $(1 \le r \le t)$ (in $\sum^{(5)}$), and $p_{g_r} | m^{(2)} - m_{g_r}$ $(1 \le r \le u)$ (in $\sum^{(6)}$). We also need that $m^{(2)} \equiv m^{(3)} \bmod F_2$ (in $\sum^{(5)}$). Finally we write $E(n) = yF_3$ (in $\sum^{(10)}$). The rearrangements and substitutions lead to the following estimate.

$$\sum_{n\in\mathcal{S}_k(x;A_1,\dots,A_{k+8})} N_{1,1}(n,C,\tau) \le \sum_{(m_1,\dots,m_k)}^{(1)} \sum_{m^{(3)}}^{(2)} \sum_{(p_{j_1},\dots,p_{j_t})}^{(3)} \sum_{F_2}^{(4)} \sum_{m^{(2)}}^{(5)} \sum_{(p_{g_1},\dots,p_{g_u})}^{(6)} \sum_{F_3}^{(7)} \sum_{l}^{(8)} \sum_{(p_{h_1},\dots,p_{h_v})}^{(9)} \sum_{y}^{(10)} 1,$$

where the ranges of summation are as follows:

in $\sum^{(1)}$ over $(m_1, m_2, \dots, m_k)$ with $U/2 \le |m_j| \le U$ $(1 \le j \le k)$;

in $\sum^{(2)}$ over $m^{(3)}$ with $|m^{(3)}| \le 2^{s+1}\eta C \prod_{1\le r\le s} A_{j_r}$;

in $\sum^{(3)}$ over $(p_{j_1},\dots,p_{j_t})$ with $A_{j_r} \le p_{j_r} \le 2A_{j_r}$, $p_{j_r} | m^{(3)} - m_{j_r}$ $(1 \le r \le t)$;

in $\sum^{(4)}$ over F_2 with $\prod_{t+1\le r\le s} A_{j_r} \le F_2 \le \prod_{t+1\le r\le s}(2A_{j_r})$;
in $\sum^{(5)}$ over $m^{(2)}$ with $(1/8)UA_1\dots A_{k+6}A_{k+7}^{-3} < |m^{(2)}| \le 2^{k+6}UA_1\dots A_{k+6}A_{k+7}^{-3}$, $m^{(2)} \equiv m_{j_r} \bmod p_{j_r}$ $(1\le r\le t)$, $m^{(2)} \equiv m^{(3)} \bmod F_2$;
in $\sum^{(6)}$ over $(p_{g_1},\dots,p_{g_u})$ with $p_{g_r}|m^{(2)} - m_{g_r}$ $(1\le r\le u)$;
in $\sum^{(7)}$ over F_3 with $C < F_3 \le 2C$;
in $\sum^{(8)}$ over l with $l \le x$, $l \equiv m_j \bmod p_j$ $(1\le j\le k)$, $l \equiv m^{(3)} \bmod F_3$, $l \equiv m^{(2)} \bmod p_1\dots p_k F_2$;
in $\sum^{(9)}$ over $(p_{h_1},\dots,p_{h_v})$ with $p_{h_r}|l - m^{(2)}$ $(1\le r\le v)$;
in $\sum^{(10)}$ over y with $y \le A_{k+8}/F_3$;

The estimate of this multiple sum is elementary, though lengthy. We use the fact that $l - m^{(2)}$ has at most $\log x$ prime factors and obtain

$$\sum_{(p_{h_1},\dots,p_{h_v})}^{(9)} \sum_{y}^{(10)} 1 \ll (\log x)^v A_{k+8} C^{-1}.$$

The number of l-values in the sum $\sum^{(8)}$ is by the Chinese Remainder Theorem $\ll x(A_1\dots A_k)^{-1}F_2^{-1}F_3^{-1}$. Thus we have

$$\sum^{(7)}\dots\sum^{(10)} 1 \ll x(\log x)^v A_{k+8}(A_1\dots A_k)^{-1}F_2^{-1}C^{-1}.$$

The number of u-tuplets $(p_{g_1},\dots,p_{g_u})$ in $\sum^{(6)}$ is again $\ll (\log x)^u$. The number of $m^{(2)}$-values in $\sum^{(5)}$ is by the Chinese Remainder Theorem

$$\ll 2^{k+6}UA_1\dots A_{k+6}A_{k+7}^{-3}(A_{j_1}\dots A_{j_s})^{-1}F_2^{-1}.$$

The number of t-tuplets $(p_{j_1},\dots p_{j_t})$ in $\sum^{(3)}$ is $\ll (\log x)^t$. In $\sum^{(2)}$ we estimate the number of $m^{(3)}$ by $\ll 2^{s+1}\eta C\prod_{1\le r\le s} A_{j_r}$. This estimate is sufficient as long as this bound is $\gg 1$ which is guaranteed by (4.17). Finally, the number of k-tuplets $(m_1,\dots,m_k)$ is $\ll U^k$. Combining these estimates leads to the bound

$$\sum_{n\in S_k(x;A_1,\dots,A_{k+8})} N_{1,1}(n,C,\tau) \ll \eta x(\log x)^{2k}U^{k+1}A_{k+1}\dots A_{k+6}A_{k+7}^{-3}A_{k+8}. \tag{4.20}$$

We conclude (see (2.8)) that

$$N_{1,1}(n,C,\tau) \ll \exp(-4(\log\log x)^2)U^{k+1}A_{k+1}\dots A_{k+6}A_{k+7}^{-3}A_{k+8}$$

for all $n \in \mathcal{S}_k(x, A_1, \ldots A_{k+8})$ except those from a set of cardinality $\ll x\exp(-4(\log\log x)^2)$. By summing over all $\ll (\log x)^{k+9}$ $(k+9)$-tuplets $(A_1, \ldots, A_{k+8}, C)$ we get

$$N_{1,1}(n, \tau) \ll |\mathcal{J}_k(n)| \exp(-2(\log\log x)^2). \tag{4.21}$$

for all $n \in \mathcal{S}_k(x)$ with at most $x\exp(-2(\log\log x)^2)$ exceptions.

We now estimate $\sum_{n\in\mathcal{S}_k(x;A_1,\ldots,A_{k+8})} N_{1,2}(n, C, \tau)$. The equation $m^{(3)} = m_{j_r}$ for *one* r with $1 \le r \le t$ obviously implies $m^{(3)} = m_{j_r}$ for *all* r with $1 \le r \le t$ because of the small size of the m_j's (see (4.11)). The estimate closely follows the estimate for $\sum N_{1,1}(n, C, \tau)$. The main differences are as follows: In the summation over $(m_1, \ldots, m_k)$ we must add the condition $m_{j_1} = m_{j_2} = \cdots = m_{j_t}$, whereas the condition $p_{j_r} | m^{(3)} - m_{j_r}$ $(1 \le r \le t)$ becomes redundant since $m^{(3)} - m_{j_r} = 0$. Also, the summation over $m^{(3)}$ is omitted since $m^{(3)} = m_{j_1} = \cdots = m_{j_t}$. We get

$$\sum_{n\in\mathcal{S}_k(x;A_1,\ldots,A_{k+8})} N_{1,2}(n, C, \tau) \le \sum_{(p_{j_1},\ldots,p_{j_t})}^{(1)} \sum_{(m_1,\ldots,m_k)}^{(2)} \sum_{F_2}^{(3)} \sum_{m^{(2)}}^{(4)} \sum_{(p_{g_1},\ldots,p_{g_u})}^{(5)} \sum_{F_1}^{(6)} \sum_{l}^{(7)} \sum_{y}^{(8)} 1,$$

where the ranges of summation are as follows:

in $\sum^{(1)}$ over $(p_{j_1}, \ldots, p_{j_t})$ with $A_{j_r} \le p_{j_r} \le 2A_{j_r}$ $(1 \le r \le t)$;

in $\sum^{(2)}$ over $(m_1, \ldots, m_k)$ with $|m_j| \le U_j$ $(1 \le j \le k)$ and $m_{j_1} = m_{j_2} = \cdots = m_{j_k}$;

in $\sum^{(3)}$ over F_2 with $\prod_{t+1\le r\le s} A_{j_r} \le F_2 \le \prod_{t+1\le r\le s}(2A_{j_r})$;

in $\sum^{(4)}$ over $m^{(2)}$ with $UV/2 \le |m^{(2)}| \le UV$, $m^{(2)} \equiv m_{j_r} \bmod p_{j_r}$ $(1 \le r \le t)$, $m^{(2)} \equiv m_{j_r} \bmod F_2$;

in $\sum^{(5)}$ over $(p_{g_1}, \ldots, p_{g_u})$ with $A_{g_r} \le p_{g_r} \le 2A_{g_r}$, $p_{g_r} | m^{(2)} - m_{g_r}$ $(1 \le r \le u)$;

in $\sum^{(6)}$ over F_3 with $C \le F_3 \le 2C$;

in $\sum^{(7)}$ over $l \le x$ with $l \equiv m_j \bmod p_j$ $(1 \le j \le k)$, $l \equiv m^{(2)} \bmod p_1 \ldots p_k$, $l \equiv m^{(3)} \bmod F_3$, $l \equiv m_{j_r} \bmod F_2$ $(1 \le r \le u)$;

in $\sum^{(8)}$ over $y \le A_{k+8}/F_1$.

The estimate is elementary as for $N_{1,1}$, and we therefore omit the details. One obtains

$$\sum_{n\in\mathcal{S}_k(x;A_1,\ldots,A_{k+8})} N_{1,2}(n, C, \tau) \ll x(\log x)^{3k} A_{k+1} \ldots A_{k+6} A_{k+7}^{-3} A_{k+8} U^{k-t+2}.$$

We conclude that if $t > 1$ then

$$N_{1,2}(n,\tau) \ll |\mathcal{J}_k(n)| \exp(-(\log\log x)^2) \tag{4.22}$$

holds for all $n \in \mathcal{S}_k(x)$ with at most $x\exp(-2(\log\log x)^2)$ exceptions. The case $t = 1$ corresponds to the divisors $d_0 = D_j t_j^{-1}$ $(t_j|E)$ and has already been settled in the construction of $\mathcal{J}_k(n)$.

The case of pairs (n, d_0) with $F_1F_2 \geq UV$ is treated in a completely analogous manner. We omit the details.

We now observe (4.15) and the fact that there are at most 2^{k+7} possible divisor patterns. From (4.21), (4.22), and the analogous estimates for $N_{2,i}(x)$ we conclude that for all $n \in \mathcal{S}_k(x)$ with at most $x\exp(-(\log\log x)^2)$ exceptions the set of $l \in \mathcal{J}_k(n)$ that do not satisfy any of the congruences (4.11iii) for $d_0 \notin \mathcal{P}(n) \cup \{nt_0^{-1} : t_0|n,\ t_0 \leq (2\eta)^{-1}\}$ is non-empty. We denote this set by $\mathcal{L}_k(n)$. We form $\bigcup_{l\in\mathcal{L}_k(n)} S_l$ and remove from certain intervals S_l a subinterval of length $\leq 2\eta^2 t_0 n^{-1}$ to ensure inequality (4.11iii) for α from the remainder also for divisors $d_0 = nt_0^{-1}$ with $t_0 \leq (2\eta)^{-1}$. This proves Lemma 4.

5. Conclusion

For $n \in \mathcal{S}_k(x)$ with non-empty $\mathcal{M}(n)$ we choose $\alpha \in \mathcal{M}(n)$ and set $z = e(\alpha)$, where $e(\alpha) = \exp(2\pi i\alpha)$. For $1 \leq j \leq k$ we set $\alpha ED_j = Y_j + \rho_j$, where Y_j is an integer and $|\rho_j| < 1/2$. Taylor's theorem gives $e(\alpha ED_j) = 1 + \rho_j + O(\rho_j^2)$ and thus by inequality (2.10i)

$$\mu(n/ED_j)\log|1 - e(\alpha ED_j)| \geq (1 - k^{-2})\log p_j. \tag{5.1}$$

In the same manner we get from (2.10ii) that

$$\mu(n/dE)\log|1 - e(\alpha dED)| = (1 + O(k^{-2}))\log d. \tag{5.2}$$

for all $d|p_{k+1\ldots}p_{k+7}$ with $\omega(d) < 4$.

We have

$$\begin{aligned}\sum_{d|p_{k+1}\cdots p_{k+7}} &\mu(n/dE)\log|1 - e(\alpha dE)| \\ &= \left(-4 + 3\binom{7}{1} - 2\binom{7}{2} + \binom{7}{3}\right)\log p_{k+7}(1+\tau) \\ &\geq (10 + \tau)\log p_{k+7}\end{aligned}$$

with $|\tau| \leq 110\epsilon_0$. This together with (5.1) gives that

$$|\Phi_n(z)| \geq np_{k+7}(n)^2. \tag{5.3}$$

The Theorem follows from Lemmas 1,3, and 4, and from (5.3).

REFERENCES

[1] P. Bateman, C. Pomerance, and R. C. Vaughan, On the size of the ocefficients of the cyclotomic polynomial, Coll. Math. Soc. J. Bolyai (1981), 171–202.

[2] N. G. de Bruijn, On the number of positive integers $\leq x$ and free of prime factors $> y$, Nederl. Akad. Wetensch. Proc. Ser. A **54** (1951), 50–60.

[3] P.D.T.A. Elliott, *Probabilistic Number Theory I,II*, Springer-Verlag,, New York, 1979/1980.

[4] H. Halberstam, H.-E. Richert, *Sieve Methods*, Academic Press, London, 1974.

[5] H. Maier and G. Tenenbaum, On the set of divisors of an integer, Invent. Math. **76** (1984), 121–128.

Helmut Maier
Department of Mathematics
University of Georgia
Athens, GA 30602

The Rudin-Shapiro Sequence, Ising Chain, and Paperfolding

MICHEL MENDÈS FRANCE

To Paul Bateman, his wife Felice, and his daughter Sally

Abstract

The Rudin-Shapiro sequence appears both in Fourier Analysis and Number Theory. The Ising chain is a crude model for magnetic substance and plays a fundamental rôle in Statistical Mechanics. The study of patterns of folds on a sheet of paper is linked to many domains including Number Theory and Dynamical Systems. These three concepts are different facets of one and the same object.

Introduction

To simplify their equations, physicists often choose the speed of light and Planck's constant as unity. I have heard a joke according to which they also take $\sqrt{-1}$ as unity. This is of course absurd. And yet, if one was allowed to do so, Number Theory would just be a branch of Physics. I hope to convince the reader in the following pages that this is indeed so, that Number Theory is imaginary temperature Physics.

We shall give no detailed proofs since they appear in previous joint articles written with J. P. Allouche and G. Tenenbaum, [1], [2], [14], [15].

1. The Rudin-Shapiro Sequence

Consider the exponential sum

$$S_N(x) = \sum_{n=0}^{N-1} \pm e^{2i\pi n x}$$

where the coefficients form an arbitrary sequence of ± 1 . Clearly

$$\int_0^1 |S_N(x)|^2 \, dx = N$$

so that the L^2-average over the range $(0,1)$ is $\sqrt{N}$. Hence

$$\max_{0 \le x \le 1} |S_N(x)| \ge \sqrt{N}.$$

This simple observation led P. Erdös, D. Newman and R. Salem to ask whether there exists a sequence of ± 1 for which the above maximum is of the order $\sqrt{N}$. In his master thesis, H.S. Shapiro [20] gives an answer to the question by constructing an infinite ± 1 sequence $(r(n))$ for which

$$|S_N(x)| \le (2 + 2\sqrt{2})\sqrt{N} \tag{1}$$

for all N and all x. Years later, W. Rudin [18] studied the same sequence. It is defined as follows.

Consider the sequence of real polynomials

$$\begin{aligned} P_0 &= Q_0 = 1 \\ P_{n+1} &= P_n + X^{2^n} Q_n, \qquad n \ge 0 \\ Q_{n+1} &= P_n - X^{2^n} Q_n \end{aligned}$$

Clearly the limit

$$P_\infty = \lim_{n \to \infty} P_n$$

exists and

$$P_\infty = \sum_{n=0}^{\infty} r(n) X^n$$

where the coefficients $r(n) = \pm 1$ are the Rudin-Shapiro elements.

The proof of inequality (1) is then very simple. In polynomials P_n and Q_n choose $X = e^{2i\pi x}$. Then

$$\begin{aligned} |P_n|^2 + |Q_n|^2 &= 2(|P_{n-1}|^2 + |Q_{n-1}|^2) \\ &= 2^{n+1}; \end{aligned}$$

hence

$$|P_n| \leq \sqrt{2} \cdot \sqrt{2^n}.$$

But $P_n = S_{2^n}(x)$ so that

$$|S_{2^n}(x)| \leq \sqrt{2} \cdot \sqrt{2^n}.$$

Positive integers N are sums of powers of 2 . The above inequality can therefore be extended to $S_N(x)$ if one is willing to replace $\sqrt{2}$ by $2 + 2\sqrt{2}$.

The constant factor in inequality (1) can actually be lowered to $(2 + \sqrt{2})$ (see [15]) and recently B. Saffari [19] managed to improve it to $(2 + \sqrt{2})\sqrt{3/5}$. It remains to find the optimal constant but we shall not be concerned here by this problem.

J. Brillhart and L. Carlitz [4] observed that if

$$n = \sum_{q=0}^{\infty} e_q(n)\, 2^q, \qquad e_q(n) = 0 \text{ or } 1$$

is the binary expansion of the positive integer n (a finite sum) , then

$$r(n) = \exp i\pi \sum_{q=0}^{\infty} e_q(n) e_{q+1}(n).$$

This is a crucial remark which, as we shall see, links the Rudin-Shapiro sequence to the Ising chain.

2. The Ising Model

The Ising model is a crude model for magnetic susbtance and plays a central rôle in Statistical Mechanics (see R. J. Baxter [3], B. A. Cipra [5], B. M. McCoy, T. T. Wu [13], C. Thompson [21], [22]).

Consider a D-dimensional cubic lattice. A volume V encloses approximately V vertices (we use the same symbol for the set and its measure). See figure 1.

Typically V should be of the order 10^{23} (Avogadro's number). At each vertex $P \in V$ there is an atom with spin $\sigma_P = \pm 1$. The set

$$\sigma = (\sigma_P = \pm 1 \mid P \in V) \in \{-1, +1\}^V$$

is called a configuration. There are thus 2^V different configurations.

By definition, the energy or Hamiltonian of a given configuration σ is

$$\mathcal{H}(\sigma) = -J \sum_{P,Q \in V}{}^{*} \sigma_P \sigma_Q - H \sum_{P \in V} \sigma_P.$$

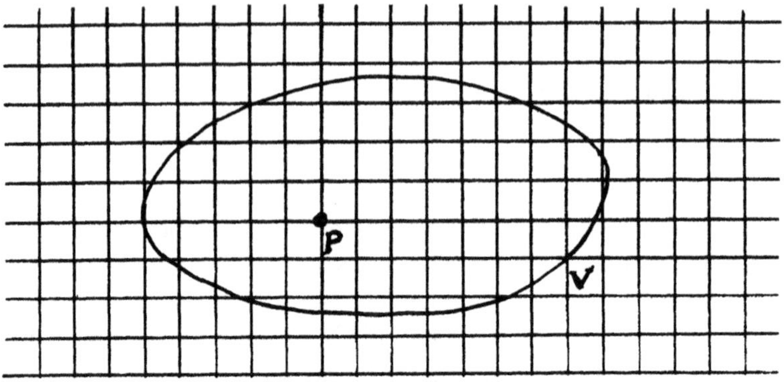

Fig. 1

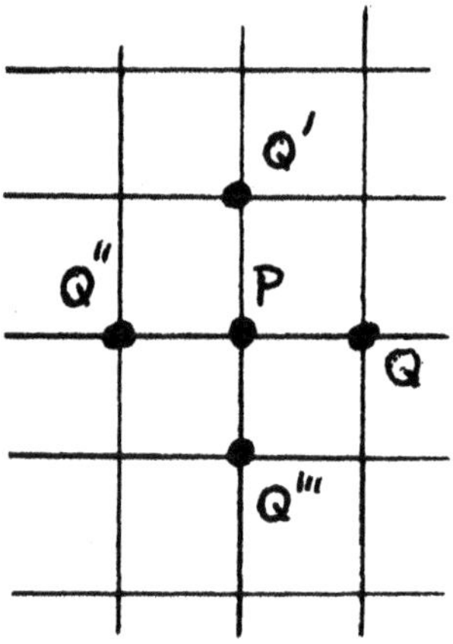

Fig. 2

The first sum is extended to all couples P, Q such that P and Q are neighbours. A given site P has $2D$ neighbours (figure 2) .

In the above formula J is a real positive parameter called the coupling constant and H is a real parameter called the external field. J and H are fixed.

The axioms of physics tell us that the system is in equilibrium if its energy is minimal. Suppose $H > 0$. Then $\mathcal{H}(\sigma)$ is minimal when for all $P \in V$, $\sigma_P = +1$. If $H < 0$, then $\mathcal{H}(\sigma)$ is minimal when $\sigma_P = -1$. Hence equilibrium is attained when all the spins point in the direction of the external field. This is exactly what one would expect of a magnetic susbtance.

The Ising model is however more subtle in that it takes account of the (absolute) temperature $T > 0$. We are told that the probability of a given configuration σ is

$$p_T(\sigma) = \frac{1}{Z} \exp(-\beta \mathcal{H}(\sigma)), \qquad \beta = \frac{1}{T}$$

where

$$Z = Z(\beta, V) = \sum_{\sigma' \in \{-1,+1\}^V} \exp(-\beta \mathcal{H}(\sigma')).$$

Z is known as the "partition function" .

Let us now justify (illustrate ?) the above definitions. When T decreases to 0 , it is easily seen that

$$p_T(\sigma) \longrightarrow \begin{cases} 1 & \text{if } \sigma \text{ minimizes } \mathcal{H}(\sigma), \\ 0 & \text{if not.} \end{cases}$$

At temperature 0 , the system has probability 1 to be in the equilibrium state. Order prevails.

If now T increases to infinity, then for all $\sigma \in \{-1, +1\}^V$,

$$p_T(\sigma) \longrightarrow 2^{-V}.$$

At high temperature, all configurations have equal probability. The system is chaotic. The behaviour of the system with respect to T is thus consistent with our intuitive vision of reality: as T varies from 0 to infinity we go from order to disorder.

Physicists are particularly interested in the dependence of Z with respect to T , especially for infinite V (remember $V = 10^{23}$). It is therefore important to compute $Z(\beta, V)$ and this turns out to be a very difficult problem when the dimension is larger than 1. In the mid 40's, L. Onsager [17] managed to show that for $D = 2$ and $H = 0$, the function

$$\beta \longmapsto \psi(\beta) = \lim_{V \to \infty} \frac{1}{V} \log Z(\beta, V)$$

has a singularity for some $\beta > 0$. This extremely important result which proves the existence of phase transitions gained him the Nobel prize. Later R. J. Baxter [3] found a very surprising and deep relationship between the Ising model and Ramanujan-type identities which enabled him to solve the so-called hard hexagon model.

At the time of writing (May, 1989) rumors are coming from the Soviet Union that the 2-dimensional Ising model with external field ($H \neq 0$) has just been solved. The situation for $D \geq 3$ is to this day intractable even though phase transitions are known to exist.

The link I wish to discuss between the Ising model and the Rudin-Shapiro sequence fortunately only involves the one-dimensional case. The next paragraph is devoted to this simple case.

3. The Ising Chain

We consider a chain with N sites $0, 1, 2, \ldots, N-1$. The volume V becomes the interval $[0, N[$. (the linear model) or $\mathbf{Z}$ modulo N (the cyclic model) .

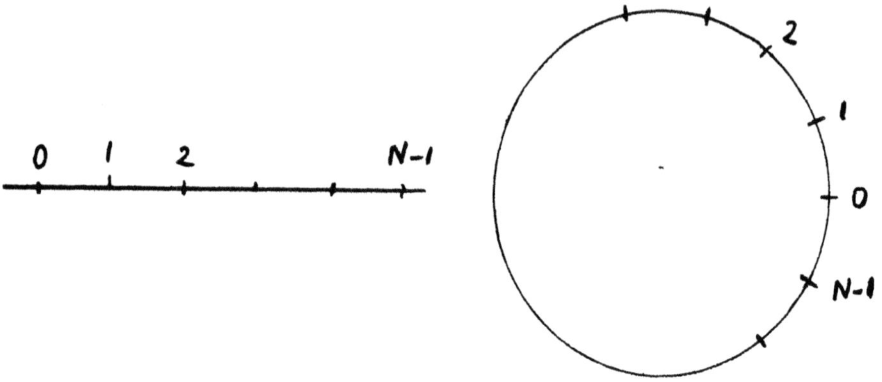

Fig. 3

The Hamiltonian is

$$\mathcal{H}(\sigma) = -J \sum_{q=0}^{N-1} \sigma_q \sigma_{q+1} - H \sum_{q=0}^{N-1} \sigma_q .$$

In the linear model we assume $\sigma_N = +1$. In the cyclic model $\sigma_N = \sigma_0$. As N goes to infinity , both models behave similarly.

Computing the partition function Z_c of the cyclic model is simple:

$$\begin{aligned} Z_c(\beta, N) &= \sum_{\sigma \in \{-1,+1\}^N} \exp \beta \left(J \sum_{q=0}^{N-1} \sigma_q \sigma_{q+1} + H \sum_{q=0}^{N-1} \sigma_q \right) \\ &= \sum_{\sigma \in \{-1,+1\}^N} \prod_{q=0}^{N-1} \exp \beta \left(J \sigma_q \sigma_{q+1} + \frac{1}{2} H(\sigma_q + \sigma_{q+1}) \right). \end{aligned}$$

Put

$$L(\delta, \delta') = \exp \beta \left(J \delta \delta' + \frac{1}{2} H(\delta + \delta') \right)$$

where δ and δ' take values ± 1 , and consider the 2×2 matrix (known as the transfer matrix)

$$L = \begin{pmatrix} L(1,1) & L(1,-1) \\ L(-1,1) & L(-1,-1) \end{pmatrix} = \begin{pmatrix} \exp \beta(J+H) & \exp(-\beta J) \\ \exp(-\beta J) & \exp \beta(J-H) \end{pmatrix}.$$

Then

$$\begin{aligned} Z_c(\beta, N) &= \sum_{\sigma \in \{-1,+1\}^N} L(\sigma_0, \sigma_1) L(\sigma_1, \sigma_2) \cdots L(\sigma_{N-1}, \sigma_0) \\ &= \sum_{\sigma_0 \in \{-1,+1\}} L_N(\sigma_0, \sigma_0) \end{aligned}$$

where $L_N(\sigma_0, \sigma_0)$ is the (σ_0, σ_0) element of the matrix L^N . Hence

$$Z_c(\beta, N) = Trace\, L^N = \lambda_1^N + \lambda_2^N$$

where λ_1 and λ_2 are the eigenvalues of the matrix L:

$$\left.\begin{matrix} \lambda_1 \\ \lambda_2 \end{matrix}\right\} = e^{\beta J} \cosh \beta H \pm \left(e^{2\beta J} \cosh^2 \beta H - 2 \sinh 2\beta J \right)^{1/2}.$$

This technique of computing Z_c is essentially due to H.A. Kramers and G.H. Wannier [9] . (Notice incidentally that

$$\lim_{N \to \infty} \frac{1}{N} \log Z(\beta, N) = \log \lambda_1$$

is continuous and differentiable with respect to β . There are no phase transitions in the one dimensional model.)

4. Back to Number Theory

Let $n < 2^N$ be an integer . Consider its binary expansion

$$n = \sum_{q=0}^{N-1} e_q(n) 2^q, \qquad e_q(n) = 0,\, 1.$$

Define

$$\sigma_q(n) = 1 - 2e_q(n) = \pm 1.$$

When n runs through the interval $[0, 2^N[$, the vector

$$\sigma = \sigma(n) = (\sigma_0(n),\, \sigma_1(n),\, \cdots,\, \sigma_{N-1}(n))$$

ranges through the set $\{-1, +1\}^N$. The mapping $n \longmapsto \sigma(n)$ is a one to one mapping of $[0, 2^N[$ onto the family of Ising configurations. The Hamiltonian of the configuration $\sigma(n)$ will now be denoted

$$\mathcal{H}(\sigma(n)) = \mathcal{H}(n).$$

Then

$$\mathcal{H}(n) = -J\sum_{q=0}^{N-1}\sigma_q(n)\sigma_{q+1}(n) - H\sum_{q=0}^{N-1}\sigma_q(n)$$

where $\sigma_N(n) = 1$ (because $n < 2^N$). The Ising chain we are considering is linear. Reverting to the binary digits $e_q(n)$,

$$\begin{aligned}\mathcal{H}(n) &= -J\sum_{q=0}^{N-1}(1-2e_q(n))(1-2e_{q+1}(n)) - H\sum_{q=0}^{N-1}(1-2e_q(n))\\ &= -(J+H)N - 4J\sum_{q=0}^{N-1}e_q(n)e_{q+1}(n) + (4J+2H)\sum_{q=0}^{N-1}e_q(n).\end{aligned}$$

The partition function Z is thus

$$\begin{aligned}Z(\beta,N) &= \sum_{q=0}^{2^N-1}\exp(-\beta\mathcal{H}(n))\\ &= e^{\beta(J+H)N}\sum_{q=0}^{2^N-1}e^{\beta\left[4J\sum_{q=0}^{N-1}e_q(n)e_{q+1}(n)-(4J+2H)\sum_{q=0}^{N-1}e_q(n)\right]}.\end{aligned}$$

At this point we start diverging from classical physics by allowing imaginary temperatures. We choose

$$\beta = i, \quad J = \frac{1}{4}\pi, \quad H = -\frac{1}{2}(\alpha+\pi), \quad \alpha\in\mathbf{R}$$

Then

$$Z(i,N) = e^{i\gamma N}\sum_{q=0}^{2^N-1}r(n)e^{2i\pi\alpha s(n)}$$

where $\gamma = -\frac{1}{4}(2\alpha+\pi)$ is a real constant, where $r(n)$ is the Rudin-Shapiro sequence and where $s(n)$ is the sum of the binary digits of n .

In paragraph 3 we learned how to compute $Z_c(i,N)$:

$$\begin{aligned}Z_c(i,N) &= e^{i\pi N/4}\left[-\sin\frac{\alpha}{2} + i(2-\sin^2\frac{\alpha}{2})^{1/2}\right]^N\\ &\quad + e^{i\pi N/4}\left[-\sin\frac{\alpha}{2} - i(2-\sin^2\frac{\alpha}{2})^{1/2}\right]^N.\end{aligned}$$

It is not hard to relate Z to Z_c (see for example [1]), and to deduce

$$|Z(i,N)| \le \sqrt{2}\sqrt{2^N},$$

and finally

$$|\sum_{n=0}^{N-1} r(n)e^{2i\pi\alpha s(n)}| \leq (2+\sqrt{2})\sqrt{N}.$$

This result is to be compared with the classical Rudin-Shapiro inequality which we recall from paragraph 1:

$$|\sum_{n=0}^{N-1} r(n)e^{2i\pi\alpha n}| \leq (2+\sqrt{2})\sqrt{N}.$$

At this point we ask whether these two inequalities are not special cases of a more general inequality. We now answer this question.

In 1968, A. O. Gelfond [8] introduced the notion of 2-multiplicative sequences f:

$$f(2^n+m) = f(2^n)\cdot f(m), \qquad \text{for all } n \geq 0 \text{ and } 0 \leq m < 2^n.$$

It is easy to see that a unimodular sequence f is 2-multiplicative if and only if there exists a real sequence $c = (c_0, c_1, c_2, \cdots)$ such that

$$f(n) = \exp 2i\pi \sum_{q=0}^{\infty} c_q e_q(n)$$

where as before $e_0(n), e_1(n), \cdots$ are the binary digits of n. Clearly $\exp 2i\pi\alpha n$ and $\exp 2i\pi\alpha s(n)$ belong to Gelfond's family. Incidentally, $\exp i\pi s(n)$ is the celebrated Thue-Morse sequence.

By modifying our approach, J.P. Allouche and I were able to establish the following theorem [2] which indeed contains both previous inequalities.

Theorem. *If $r(n)$ denotes the Rudin-Shapiro sequence and if $f(n)$ is an arbitrary unimodular ($|f(n)| = 1$) 2-multiplicative sequence , then for all $n \in \mathbf{N}$*

$$|\sum_{n=0}^{N-1} r(n)f(n)| \leq (2+\sqrt{2})\sqrt{N}.$$

5. Paperfolding

A sheet of paper is folded in two over and over again.

By unfolding a sheet of paper three times folded one observes a sequence of $2^3 - 1$ creases

$$\vee\ \vee\ \wedge\ \vee\ \vee\ \wedge\ \wedge.$$

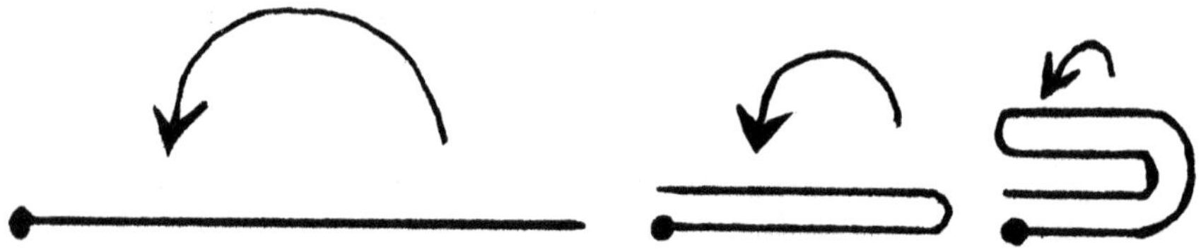

Fig. 4

After n folds, the sequence of creases has length $2^n - 1$. Let us denote by s_n this finite sequence of V's and Λ's . It is quite obvious that

$$s_{n+1} = s_n \vee \tilde{s}_n$$

where $\tilde{s}_n$ is the reverse of s_n in which the symbols V and Λ are permuted. As n tends to infinity , the sequence s_n converges to an infinite sequence

$$\vee \vee \wedge \vee \vee \wedge \wedge \vee \vee \vee \wedge \wedge \vee \wedge \wedge \cdots$$

which we call the paperfolding sequence.

At this point I believe it is interesting to mention a curious result of J. Loxton and A.J.van der Poorten [10], [11] even though we shall not use it. Replace the symbols V and Λ repectively by 0 and 1 . The real number whose binary expansion is the paperfolding sequence is transcendental. Extensions of this result are discussed in [16].

We now come back to our main concern. Replace V by +1 and Λ by −1. The sequence then reads

$$+1, +1, -1, +1, +1, -1, -1, \cdots.$$

Let $f(n)$ be the nth sign ($f(1) = 1,\ f(2) = 1,\ f(3) = -1,\ \cdots$).

Suppose we unfold to the angle $\alpha \in [0, \pi]$ the infinitely folded sheet of paper (the sheet has infinite length and the distance between two adjacent

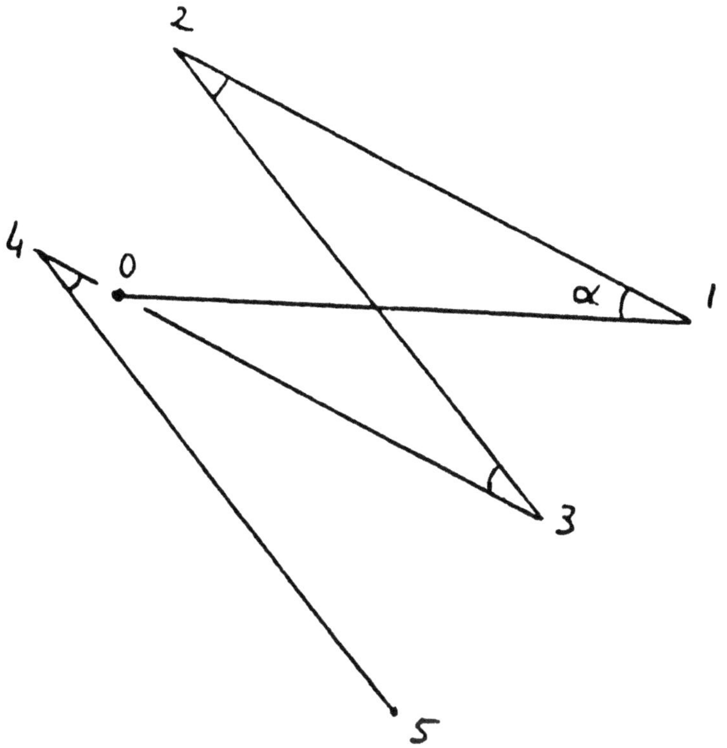

Fig. 5

creases is taken as unity). Call $\Gamma(\alpha)$ the infinite broken line formed by the edge of the sheet (see figure 5).

For $\alpha = \pi/2$ we obtain the well known dragon curve discovered by Ch. Davis and D.Knuth [6]. They show that it is self-avoiding (figure 6).

Coming back to the general case, let $z = z(m,\alpha) = x + iy$ be the coordinates of the mth vertex of $\Gamma(\alpha)$ (the first side has its origin at 0 and is supported by the positive x-axis). Clearly

$$z(m,\alpha) = \sum_{n=0}^{m-1} \exp i(\pi - \alpha) \sum_{j=1}^{n} f(j).$$

Theorem. *For all* $n < 2^N$

$$\sum_{j=1}^{n} f(j) = -\frac{1}{2} \sum_{q=0}^{N-1} \sigma_q(n)\sigma_{q+1}(n) + \frac{1}{2}N$$

Fig. 6

so that for $J = -1/2$ *and* $H = 0$

$$z(2^N, \alpha) = Z(i(\pi - \alpha), N) \exp i \frac{\pi - \alpha}{2} N.$$

(The proof of this result appears in [14] .)

The paperfolding broken line is thus the realization of an Ising chain with imaginary temperature. The calculation in paragraph 3 with $H = 0$, or better, direct computation shows that

$$Z(\beta, N) = (2 \cosh \beta J)^N;$$

hence

$$|z(2^N, \alpha)| = |2 \cosh i \frac{\pi - \alpha}{2}|^N = |2 \sin \frac{\alpha}{2}|^N.$$

Therefore

$$\lim_{N \to \infty} |z(2^N, \alpha)| = \begin{cases} 0 & \text{if } 0 \leq \alpha < \pi/3 \\ 1 & \text{if } \alpha = \pi/3 \\ \infty & \text{if } \pi/3 < \alpha \leq \pi \end{cases}$$

and

$$\limsup_{n \to \infty} |z(n, \alpha)| = \begin{cases} 0 & \text{if } 0 \leq \alpha < \pi/3 \\ \infty & \text{if } \pi/3 \leq \alpha \leq \pi \end{cases}.$$

The diameter $\Delta(\alpha)$ of $\Gamma(\alpha)$ is defined as

$$\Delta(\alpha) = \sup_{n,m} |z(n, \alpha) - z(m, \alpha)|$$

so that $\Delta(\alpha)$ stays finite if and only if $\alpha < \pi/3$.

At the time of writing, I do not know whether the function $\alpha \longmapsto \Delta(\alpha)$ is continuous or not in the interval $[0, \pi/3[$. I do not know whether it is increasing . Quite trivially in this interval

$$\Delta(\alpha) \leq \frac{1}{1 - 2 \sin \frac{\alpha}{2}}.$$

In December 1988, T. Kamae with whom I discussed these problems, sent me a letter in which he gives a partial answer to these questions. $\Delta(\alpha)$ is indeed continuous at all points with possible exceptions at those α such that α/π is a rational number p/q where p and q are odd. He also proves that if α/π is irrational , then the closure of $\Gamma(\alpha) \subset \mathcal{R}^2$ is a disk centered at the origin.

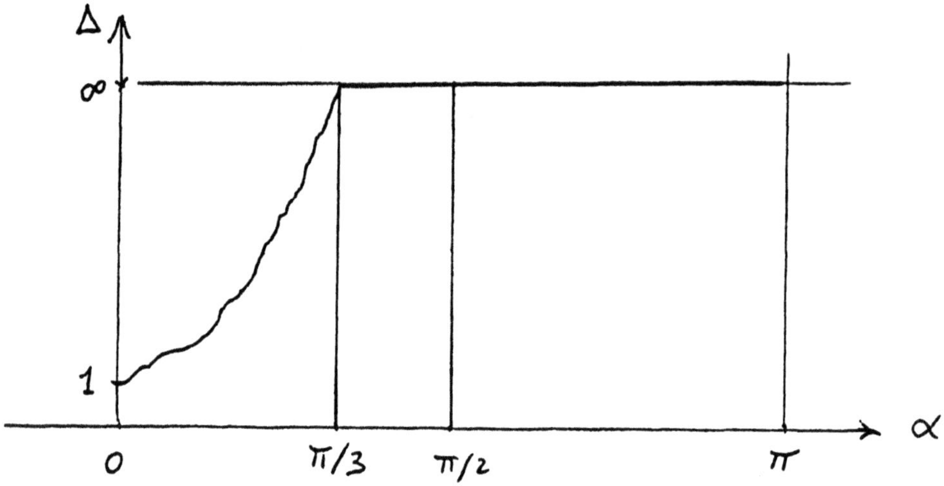

Fig. 7

6. A Final Remark And A Final Problem

If one looks at the graph $(\alpha,\ \Delta(\alpha))$ (see figure 7), one sees that as α increases from 0 to π , it crosses two critical angles. At $\pi/3$ the diameter Δ "explodes" and at $\pi/2$ the curve $\Gamma(\alpha)$ ceases abruptly to be self-intersecting.

Identifying the angle α with a temperature , the curve $\Gamma(\alpha)$ mimics the three states of matter. For $0 \leq \alpha < \pi/3$, $\Gamma(\alpha)$ is very densely packed on itself suggesting a solid state. At $\pi/3$, $\Gamma(\alpha)$ "melts" and stays liquid up to $\pi/2$. At $\pi/2$ and beyond $\Gamma(\alpha)$ is gaseous (liquid boils at 90°).

To conclude I would like to suggest a final problem. We have seen that the set

$$\left\{\gamma \in [0, 2\pi] \mid \limsup_{N\to\infty} \left|\sum_{n=0}^{N-1} \exp i\gamma\mathcal{H}(n)\right| < \infty\right\}$$

is the interval $]\pi - \pi/3, \pi + \pi/3[$. What can be said about those integer sequences $\mathcal{K}(n)$ for which the set

$$E(\mathcal{K}) = \left\{\gamma \in [0, 2\pi] \mid \limsup_{N\to\infty} \left|\sum_{n=0}^{N-1} \exp i\gamma\mathcal{K}(n)\right| < \infty\right\}$$

contains a nontrivial interval ?

Maybe one should restrict the question to special kinds of sequences. For example let $f(n)$ be an arbitrary sequence of ± 1 . Define

$$\mathcal{K}(n) = \mathcal{K}_f(n) = \sum_{j=1}^{n} f(j).$$

Then $\pi \in E(\mathcal{K})$. When $f(n)$ is the paperfolding sequence or generalized paperfolding sequence (see [14]), then

$$E(\mathcal{K}) =]\pi - \pi/3, \pi + \pi/3[.$$

Another case (trivial) occurs when $f(n)$ is the constant sequence $+1$. Then

$$E(\mathcal{K}) =]0, 2\pi[.$$

The problem is to characterize those f for which $E(\mathcal{K}_f)$ contains a nontrivial interval.

REFERENCES

[1] J.P. Allouche, M. Mendès France, Suite de Rudin-Shapiro et modèle d'Ising, Bull. Soc. Math. France **113** (1985), 273–283.

[2] J.P. Allouche, M. Mendès France, On an extremal property of the Rudin-Shapiro sequence, Mathematika **32** (1985), 33–38.

[3] R.J. Baxter, *Exactly Solved Models in Statistical Mechanics*, Academic Press, 1982.

[4] J.Brillhart, L. Carlitz, Note on the Shapiro polynomials, Proc. Amer. Math. Soc. **25** (1970), 114–118.

[5] B.A. Cipra, An introduction to the Ising Model, Amer. Math. Monthly **94** (1987), 937–959.

[6] Ch. Davis, D. Knuth, Number representations and dragon curves, J. Recreational Math. **3** (1970), 61–81; 133-149.

[7] M.Dekking, M. Mendès France, A.J.van der Poorten, Folds !, Math. Intelligencer **4** (1982), 130–138; 173–181; 190–195.

[8] A.O.Gelfond, Sur les nombres qui ont des propriétés additives et multiplicatives données, Acta Arith. **13** (1968), 259–265.

[9] H.A. Kramers, G.H. Wannier, Statistics of the two-dimensional ferromagnet, Phys. Rev. **60** (1941), 252–262; 263–276.

[10] J. Loxton, A.J. van der Poorten, Arithmetic properties of the solutions of a class of functional equations, V. reine angew. Math. **330** (1982), 159–172.

[11] J. Loxton, A.J. van der Poorten, Arithmetic properties of automata: regular sequences, V. reine angew. Math. **392** (1988), 57–69.

[12] S.-K. Ma, *Statistical Mechanics*, World Scientific, 1985.

[13] B.M.McCoy, T.T.Wu, *The Two-Dimensional Ising Model*, Harvard University Press, 1973.

[14] M.Mendès France, *The inhomogeneous Ising chain and paperfolding, in Physics and Number Theory*, Springer-Verlag (to appear).

[15] M.Mendès France, G. Tenenbaum, Dimension des courbes planes, papiers pliés et suites de Rudin-Shapiro, Bull. Soc. Math. France **109** (1981), 207–215.

[16] M.Mendès France, A.van der Poorten, Arithmetic and analytic properties of paperfolding sequences (dedicated to K. Mahler), Bull. Austral. Math. Soc. **24** (1981), 123–131.

[17] L.Onsager, Crystal statistics. A Two-Dimensional Model with an Order-Disorder transition, Phys. Rev **65** (1944), 117–149.

[18] W. Rudin, Some theorems on Fourier coefficients, Proc. Amer. Math. Soc. **10** (1959), 855–859.

[19] B.Saffari, Une fonction extrémale liée à la suite de Rudin-Shapiro, C.R.Acad.Sc.Paris **303** (1986), 97–100.

[20] H.S.Shapiro, *Extremal problems for polynomials and power series*, Thesis MIT, 1951.

[21] C.Thompson, *Mathematical Statistical Mechanics*, Princeton University Press, 1972.

[22] C.Thompson, *Classical Equilibrium Statistical Mechanics*, Oxford University Publications, 1988.

Michel Mendès France
Dept. Math.
Université Bordeaux I
F-33405 Talence Cedex
FRANCE

On Binomial Equations over Function Fields And a Conjecture of Siegel

J. MUELLER

Dedicated to Paul Bateman on his 70th birthday

§1. Introduction

Let $f(x, y)$ be a polynomial in variables x and y and with rational integral coefficients. In this note we study the diophantine equation $f(x, y) = 0$, in the case in which f is not homogeneous. Siegel [5] conjectured in 1929 that the number of integral solutions of $f(x, y) = 0$, where the curve defined by the equation is irreducible and of positive genus, may be bounded in terms of of the number of monomials of f. The conjecture in this form is too strong to be true (for further discussions on this topic, we refer the readers to the introduction of [4]); nevertheless, a modified version of this conjecture, restricted to the Thue equation $F(x, y) = h$, where F is homogeneous of degree $r \geq 3$, has been proved by Mueller and Schmidt [4]. Unfortunately, their method cannot be applied to general inhomogeneous equations. Therefore, as of now this conjecture appears to be inaccesible over a number field. The object of this paper is to give a partial verification of Siegel's conjecture over function fields. Our main result is the following

Theorem. *Let k be an algebraically closed field of characteristic 0, and let K/k be a function field of genus g. Suppose $a, b \in K^*(= K - \{0\})$ and either $a \notin K^{*^r}$ or $b \notin K^{*^t}$. Then*

$$ax + by = 1 \tag{1.1}$$

has at most two solutions $(x, y) \in K^{*^r} \times K^{*^t}$ provided

$$\min(r, t) > 120 + 40g \tag{1.2}$$

Research supported in part by NSF Grant NSF-DMS-8808398.

The basic idea of our proof is to assume that (1.1) has 3 distinct solutions and then use an inequality to show that (1.2) cannot hold. This inequality (see (2.2)) is the function field analogue of the well known *abc*-conjecture of Masser and Oesterlé . We are thus led to speculate that the truthfulness of Siegel's conjecture may yet be another consequence of the celebrated *abc*-conjecture.

Finally, we would like to mention that the Theorem in this paper is a generalization of the Main Theorem in [3] which deals with the case $r = t$ in (1.1); the constant there is $30 + 20g$. As it turned out, the result of this paper is not an automatic generalization of [3]. The new ingredient in this paper is in the way we handle the additional complications which appear in comparing the heights of various sets. Such difficulty was not present in [3]. We have not tried to obtain the best possible results and it is very likely that the constant $120 + 40g$ in the Theorem could be substantially improved.

I would like to thank Professor Wolfgang Schmidt for several helpful conversations on this subject.

§2. Preliminaries

Let $W = \{w_1, \cdots, w_n\}$ with $w_i \in K^*$. The height of $w_1, \cdots, w_n$ is defined to be

$$H(W) = - \sum_{v \in \mathcal{M}_k} v(W), \tag{2.1}$$

where $v(W) = \min(v(w_1), \cdots, v(w_n))$ and v runs through the set of valuations $\mathcal{M}_k$ of K/k with the rational integers as its value group.

Definition: We say W is **dependent** if $w_1 + \cdots + w_n = 0$, and W is **non-degenerate** if no proper subset of W is dependent. We say w is **minimal** if W is **dependent and non-degenerate.**

Our basic tool in the proof of the Theorem is the following fundamental inequality which was first formulated by Mason and later generalized as well as sharpened by Brownawell and Masser [1]:

$$H(W) \leq \frac{1}{2}(n-1)(n-2)(|S| + \max(0, 2g-2)), \tag{2.2}$$

where W is minimal, $n \geq 3$, and S is a finite subset of $\mathcal{M}_k$ such that every element of W is an S-unit.

For a proof of (2.2), which is the function field analogue of the *abc*-conjecture, we refer the readers to [1, Theorem B].

The following basic facts about $H(W)$ can easily be verified from (2.1) and the sum formula $\sum_{v \in \mathcal{M}_k} v(w) = 0$, for w in K^*.

We have

$$H(W) \geq 0, \tag{2.3}$$

$$H(W') \leq H(W) \quad \text{if} \quad W' \subset W, \tag{2.4}$$

and

$$H(uW') = H(W), \quad \forall u \in K^* \tag{2.5}$$

Moreover, for any $w_1, w_2, w_3 \in K^*$, we have

$$H(1, w_1 w_2) \leq H(1, w_1) + H(1, w_2), \tag{2.6}$$

and

$$H(w_1, w_2, w_3) \leq H(w_1, w_2) + H(w_2, w_3). \tag{2.7}$$

We remark that (2.6) and (2.7) are immediate consequences of the following inequalities:

$$\min(0, v(w_1)) + \min(0, v(w_2)) \leq \min(0, v(w_1 w_2)),$$

and

$$\min\left(0, v\left(\frac{w_1}{w_2}\right)\right) + \min\left(0, v\left(\frac{w_3}{w_2}\right)\right) \leq \min\left(0, v\left(\frac{w_1}{w_2}\right), v\left(\frac{w_3}{w_2}\right)\right)$$

Definition: Let $w_1, w_2 \in K^*$. We say w_1, w_2 are **proportional** if $w_1/w_2 \in k$. Two distinct solutions (x_1, y_1) and (x_2, y_2) are said to be **non-proportional** if either $x_1/x_2 \notin k$ or $y_1/y_2 \notin k$.

Proposition 1 [3, Lemma 1]. *Let $W = \{w_1, \cdots, w_n\}$ with $n \geq 2$ and $w_i \in K^*$. Then $H(W) = 0$ iff the elements of W are pairwise proportional.*

Proposition 2 [3, Lemma 3]. *Suppose either $a \notin K^{*^r}$ or $b \notin K^{*^t}$ in (1.1), then any two distinct solutions of (1.1) are non-proportional.*

From now on, we assume that (1.1) has three distinct solutions $(x_i, y_i) \in K^{*^r} \times K^{*^t}, i = 1, 2, 3$. Set

$$X = \{x_1, x_2, x_3\} \qquad \text{and} \qquad Y = \{y_1, y_2, y_3\} \tag{2.9}$$

Then Propositions 1 and 2 imply that either

$$H(X) \neq 0 \qquad \text{or} \qquad H(Y) \neq 0 \tag{2.10}$$

Let

$$\mathcal{U} = \{x_1 y_2, x_2 y_3, x_3 y_1, -x_1 y_3, -x_2 y_1, -x_3 y_2\} \tag{2.11}$$

This set $\mathcal{U}$ is the basic set we will be working with. We remark that $\mathcal{U}$ is a dependent set, that is

$$x_1 y_2 - x_1 y_3 + x_2 y_3 - x_2 y_1 + x_3 y_1 - x_3 y_2 = 0. \tag{2.12}$$

To see (2.12) we note that the left-hand side of (2.12) is the determinant of the solution matrix of the linear equations $ax_i + by_i = 1, i = 1, 2, 3$. Since the coefficients a and b are non-zero, this determinant must vanish.

If $\mathcal{U}$ is non-degenerate, then it is a minimal set. In the next proposition, we will show that if $\mathcal{U}$ is degenerate, there is always a subset of $\mathcal{U}$ which is minimal.

Proposition 3. *Suppose $\mathcal{U}$ is degenerate where $\mathcal{U}$ is given by (2.11). Then there is a subset of $\mathcal{U}$ containing either 3 or 4 elements which is minimal.*

Proof: Suppose $\mathcal{U}$ is degenerate, then $\mathcal{U}$ has a proper subset which is a dependent set. Denote this set by W and its complement by $\bar{W}$. Then $|\bar{W}| \neq 1$ or 5, since the monomials of $\mathcal{U}$ are non-zero. Suppose $|W| = 3$ and suppose W has a proper subset W' which is dependent, then either $|W'| = 1$ or $|\bar{W}'| = 1$ where $\bar{W}'$ is the complement of W' in W . But we have seen that this is impossible. Therefore W has no such proper subset and hence W is non-degenerate. Next, suppose $|W| = 4$. If W has a dependent, proper subset W', then $|W'| = 1, 2$ or 3. Since the cases $|W'| = 1$ or 3 are impossible, to show W is non-degenerate we only need to show $|W'| \neq 2$. Suppose $|W'| = 2$. Then $\mathcal{U}$ is divided into three disjoint, dependent subsets $W', \bar{W}'$ and $\bar{W}$ with $|W'| = |\bar{W}'| = |\bar{W}| = 2$. Let $W' = \{u, v\}$ and $u = x_1y_2$. If $v = -x_1y_3$ or $v = -x_3y_2$, then from $u + v = 0$ we get $y_2 = y_3$ or $x_1 = x_3$ which gives two equal solutions, contradicting our assumption that the solutions are distinct. If $v = -x_2y_1$, then $x_1/y_1 = x_2/y_2$ which together with (1.1) again yields two equal solutions. Hence u and v must be monomials in $\mathcal{U}$ with the same sign. Similarly, one can show the monomials in $\bar{W}'$ or $\bar{W}$ must have the same sign. But this implies that the number of positive and negative monomials in $\mathcal{U}$ must be even, contradicting (2.11). Therefore, $|W'| \neq 2$ and W is non-degenerate. This completes the proof of Proposition 3.

Proposition 4. *Let X, Y and $\mathcal{U}$ be given by (2.9) and (2.11). If W is a minimal subset of $\mathcal{U}$ with $|W| = 4$. Then*

$$H(W) \geq \frac{1}{2} \max(H(X), H(Y)) \tag{2.13}$$

We remark that since $W \subset \mathcal{U}$, (2.4) and (2.13) imply that

$$H(\mathcal{U}) \geq \frac{1}{2} \max(H(X), H(Y)) \tag{2.14}$$

Proof: Let $I = \{1, 2, 3\}$ and let i, j, k be distinct elements of I. Suppose W is a minimal subset of $\mathcal{U}$ with $|W| = 4$. Then for some i and j, x_i and y_j must appear twice among the 4 monomials in W. Let $\bar{W}$ be the complement of W in $\mathcal{U}$. Then we know from Proposition 3 that the two monomials of $\bar{W}$ have the same sign. Hence we have

$$W = \{\pm x_iy_j, \pm x_jy_k, \pm x_ky_i, \mp x_iy_k\}, \quad \bar{W} = \{\mp x_jy_i, \mp x_ky_j\}. \tag{2.15}$$

For the computation of $H(W)$ that follows, we remark that the signs of the monomials will not appear. To justify this, we note that $H(w_1, w_2) = H(\pm w_1, \pm w_2)$ for any $w_1, w_2 \in K^*$.

From (2.15) we first get

$$H(W) \geq H(x_j y_k, x_i y_k) = H(x_j, x_i), \tag{2.16}$$

and

$$H(W) \geq H(x_i y_j, x_i y_k) = H(y_j, y_k) \tag{2.17}$$

From the dependency of $\bar{W}$ we find that

$$\frac{x_k{}^2}{x_i{}^2} = \frac{x_j y_k}{x_i y_k} \cdot \frac{x_k y_i}{x_i y_j} \cdot \frac{x_k y_j}{x_j y_i} = \frac{-x_j y_k}{x_i y_k} \cdot \frac{x_k y_i}{x_i y_j} \tag{2.18}$$

and

$$\frac{y_i{}^2}{y_k{}^2} = \frac{x_i y_j}{x_i y_k} \cdot \frac{x_k y_i}{x_j y_k} \cdot \frac{x_j y_i}{x_k y_j} = \frac{-x_i y_j}{x_i y_k} \cdot \frac{x_k y_i}{x_j y_k}. \tag{2.19}$$

To apply those identities, we resort to the basic properties of $H(W)$ stated in (2.3) to (2.7). For example, using (2.5) we have

$$H\left(1, \frac{x_k{}^2}{x_i{}^2}\right) = H\left(x_i{}^2, x_k{}^2\right) = 2H\left(x_i, x_k\right)$$

Hence from (2.18) and (2.19) we obtain

$$2H(x_i, x_k) \leq H(x_i y_k, x_j y_k) + H(x_i y_j, x_k y_i) \leq 2H(W), \tag{2.20}$$

and

$$2H(y_i, y_k) \leq H(x_i y_k, x_i y_j) + H(x_j y_k, x_k y_i) \leq 2H(W). \tag{2.21}$$

But (2.16) and (2.20) give

$$2H(W) \geq H(x_j, x_i) + H(x_i, x_k) \geq H(x_i, x_j, x_k) = H(X),$$

while (2.17) and (2.21) give

$$2H(W) \geq H(y_j, y_k) + H(y_i, y_k) \geq H(y_i, y_j, y_k) = H(Y).$$

The last two inequalities clearly imply (2.13).

Proposition 5. *Let X, Y and $\mathcal{U}$ be given by (2.9) and (2.11). If W is a minimal subset of $\mathcal{U}$ with $|W| = 3$ and $H(W) \geq H(\bar{W})$. Then*

$$H(W) \geq \frac{1}{4} \max(H(X), H(Y)) \tag{2.22}$$

Proof: Let i, j and k be distinct elements of $I = \{1, 2, 3\}$. Suppose W is a minimal subset $\mathcal{U}$ with $|W| = 3$. The monomials of W can be described as follows:

(I) For some i and j in I, both x_i and y_j appear twice. Then

$$W = \{\pm x_i y_j, \mp x_i y_k, \mp x_k y_j\}, \ \bar{W} = \{\pm x_j y_k, \mp x_j y_i, \pm x_k y_i\} \tag{2.23}$$

(II) For some i in I, either x_i appears twice, then

$$W = \{\pm x_i y_j, \mp x_i y_k, \pm x_k y_i\}, \ \bar{W} = \{\pm x_j y_k, \mp x_j y_i, \mp x_k y_j\}, \tag{2.24}$$

or y_i appears twice, then

$$W = \{\mp x_j y_i, \pm x_k y_i, \pm x_i y_j\}, \ \bar{W} = \{\mp x_i y_k, \pm x_j y_k, \pm x_k y_i\}, \tag{2.25}$$

(III) For any i in I, no x_i or y_i appears more than once. Then

$$W = \{\pm x_i y_j, \pm x_j y_k, \pm x_k y_i\}, \ \bar{W} = \{\mp x_i y_k, \mp x_j y_i, \mp x_k y_j\}, \tag{2.26}$$

We remark first that we may choose i, j and k so that $H(W) \geq H(\bar{W})$. Our next remark is that the signs of the monomials need not appear in the computations of $H(W)$, as in Proposition 4.

Our first object is to show that

$$H(W) \geq \frac{1}{2} \max(H(X), H(Y)) \tag{2.27}$$

where W is given by (2.23) under case (I).

In this case, we have

$$H(W) \geq H(x_i y_j, x_i y_k) = H(y_j, y_k), \ H(\bar{W}) \geq H(x_j y_k, x_j y_i) = H(y_k, y_i),$$

and

$$H(W) \geq H(x_i y_j, x_k y_j) = H(x_i, x_k), \ H(\bar{W}) \geq H(x_j y_i, x_k y_i) = H(x_j, x_k).$$

Since $H(W) \geq H(\bar{W})$, it follows that

$$2H(W) \geq H(W) + H(\bar{W}) \geq H(y_j, y_k) + H(y_k, y_i) \geq H(Y),$$

and

$$2H(W) \geq H(W) + H(\bar{W}) \geq H(x_i, x_k) + H(x_j, x_k) \geq H(X).$$

The last two inequalities clearly imply (2.27).

Next, we will show that

$$H(W) \geq \max\left(\frac{1}{4}H(X), \frac{1}{2}H(Y)\right) \tag{2.28}$$

where W is given by (2.24) under case (II).

In this case, we have

$$H(W) \geq H(x_i y_j, x_i y_k) = H(y_j, y_k), \;\; H(\bar{W}) \geq H(x_j y_k, x_j y_i) = H(y_k, y_i).$$

Since $H(W) \geq H(\bar{W})$, it then follows that

$$2H(W) \geq H(W) + H(\bar{W}) \geq H(y_j, y_k) + H(y_k, y_i) \geq H(Y). \tag{2.29}$$

Next, from both W and $\bar{W}$ in (2.24) we get the following identities

$$\frac{x_i}{x_k} = \frac{x_i y_k}{x_k y_i} \cdot \frac{x_j y_i}{x_j y_k}, \;\; \frac{x_j}{x_k} = \frac{x_j y_k}{x_k y_j} \cdot \frac{x_i y_j}{x_i y_k}. \tag{2.30}$$

As we have seen in the proof of Proposition 4, the identities in (2.30) will give

$$\begin{aligned} H(x_i, x_k) &\leq H(x_i y_k, x_k y_i) + H(x_j y_i, x_j y_k) \\ &\leq H(W) + H(\bar{W}) \leq 2H(W), \end{aligned}$$

and

$$\begin{aligned} H(x_j, x_k) &\leq H(x_j y_k, x_k y_j) + H(x_i y_j, x_i y_k) \\ &\leq H(W) + H(\bar{W}) \leq 2H(W). \end{aligned}$$

Hence

$$H(X) \leq H(x_i, x_k) + H(x_j, x_k) \leq 4H(W), \tag{2.31}$$

and (2.28) follows from (2.29) and (2.31).

We remark that if W is given by (2.25) instead of (2.24), then instead of (2.28), we will have

$$H(W) \geq \max\left(\frac{1}{2}H(X), \frac{1}{4}H(Y)\right). \tag{2.32}$$

Finally, we will show that

$$H(W) \geq \frac{3}{8} \max(H(X), H(Y)) \tag{2.33}$$

where W is given by (2.26) under case (III).

This is the trickiest case to handle. From both W and $\bar{W}$ in (2.26), we get the following identities which play a central role in the derivation of (2.33):

$$\frac{x_i{}^3}{x_j{}^3} = \frac{x_i y_j}{x_j y_k} \cdot \frac{x_k y_i}{x_j y_k} \cdot \frac{x_i y_k}{x_k y_j} \cdot \frac{x_i y_k}{x_j y_i}, \quad \frac{x_k{}^3}{x_j{}^3} = \frac{x_k y_j}{x_j y_i} \cdot \frac{x_i y_k}{x_j y_i} \cdot \frac{x_k y_i}{x_i y_j} \cdot \frac{x_k y_i}{x_j y_k}, \tag{2.34}$$

and

$$\frac{y_i{}^3}{y_k{}^3} = \frac{x_k y_i}{x_j y_k} \cdot \frac{x_i y_j}{x_j y_k} \cdot \frac{x_j y_i}{x_k y_j} \cdot \frac{x_j y_i}{x_i y_k}, \quad \frac{y_i{}^3}{y_j{}^3} = \frac{x_k y_i}{x_i y_j} \cdot \frac{x_k y_i}{x_j y_k} \cdot \frac{x_j y_i}{x_k y_j} \cdot \frac{x_i y_k}{x_k y_j}. \tag{2.35}$$

As we have seen in the proof of Proposition 4, the four identities in (2.34) and (2.35) will give

$$3 \max(H(x_i, x_j), H(x_j, x_k), H(y_i, y_k), H(y_i, y_j)) \leq 4H(W). \tag{2.36}$$

For example, we have

$$\begin{aligned} 3H(x_i, x_j) &\leq H(x_i y_j, x_j y_k) + H(x_k y_i, x_j y_k) + H(x_i y_k, x_k y_j) + H(x_i y_k, x_j y_i) \\ &\leq 2H(W) + 2H(\bar{W}) \leq 4H(W), \end{aligned}$$

and in a similar fashion, from the other three identities we have

$$3 \max(H(x_j, x_k), H(y_i, y_k), H(y_i, y_j)) \leq 4H(W).$$

Thus from (2.36) we get

$$H(X) \leq H(x_i, x_j) + H(x_j, x_k) \leq \frac{8}{3} H(W),$$

and

$$H(Y) \leq H(y_i, y_k) + H(y_i, y_j) \leq \frac{8}{3} H(W).$$

The last two inequalities clearly imply (2.33). The proof of Proposition 5 is now complete.

§3. Proof of Theorem

Let $W = \{w_1, \cdots, w_n\}$ be a minimal subset of $\mathcal{U}$ such that $H(W) \geq H(\bar{W})$ if $n = 3$. Let

$$W_1 = \left\{1, \frac{w_2}{w_1}, \cdots, \frac{w_n}{w_1}\right\}.$$

Then it is easy to see that W_1 is also a minimal set. Moreover, we have

$$H(W_1) = H(W).$$

Since the elements of W are monomials $x_i y_j$ with $x_i \in K^{*^r}$ and $y_j \in K^{*^t}$, we may pick p_i and q_i in K^* such that

$$p_i{}^r = x_i \qquad \text{and} \quad q_i{}^t = y_i, \quad i = 1, 2, 3.$$

Set

$$P = \{p_1, p_2, p_3\} \qquad \text{and} \qquad Q = \{q_1, q_2, q_3\},$$

then

$$H(X) = rH(P) \qquad \text{and} \qquad H(Y) = tH(Q),$$

where X and Y are given by (2.9).

Define for each $i, i = 1, 2, 3,$

$$S_i = \{v \in \mathcal{M}_k \mid v(p_i) > v(P)\}$$

and

$$T_i = \{v \in \mathcal{M}_k \mid v(q_i) > v(Q)\}$$

Furthermore, let

$$S = \bigcup_{i=1}^{3} (S_i \cup T_i). \tag{3.1}$$

We claim that every element in W_1 is an S-unit; let $v \notin S$ and $w \in W$, such that $w = w_h/w_1$, where $w_h, w_1 \in W$. Since $v \notin S$, we have

$$v(p_1) = v(p_2) = v(p_3) \qquad \text{and} \quad v(q_1) = v(q_2) = v(q_3).$$

Writing $w_h = p_i{}^r q_j{}^t$ and $w_1 = p_k{}^r q_l{}^t$, then

$$v(w) = v\left(\frac{w_h}{w_1}\right) = rv\left(\frac{p_i}{p_k}\right) + tv\left(\frac{q_j}{q_l}\right) = 0.$$

This shows that w is an S-unit and hence our claim is proved. We may then apply (2.2) to W_1.

Our next object is to obtain a bound on $|S|$. Since $v(p_i) = v(P)$ for $v \notin S_i$ and $v(q_i) = v(Q)$ for $v \notin T_i$, we have, for each i,

$$|S_i| = \sum_{v \in S_i} 1 \leq \sum_{v \in \mathcal{M}_k} (v(p_i) - v(P)) = - \sum_{v \in \mathcal{M}_k} v(P) = H(P), \tag{3.2}$$

and

$$|T_i| = \sum_{v \in T_i} 1 \leq \sum_{v \in \mathcal{M}_k} (v(q_i) - v(Q)) = - \sum_{v \in \mathcal{M}_k} v(Q) = H(Q). \tag{3.3}$$

Then from (3.1), (3.2) and (3.3) we get

$$|S| \leq \sum_{i=1}^{3} (|S_i| + |T_i|) \leq 3(H(P) + H(Q)) \leq 6 \max(H(P), H(Q)). \tag{3.4}$$

Finally we remark that since $H(W) = H(W_1)$, $H(X) = rH(P)$ and $H(Y) = tH(Q)$, we may deduce from (2.13), (2.14) and (2.22) that

$$H(W_1) \geq c(n) \min(r,t) \max(H(P), H(Q)) \tag{3.5}$$

where

$$c(3) = \frac{1}{4}, \qquad c(4) = c(6) = \frac{1}{2}.$$

From (2.2), (3.4) and (3.5) we get

$$\begin{aligned} &c(n) \min(r,t) \max(H(P), H(Q)) \\ &\quad \leq \frac{1}{2}(n-1)(n-2)(6 \max(H(P), H(Q)) + \max(2g-2, 0)) \end{aligned} \tag{3.6}$$

Since $\max(H(P), H(Q)) \geq 1$, we may divide both sides of (3.6) by this quantity and get $\min(r,t) \leq t(n)$ where

$$t(3) = 24 + 8g, \quad t(4) = 36 + 12g, \quad t(6) = 120 + 40g.$$

We have shown that when $\min(r,t) > 120 + 40g$, (1.1) cannot have more than two distinct solutions. This completes the proof of the our Theorem.

REFERENCES

[1] W.D.Brownawell and D.W.Masser, Vanishing sums in function fields, Math. Proc. Camb. Phil. Soc. **100** (1986), 427–434.

[2] R.C.Mason, *Diophantine equations over function fields*, London Math. Soc. Lecture Notes, Vol. 96, Cambridge University Press, 1984.

[3] J. Mueller, Binomial Thue's equation over function fields, Compositio Math. (to appear).

[4] J. Mueller and W.M.Schmidt, Thue's equation and a conjecture of Siegel, Acta Math. **160** (1988), 207–247.

[5] C.L. Siegel, Über einige Anwendungen diophantischer Approximationen, Abh. Preuss. Akad. Wiss. Phys.-math. Kl. **1** (1929).

J. Mueller
Department of Mathematics
Fordham University
Bronx, NY 10458
U. S. A.

Best Possible Results on the Density of Sumsets

MELVYN B. NATHANSON

Dedicated to Paul Bateman on his 70th birthday

Abstract

Mann and Kneser obtained lower bounds for the Shnirel'man density and lower asymptotic density of the sum of a finite number of sequences of non-negative integers. In this paper, special integer sequences are constructed to prove that these results are best possible.

1. Introduction

Let $\mathbf{N}$ denote the set of nonnegative integers. Let $A_1, \ldots, A_h$ be subsets of $\mathbf{N}$. The *sumset* $A_1 + \cdots + A_h$ is the set of all integers of the form $a_1 + \cdots + a_h$, where $a_i \in A_i$ for $i = 1, \ldots, h$. A *sum of rank* r of the sets $A_1, \ldots, A_h$ is a sumset of the form $A_{i(1)} + \cdots + A_{i(r)}$, where $1 \leq i(1) < \cdots i(r) \leq h$. Clearly, there are $\binom{h}{r}$ sums of rank r of any h sets.

Let A be a subset of $\mathbf{N}$. Denote by $A(x)$ the number of positive elements of A not exceeding x. The *Shnirel'man density* of A is defined by

$$\delta(A) = \inf\left\{\frac{A(n)}{n} \mid n = 1, 2, 3, \ldots\right\}.$$

The lower asymptotic density of A is defined by

$$d_L(A) = \liminf\left\{\frac{A(n)}{n} \mid n = 1, 2, 3, \ldots\right\}.$$

Clearly, $0 \leq \delta(A) \leq d_L(A) \leq 1$. If $A' \subseteq A$, then $\delta(A') \leq \delta(A)$ and $d_L(A') \leq d_L(A)$.

Mann [7] proved the following fundamental inequality for the addition of sets of nonnegative integers.

Research supported in part by grants 6-67337 and 6-68344 from the PSC-CUNY Research Award Program of the City University of New York.

Mann's Theorem. *Let A_1, A_2 be sets of nonnegative integers with $0 \in A_i$ for $i = 1, 2$. If*

$$A_1(m) + A_2(m) \geq \mu m$$

for $m = 1, 2, \ldots, n$ and $m \notin A_1 + A_2$, then

$$(A_1 + A_2)(n) \geq \min(1, \mu)n. \tag{1}$$

In particular, if $\delta(A_i) = \alpha_i$ for $i = 1, 2$, then

$$\delta(A_1 + A_2) \geq \min(1, \alpha_1 + \alpha_2). \tag{2}$$

Corollary. *Let $h \geq 2$, and let $A_1, \ldots, A_h$ be sets of nonnegative integers with $0 \in A_i$ for $i = 1, \ldots, h$. Let $\delta(A_i) = \alpha_i$ for $i = 1, \ldots, h$. Then*

$$\delta(A_1 + \cdots + A_h) \geq \min(1, \alpha_1 + \cdots + \alpha_h). \tag{3}$$

Let $1 \leq r \leq h$. Then

$$\sum_S \delta(S) \geq \binom{h-1}{r-1} \min(1, \alpha_1 + \cdots + \alpha_h), \tag{4}$$

where the summation runs over all sums of rank r of the sets $A_1, \ldots, A_h$.

Proof: Inequality (3) follows immediately from (2) by induction on h.

Let $\sigma = \alpha_1 + \cdots + \alpha_h$. Suppose that $\sigma \leq 1$. Let $\sum'$ denote the sum over the $\binom{h}{r}$ subsets of cardinality r chosen from $\{1, \ldots, h\}$. Note that for each $i \in \{1, \ldots, h\}$ there are exactly $\binom{h-1}{r-1}$ such subsets that contain the integer i. Let S denote a sum of rank r of the sets $A_1, \ldots, A_h$. Since $\sigma \leq 1$, it follows from (3) that

$$\begin{aligned}
\sum_S \delta(S) &= \sum{}' \delta(A_{i(1)} + \cdots + A_{i(r)}) \\
&\geq \sum{}' (\alpha_{i(1)} + \cdots + \alpha_{i(r)}) \\
&= \sum_{i=1}^{h} \binom{h-1}{r-1} \alpha_i = \binom{h-1}{r-1} \sigma \\
&= \binom{h-1}{r-1} \min(1, \alpha_1 + \cdots + \alpha_h).
\end{aligned}$$

Now suppose that $\sigma = \alpha_1 + \cdots + \alpha_h > 1$. Define $\beta_i = \alpha_i / \sigma$ for $i = 1, \ldots, h$. Then $0 \leq \beta_i \leq \alpha_i \leq 1$ and $\beta_1 + \cdots + \beta_h = 1$. For $i = 1, \ldots, h$, there exists a subset B_i of A_i with $\delta(B_i) = \beta_i$. (For completeness, a proof of this fact appears as Lemma 1 in the next section.) Let $T = B_{i(1)}+\cdots+B_{i(r)}$ be a sum of rank r of the sets $B_1, \ldots, B_h$, and let $S = A_{i(1)} + \cdots + A_{i(r)}$ be the corresponding sum of rank r of the sets $A_1, \ldots, A_h$. Since $B_i \subseteq A_i$, it follows that $T \subseteq S$ and $\delta(T) \leq \delta(S)$. Therefore,

$$\sum_S \delta(S) \geq \sum_T \delta(T) \geq \binom{h-1}{r-1}(\beta_1 + \cdots + \beta_h)$$
$$= \binom{h-1}{r-1} \min(1, \alpha_1 + \cdots + \alpha_h).$$

This completes the proof.

Dyson [2] generalized Mann's inequality (1) for two summands to r-fold sums of the form (4).

Dyson's Theorem. *Let $h \geq 2$, and let $A_1, \ldots, A_h$ be sets of nonnegative integers with $0 \in A_i$ for $i = 1, \ldots, h$. If*

$$A_1(m) + \cdots + A_h(m) \geq \mu m \tag{5}$$

for $m = 1, \ldots, n$, then

$$\sum_S S(n) \geq \binom{h-1}{r-1} \min(1, \mu)n, \tag{6}$$

where the summation runs over all sums of rank r of the sets $A_1, \ldots, A_h$.

Mann's theorem on the Shnirel'man density of sumsets was extended by Kneser [5] to the case of the lower asymptotic density of sumsets.

Kneser's Theorem. *Let $h \geq 2$ and let $A_1, \ldots, A_h$ be sets of nonnegative integers with $0 \in A_i$ for $i = 1, \ldots, h$. Then either*

$$d_L(A_1 + \cdots + A_h) \geq d_L(A_1) + \cdots + d_L(A_h) \tag{7}$$

or there exists an integer $g \geq 1$ and sets $A_1^{(g)}, \ldots, A_h^{(g)}$ such that each $A_i^{(g)}$ contains A_i and is a union of congruence classes modulo g, the sumsets $A_1+\cdots+A_h$ and $A_1^{(g)}+\cdots+A_h^{(g)}$ coincide for all sufficiently large integers, and

$$d_L(A_1 + \cdots + A_h) \geq d_L(A_1) + \cdots + d_L(A_h) - \frac{h-1}{g}.$$

Lepson[6] and Cheo[1] proved that Mann's lower bound (2) for the Shnirel'man density of the sum of two sets is best possible in the sense that if $\alpha_1, \alpha_2 \in [0,1]$ and $\alpha_1 + \alpha_2 \leq \beta \leq 1$, then there exist sets A_1 and A_2 of nonnegative integers such that $\delta(A_i) = \alpha_i$ for $i = 1, 2$ and $\delta(A_1+A_2) = \beta$. In this paper I shall use Dyson's Theorem to show that the lower bounds (3) and (7) are also best possible. I shall also prove that the lower bound (4) is sharp.

Notation. For real numbers x and y, let $[x, y]$ denote the set of integers n such that $x \leq n \leq y$. Let $\langle x \rangle$ denote the least integer n such that $n \geq x$.

2. Two Lemmas

The following lemma concerns the Shnirel'man densities of the subsets of a set of nonnegative integers. It should be noted that Grekos [3] and Grekos and Volkmann [4] have made an extensive study of the upper and lower asymptotic densitities of the subsets of a set of nonnegative integers.

Lemma 1. *Let A be a set of nonnegative integers with $0 \in A$ and $\delta(A) = \alpha > 0$. Let $0 \leq \alpha' \leq \alpha$. There exists a subset A' of A with $0 \in A'$ and $\delta(A') = \alpha'$.*

Proof: If $\alpha' = 0$, let $A' = \{0\}$. Now suppose that $\alpha' > 0$. Note that $\alpha \geq \alpha' > 0$ implies that $1 \in A$. Let $1 = a_1 < a_2 \cdots$ be the positive elements of A listed in strictly increasing order. I shall construct a decreasing sequence of sets $A = A_0 \supseteq A_1 \supseteq A_2 \supseteq \cdots$ such that $\delta(A_k) \geq \alpha'$ for all $k \geq 0$ and $A' = \bigcap_{k=0}^{\infty} A_k$ satisfies $\delta(A') = \alpha'$.

Let $A_0 = A$. Then $\delta(A_0) = \alpha \geq \alpha'$. Let $k \geq 1$, and suppose that we have constructed set $A_0, A_1, \ldots, A_{k-1}$ with $A_0 \supseteq A_1 \supseteq \cdots \supseteq A_{k-1}$ and $\delta(A_i) \geq \alpha'$ for $i = 0, 1, \ldots, k-1$. Let $B_k = A_{k-1} \setminus \{a_k\}$. Define the set A_k by

$$A_k = \begin{cases} B_k & \text{if } \delta(B_k) \geq \alpha', \\ A_{k-1} & \text{if } \delta(B_k) < \alpha'. \end{cases}$$

Then $A_{k-1} \supseteq A_k$. Since $\delta(A_{k-1}) \geq \alpha'$, it follows that $\delta(A_k) \geq \alpha'$. Let $A' = \bigcap_{k=0}^{\infty} A_k$.

Let $n \geq 1$. Choose $a_k \in A$ such that $a_k \leq n < a_{k+1}$. Then $A'(n) = A_k(n) \geq \alpha' n$, and so $\delta(A') \geq \alpha'$.

Suppose that $\delta(A') = \beta > \alpha'$. Since the set A is infinite, there exists an integer $a_t \in A'$ such that $a_t \geq 1/(\beta - \alpha')$. Clearly, $a_t \in A_t$. It follows that $B_t = A_t \setminus \{a_t\} \neq A_t$ and

$$\delta(B_t) < \alpha'. \tag{$*$}$$

Let $n \geq 1$. If $1 \leq n < a_t$, then $B_t(n) = A_{t-1}(n) \geq \alpha' n$. If $n \geq a_t$, then $B_t(n) = A_t(n) - 1$. Since $a_t \geq 1/(\beta - \alpha')$, it follows that $\beta - 1/a_t \geq \alpha'$ and

$$\begin{aligned} \frac{B_t(n)}{n} &= \frac{A_t(n) - 1}{n} \geq \frac{A'(n) - 1}{n} \\ &\geq \beta - \frac{1}{n} \geq \beta - \frac{1}{a_t} \geq \alpha'. \end{aligned}$$

Therefore, $\delta(B_t) \geq \alpha'$, which contradicts ($*$). Thus, $\delta(A') = \alpha'$. This completes the proof.

Lemma 2. *Let $0 \leq \alpha \leq 1$. Let $1 = M_1 < M_2 < \cdots$ be a strictly increasing sequence of integers such that $M_{k+1}/M_k \rightarrow \infty$. Choose $\alpha_k \in [0,1]$ such that $\alpha_k \geq \alpha$ for all $k \geq 1$ and $\alpha_k = \alpha$ for infinitely many k. Define integers a_k by $a_k = \langle \alpha_k(M_{k+1} - M_k)\rangle$. Define the set A of nonnegative integers by*

$$A = \{0,1\} \cup \bigcup_{k=1}^{\infty} [M_k + 1, M_k + a_k].$$

Then $\delta(A) = d_L(A) = \alpha$.

Proof: The first step is to show that $A(M_k) \geq \alpha M_k$ for all $k \geq 1$. Clearly, $A(M_1) = A(1) = 1 \geq \alpha = \alpha M_1$. Let $k \geq 2$. Then

$$\begin{aligned} A(M_k) &= 1 + a_1 + \cdots + a_{k-1} \\ &\geq 1 + \sum_{i=1}^{k-1} \langle \alpha_i(M_{i+1} - M_i)\rangle \geq 1 + \alpha \sum_{i=1}^{k-1} (M_{i+1} - M_i) \\ &= 1 + \alpha(M_k - 1) \geq \alpha M_k. \end{aligned}$$

Thus, $A(M_k)/M_k \geq \alpha$ for all $k \geq 1$.

Let $m \geq 2$, $m \neq M_k$. If $M_k < m \leq M_k + a_k$, let $t = m - M_k$. Then

$$\frac{A(m)}{m} = \frac{A(M_k + t)}{M_k + t} = \frac{A(M_k) + t}{M_k + t} \geq \frac{A(M_k)}{M_k} \geq \alpha.$$

If $M_k + a_k < m < M_{k+1}$, then

$$\frac{A(m)}{m} = \frac{A(M_{k+1})}{m} > \frac{A(M_{k+1})}{M_{k+1}} \geq \alpha.$$

Therefore, $A(m)/m \geq \alpha$ for all $m \geq 1$, and so

$$\alpha \leq \delta(A) \leq d_L(A).$$

To obtain an upper bound for the densities, recall that $M_{k+1}/M_k \rightarrow \infty$ and that $\alpha_k = \alpha$ for infinitely many k. Since

$$A(M_{k+1}) \leq M_k + a_k < M_k + \alpha_k(M_{k+1} - M_k) + 1,$$

it follows that, if $\alpha_k = \alpha$, then

$$\frac{A(M_{k+1})}{M_{k+1}} < \alpha + \frac{(1 - \alpha_k)M_k + 1}{M_{k+1}}$$

and so $d_L(A) \leq \alpha$. Therefore, $\delta(A) = d_L(A) = \alpha$. This completes the proof.

3. Main results

The following result shows that inequalities (3) and (7) are best possible.

Theorem 1. *Let $\alpha_1, \dots, \alpha_h \in [0,1]$, and let $\sigma = \alpha_1 + \cdots + \alpha_h \leq 1$. Let $\sigma \leq \beta \leq 1$. Then there exist sets $A_1, \dots, A_h$ of nonnegative integers with $0 \in A_i$ and $\delta(A_i) = d_L(A_i) = \alpha_i$ for $i = 1, \dots, h$, such that*

$$\delta(A_1 + \cdots + A_h) = d_L(A_1 + \cdots + A_h) = \beta.$$

Proof: Define $\beta_i \in [0,1]$ by

$$\beta_i = \beta - \sum_{\substack{1 \leq j \leq h \\ j \neq i}} \alpha_j.$$

Then $\alpha_i \leq \beta_i \leq \beta$ for $i = 1, \dots, h$. Let $1 = M_1 < M_2 < \cdots$ be a strictly increasing sequence of integers such that $M_{k+1}/M_k \to \infty$. Let $\mathbf{N}^* = W_1 \cup W_2 \cup \cdots \cup W_h$ be a partition of the positive integers into h pairwise disjoint, infinite sets. For $i = 1, \dots, h$, define the sequences $\{\alpha_{ik}\}_{k=1}^{\infty}$ of real numbers by

$$\alpha_{ik} = \begin{cases} \alpha_i & \text{if } k \notin W_i, \\ \beta_i & \text{if } k \in W_i. \end{cases}$$

Then $\alpha_{ik} \geq \alpha_i$ for all $k \geq 1$ and $\alpha_{ik} = \alpha_i$ for infinitely many k. Moreover,

$$\alpha_{1k} + \alpha_{2k} + \cdots + \alpha_{hk} = \beta$$

for all $k \geq 1$. Define $a_{ik} = \langle \alpha_{ik}(M_{k+1} - M_k) \rangle$. Use Lemma 2 to construct the h sets A_i from the sequences $\{M_k\}_{k=1}^{\infty}$, $\{\alpha_{ik}\}_{k=1}^{\infty}$, and $\{a_{ik}\}_{k=1}^{\infty}$. Then $\delta(A_i) = d_L(A_i) = \alpha_i$ for $i = 1, \dots, h$.

I shall now show that the sets $A_1, \dots, A_h$ satisfy condition (5) of Dyson's theorem. For $k \geq 1$,

$$\begin{aligned} A_i(M_k) &= 1 + a_{i,1} + a_{i,2} + \cdots + a_{i,k-1} \\ &\geq 1 + \sum_{j=1}^{k-1} \alpha_{ij}(M_{j+1} - M_j) \end{aligned}$$

and so

$$\begin{aligned} A_1(M_k) + \cdots + A_h(M_k) &\geq h + \sum_{i=1}^{h} \sum_{j=1}^{k-1} \alpha_{ij}(M_{j+1} - M_j) \\ &= h + \sum_{j=1}^{k-1} \sum_{i=1}^{h} \alpha_{ij}(M_{j+1} - M_j) \\ &= h + \beta(M_k - 1) > \beta M_k. \end{aligned}$$

Let m be a positive integer satisfying $M_k < m < M_{k+1}$ for some $k \geq 1$. Let $t = m - M_k$. Let $a^* = \max\{a_{ik} \mid i = 1, \ldots, h\}$. If $t \leq a^*$, then

$$\begin{aligned} A_1(m) + \cdots + A_h(m) &\geq A_1(M_k) + \cdots + A_k(M_k) + t \\ &\geq \beta M_k + t \geq \beta(M_k + t) = \beta m. \end{aligned}$$

If $t > a^*$, then $A_i(m) = A_i(M_{k+1})$ for $i = 1, \ldots, h$, and

$$A_1(m) + \cdots + A_h(m) = A_1(M_{k+1}) + \cdots + A_h(M_{k+1}) > \beta M_{k+1} \geq \beta m.$$

Thus, (5) holds for all $m \geq 1$. Applying (6) with $r = h$, we obtain

$$(A_1 + \cdots + A_h)(n) \geq \beta n$$

for all $n \geq 1$, and so

$$\beta \leq \delta(A_1 + \cdots + A_h) \leq d_L(A_1 + \cdots + A_h).$$

To get an upper bound for the densities, observe that

$$\begin{aligned} (A_1 + \cdots + A_h)(M_{k+1}) &\leq hM_k + a_{1k} + \cdots + a_{hk} \\ &\leq hM_k + h + (\alpha_{1k} + \cdots + \alpha_{hk})(M_{k+1} - M_k) \\ &= hM_k + h + \beta(M_{k+1} - M_k). \end{aligned}$$

Therefore,

$$\frac{(A_1 + \cdots + A_h)(M_{k+1})}{M_{k+1}} \leq \beta + \frac{(h - \beta)M_k + h}{M_{k+1}}.$$

Since $M_k/M_{k+1} \to 0$, it follows that $d_L(A_1 + \cdots + A_h) \leq \beta$. This completes the proof.

The next result shows that the lower bound in (4) is sharp.

Theorem 2. *Let $0 \leq \alpha_i \leq 1$ for $i = 1, \ldots, h$, and let $\alpha_1 + \cdots + \alpha_h \leq 1$. Then there exist sets $A_1, \ldots, A_h$ of nonnegative integers such that $\delta(A_i) = d_L(A_i) = \alpha_i$ for $i = 1, \ldots, h$, and, for $r = 1, \ldots, h$,*

$$\sum_S \delta(S) = \binom{h-1}{r-1}(\alpha_1 + \cdots + \alpha_h),$$

where the summation runs over all sums of rank r of the sets $A_1, \ldots, A_h$.

Proof: Let $\alpha_{ik} = \alpha_i$ for $i = 1, \ldots, h$ and all $k \geq 1$. Construct sets $A_1, \ldots, A_h$ as in the proof of Theorem 1. If $1 \leq i_1 < i_2 < \cdots < i_r \leq h$, then

$$d_L(A_{i_1} + \cdots + A_{i_r}) = \delta(A_{i_1} + \cdots + A_{i_r}) = \alpha_{i_1} + \cdots + \alpha_{i_r}.$$

It follows exactly as in the proof of the Corollary to Mann's Theorem that

$$\sum_S d_L(S) = \sum_S \delta(S) = \binom{h-1}{r-1}(\alpha_1 + \cdots + \alpha_h).$$

This completes the proof.

Theorem 3. *Let $0 \leq \alpha_i \leq 1$ for $i = 1, \ldots, h$, and let $\alpha_1 + \cdots + \alpha_h \leq \mu \leq 1$. Then there exist sets $A_1, \ldots, A_h$ of nonnegative integers such that $\delta(A_i) = \alpha_i$ and, for $r = 1, \ldots, h$,*

$$\inf \sum_S \frac{S(m)}{m} = \liminf \sum_S \frac{S(m)}{m} = \binom{h-1}{r-1}\mu,$$

where the summation runs over all sums of rank r of $A_1, \ldots, A_h$.

Proof: Construct the sets A_i exactly as in the proof of Theorem 1. Then (5) holds for all $m \geq 1$, and so, by Dyson's Theorem, $\sum_S S(n) \geq \binom{h-1}{r-1}\mu n$ for $r = 1, \ldots, h$ and for all $n \geq 1$.

Let $S = A_{i_1} + \cdots + A_{i_r}$ be a sum of rank r. Then

$$\begin{aligned} S(M_{k+1}) &\leq rM_k + \sum_{j=1}^{r} a_{i_j,k} \\ &\leq rM_k + r + \left(\sum_{j=1}^{r} \alpha_{i_j,k}\right)(M_{k+1} - M_k). \end{aligned}$$

It follows that

$$\sum_S S(M_{k+1}) \leq r\binom{h}{r}(M_k + 1) + \binom{h-1}{r-1}\mu(M_{k+1} - M_k).$$

Since $M_k/M_{k+1} \to 0$, it follows that

$$\liminf \frac{1}{n}\sum_S S(n) \leq \binom{h-1}{r-1}\mu.$$

This completes the proof.

It is an open problem to determine if, under the conditions of Theorem 3, there exist sets $A_1, \ldots, A_h$ such that $\delta(A_i) = \alpha_i$ for $i = 1, \ldots, h$ and $\sum_S \delta(S) = \binom{h-1}{r-1}\mu$ for $r = 1, \ldots, h$.

REFERENCES

[1] L. Cheo, A remark on the $\alpha + \beta$ theorem, Proc. Amer. Math. Soc. **3** (1952), 175–177.

[2] F. J. Dyson, A theorem on the densities of sets of integers, J. London Math. Soc. **20** (1945), 8–14.

[3] G. Grekos, Répartition des densités des sous-suites d'une suite d'entiers, J. Number Theory **10** (1978), 177–191.
[4] G. Grekos and B. Volkman, On densities and gaps, J. Number Theory **26** (1987), 129–148.
[5] M. Kneser, Abschätzung der asymptotischen Dichte von Summenmengen, Math. Z. **58** (1953), 459–484.
[6] B. Lepson, Certain best possible results in the theory of Schnirelmann density, Proc. Amer. Math. Soc. **1** (1950), 592–594.
[7] H. B. Mann, A proof of the fundamental theorem on the density of sums of sets of positive integers, Annals Math. **43** (1942), 523–527.

Melvyn B. Nathanson
Provost and Vice President
for Academic Affairs
Lehman College (CUNY)
Bronx, New York 10468

Some Powers of The Euler Product

MORRIS NEWMAN

For Paul Bateman, in friendship

The coefficients of the rth power of the Euler product for r even, $0 < r \leq 24$, are a natural generalization of the Ramanujan τ-function, and satisfy similar recurrence formulas. These coefficients are studied here and an analogue of Lehmer's question on the non-vanishing of the τ-function is stated and proved for $r = 2, 4, 6, 8, 10, 14$. In addition, certain congruences for these coefficients are proved, and the results of extensive numerical investigations are given.

The Euler product is given by

$$\varphi(x) = \prod_{n=1}^{\infty}(1 - x^n) = \sum_{n=-\infty}^{\infty} (-1)^n x^{(3n^2+n)/2}.$$

If r is an integer, we set

$$\varphi(x)^r = \sum_{n=0}^{\infty} p_r(n)x^n.$$

The Ramanujan τ-function is then given by $\tau(n) = p_{24}(n-1)$. Ramanujan introduced this function in [13], and conjectured that it is a multiplicative arithmetic function satisfying the relationship

$$\tau(np) = \tau(n)\tau(p) - p^{11}\tau(n/p)$$

where p is a prime and $\tau(m)$ is defined to be 0 if m is not an integer. This conjecture was subsequently proved by Mordell, in [6] .

Let r be an integer, and let S_r be the set of positive integers n such that each prime divisor p of n satisfies $r(p-1) \equiv 0 \mod 24$. In the same paper, Ramanujan introduced the function $\tau_r(n)$ defined by

$$\tau_r(n) = p_r(r(n-1)/24), \qquad r(n-1) \equiv 0 \mod 24$$

Ramanujan conjectured and Mordell proved that if r is an even divisor of 24, then $\tau_r(n)$ is a multiplicative arithmetic function on S_r and satisfies the relationship

$$\tau_r(np) = \tau_r(n)\tau_r(p) - p^{(r-2)/2}\tau_r(n/p)$$

where p is a prime belonging to S_r, and $\tau_r(m)$ is defined to be 0 if m is not an integer. This result was subsequently extended by the author in [9] to include all even r satisfying $0 < r \leq 24$.

In connection with the Ramanujan function, Lehmer asked in [4] whether $\tau(n)$ ever vanishes. The conjecture that it does not is as yet unproved, although it is supported by massive amounts of computation. The same conjecture can be made for the generalized functions $\tau_r(n)$ defined above. We will show in this note that $\tau_r(n)$ does not vanish if n belongs to S_r, and r is limited to the values $2, 4, 6, 8, 10, 14$. No information is available in the remaining cases; namely $r = 12, 16, 18, 20, 22, 24$. However, for these cases, congruential results are available and will be proved.

Perhaps more fundamental than $\varphi(x)$ is the Dedekind η-function, defined by

$$\eta(\tau) = x^{1/24}\varphi(x), \qquad x = \exp(2\pi i\tau), \quad \Im m(\tau) > 0$$

a modular form of dimension $-1/2$ with well-studied properties. There has been a revival of interest in the powers of the Dedekind η-function ever since the work of MacDonald [5], who found identities for the coefficients of the dth power of the Euler product whenever d is the dimension of a simple Lie algebra. It is still not understood why such identities exist. There is some arcane connection between the affine root systems of Lie theory and the classical theory of modular forms.

We state the results discussed above as a theorem.

Theorem 1 (Ramanujan, Mordell, Newman). *Suppose that r is even, $0 < r \leq 24$. Let the positive integer n satisfy $r(n-1) \equiv 0 \mod 24$. Then if p is a prime such that $r(p-1) \equiv 0 \mod 24$,*

$$\tau_r(np) = \tau_r(n)\tau_r(p) - p^{(r-2)/2}\tau_r(n/p)$$

and $\tau_r(n)$ is a multiplicative arithmetic function of n on S_r .

An interesting consequence of this theorem is that if $r > 2$, then $\tau_r(n) \equiv 0 \mod p$ whenever $n \equiv 0 \mod p$, provided that $\tau_r(p) \equiv 0 \mod p$. It is therefore of interest to determine those primes p for which $\tau_r(p)$ is divisible by p. It turns out that this cannot happen for $r = 2, 4, 6, 8, 10, 14$. This remark (which is proved below) is crucial in settling the question of the vanishing of $\tau_r(n)$. However , it *can* happen for $r = 12, 16, 18, 20, 22, 24$, and a computer search was made to determine primes for which this occurs.

We must have available the explicit formulas for the coefficients for the values of r under consideration. The following result furnishes these formulas. Relevant references as to the source of these identities are given in the bibliography.

Theorem 2 (Klein-Fricke, Ramanujan, Winquist, Dyson, MacDonald, Atkin). *The following formulas are valid:*

$$p_2(n) = \sum (-1)^b, \tag{2.1}$$

summed over all solutions of $a^2 + 9b^2 = 12n + 1$, $a \equiv 1 \mod 3$,

$$p_4(n) = \sum c, \tag{2.2}$$

summed over all solutions of $c^2 + 3d^2 = 6n + 1$, $c \equiv 1 \mod 3$,

$$p_6(n) = \sum (a^2 - 4b^2), \tag{2.3}$$

summed over all solutions of $a^2 + 4b^2 = 4n + 1$, $a > 0$,

$$p_8(n) = \frac{1}{8} \sum c(c^2 - 9d^2), \tag{2.4}$$

summed over all solutions of $c^2 + 3d^2 = 12n + 4$, $c \equiv d \equiv 2 \mod 3$,

$$p_{10}(n) = \frac{-1}{3} \sum ab(a^2 - 4b^2), \tag{2.5}$$

summed over all solutions of $a^2 + 4b^2 = 12n + 5$, $a \equiv b \mod 3$, $a > 0$,

$$p_{14}(n) = \frac{1}{30} \sum cd(c^2 - d^2)(c^2 - 9d^2), \tag{2.6}$$

summed over all solutions of $c^2 + 3d^2 = 12n + 7$, $c \equiv 2 \mod 6$, $d \equiv 1 \mod 6$.

From these formulas it is easy to derive the following:

Theorem 3. *Let p be a prime such that $r(p-1) \equiv 0 \mod 24$. Then the following identities for $\tau_r(p)$ are valid:*

$$\tau_2(p) = 2(-1)^b, \qquad \text{where } p = a^2 + 9b^2, \tag{3.1}$$

$$\tau_4(p) = 2c, \qquad \text{where } p = c^2 + 3d^2,\ c \equiv 1 \mod 3, \tag{3.2}$$

$$\tau_6(p) = 2(a^2 - 4b^2), \qquad \text{where } p = a^2 + 4b^2, \tag{3.3}$$

$$\tau_8(p) = 2c(c^2 - 9d^2), \quad \text{where } p = c^2 + 3d^2,\ c \equiv 1 \mod 3, \tag{3.4}$$

$$\tau_{10}(p) = \begin{cases} 2(a^4 - 216a^2b^2 + 1296b^4), & \text{where } p = a^2 + 36b^2, \\ -2(81a^4 - 216a^2b^2 + 16b^4), & \text{where } p = 9a^2 + 4b^2, \end{cases} \tag{3.5}$$

$$\tau_{14}(p) = -2(a^6 - 180a^4b^2 + 2160a^2b^4 - 1728b^6), \tag{3.6}$$

$$\text{where } p = a^2 + 12b^2,\ a \equiv 1 \mod 4.$$

Proof: The proof depends only on the fact that if a prime p has a representation as $a^2 + b^2$ or as $a^2 + 3b^2$, then it is essentially unique . It is straightforward, and will be omitted.

From the above, we can deduce

Theorem 4. *Suppose that $r = 2, 4, 6, 8, 10, 14$, and that p is a prime such that $r(p-1) \equiv 0 \mod 24$. Then $\tau_r(p)$ is not divisible by p.*

Proof: The proof is a direct consequence of the identities above and offers no difficulty. We give the proof for one case only; namely $r = 14$.

We have from (3.6) that $a^2 \equiv -12b^2 \mod p$, and substituting for a^2 in this formula, we find that $\tau_{14}(p) \equiv 2^{12}3^3b^6 \mod p$, which is not divisible by p, since $p \equiv 1 \mod 12$ and p and b are relatively prime. The other cases are proved in similar fashion.

We now state the main result of this note, which supplies an answer to the generalization of Lehmer's question about the Ramanujan τ-function:

Theorem 5. *Let $r = 2, 4, 6, 8, 10, 14$. Suppose that n belongs to S_r. Then $\tau_r(n) \neq 0$.*

Proof: Suppose first that $r > 2$. Let p be any prime belonging to S_r and let k be any positive integer. Then Theorem 1 implies that

$$\tau_r(p^k) = \tau_r(p)\tau_r(p^{k-1}) - p^{(r-2)/2}\tau_r(p^{k-2}).$$

which in turn implies that

$$\tau_r(p^k) \equiv \tau_r(p)^k \mod p, \qquad k \geq 1$$

Now suppose that $\tau_r(n) = 0$. Then for some prime p dividing n and some positive integer k, $\tau_r(p^k) = 0$, since $\tau_r(n)$ is arithmetically multiplicative on the set S_r. It then follows from the formula above that $\tau_r(p) \equiv 0 \mod p$. But this is false, by Theorem 4. Hence the initial assumption that $\tau_r(n) = 0$ is false, and the proof is complete for $r > 2$.

Now suppose that $r = 2$. Then the formula (3.1) above states that

$$\tau_2(p) = 2(-1)^b, \qquad \text{where } p = a^2 + 9b^2.$$

The recurrence formula of Theorem 1 now implies that if k is any nonnegative integer, then

$$\tau_2(p^k) = (-1)^{bk}(k+1),$$

which is not 0. This proves the result for $r = 2$ and completes the proof.

If all that is assumed is that $r(n-1) \equiv 0 \mod 24$, then $\tau_r(n)$ can certainly vanish. It is quite easy to show that $\tau_r(n) = 0$ whenever n is exactly divisible by an odd power of a prime p such that $r(p+1) \equiv 0 \mod 24$, where now $r = 2, 4, 6, 8, 10, 14, 26$. The proof follows from the explicit formulas of Theorem 2 and from the results of the author in [7].

Whether Theorem 5 is true in the remaining cases ($r = 12, 16, 18, 20, 22, 24$) is not known. In these cases a computer search was made for values of p such that $\tau_r(p) \equiv 0 \mod p$. The results are summarized in the following table:

r	p	bound
12	3,19,11003,12197,139361	< 399000
16	13	< 299000
18	5,541	< 266000
20	7	< 239000
22	61	< 163000
24	2,3,5,7,2411	< 300000

TABLE 1: Values of r and p such that $\tau_r(p) \equiv 0 \mod p$, where p is a prime

The only oddity exhibited here is the relatively large number of instances for $r = 12$. It is not clear if there is something more than coincidence here.

Explicit formulas involving sigma-functions for $p_{12}(n)$, $p_{24}(n)$ have been given by Ramanujan, Schoeneberg, Van der Pol, Niebur, and others (references are given in the bibliography). For example, Schoeneberg has shown that

$$\tau_{12}(n) = n\{2\sigma(n) - \sigma_3(n)\} + 48 \sum_{k=1}^{(n-1)/2} (n-2k)\sigma(n-2k)\sigma'(k),$$

where $\sigma_r(n)$ is the sum of the rth powers of the divisors of n, and $\sigma'(n)$ is the sum of the odd divisors of n. This implies that if p is a prime > 2 then

$$\tau_{12}(p) \equiv -96 \sum_{k=1}^{(p-1)/2} k\sigma(p-2k)\sigma'(k) \qquad \mod p.$$

Similarly, Niebur has shown that

$$\tau(n) = n^4\sigma(n) - 24 \sum_{k=1}^{n-1} (35k^4 - 52k^3 n + 18k^2 n^2)\sigma(k)\sigma(n-k).$$

This in turn implies that if p is a prime, then

$$\tau(p) \equiv -1680 \sum_{k=1}^{(p-1)/2} k^4 \sigma(k)\sigma(p-k) \mod p.$$

Unfortunately , it seems difficult (if not impossible) to extract any significant information about the behaviour of these functions modulo a prime from the above.

Table 1 yields the following congruential result, which was mentioned previously:

Theorem 6. *Let r, p have the values given in Table 1. Then*

$$\tau_r(n) \equiv 0 \mod p, \quad \textit{whenever } n \equiv 0 \mod p.$$

The proof follows directly from the recurrence formula of Theorem 1, and will be omitted.

In connection with Lehmer's question about the τ-function, it is of interest to look for instances of the vanishing of $p_r(n)$. The following table summarizes the result of extensive machine computation:

r	z	r	z
1	3	14	4
2	7	15	53
3	2	16	none < 500000
4	9	17	none < 500000
5	1560	18	none < 500000
6	5	19	none < 500000
7	28017	20	none < 500000
8	3	21	none < 500000
9	none < 500000	22	none < 500000
10	6	23	none < 500000
11	none < 500000	24	none known
12	"	25	none < 100000
13	"	26	9

TABLE 2: z = first value of n for which $p_r(n) = 0$

The values for $r = 5$ and $r = 7$ were first found by Atkin.

Once a value of n has been found such that $p_r(n) = 0$, where $1 \le r \le 24$, or $r = 26$, a result due to the author [8] implies that there are infinitely many values of n such that $p_r(n) = 0$. For example, the fact that $p_5(1560) = 0$ implies that

$$p_5(1560n^2 + 5(n^2-1)/24) = 0, \quad \text{provided that } (n,6) = 1.$$

REFERENCES

[1] A.O.L.Atkin, various private communications.

[2] F.J.Dyson, Missed opportunities, Bull. Amer. Math. Soc. **78** (1972), 635–652.

[3] F. Klein and R. Fricke, Vorlesungen über die Theorie der elliptischen Modulfunktionen, vol. 2, Teubner, Leipzig, 1892, pp. 373–377.

[4] D.H.Lehmer, The vanishing of Ramanujan's function $\tau(n)$, Duke Math. J. **14** (1947), 429–433.

[5] I.G. MacDonald, Affine root systems and Dedekind's η-function, Invent. Math. **15** (1972), 91–143.

[6] L.J. Mordell, On Mr. Ramanujan's empirical expansions of modular functions, Proc. Cambridge Philos Soc. **19** (1917), 117–124.

[7] M. Newman, An identity for the coefficients of certain modular forms, J. London Math. Soc. **30** (1955), 488–493.

[8] ———, Further identities and congruences for the coefficients of modular forms, Can. J. Math. **10** (1958), 577–586.

[9] ———, Remarks on some modular identities, Trans Amer. Math. Soc. **73** (1952), 313–320.

[10] D. Niebur, A formula of Ramanujan's τ-function, Illinois J. Math **19** (1975), 448–449.

[11] B. van der Pol, On a non-linear partial differential equation satisfied by the logarithm of the Jacobian theta-functions, with arithmetical applications I, II, Indag. Math. **13** (1951), 261–284.

[12] B. Schoeneberg, Über die Diskriminante der elliptischen Funktionen und ihre Quadratwurzel, Math. Z. **65** (1956), 16–24.

[13] S. Ramanujan, On certain arithmetic functions, Trans. Cambridge Philos. Soc. **22** (1916), 159–184.

[14] L. Winquist, An elementary proof of $p(11m+6) \equiv 0 \pmod{11}$, J. Combinatorial Theory **6** (1969), 56–59.

Morris Newman
Department of Mathematics
University of California
Santa Barbara, CA 93106

A Divergent Argument Concerning Hadamard Roots Of Rational Functions

A. J. VAN DER POORTEN

To Paul Bateman

1. Introduction

A compelling conjecture attributed variously to Pisot and Schutzenberger suggests that $\sum a_h X^h$ is the Taylor expansion of a rational function provided only that the a_h all belong to some field finitely generated over $\mathbf{Q}$ and there is a nonzero polynomial f so that $\sum f(a_h)X^h$ represents a rational function – of course only 'coherent' choices of the a_h yielding $f(a_h)$ will do. In part because the sequence of Taylor coefficients is a recurrence sequence (satisfying a linear homogeneous recurrence relation with constant coefficients), the following is a plain language example of the conjecture: A sequence of rationals (a_h) is a recurrence sequence if (a_h^3) is a recurrence sequence. Conversely, $\sum \sqrt{h}X^h$ is not rational because the $\sqrt{h}$ do not all belong to a field finitely generated over $\mathbf{Q}$. The conjecture can be proved in a 'generic' case, but seems inaccessible in general. Breaching the tradition that one should only sketch proofs one believes to be correct, I manipulate formal series divergent everywhere to 'verify' the conjecture and to illustrate the apparent obstructions to a convincing proof.

Research partially supported by the Australian Research Council.

2. Generalised Power Sums, Rational Functions and Recurrence Sequences

A *generalised power sum* $a(h)$, $h = 0,1,2,\ldots\ldots$ is an expression of the shape

$$a(h) = \sum_{i=1}^{m} A_i(h)\alpha_i^h \;,\; h = 0,1,2,\ldots\ldots \tag{1}$$

with *roots* α_i, $1 \le i \le m$, distinct non-zero numbers, and *coefficients* $A_i(h)$ polynomials respectively of degree $n_i - 1$, for positive integers n_i, $1 \le i \le m$. The generalised power sum $a(h)$ is said to have *order*

$$n = \sum_{i=1}^{m} n_i \,.$$

Set

$$s(X) = \prod_{i=1}^{m}(1 - \alpha_i X)^{n_i} = 1 - s_1 X - \cdots - s_n X^n \,. \tag{2}$$

Then the sequence (a_h) with $a_h = a(h)$, $h = 0,1,2,\ldots\ldots$ satisfies the linear homogeneous recurrence relation

$$a_{h+n} = s_1 a_{h+n-1} + \cdots + s_n a_h \,,\; h = 0,1,2,\ldots. \tag{3}$$

To see this let $E : f(h) \mapsto f(h+1)$ be the shift operator and $\Delta = E - 1$ the difference operator. Then

$$(E - \alpha)\big(A_i(h)\alpha_i^h\big) = (\Delta A_i(h))\alpha_i^{h+1}$$

and since $\Delta A_i(h)$ has lower degree than does A_i, by linearity of E and induction it is plain that

$$\prod_{i=1}^{m}(E - \alpha_i)^{n_i}$$

annihilates the sequence (a_h) as asserted. Thus generalised power sums are interesting in that they coincide with the sequences satisfying the recurrence relations (3). It follows that there is a polynomial $r(x)$, of degree less than n, so that the power series

$$\sum_{h=0}^{\infty} a_h X^h = \frac{r(X)}{s(X)} \tag{4}$$

is a rational function; to see this multiply by $s(X)$ and note the recurrence relation.

Conversely given a rational function as above, with $\deg r < \deg s$, a partial fraction expansion yields

$$\frac{r(X)}{s(X)} = \sum_{i=1}^{m}\sum_{j=1}^{n_i} \frac{r_{ij}}{(1-\alpha_i X)^j} = \sum_{h=0}^{\infty}\left(\sum_{i=1}^{m}\sum_{j=1}^{n_i} r_{ij}\binom{h+j-1}{j-1}\alpha_i^h\right) X^h$$

and the coefficients of X^h, $h = 0, 1, 2, \ldots\ldots$ are indeed the values of a generalised power sum as described.

Accordingly, results on generalised power sums are equivalent to corresponding results for the Taylor coefficients of rational functions.

A sequence (a_h) satisfying a relation (3) is often called a *recurrence sequence* (or *linearly recursive sequence*) of *order* n; the polynomial $X^n s(X^{-1})$ reciprocal to the polynomial (2) is called the *characteristic* or *companion polynomial* of the recurrence sequence. Our "roots" α_i are the distinct zeros of the companion polynomial. The archetypal example of a recurrence sequence is of course the celebrated Fibonacci sequence (f_h) defined by

$$f_{h+2} = f_{h+1} + f_h\,,\ h = 0, 1, 2, \ldots \text{ with } f_0 = 0\,, f_1 = 1\,;$$

and generated by

$$\frac{X}{1 - X - X^2} = \sum_{h=0}^{\infty} f_h X^h\,.$$

The expression (1) for the $a_h = a(h)$ as a generalised power sum provides a well known formula for the terms of the recurrence sequence. One obtains a less well known formula from directly expanding (4). In terms of the given *initial values* $a_0, a_1, \ldots, a_{n-1}$ of (a_h) one has

$$r(X) = \sum_{j=0}^{n-1}\left(a_j - \sum_{i=1}^{j} s_i a_{j-i}\right) X^j\,,$$

and

$$s(X)^{-1} = \sum_{h=0}^{\infty}\ \sum_{j_1+2j_2\cdots+nj_n=h} \frac{(j_1+j_2\cdots+j_n)!}{j_1!\ldots j_n!} s_1^{j_1}\cdots s_n^{j_n} X^h\,.$$

For the Fibonacci numbers this yields (with the usual conventions for interpreting the combinatorial symbol)

$$f_{h+1} = \sum_j \binom{h-j}{j}\,.$$

2. Exponential Polynomials: Complex and p-Adic

The field $\mathbf{F}$ of definition of a given generalised power sum is finitely generated over the field of rationals $\mathbf{Q}$ and, indeed, the a_h all belong to a subring R of $\mathbf{F}$ finitely generated (of finite type) over $\mathbf{Z}$. This says, exactly: There is a finite number, say t, of algebraically independent transcendentals $x = (x_1, \ldots, x_t)$ and a y algebraic over $\mathbf{Q}(x)$ so that $\mathbf{F} = \mathbf{Q}(x)[y]$. Further, for $j = 1, 2, \ldots, g$, say, there are polynomials $U_j \in \mathbf{Z}[y; x]$ and $V_j \in \mathbf{Z}[x]$ so that R is the ring $\mathbf{Z}[U_1(y;x)/V_1(x), \ldots, U_g(y;x)/V_g(x)]$. In the case $t = 0$ we have $\mathbf{F} = \mathbf{K}$, an algebraic number field, and R a subring (usually referred to as a ring of S-integers) in that field.

Plainly, we may view $\mathbf{F}$ as a subfield of $\mathbf{C}$. Then a generalised power sum is the restriction to the nonnegative integers of an *exponential polynomial*

$$a(z) = \sum_{i=1}^{m} A_i(z) e^{z \log \alpha_i} \, , z \in \mathbf{C} \, .$$

Note, however, that the continuation is not well defined because we are free to choose the branches of the $\log \alpha_i$.

It turns out that there are infinitely many primes (indeed, a set of positive density) so that a given generalised power sum can be suitably embedded in the field of p-adic rationals $\mathbf{Q}_p$ and analytically continued to exponential polynomials on $\mathbf{C}_p$, the algebraic closure of the completion of $\mathbf{Q}_p$. Cassels [3] provides an elegant description. There are two steps in the embedding process, the first of which provides a notion of *specialisation* of a generalised power sum which is used in lifting results from the number field to the general case.

It is straightforward to see that each element ϕ in a field $\mathbf{F} = \mathbf{Q}(x)[y]$, containing the terms

$$a(h) = \sum_{i=1}^{m} A_i(h) \alpha_i^h \, , \; h = 0, 1, 2, \ldots\ldots$$

of the generalised power sum a , has a representation

$$\phi = U_\phi(y; x)/V_\phi(x) \, ,$$

with $U_\phi \in \mathbf{Z}[y; x]$ and $V_\phi \in \mathbf{Z}[x]$, say relatively prime to the set of coefficients of U_ϕ and with its set of coefficients relatively prime over $\mathbf{Z}$. We may then refer to $V_\phi \in \mathbf{Z}[x]$ as *the* denominator of ϕ. Denote the defining polynomial of y over $\mathbf{Z}[x]$ by $F[x](Y)$, and suppose that it is of degree r .

Cassels' idea is to introduce a finite set Γ of elements of $\mathbf{F}$ with the property that whenever $\gamma \in \Gamma$ and $\gamma \neq 0$ then also $\gamma^{-1} \in \Gamma$. It will be convenient to always require that Γ contains the discriminant and leading and trailing coefficients of $F[x](Y)$. Set

$$V_\Gamma(x) = \prod_{\gamma\in\Gamma} V_\gamma(x).$$

It follows by induction on t that there are infinitely many t-tuples of rational integers $c = (c_1, \ldots, c_t)$ so that $V_\Gamma(c) \neq 0$. Whenever $V_\Gamma(c) \neq 0$, we refer to a map $x \mapsto c$, together with an induced map $y = y(x) \mapsto y(c)$ with $y(c)$ some zero of $F[c](Y)$, as a Γ-specialisation of $\mathbf{F}$. (This is an abuse of language; we specialise only the elements of a subring of $\mathbf{F}$.)

I allege that if $\gamma = \gamma(y(x); x) \in \Gamma$, its Γ-specialisation $\gamma(y(c); c)$ is an element of an algebraic number field $\mathbf{K} = \mathbf{Q}(c)[y(c)]$ of degree at most r over $\mathbf{Q}$. But this is clear. Trivially, $\mathbf{Q}(c) = \mathbf{Q}$ and $y(c)$ is a zero of a polynomial $F[c](Y)$ of degree r over $\mathbf{Q}$. Moreover, if $\gamma \neq 0$, the specialisation of γ is nonzero. For, by the condition on the sets Γ, the element γ^{-1} also belongs to Γ and thus also has an image in $\mathbf{K}$ under the specialisation.

I now turn to the second step of the *p-adification* process.

One notes that, having selected a Γ-specialisation $x \mapsto c$, there are infinitely many rational primes p so that both $V_\Gamma(c) \not\equiv 0 \bmod p$ and so that the reduction of the irreducible factor of the polynomial $F[c](Y)$ with $y(c)$ a zero, viewed as a polynomial over $\mathbf{F}_p$, has a linear factor $Y - \overline{y(c)}$. The first condition excludes just finitely many primes and the second condition is satisfied by all those primes p with a prime ideal factor of degree 1 in the number field $\mathbf{K} = \mathbf{Q}(c)[y(c)]$. By the Tchebotarev density theorem one is left with a set of *admissible* primes of positive density in the set of all primes.

Now select t algebraically independent elements $\xi = (\xi_1, \ldots, \xi_t)$ of $\mathbf{Q}_p$ subject to $\xi_i \equiv c_i \bmod p$, $i = 1, \ldots, t$, as one may since $\mathbf{Q}_p$ has uncountable transcendence degree over $\mathbf{Q}$. Then, by Hensel's lemma, there is an element η of $\mathbf{Q}_p$ with $\eta \equiv \overline{y(c)} \bmod p$ and $F[\xi](\eta) = 0$ in $\mathbf{Q}_p$. By the remarks above, the map $(y; x) \mapsto (\eta; \xi)$ yields an embedding of $\mathbf{F}$ into $\mathbf{Q}_p$ under which nonzero elements of Γ become units in $\mathbf{Q}_p$. One obtains such an embedding for each p admissible with respect to the selected Γ-specialisation and the given polynomial F.

Given a generalised power sum

$$a(h) = \sum_{i=1}^{m} A_i(h)\alpha_i^h, \quad h = 0, 1, 2, \ldots\ldots$$

select Γ so that it contains the roots α_i. Then there are admissible p for which one obtains an embedding of the generalised power sum into $\mathbf{Q}_p$ so that the α_i become units in $\mathbf{Q}_p$. (It is convenient, and is a harmless abuse of notation, to fail to indicate that elements once in $\mathbf{F}$ are now in $\mathbf{Q}_p$).

Thus for each i we have $\alpha_i^{p-1} \equiv 1 \bmod p$, whence the p-adic logarithms

$$\log_p \alpha_i^{p-1} = \log_p \left(1-(1-\alpha_i^{p-1})\right)$$

are defined, and satisfy $\operatorname{ord}_p(\log_p \alpha_i^{p-1}) \geq 1$. Finally, we recall that the p-adic exponential $\exp_p t$ converges for $t \in \mathbf{C}_p$ with $\operatorname{ord}_p t > 1/(p-1)$. Since p is fixed in the course of any paragraph, below we may omit the subscripts p.

With all this, one obtains p-adic analytic functions

$$a_{p,r}(t) = \sum_{i=1}^{m} A_i(r+(p-1)t)\alpha_i^r \exp\left(t \log \alpha_i^{p-1}\right) \quad , \ r = 0,1,\ldots,p-2 \,,$$

converging for $t \in \mathbf{C}_p$ with $\operatorname{ord} t > -1 + 1/(p-1)$ and analytically continuing the given generalised power sum in the sense that $a_{p,r}(h) = a(r+(p-1)h)$ for $0 \leq r < p-1$ and $h = 0,1,2,\ldots\ldots.$

4. A Criterion for Rationality

We have seen that $\sum a_h X^h$ represents a rational function vanishing at ∞ if and only if, for some n, there is a recurrence relation $a_{h+n} = s_1 a_{h+n-1} + \cdots + s_n a_h$ for all $h = 0,1,\ldots$. Thus, necessarily, the Kronecker-Hankel determinants

$$K_N(a) = \begin{vmatrix} a_0 & a_1 & \cdots & a_N \\ a_1 & a_2 & \cdots & a_{N+1} \\ \vdots & \vdots & \ddots & \vdots \\ a_N & a_{N+1} & \cdots & a_{2N} \end{vmatrix}$$

vanish for $N = n, n+1, \ldots$. This condition is, not altogether obviously, also sufficient. To see that, suppose $K_{n-1}(a) \neq 0$ but $K_n(a) = 0$. On setting $b_h = a_{h+n} - s_1 a_{h+n-1} - \cdots - s_n a_h$, with certain constants $s_1, \ldots, s_n$, we therefore have $b_h = 0$ for $h = 0,1,\ldots,n$. But

$$K_{n+1}(a) = \begin{vmatrix} a_0 & a_1 & \cdots & a_{n-1} & 0 & 0 \\ a_1 & a_2 & \cdots & a_n & 0 & 0 \\ \vdots & \vdots & \ddots & \vdots & \vdots & \vdots \\ a_{n-1} & a_n & \cdots & a_{2n-2} & 0 & 0 \\ 0 & 0 & \cdots & 0 & 0 & b_{n+1} \\ 0 & 0 & \cdots & 0 & b_{n+1} & b_{n+2} \end{vmatrix} = -b_{n+1}^2 K_{n-1}(a) \,.$$

Thus $K_{n+1}(a) = 0$ implies $b_{n+1} = 0$, and, by induction, $K_N(a) = 0$ for $N = n, n+1, \ldots$ entails $b_h = 0$ for $h = 0,1,\ldots$.

As suggested by Bombieri [1], it turns out to be appropriate to define the height $\mathrm{H}(a)$ of a sequence (a_h) of elements a_h of a number field $\mathbf{K}$ by

$$\log \mathrm{H}(a) = ([\mathbf{K} : \mathbf{Q}])^{-1} \limsup_{h\to\infty} h^{-1} \sum_v \max_{0\le i\le h} \log |a_i|_v ,$$

with the sum over all appropriately normalised values of $\mathbf{K}$. The definition implies that the height of a sequence is invariant under multiplication by a nonzero element of $\mathbf{K}$; by the product formula, the sequences (a_h) and (ca_h) have the same height. Our purpose is to attach a height to the sequence of coefficients of a power series $\sum a_h X^h \in \mathbf{K}[[X]]$. This is achieved, felicitously, by the given definition. Plainly, the invariance under multiplication by nonzero algebraic constants is desirable. Moreover, the nonarchimedean values progressively pick up the lowest common multiple of the denominators of $a_0, \ldots, a_h$, so that the height is a suitable arithmetic measure of the growth of the sequence. The geometric progression $(1, \alpha, \alpha^2, \ldots\ldots)$ has height $\mathrm{H}(\alpha)$; the harmonic sequence $(1, 1/2, 1/3, \ldots\ldots)$ has height e.

Suppose that (a_h) is a sequence of elements of a number field $\mathbf{K}$ and has height $\mathrm{H}(a) = A < \infty$. It will be convenient to define the height $\mathrm{H}(K(a))$ of the sequence of Kronecker-Hankel determinants by

$$\log \mathrm{H}(K(a)) = ([\mathbf{K} : \mathbf{Q}])^{-1} \limsup_{h\to\infty} h^{-2} \sum_v \max_{0\le N\le h} \log |K_N(a)|_v ;$$

note that this is not the same definition of height of a sequence where one multiplies just by h^{-1}. Then, but (*cf* [4]) this is not as obvious as may seem at first, $\mathrm{H}(K(a)) = \mathrm{H}(a) = A$.

Let Δ be the forward difference operator, here acting on the subscript of a. Manipulation of rows and columns in the Kronecker-Hankel determinant shows that

$$K_N(a) = \left|\Delta^{i+j} a_0\right|_{0\le i,j<N} .$$

Let $x(t) = \sum x_h t^h$ be a power series with coefficients in $\mathbf{C}_p$ and converging on the disc $\{t \in \mathbf{C}_p : \operatorname{ord} t > -c + 1/(p-1)\}$, some $c > 1/(p-1)$. Then $\limsup_{h\to\infty} h^{-1} \operatorname{ord} x_h = c - 1/(p-1)$. One has

$$\frac{\Delta^k x(0)}{k!} = \sum_{h=k}^{\infty} x_h S(h,k),$$

where the integers $S(h,k)$ are the Stirling numbers of the second kind. It follows, recalling $\operatorname{ord} k! = (k - \sigma(k))/(p-1)$, where $\sigma(k)$ is the sum of the p-adic digits of k, that

$$\limsup_{k\to\infty} k^{-1} \operatorname{ord} \Delta^k x(0) \le \lim_{k\to\infty} k^{-1} \operatorname{ord} k! + \limsup_{h\to\infty} h^{-1} \operatorname{ord} x_h = c .$$

On the other hand,

$$x_h = \sum_{k=h}^{\infty} \frac{\Delta^k x(0)}{k!} s(h,k),$$

where the integers $s(h,k)$ are the Stirling numbers of the first kind. This yields

$$c = \lim_{k\to\infty} k^{-1}\operatorname{ord} k! + \limsup_{h\to\infty} h^{-1}\operatorname{ord} x_h \le \limsup_{k\to\infty} k^{-1}\operatorname{ord}\Delta^k x(0);$$

and we have proved

$$\limsup_{k\to\infty} k^{-1}\operatorname{ord}\Delta^k x(0) = c.$$

But we saw that recurrence sequences yield maps $h \mapsto a(r+(p-1)h)$ that can be analytically continued to maps on the disc $\{t \in \mathbf{C}_p : \operatorname{ord} t > -1 + 1/(p-1)\}$. That is the context in which the following criterion is useful:

Theorem (A criterion for rationality). *Let (a_h) be a sequence of elements of a number field $\mathbf{K}$ with finite height $\mathrm{H}(a) = A$. Suppose there is a set $\mathcal{P}$ of rational primes p for which the $p-1$ maps, with $0 \le r < p-1$,*

$$h \mapsto a_{r+(p-1)h}$$

may be analytically continued to maps on the disc $\{t \in \mathbf{C}_p : \operatorname{ord} t > -1 + 1/(p-1)\}$. If

$$\prod_{p\in\mathcal{P}} p^{1/(p-1)} > A^{[\mathbf{K}:\mathbf{Q}]}$$

then $\sum a_h X^h$ is a rational function.

Proof: If Δ_{p-1} is the difference operator $\Delta_{p-1} : f(h) \mapsto f(h+p-1) - f(h)$, then a generalision of a remark above yields

$$K_N(a) = \left|\Delta_{p-1}^{\lfloor i/(p-1)\rfloor + \lfloor j/(p-1)\rfloor} a_{r+s}\right|_{\substack{r\equiv i,\, s\equiv j \pmod{p-1} \\ 0\le r,s<p-1}}$$

and we can verify that the data of the criterion implies: for $p \in \mathcal{P}$,

$$\liminf_{N\to\infty} N^{-2}\operatorname{ord}_p K_N(a) \ge \frac{1}{p-1}.$$

Now let T be the set of all places above the primes $p \in \mathcal{P}$. Then, by Liouville's theorem, namely,

For any subset T of the places v of $\mathbf{K}$, either $\alpha = 0$ or

$$\sum_v \log|\alpha|_v = 0 \text{ implies}$$

$$\sum_{v\in T} \log|\alpha|_v = -\sum_{v\notin T} \log|\alpha|_v$$
$$\geq -\sum_{v\notin T} \log^+|\alpha|_v \geq -\sum_v \log^+|\alpha|_v = -[\mathbf{K}:\mathbf{Q}]\log \mathrm{H}(\alpha),$$

we have, for all sufficiently large N, either $K_N(a) = 0$ or

$$-N^{-2}\sum_{v\in T}\log|K_N(a)|_v \lesssim \sum_{p\in\mathcal{P}} \frac{1}{p-1}\log p$$
$$< [\mathbf{K}:\mathbf{Q}]N^{-2}\log \mathrm{H}(K_N(a)) \lesssim \log A.$$

This is just a restatement of the criterion and establishes its validity.

5. Algebraic Functions of Exponential Polynomials

Let $a^{(1)}(z)$, $a^{(2)}(z)$, $\ldots$, $a^{(r)}(z)$ be exponential polynomials. It is a theorem of Ritt [6] that *an entire function $Y(z)$ satisfying an equation*

$$F(Y;z) = Y^r + a^{(1)}(z)Y^{r-1} + \cdots\cdots + a^{(r-1)}(z)Y + a^{(r)}(z) = 0$$

is an exponential polynomial.

Ritt's result and argument is more general but its principles are contained in the following sketch: Suppose firstly that the frequencies ω_{ij} of the exponential polynomials comprising the data

$$a^{(i)}(z) = \sum_j A_{ij}(z)\exp(z\omega_{ij}),$$

for $i = 1, 2, \ldots, r$ all are real. Then $Y(z)$ has a series expansion

$$Y(z) = \sum t_h(z)\exp(z\tau_h)$$

where, for some integer $d \leq r$, $(d\tau_h)$ is a monotonic increasing sequence of $\mathbf{Z}$-linear combinations of the given frequencies and each t_h is an algebraic function in z. More generally, if the given frequencies are arbitrary then the $d\tau_h$ may be taken to be $\mathbf{Z}$-linear combinations of the real parts of the given frequencies and each t_h expands to a series

$$t_h(z) = \sum s_{hk}(z)\exp(iz\sigma_{hk})$$

where, for some integer $d_h \le r$, $(d_h\sigma_{hk})$ is a monotonic increasing sequence of **Z**-linear combinations of the imaginary parts of the given frequencies and each s_{hk} is an algebraic function in z. The proof of the allegations comprising this sketch is little different from the more familiar argument that yields the Puiseux expansion of an algebraic function in one variable. The point in ordering the frequencies as described is that Ritt can show that his expansion converges in some sector.

Ritt proves that a series

$$Y(z) = \sum\sum s_{hk}(z)\exp(iz\sigma_{hk})\exp(z\tau_h),$$

with frequencies as described, is meromorphic if and only if it is a quotient of exponential polynomials. The assertion commencing this section then follows from Ritt's Quotient Theorem [7], whereby an entire quotient of exponential polynomials is itself, up to possible division by a polynomial, an exponential polynomial.

6. Hadamard Roots of Rational Functions

It is obvious to undergraduates[1] that if $\sum a_hX^h$ and $\sum b_hX^h$ are rational functions then so is $\sum a_hb_hX^h$. It follows that if $\sum a_hX^h$ is rational then $\sum f(a_h)X^h$ is rational for every polynomial f.

My remarks concern possible converse results. If $\sum f(a_h)X^h$ is rational, where f is a polynomial, what can be said of $\sum a_hX^h$?

Actually, that can be readily answered: Without more data, very little indeed can be said. The correct question is, therefore, whether there are some simple but evidently necessary additional conditions which suffice to entail the rationality of $\sum a_hX^h$.

But, in order that $\sum a_hX^h$ possibly be rational it is certainly necessary that the a_h all lie in some finitely generated extension field of the field of rationals **Q**. If $\sum a_hX^h$ is rational then a_h is given by a generalised power sum $a(h)$ and its values evidently have the cited property. Moreover, we should recall that the a_h are presented as roots of equations $b_h = f(a_h)$. Thus, at best, we may make allegations about *some* sequence of roots rather than about a *given* sequence of roots. In all, these considerations lead to the following conjecture:

Conjecture. *Let f be a polynomial and $\sum f(a'_h)X^h$ a power series representing a rational function. If the a'_h all belong to a field finitely generated*

---

[1] It is obvious to us because the product of generalised power sums is of course once again a generalised power sum. Most undergraduates will, unfortunately, simply report that the product of rational functions is once again a rational function, believing the Hadamard product to be the ordinary product.

over $\mathbf{Q}$ *then there is a sequence* (a_h) *with* $f(a_h) = f(a'_h)$ *for all* $h = 0, 1, 2, \ldots$ *so that* $\sum a_h X^h$ *is a rational function.*

This conjecture is not just wishful thinking. Rumely and I [9] have established it in a special, but arguably a 'generic' case: Namely, we deal with $f(a_h) = a_h^k = b_h$ subject to the roots (which we may have reordered without loss of generality) $\beta_1, \ldots, \beta_m$ of the given recurrence sequence (b_h) satisfying

$$|\beta_1|_v > |\beta_2|_v \geq \cdots\cdots \geq |\beta_m|_v \,.$$

That is, at some absolute value v, (b_h) has a *unique* maximal root. It turns out that, for the special problem cited, we can also deal with a unique minimal root. Graham Everest has explained to me that under the dominant root condition the general problem should be similarly accessible. Given this, it seems that the conjecture can be proved for 'almost all' given recurrence sequences (b_h). A brief history of these matters is summarised by:

In work of Pisot it's been shown
That Hadamard's secret is blown;
The remarks of Cantor
Are quite without flaw,
And should have become better known.

In an analogous context Cantor[2][2] had shown that problems of the present kind are accessible in the dominant root case. Pisot had dealt with the Hadamard k th root problem, showing that $\sum b_h X^h = \sum a_h X^h$ is rational if $\sum a_h^k X^h$ is rational, subject to the $a_h \in \mathbf{Z}$ and the dominant root β_1 having multiplicity one, that is, with the leading coefficient B_1 constant.

The conjecture is a special case of the arithmetic analogue of Ritt's result of §5. Since I cannot settle the conjecture it seems wiser[3] to study the yet more general problem which is the arithmetic analogue of the theorem of Ritt of the previous section.

Generalised Conjecture. *Let* $b^{(1)}(h), b^{(2)}(h), \ldots, b^{(r)}(h)$ *be generalised power sums and let* $F(Y;h)$ *be a polynomial in* Y

$$F(Y;h) = Y^r + b^{(1)}(h)Y^{r-1} + \cdots\cdots + b^{(r-1)}(h)Y + b^{(r)}(h)$$

with generalised power sum coefficients. If there is a sequence (a'_h) *with all its elements belonging to a field finitely generated over* $\mathbf{Q}$ *so that* $F(a'_h; h) =$

[2] At the meeting, Knopp chose to criticise me by commenting that he saw a defect in my attempt to rhyme 'flaw' and 'Cantor'. I sadly replied (in my best Southern accent): "Oh! Marvin ... ".

[3] Following the principle that one's wisdom increases according to the amount one knows one doesn't know.

0 for $h = 0, 1, 2, \ldots$ then there is a sequence (a_h) with $F(a_h; h) = 0$ for $h = 0, 1, 2, \ldots$ so that $\sum a_h X^h$ is a rational function.

I should mention that we now know the arithmetic analogue of Ritt's Quotient Theorem. In [10] I show that if both $\sum c_h X^h$ and $\sum b_h X^h$ are rational and (a'_h) is a sequence of elements all belonging to a ring finitely generated (of finite type) over $\mathbf{Z}$ and satisfying $a'_h b_h = c_h$ for $h = 0, 1, 2, \ldots$ then there is a sequence (a_h) with $a_h b_h = c_h$ for $h = 0, 1, 2, \ldots$ so that $\sum a_h X^h$ is a rational function. Thus if exponential polynomials $c(z)$ and $b(z)$ have, for example, the property that the quotients $c(h)/b(h)$, $h = 0, 1, 2, \ldots$ all are integers, then, for some analytic continuation to $\mathbf{C}$, the quotient $c(z)/b(z)$ is an exponential polynomial.

7. The Divergent Argument

In the proof of the Hadamard Quotient Theorem I p-adify the data as described at §3 and use the rationality criterion detailed in §4. However, attempts to apply similar techniques to the Hadamard root problem seem to fail. There are a number of difficulties. But the most striking problem is that there does not seem to be a method, other than in a dominant root situation, to specify that one is studying a *coherent*[4] sequence $a(h)$, $h = 0, 1, 2, \ldots$ of roots. For example, $\sum X^h$ has itself as a rational Hadamard square root. But whilst all the series $\sum \pm X^h$ are Hadamard square roots of $\sum X^h$, almost all of them are are not even continuable beyond the unit circle. The exceptions are rational and are those Hadamard square roots for which the signs are chosen 'consistently'; in the present example, so that the sequence of signs is periodic. There is no analogous coherency difficulty in the quotient problem.

My divergent argument is directed at attaining coherency. Given the data $F(Y; h) = 0$, $h = 0, 1, 2, \ldots$ observe that no generality is lost in supposing that the given polynomial $F(Y; h)$ is irreducible in the ring of polynomials in Y over the ring of generalised power sums. Now obtain a series expansion

$$Y(h) = \sum q_i(h)\theta_i^h$$

in a manner similar to that sketched at §5. One can always arrange that the sequence θ_i of roots be monotonic decreasing in absolute value. Indeed, and this seems the appropriate approach, one can arrange that the sequence is monotonic decreasing in g-adic pseudo-value (see Mahler [5] for the appropriate notions), where g is some product of the prime ideals dividing the roots appearing in the data. In the dominant root case the sequence will be strictly monotonic decreasing in absolute value and the series actually converges for all sufficiently large integers h.

[4] The relevance of coherence was pointed out to me by the late Philippe Robba.

If the generalised power sums of the data all have constant coefficients (all their roots have multiplicity one) then the q_i all are constants. It will be convenient to restrict ourselves to that case for the remaining remarks. As matters are phrased here, in general the $q_i(h)$ must be supposed algebraic functions of h. However, Rumely and I [9] show, unconditionally for the k th root problem, that one may indeed assume, without loss of generality, that the coefficients of the data are constant. Equivalently, we prove that the coefficients appearing in our 'Puiseux expansion' all are rational functions in h.

I now define the purportedly coherent sequence (a_h) by the formal expressions $Y(h) = a_h$ and as in §4, form the Kronecker-Hankel determinants

$$K_N(a) = \begin{vmatrix} a_0 & a_1 & \cdots & a_N \\ a_1 & a_2 & \cdots & a_{N+1} \\ \vdots & \vdots & \ddots & \vdots \\ a_N & a_{N+1} & \cdots & a_{2N} \end{vmatrix}$$

$$= \left| \Delta_{p-1}^{\lfloor i/(p-1)\rfloor + \lfloor j/(p-1)\rfloor} a_{r+s} \right|_{\substack{r\equiv i,\, s\equiv j \pmod{p-1} \\ 0\le r,s<p-1}} .$$

Since the θ_i are monomials in the roots of the data there are plenty of primes p so that, formally, *each term* of the series giving an entry in the determinant on the right is highly divisible by p. That is no more than iteration of the congruence

$$\theta^{p-1} \equiv 1 \pmod{p}.$$

I then claim that the entry itself, that is, the formal infinite sum, has high p-adic order.

It is not clear to me that this argument can be sustained. It entails, regardless of the arithmetic preconditions, that there always are infinitely many primes p for which an algebraic function of generalised power sums has p-adic analytic continuations converging p-adically in 'large' discs (to be precise, for which $\log_p |t|_p < 1 - 1/(p-1)$). I have convinced myself that this is not necessarily false, but not that it is in fact so. In any event, it is an independent question that seems worthy of study and that may be more accessible than the principal conjecture.

My divergent argument proceeds by noticing that, for given h, zeros a_h of $F(Y(h); h)$ are certainly bounded by the data. If the a_h lie in a number field this yields the vanishing of the Kronecker-Hankel determinants once N is sufficiently large. The general result would then follow by lifting the result from the number field case. A detailed description of such a lifting argument is provided by Rumely [8].

The argument appropriate to the dominant root case is rather different; see [9] for a concluding argument likely to be appropriate. There is no need to manipulate divergent series nor to attempt to obtain arithmetic properties of sums of series from arithmetic properties of their terms. As suggested, I believe that the problems that may arise from nonconstant coefficients $q_i(h)$ should be tractable without the introduction of radically new ideas.

8. Concluding Remarks

My purpose in sketching an argument which is, as it stands, unsustainable is, of course, to take advantage of an opportunity to listen carefully to an exposition of that argument. I have listened but have not learnt as much as I hoped I would. Most of all, I have not succeeded in my real goal, which was to provide a valid proof of the conjectures (a convergent argument, so to speak) in this, the written version of my talk.

However, I have had an opportunity to state the relevant conjectures and to give a flavour of the methods and principles that appear necessary to deal with problems of the present sort. Remarking on a 40 minute talk of mine at Oberwolfach some years ago (on roughly the present subject) the poet Martin Huxley wrote:

Matters of specialisation
And Hadamard multiplication
Were dealt with by Alf
In an hour and a half
By clever p–adification.

I have probably been excessively verbose again. But I wrote:

For Bateman, Whom we've all known for ages
First chairman, then one of the sages.
Without undue urgence
Or thought for convergence
I offer these miserable pages.

References

[1] E. Bombieri, 'On G-functions', in H. Halberstam and C. Hooley, eds., *Recent progress in analytic number theory*, Academic Press (1981), Chapter 24, Vol. 2, 1–67

[2] David G. Cantor, 'On arithmetic properties of the Taylor series of rational functions II', *Pacific J. Math.* **41** (1972), 329–334

[3] J. W. S. Cassels, 'An embedding theorem for fields', *Bull. Austral. Math. Soc.* **14** (1976) 193–198; Addendum: *ibid* **14** (1976) 479–480

[4] V. Laohakosol, J. H. Loxton and A. J. van der Poorten, 'Integer p-adic functions', *Macquarie Math. Reports*, **87-0008** (June, 1987) = *Coll. Math. Soc. János Bolyai*(Budapest, 1987)

[5] Kurt Mahler, *p-Adic Numbers and Their Functions*, (2nd edition of "Introduction to p-adic Numbers and Their Functions"), Cambridge Tracts in Mathematics **76**, (Cambridge University Press, Cambridge New York, 1981), 320pp

[6] J. F. Ritt, 'Algebraic combinations of exponentials', *Trans. Amer. Math. Soc.* **31** (1929), 654–679

[7] J. F. Ritt, 'On the zeros of exponential polynomials', *Trans. Amer. Math. Soc.*, **31** (1929), 680–686

[8] Robert S. Rumely, 'Notes on van der Poorten's proof of the Hadamard Quotient Theorem: Part II', in Catherine Goldstein ed., *Sém. Théorie des Nombres* Paris 1986-87, Progress in Mathematics **75** (Birkhäuser, Boston Basel, 1989), 383–409

[9] Robert S. Rumely and A. J. van der Poorten, 'A note on the Hadamard kth root of a rational function', *J. Austral. Math. Soc. (Ser. A)* **43** (1987), 314–327

[10] Alfred J. van der Poorten, 'Solution de la conjecture de Pisot sur le quotient de Hadamard de deux fractions rationnelles', *C. R. Acad. Sc. Paris* **306** Série 1, (1988), 97–102

Alf van der Poorten
School of Mathematics, Physics, Computing and Electronics
Macquarie University, NSW 2109
Australia
alf@mqcomp.mqcs.mq.oz

Diagonalizing Eisenstein Series. I

ROBERT A. RANKIN

Dedicated to Paul Bateman on his 70th birthday

1. Introduction

In this paper we consider the action of Hecke operators T_n $(n \in \mathbb{N})$, and their adjoint operators T_n^*, on Eisenstein series belonging to the group $\Gamma_0(N)$, and having integral weight $k > 2$ and arbitrary character χ modulo N. It is shown that the space $\mathcal{E}_k(\chi)$ spanned by these Eisenstein series splits up into a number of subspaces $\mathcal{E}_k(\chi, t)$, where t is a divisor of N, each being invariant under the operators T_n and T_n^* with $(n, N) = 1$. If χ is a primitive character modulo N, this holds also for T_n with $(n, N) > 1$, but this need not be true for general χ modulo N. A basis of modular forms that are eigenfunctions for T_n with $(n, N) = 1$ is constructed for each appropriate t and explicit evaluations of $G_L|T_n$ are given for each Eisenstein series G_L $(L \in \Gamma(1))$ and any positive integer n prime to N, or any n that is a prime divisor of N, the results being particularly simple when N is squarefree. The corresponding results for $G_L|T_n^*$ when $(n, N) > 1$ will be given in a subsequent paper.

Although over the last 70 years much attention has been devoted to the diagonalization of cusp forms, by authors such as Mordell, Hecke, Petersson, Atkin, Lehner, Li and others, very little work has been done on the corresponding problem for Eisenstein series, possibly because a finite inner product of pairs of such series cannot be defined in all cases. However, the problem was considered by Hecke in 1937 and further results were obtained by Petersson in 1948; see, in particular, Satz 44 of [2] and Satz 8 of [3].

The starting point of Hecke's work was his explicit determination of the Fourier coefficients of Eisenstein series. By taking linear combinations of the coefficients with characters χ_1 and χ_2 to the moduli N/t_1 and N/t_2, he obtained modular forms whose associated Dirichlet series had Euler factors for each prime $p \nmid N$, namely those Euler products occurring in the product of L-functions

$$L(s - k + 1, \chi_1)L(s, \chi_2); \tag{1.1}$$

see §5 below. However, the associated modular forms are not necessarily linearly independent, no doubt because the precise conditions to be satisfied by the characters χ_1 and χ_2 and divisors t_1 and t_2 are not specified. Hecke was aware that his results did not in general extend to T_n with $(n, N) > 1$, since he illustrated this fact in a particular case when $N = q^3$ (q prime); see Satz 45a of [2]. In the present paper I employ a different method to make explicit the relationship between Hecke's characters χ_1, χ_2 and the character χ and divisor t mentioned in the first paragraph of this section.

It should be mentioned that the *divisor* t of N occurring in $\mathcal{E}_k(\chi, t)$ has no connection with Hecke's use of the word, which relates to modular forms belonging to the subgroup $\Gamma_0(N) \cap \Gamma^0(N/t)$. In what follows we confine our attention to $\Gamma_0(N)$ (*i.e.* we take Hecke's divisor t to be N), but there is no doubt that the analysis could be extended to the general case without difficulty. Moreover, as stated above, we shall assume that $k > 2$; the case $k = 2$ could be included by use of the well known Hecke limiting process.

2. Notation

We write

$$I = \begin{pmatrix} 1 & 0 \\ 0 & 1 \end{pmatrix}, \quad U = \begin{pmatrix} 1 & 1 \\ 1 & 0 \end{pmatrix}, \quad W = \begin{pmatrix} 1 & 0 \\ 1 & 1 \end{pmatrix} \tag{2.1}$$

and

$$S = \begin{pmatrix} \alpha & \beta \\ \gamma & \delta \end{pmatrix}, \qquad L = \begin{pmatrix} A & B \\ C & D \end{pmatrix}, \tag{2.1}$$

where both S and L belong to $\mathrm{SL}(2, \mathbb{Z}) = : \Gamma(1)$.

Throughout both N and k will be fixed positive integers and we write

$$\Gamma = \Gamma_0(N) = \{S \in \Gamma(1) : \ \gamma \equiv 0 \pmod N\} \tag{2.3}$$

and assume that $k > 2$.

For any character χ modulo N we write $\mathcal{M}(\chi)$ and $\mathcal{C}(\chi)$, respectively, for the vector spaces of entire modular forms and cusp forms of weight k and character χ belonging to Γ. Moreover, if

$$T = \begin{pmatrix} a & b \\ c & d \end{pmatrix}, \tag{2.4}$$

is any matrix with real entries and positive determinant $\det T$, we define

$$f(z)|T = (cz+d)^{-k}(\det T)^{k/2} f\{(az+b)/(cz+d)\}, \tag{2.5}$$

for any $f \in \mathcal{M}(\chi)$.

We shall be much concerned with (parabolic) cusps, which we write in the form $L^{-1}\infty$, where $L \in \Gamma(1)$, as above. It is easily seen that the cusps $L_1^{-1}\infty$ and $L_2^{-1}\infty$ are congruent modulo Γ if and only if

$$S := L_1^{-1} U^r L_2 \in \Gamma \quad \text{for some} \quad r \in \mathbb{Z}. \tag{2.6}$$

We write (2.6) as $L_1 \sim L_2$. The corresponding equivalence class $[L]$ we call a *cusp class.*

The *width* n_L of the cusp $L^{-1}\infty$ is defined to be the smallest positive integer n such that $L^{-1}U^n L \in \Gamma$. It depends on $[L]$ and is easily evaluated as

$$n_L = t_2/(t_1, t_2), \tag{2.7}$$

where

$$t_1 = t_1(L) = (N, C) \qquad \text{and} \qquad N = t_1 t_2. \tag{2.8}$$

We call t_1 the *divisor* of L and write $t_1 = \operatorname{div} L$. It is convenient to write

$$C_1 = C/t_1. \tag{2.9}$$

The corresponding *cusp parameter* $\kappa = \kappa_L(\chi)$ is defined by

$$e(\kappa_L) = \chi(L^{-1}U^{n_L}L) = \chi(1 - n_L CD) \qquad (0 \le \kappa_L < 1), \tag{2.10}$$

where, as usual

$$e(z) = e^{2\pi i z} \quad (z \in \mathbb{C}). \tag{2.11}$$

From (2.6) it follows that κ_L depends only on $[L]$ and the character χ. We also put

$$h = h_L = (t_1, t_2). \tag{2.12}$$

3. Characters

For any character χ modulo N we write $N(\chi)$ for its conductor, so that

$$\chi = \chi^* \chi_0, \tag{3.1}$$

where χ^* is the associated primitive character with modulus $N(\chi)$ and χ_0 is the principal character modulo N; similarly for other characters to various moduli.

Lemma 3.1. *For any character χ modulo N and $L \in \Gamma(1)$, $\kappa_L = 0$ if and only if $N(\chi)$ divides N/h_L.*

Proof: By (2.10),

$$e(\kappa_L) = \chi(1 - x_L N/h_L), \tag{3.2}$$

where $x_L = CD/t_1 = C_1 D$ and $(x_L, h_L) = 1$.

Now suppose that $\chi(1 - xN/h) = 1$ for some x prime to $h = h_L$ and take any y with $(y, h) = 1$, so that, for some positive integer r,

$$y \equiv rx \pmod{h}.$$

Then,

$$\chi(1 - yN/h) = \chi(1 - rxN/h) = \chi^r(1 - xN/h) = 1,$$

since N divides $(N/h)^2$, *i.e.* $h^2 | N$. Now take any z prime to N, so that

$$\chi(z - N/h) = \chi(z)\chi(1 - z'N/h) = \chi(z),$$

where $zz' \equiv 1 \pmod{N}$. This shows that $N(\chi)$ divides N/h.

Conversely, if this holds, it is obvious from (3.2) that $\kappa_L = 0$.

When $\kappa_L = 0$ we say that L, $[L]$ and $L^{-1}\infty$ are *unbranched* for χ, or merely, when no confusion can arise, unbranched. Conversely, if $\kappa_L \neq 0$, L is said to be *branched*. It is clear that, if L is unbranched, so is every matrix of the same divisor t_1, so that we may say that t_1 is unbranched.

It is convenient to write $\bmod \epsilon$ for the modulus of a character ϵ, so that we always have $N(\epsilon) \mid \bmod \epsilon$. As usual, we write $q^r \| n$ to denote that q^r is the highest power of q that divides n.

Lemma 3.2. *Suppose that $N(\chi)$ divides N/h, where $h = (t_1, t_2)$, $t_1 t_2 = N$ and $\bmod \chi = N$. Then there exist characters χ_1 and χ_2 with moduli t_1 and t_2, respectively, such that*

$$\chi = \chi_1 \chi_2. \tag{3.3}$$

Proof: The integer h is a product of powers p^γ of primes p. If $p^\alpha \| t_1$ and $p^\beta \| t_2$ then $\gamma = \min(\alpha, \beta)$ and we factorize h as $h = h_0 h_1 h_2$, where (i) h_0 consists of powers p^γ for which $\gamma = \alpha = \beta$, (ii) h_1 consists of powers for which $\gamma = \alpha < \beta$, and (iii) h_2 consists of powers for which $\gamma = \beta < \alpha$. Then

$$\frac{N}{h} = h_0 \cdot \frac{t_1}{h_0 h_1} \cdot \frac{t_2}{h_0 h_2} = N_0 \cdot N_1 \cdot N_2,$$

say, these three factors being coprime in pairs.

Accordingly, since $N(\chi)$ divides N/h, it follows that we can write

$$\chi^* = \psi_0 \psi_1 \psi_2, \qquad \chi = \chi^* \chi_0,$$

where $\bmod \psi_j = N_j$ $(j = 0, 1, 2)$ and χ_0 is the principal character modulo N. Now put

$$\chi_1 = \psi_0\psi_1\phi_1, \qquad \chi_2 = \psi_2\phi_2, \tag{3.4}$$

where ϕ_1 and ϕ_2 are the principal characters with modulo t_1 and t_2, respectively. It is clear that (3.3) holds.

Note that, when $h_0 > 1$, the decomposition is not unique. But $N(\chi_1)$ divides t_1/h and $N(\chi_2)$ divides $t_2/(h_0h_2)$; subject to these conditions, the decomposition is unique.

Lemma 3.3. *Let $N = t_1t_2$ and assume that, for some prime factor q of N,*

$$q^\alpha \| t_1 \quad (\alpha \geq 1), \qquad q^\beta \| t_2 \quad (\beta \geq 0). \tag{3.5}$$

Write

$$h = (t_1, t_2), \qquad h' = (t_1/q, qt_2). \tag{3.6}$$

Then

$$h' = qh \ \ (\beta \leq \alpha - 2), \ \ h' = h \ \ (\beta = \alpha - 1), \ \ h' = h/q \ \ (\beta \geq \alpha). \tag{3.7}$$

Now assume that both h and h' divide $N/N(\chi)$, where χ has modulus N. Then

$$N(\chi) \ \text{divides} \ N/\{h, h'\}, \tag{3.8}$$

where $\{h, h'\}$ is the least common multiple of h and h'.

Further, let χ be factored as in Lemma 3.2 in the form

$$\chi = \chi_1\chi_2 = \chi_1'\chi_2', \tag{3.9}$$

where χ_1, χ_2, χ_1' and χ_2' have moduli t_1, t_2, t_1/q and qt_2, respectively. Let ϕ_1, ϕ_2, ϕ_1', and ϕ_2' be the principal characters to the same moduli, and write

$$\chi = \epsilon_q\epsilon_0, \tag{3.10}$$

where $\bmod \epsilon_q$ is a power of q and $\bmod \epsilon_0$ is the greatest factor of N prime to q.

Then there exist primitive characters η_1 and η_2 such that $N(\eta_1)|t_1$, $N(\eta_2)|t_2$ and $N(\eta_1)N(\eta_2)$ is prime to q, for which

$$\begin{aligned}
&\chi_1 = \epsilon_q\eta_1\phi_1, \ \ \chi_2 = \eta_2\phi_2 && (\beta \leq \alpha - 2),\\
&\chi_1' = \epsilon_q\eta_1\phi_1', \ \ \chi_2' = \eta_2\phi_2' && (\beta \leq \alpha - 2),\\
&\chi_1 = \epsilon_q\eta_1\phi_1, \ \ \chi_2 = \eta_2\phi_2 && (\beta = \alpha - 1 \text{ or } \alpha),\\
&\chi_1' = \eta_1\phi_1', \ \ \chi_2' = \epsilon_q\eta_2\phi_2' && (\beta = \alpha - 1 \text{ or } \alpha),\\
&\chi_1 = \eta_1\phi_1, \ \ \chi_2 = \epsilon_q\eta_2\phi_2 && (\beta \geq \alpha + 1),\\
&\chi_1' = \eta_1\phi_1', \ \ \chi_2' = \epsilon_q\eta_2\phi_2' && (\beta \geq \alpha + 1).
\end{aligned}$$

Proof: By considering the different cases, (3.7) is easily established and (3.8) follows from the fact that $\{h, h'\}$ must divide $N/N(\chi)$. The remainder of the Lemma is obvious when ϵ_q is a principal character, so that we now assume that this is not the case.

Because of the principal characters ϕ_1 and ϕ_2 in (3.4), we confine our attention to the characters $\psi_0\psi_1$ and ψ_2, noting that their moduli are coprime. It is easily checked that, when $\beta \leq \alpha$, $q\nmid$ mod ψ_2, so that ϵ_q must occur as a factor of $\psi_0\psi_1$ and therefore of χ_1; but if $\beta > \alpha$, then $q\nmid$ mod $\psi_0\psi_1$ and so ϵ_q is not a factor of $\psi_0\psi_1$. A similar situation holds in all the other cases. Observe that the characters η_1 and η_2 are the same throughout.

4. Eisenstein Series

By Theorem 5.1.2 of [4], the general Poincaré series of weight $k > 2$ and character χ for Γ is defined by

$$G_L(z, m, \chi) = \sum_{T \in \mathcal{R}_L} \bar{\chi}(T)(LT : z)^{-k} e\{(m + \kappa_L)LT(z)/n_L\}, \qquad (4.1)$$

where $L \in \Gamma(1)$ and $L : z = Cz + D$. Here $\mathcal{R}_L$ is an arbitrary right transversal of the group $< -I, L^{-1}U^{n_L}L >$ generated by $-I$ and $L^{-1}U^{n_L}L$ in Γ; G_L depends, of course, on k but we suppress this dependence since k is constant throughout. In order that G_L may not vanish identically, we require that $\chi(-1) = (-1)^k$.

The series (4.1) is an Eisenstein series if and only if $m = \kappa_L = 0$ and then takes the value 1 at the cusp $L^{-1}\infty$. If $L_1 \sim L_2$, it is easily seen, for example by Lemma 3.1 of [5], that

$$G_{L_2}(z, 0, \chi) = \chi(S)G_{L_1}(z, 0, \chi). \qquad (4.2)$$

where S is given by (2.6).

Lemma 4.1. *Let*

$$L_1 = \begin{pmatrix} A_1 & B_1 \\ C_1 & D_1 \end{pmatrix}, \quad L_2 = \begin{pmatrix} A_2 & B_2 \\ C_2 & D_2 \end{pmatrix},$$

where $div\, L_1 = \tau_1$, $div\, L_2 = \tau_2$ *and write*

$$C_1 = \tau_1 C_1', \qquad C_2 = \tau_2 C_2', \qquad (4.3)$$

Then $L_1 \sim L_2$ *if and only if* $\tau_1 = \tau_2 = t_1$ *(say), and*

$$h \mid (A_1C_2' - A_2C_1'), \qquad (4.4)$$

where, as usual, $N = t_1t_2$ and $h = (t_1, t_2)$. Moreover, when (4.4) holds and also h divides $N/N(\chi)$, then

$$\chi(S) = \bar{\chi}(L_1)\chi(L_2), \tag{4.5}$$

where S is given by (2.6) and

$$\chi(L_j) = \chi_1(D_j)\bar{\chi}_2(C_j') \quad (j = 1, 2), \tag{4.6}$$

where χ_1 and χ_2 are defined as in Lemma 3.2.

Proof: Write $\tau_1\tau_1' = \tau_2\tau_2' = N$. Then it is easily seen that $S = L_1^{-1}U^rL_2 \in \Gamma$ if and only if

$$rC_1'C_2'\tau_1\tau_2 \equiv \tau_2C_2'A_1 - \tau_1C_1'A_2 \pmod{N}, \tag{4.7}$$

from which we deduce that $\tau_1|\tau_2C_2'A_1$ and therefore, as $L_1 \in \Gamma(1)$, $\tau_1|\tau_2C_2'$. This is equivalent to $\tau_2'|\tau_1'C_2'$, and therefore, since $(\tau_2', C_2') = 1$, we have $\tau_2'|\tau_1'$. Similarly $\tau_1'|\tau_2'$ and therefore $\tau_1' = \tau_2'$ and $\tau_1 = \tau_2$. The congruence (4.7) now becomes

$$rC_1'C_2't_1 \equiv A_1C_2' - A_2C_1' \pmod{t_2}$$

and this is soluble if and only if (4.4) holds. We then have, by (2.2),

$$\begin{aligned}\delta &= A_1D_2 - C_1B_2 - rC_1D_2 \\ &\equiv A_1D_2 - C_1B_2 - C_2^*D_2rC_1'C_2't_1 \pmod{N},\end{aligned}$$

where $C_2'C_2^* \equiv 1 \pmod{t_2}$. From this we deduce that

$$\delta \equiv A_1D_2 \pmod{t_1}, \quad \delta \equiv C_1'C_2^* \pmod{t_2}.$$

This gives

$$\chi(S) = \chi(\delta) = \chi_1(A_1D_2)\chi_2(C_1'C_2^*).$$

from which (4.5) follows by (4.6).

Observe that the definition (4.6) agrees with the existing meaning of $\chi(L)$ when $L \in \Gamma$, since then $t_1 = N$ and $t_2 = 1$.

It follows from Lemma 4.1 that each cusp class is associated with a unique divisor t_1 of N.

Lemma 4.2. *The number of different cusp classes $[L]$ of divisor t_1 is $\phi(h)$, where ϕ is Euler's function and $h = (t_1, t_2)$. A representative set of such classes is the set of matrices W^{ut_1}, where u runs through a set of $\phi(h)$ different integers prime to t_2 and such that no two different values are congruent modulo h.*

Proof: From (4.4) and (2.9) it follows that, if $\operatorname{div} L = t_1$, then $L \sim W^{ut_1}$, if and only if $u \equiv C_1D \pmod{h}$ and we note that $(C_1D, h) = 1$; u is necessarily prime to t_2.

We write $\mathcal{K}(t_1)$ to denote a set of $\phi(h)$ numbers u satisfying the conditions stated in the enunciation of the lemma, and call it a *cusp index set* (of divisor t_1); so clearly is $q\mathcal{K}(t_1)$ for any q prime to t_2.

Lemma 4.3. *The number of incongruent cusps for* Γ *is*

$$\sigma(N) = \sum_{t_1|N} \phi(h) \quad (h = (t_1, t_2)). \tag{4.8}$$

The number of unbranched incongruent cusps for Γ *is*

$$\sigma(N,\chi) = \sum_{\substack{t_1 \mid N \\ t_1 \text{ unbranched}}} \phi(h). \tag{4.9}$$

The result (4.8) is known; see [1, p.240] or [7, p.102]. Clearly $\sigma(N,\chi)$ is the number of essentially different Eisenstein series for Γ of weight $k > 2$. For unbranched L we now define

$$G(L,\chi;z) = G(L,\chi)(z) = \overline{\chi}(L)G_L(z,0,\chi). \tag{4.10}$$

We have

Lemma 4.4. *For each* L *in an unbranched cusp class* $[L]$, $G(L,\chi;z)$ *takes the same value.*

This is clear from (4.2) and (4.6).

We write $\mathcal{E}(\chi)$ for the subspace spanned by the $\sigma(N,\chi)$ linearly independent Eisenstein series in $\mathcal{M}(\chi)$. Since Petersson's inner product (f,g) can be defined whenever fg is a cusp form, it follows that $\mathcal{E}(\chi)$ is orthogonal to $\mathcal{C}(\chi)$, by Theorem 5.2.2 of [4], which also remains valid when the cusp form f occurring in the enunciation is replaced by $G_K(z;m;\Gamma,k,v)$, provided that $K \not\sim L$. Thus the $\sigma(N,\chi)$ different Eisenstein series are orthogonal in pairs.

We also write $\mathcal{E}(\chi,t_1)$ for the subspace of $\mathcal{E}(\chi)$ generated by those $G(L,\chi)$ for which div $L = t_1$, where t_1 is unbranched. All these spaces depend, of course, on the weight k.

5. Hecke operators

For any $f \in \mathcal{M}(\chi)$ we have

$$f|T_n = n^K \sum_{ad=n} \sum_{r=1}^{d} \chi(a) f|J_d U^r J_a^{-1} \tag{5.1}$$

where $n \in \mathbb{N}$, and for any $m \in \mathbb{N}$,

$$J_m = \begin{pmatrix} 1 & 0 \\ 0 & m \end{pmatrix}, \qquad K = \frac{1}{2}k - 1, \tag{5.2}$$

the Hecke stroke operator on the right of (5.1) being defined by (2.5).

Similarly, when n is composed entirely of prime factors of N,

$$f \mid T_n^* = n^K \sum_{r=1}^{n} f|J_n^{-1} W^{-rN} = f|H_N T_n H_N^{-1}, \tag{5.3}$$

where H_N is the Fricke involution

$$H_N = \begin{pmatrix} 0 & -1 \\ N & 0 \end{pmatrix}; \tag{5.4}$$

otherwise, we have

$$f \mid T_n^* = \bar{\chi}(n) f|T_n \quad \text{when } (n, N) = 1. \tag{5.5}$$

On the Hilbert Space $\mathcal{C}(\chi)$, T_n^* is the adjoint operator to T_n, but it also acts on $\mathcal{M}(\chi)$ and, like T_n, maps $\mathcal{M}(\chi)$ into itself. Moreover, since

$$T_m T_n = \sum_{d \mid (m,n)} d^{k-1} \chi(d) T_{mn/d^2}, \tag{5.6}$$

we also have

$$T_n^* T_m^* = \sum_{d \mid (m,n)} d^{k-1} \bar{\chi}(d) T_{mn/d^2}^* \tag{5.7}$$

for any positive integer m and n.

Because of the multiplicative properties (5.6) and (5.7) it is only necessary to define T_n and T_n^* when n is a prime p. Thus, for example,

$$f(z) \mid T_p = \frac{1}{p} \sum_{r=1}^{p} f(J_p U^r z) + p^{k-1} \chi(p) f(J_p^{-1} z) \tag{5.8}$$

and

$$f(z) \mid T_p^* = p^K \sum_{r=1}^{p} f(z) | W^{-rN/p} J_p^{-1} \qquad (p|N). \tag{5.9}$$

See, for example, [5] or [6].

Lemma 5.1. *Each of the spaces $\mathcal{C}(\chi)$ and $\mathcal{E}(\chi)$ is invariant under the operators T_n and T_n^* $(n \in \mathbb{N})$.*

Proof: This is well known for $\mathcal{C}(\chi)$. Moreover, if $f \in \mathcal{E}(\chi)$ and $g \in \mathcal{C}(\chi)$, then

$$(f|T_n, g) = (f, g|T_n^*)$$

as is easily verified, for example, by Theorem 6.1 of [6]. Hence $(f|T_n, g) = 0$, and $(f|T_n^*, g) = 0$, similarly. This completes the proof.

We now investigate the action of Hecke operator on Eisenstein series. By Theorem 9.3.1 of [4] we deduce that, when $(n, N) = 1$ and $\kappa_L = 0$,

$$G_L(z, 0, \chi) \,|\, T_n = \sum_{d|n} (n/d)^{k-1} G_{R_a L_n}(z, 0, \chi), \tag{5.10}$$

where

$$R_a L_n = \begin{pmatrix} a_1 A & dB \\ d_1 C & aD \end{pmatrix}, \tag{5.11}$$

and $a_1 a \equiv d_1 d \equiv 1 \pmod N$.

When $\operatorname{div} L = t_1$, then $\operatorname{div} R_a L_n = t_1$, and, by (4.6)

$$\chi(R_a L_n) = \chi_1(a)\chi_2(d)\chi(L) \tag{5.12}$$

for unbranched t_1. Hence we have

Theorem 5.2. *let L be unbranched with divisor t_1. Then for $(n, N) = 1$,*

$$G(L, \chi; z) \,|\, T_n = \sum_{n=ad} \chi_1(a)\chi_2(d) a^{k-1} G(R_a L_n, \chi; z). \tag{5.13}$$

It follows that $\mathcal{E}(\chi, t_1)$ is invariant under the operator T_n for $(n, N) = 1$.

Now let ϵ be any character modulo h, where $h = (t_1, t_2)$. For unbranched t_1 put

$$f_\epsilon(z, t_1) = \sum_{u \in \mathcal{K}(t_1)} \epsilon(u) G(W^{ut_1}, \chi; z). \tag{5.14}$$

Note that, for $L = W^{ut_1}$, $R_a L_n \sim W^{ad_1 ut_1}$ and observe that $ad_1 \mathcal{K}(t_1)$ is also a cusp index set of divisor t_1. Hence, for $(n, N) = 1$,

$$\begin{aligned} f_\epsilon(z, t_1) \,|\, T_n &= \sum_{u \in \mathcal{K}(t_1)} \epsilon(u) \sum_{n=ad} \chi_1(a)\chi_2(d) a^{k-1} G(R_a L_n, \chi; z) \\ &= \sum_{n=ad} \chi_1(a)\epsilon(a_1 d)\chi_2(d) a^{k-1} \sum_{u \in \mathcal{K}(t_1)} \epsilon(uad_1) G(R_a L_n, \chi; z) \\ &= f_\epsilon(z, t_1) \sum_{n=ad} a^{k-1} \chi_1(a)\bar{\epsilon}(a)\epsilon(d)\chi_2(d). \end{aligned} \tag{5.15}$$

Thus $f_\epsilon(z, t_1)$ is an eigenform for the operator T_n when $(n, N) = 1$, the corresponding eigenvalue being

$$\lambda(n; t_1, \epsilon, \chi) = \sum_{n=ad} a^{k-1} \chi_1(a)\chi_2(d)\bar{\epsilon}(a)\epsilon(d). \tag{5.16}$$

To elucidate the comment in the third paragraph of §1, it may be noted that I take $\chi_1\bar{\epsilon}$ and $\chi_2\epsilon$ to correspond to Hecke's characters χ_1 and χ_2, so that their product is χ as he claimed.

Theorem 5.3. *Let t_1 be unbranched. Then the space $\mathcal{E}(t_1,\chi)$ is spanned by the $\phi(h)$ eigenforms $f_\epsilon(z,t_1)$ defined by (5.14), where ϵ runs through the characters modulo h.*

Proof: The linear independence of the functions f_ϵ is clear, since the series $G(W^{ut_1},\chi)$ in (5.14) are linearly independent.

Note that, although the $\phi(h)$ Eisenstein series on the right of (5.14) are mutually orthogonal, this is not the case for the functions f_ϵ, since the inner product of any pair of them cannot be defined.

6. Behaviour at cusps

Let $f \in \mathcal{M}(\chi)$ and $L \in \Gamma(1)$. We consider the bahaviour of f at the cusp $L^{-1}\infty$ and, for convenience, write $M = L^{-1}$. Let

$$f_M(z) = f(z)\,|\,M = f(Mz)(M:z)^{-k}. \tag{6.1}$$

The *value* of f at the cusp $M\infty$ is defined to be

$$V(f,M) = \lim_{z\to\infty} f_M(z). \tag{6.2}$$

Lemma 6.1. *(i) If $M_2\infty = M_1\infty$, then*

$$V(f,M_2) = V(f,M_1).$$

(ii) If $L_1 \sim L_2$, where $L_1M_1 = L_2M_2 = I$, then $M_2U^rM_1^{-1} = S \in \Gamma$ for some $r \in \mathbb{Z}$ and

$$V(f,M_2) = \chi(S)V(f,M_1).$$

(iii) If $G_L(z) = G_L(z,0,\chi)$, then

$$V(G_L,L^{-1}) = \begin{cases} 1 & \text{if } \kappa_L = 0, \\ 0 & \text{if } \kappa_L \neq 0. \end{cases} \tag{6.3}$$

Moreover, if $\kappa_L = 0$ and $K \sim L$, then

$$V(G(K,\chi),L^{-1}) = V(G(L,\chi),L^{-1}) = \bar{\chi}(L). \tag{6.4}$$

Proof: Parts (i) and (ii) are straightforward from (6.1) and (6.2). From the series definition (4.1) of G_L (6.3) follows, and from (4.10) we deduce (6.4).

Lemma 6.2. *Let $F \in \mathcal{E}(\chi)$. Then*

$$F(z) = \sum_L \chi(L)V(F,L^{-1})G(L,\chi;z). \tag{6.5}$$

where the summation is over a set of $\sigma(N,\chi)$ incongruent unbranched matrices L.

This follows from (4.10) and Theorem 5.1.3 of [4].

7. The action of the operators T_q for $q \mid N$

Suppose that f is any member of $\mathcal{E}(\chi)$ and let q be a prime divisor of N. Write

$$Q = \{0, 1, \cdots, q^{-1}\}, \quad Q^* = \{1, 2, \cdots, q^{-1}\}. \tag{7.1}$$

We examine the behaviour of $f|T_q$ at an unbranched cusp $L^{-1}\infty$, where $L \in \Gamma(1)$. Without loss of generality we may take

$$L = W^{ut_1}, \tag{7.2}$$

where $t_1 = \operatorname{div} L$ and $(u, t_2) = 1$.

Write $J = J_q$ and put

$$F_q(z) = f(z) \mid T_q = \frac{1}{q} \sum_{n \in Q} f(JU^n z). \tag{7.3}$$

Then $F_q \in \mathcal{E}(\chi)$ and we are interested in finding $V(F_q, L^{-1})$.

For each $n \in Q$ we define a matrix $L_n \in \Gamma(1)$ by

$$JU^n L^{-1} = L_n^{-1} JU^m. \tag{7.4}$$

where we attempt to choose $m \in Q$ so that $L_n \in \Gamma(1)$. By (7.4) we have

$$L_n = \begin{pmatrix} 1 + mut_1 & b_n \\ uqt_1 & 1 - nut_1 \end{pmatrix}, \tag{7.5}$$

where

$$qb_n = m(1 - nut_1) - n. \tag{7.6}$$

We distinguish various cases:

(i) When $q|t_1$, we can take $m = n$ so that $b_n \in \mathbb{Z}$ and therefore $L_n \in \Gamma(1)$. If (a) $q \nmid t_2$ the $\operatorname{div} L_n = t_1$, while (b) if $q|t_2$ then $\operatorname{div} L_n = qt_1$.

(ii) Suppose that $q \nmid t_1$. We have

$$ut_1qb_n = (1 + mut_1)(1 - nut_1) - 1.$$

Thus (a) if $nut_1 \not\equiv 1 \pmod q$ we can choose $m \in Q$ to make $b_n \in \mathbb{Z}$; observe that the corresponding value of m satisfies the congruence $mut_1 \not\equiv -1 \pmod q$ and that $\operatorname{div} L_n = qt_1$.

However, if (b) $n = n_u$, where $n_u ut_1 \equiv 1 \pmod q$, then it is not possible to choose $m \in Q$ in such a way as to make b_n an integer. Observe that $n_u \neq 0$. In this case we write

$$1 - n_u ut_1 = qd \tag{7.7}$$

and take

$$Z_L = \begin{pmatrix} q & -n_u \\ ut_1 & d \end{pmatrix} \tag{7.8}$$

so that $Z_L \in \Gamma(1)$ and $\operatorname{div} Z_L = t_1$.

Lemma 7.1. *(a) If, for some $n \in Q$, $L_n \in \Gamma(1)$, then*

$$\lim_{z\to\infty} \frac{f(JU^n L^{-1}z)}{(L^{-1}:z)^k} = V(f, L_n^{-1}). \tag{7.9}$$

(b) Suppose that $q \nmid t_1$ and that $n = n_u$ is defined by (7.7). Then

$$\lim_{z\to\infty} \frac{f(JU^n L^{-1}z)}{(L^{-1}:z)^k} = q^k V(f, Z_L^{-1}). \tag{7.10}$$

Proof: (a) Write $\zeta = JU^m z$ and note that

$$L^{-1}: z = 1 - ut_1 z = L_n^{-1} : \zeta,$$

so that the left-hand side of (7.9) is

$$\lim_{\zeta\to\infty} \frac{f(L_n^{-1}\zeta)}{(L_n^{-1}:\zeta)^k}.$$

We deduce (7.9).

(b) Write $\zeta = qz$ and observe that

$$JU^n L^{-1} = qZ_L^{-1}J^{-1}$$

and that

$$Z_L^{-1} : \zeta = q - ut_1\zeta.$$

Hence the left-hand side of (7.10) follows.

Theorem 7.2. *Let $f \in \mathcal{E}(\chi)$ and let F_q be defined by (7.3), where $q|N$. Then $F_q \in \mathcal{E}(\chi)$ and, if L is given by (7.2) and is unbranched, then*

$$V(F_q, L^{-1}) = \frac{1}{q} \sum_{n\in Q} V(f, L_n^{-1}) \qquad \text{if } q|t_1 \tag{7.11}$$

and

$$V(F_q, L^{-1}) = \frac{1}{q} \sum_{\substack{n\in Q\\ n\neq n_u}} V(f, L_n^{-1}) + q^{k-1}V(f, Z_L^{-1}) \qquad \text{if } q\nmid t_1. \tag{7.12}$$

Proof: We have

$$F_q(L^{-1}z) = \frac{1}{q}\sum_{n\in Q} f(JU^n L^{-1}z)$$

and the theorem follows by Lemma 7.1.

As a corollary we have

Lemma 7.3. *Let $N = \tau_1\tau_2$ and suppose that τ_1 is unbranched for χ and that $q|N$. Then*

$$\mathcal{E}(\chi,\tau_1)\,|\,T_q \subset \mathcal{E}(\chi,\tau_1) \oplus \mathcal{E}(\chi,\tau_1/q),$$

where $\mathcal{E}(\chi,\tau_1/q)$ is the zero space if either $q\nmid\tau_1$ or τ_1/q is branched.

Proof: Suppose that $f \in \mathcal{E}(\chi,\tau_1)$. Take t_1 to be any divisor of N other than τ_1 or τ_1/q and let $\operatorname{div} L = t_1$. Then, since L_n and Z_L have divisors equal to t_1 or qt_1, it follows that all the terms on the right of (7.11) and (7.12) are zero, and the result follows.

We now obtain more explicit results by taking

$$f(z) = G(K,\chi;z), \tag{7.13}$$

where

$$K = W^{v\tau_1}, \quad (v,\tau_2) = 1 \tag{7.14}$$

and K is unbranched. Note that, if $q\nmid\tau_2$, we may assume that $q|v$. By (7.2) (with $t_1 = \tau_1$ or τ_1/q), Lemma 6.2 and Lemma 7.3, we have

$$F_q(z) = G(K,\chi;z)\,|\,T_q = \sum_L \chi(L)V(F_q,L^{-1})G(L,\chi;z), \tag{7.15}$$

where the summation is taken over all u in $\mathcal{K}(\tau_1) \cup \mathcal{K}(\tau_1/q)$, the latter summand being omitted if $q\nmid\tau_1$ or if τ_1/q is branched.

To evaluate $V(F_q,L^{-1})$ in (7.15) we use Theorem 7.2.

Let us consider first the case when $\operatorname{div} L = t_1$, where $t_1 = \tau_1$ and $q|\tau_1$. We take $L = W^{ut_1}$, where $(u,t_2) = 1$. Then, by (7.11) we get zero on the right unless $\operatorname{div} L_n = \tau_1$, which is case (i) (a) above; this we label as

(1) $q|\tau_1$, $q\nmid\tau_2$; $\operatorname{div} L = \operatorname{div} L_n = \tau_1$ $(n \in Q)$.

However, if $q\nmid\tau_1$, it is only the last term on the right of (7.12) that makes any contribution. This is case (ii) (b), which we state as

(2) $q\nmid\tau_1$, $n = n_u$; $\operatorname{div} L = \operatorname{div} Z_L = \tau_1$.

Next suppose that $q|\tau_1$ and take $t_1 = \tau_1/q$, so that $t_2 = q\tau_2$. Put $L = W^{u\tau_1/q}$. Note that case (i) (a) does not arise; nor does case (ii)(b) if $q\nmid t_1$. Thus there are two remaining cases:

(3) $q^2|\tau_1$; $\operatorname{div} L_n = \tau_1$ $(n \in Q)$, $\operatorname{div} L = \tau_1/q$

and

(4) $q|\tau_1$, $q^2\nmid\tau_1$; $\operatorname{div} L_n = \tau_1$ $(n \neq n_u)$, $\operatorname{div} L = \tau_1/q$.

These two cases occur only when τ_1/q in unbranched.

As usual, we write $h = (\tau_1,\tau_2)$ and, when $q|\tau_1$ we put

$$h' = (\tau_1/q, \tau_2 q). \tag{7.16}$$

From Lemma 3.3 (with $t_1 = \tau_1$) we see that both τ_1 and τ_1/q are unbranched when h' equals h or h/q, but τ_1/q may be branched if $h' = qh$.

We define the non-negative integers α, β and δ by

$$q^{\alpha}||\tau_1,\ q^{\beta}||\tau_2,\ q^{\delta}||N(\epsilon_q), \tag{7.17}$$

where ϵ_q is defined by (3.10). Since τ_1 is unbranched, it follows that

$$\delta \le \max(\alpha, \beta),$$

and it is easily seen that then τ_1/q is branched if and only if

$$0 \le \beta \le \alpha - 2 \ \text{ and } \ \delta = \alpha. \tag{7.18}$$

Also let

$$q_{\alpha} = \begin{cases} 1 - q^{-1} & (\alpha = 1), \\ 1 & (\alpha \ge 2). \end{cases} \tag{7.19}$$

and put

$$g = \tau_2 q^{-\beta}, \quad K_q = W^{gh\tau_1}. \tag{7.20}$$

We dispose of case (2) first, as it is the simplest. By (7.12) and (7.15)

$$F_q = q^{k-1}\chi(L)V(G(K,\chi), Z_L^{-1})G(L,\chi),$$

where we have to choose L so that $Z_L \sim K$. By Lemma 4.1, we must have $u \equiv qv \pmod{h}$. This is uniquely satisfied modulo h, with $(u, \tau_2) = 1$, by taking $u = qv + gh$, which gives $L = K^q K_q$. Thus, by (6.4) and (4.6),

$$\begin{aligned} F_q &= q^{k-1}\chi(L)\bar{\chi}(Z_L)G(L,\chi) \\ &= q^{k-1}\chi_1(q)G(K^q K_q, \chi) \quad (q \nmid \tau_1). \end{aligned} \tag{7.21}$$

We now consider the three remaining cases (1), (3) and (4), where $q|\tau_1$. We have

$$F_q = \sum\nolimits_1 + \sum\nolimits^*, \tag{7.22}$$

where, by (7.15), (7.11) and (6.4),

$$\sum\nolimits_1 = \frac{1}{q} \sum_{\operatorname{div} L = \tau_1} \chi(L) \sum_{\substack{n \in Q \\ L_n \sim K}} \bar{\chi}(L_n) G(L,\chi) \quad (q \nmid \tau_2). \tag{7.23}$$

If $q|\tau_2$, then $\sum_1 = 0$.

If τ_1/q is branched, $\sum^* = 0$. As mentioned above, this can only happen when (7.18) holds. Otherwise we have

$$\sum\nolimits^* = \frac{1}{q} \sum_{\operatorname{div} L = \tau_1/q} \chi(L) \sum_{\substack{n \in Q \\ L_n \sim K}}{}' \bar{\chi}(L_n) G(L,\chi), \tag{7.24}$$

where the dash indicates that, in case (4), the term with $n = n_u$ is omitted.
In (7.23) $L_n \sim K$ implies that

$$uq \equiv v \pmod{h}.$$

Since $q \nmid \tau_2$, we may assume that $q|v$ and take $u = v/q$. Then for each $n \in Q$ we have

$$L = W^{v\tau_1/q} = J^{-1}KJ =: K^{(0)}, \tag{7.25}$$

say. Hence

$$\sum\nolimits_1 = \frac{1}{q}\sum_{n\in Q} \bar{\chi}_2(u)\chi_2(uq)G(K^{(0)},\chi) = \chi_2(q)G(K^{(0)},\chi) \quad (q \nmid \tau_2). \tag{7.26}$$

Note that, although this has been proved on the assumption that $q \nmid \tau_2$, it holds also when $q|\tau_2$, since then $\chi_2(q) = 0$. Accordingly,

$$\sum\nolimits_1 = \chi_2(q)G(K^{(0)},\chi) \qquad (\alpha \geq 1, \beta \geq 0). \tag{7.27}$$

Now suppose that both τ_1 and τ_1/q are unbranched, so that we need to consider $\sum^*$. The condition gives

$$u \equiv v(1 + mu\tau_1/q) \pmod{h}. \tag{7.28}$$

Before considering the various cases we prove

Lemma 7.4. *Let L and L_n be given by (7.2) and (7.5), where $t_1 = \tau_1/q$ and $\operatorname{div} L = t_1$, $\operatorname{div} L_n = \tau_1$ and $L_n \in \Gamma(1)$. Then*

$$\bar{\chi}(L)\chi(L_n) = \begin{cases} \epsilon_q(u)\epsilon_q(1 - nu\tau_1/q), & \text{if } \beta = \alpha - 1 \text{ or } \alpha, \\ \epsilon_q(1 - nu\tau_1/q), & \text{if } \beta \leq \alpha - 2, \\ 1, & \text{if } \beta \geq \alpha + 1\ . \end{cases}$$

Proof: The character ϵ_q is defined in Lemma 3.3. We have, by (4.6),

$$\chi(L) = \bar{\chi}_2'(u), \ \chi(L_n) = \chi_1(1 - nu\tau_1/q)\bar{\chi}_2(u),$$

and the result follows from Lemma 3.3.

(a) Suppose first that $q \nmid \tau_2$, so that $\beta = 0$. We can assume that $q|v$ and have to choose u modulo h' to satisfy (7.28) and to be prime to $q\tau_2$.

If $\alpha \geq 2$, then $h' = qh$ and we take

$$u = v + rgh \ (r \in Q^*), \tag{7.29}$$

which gives, for each $n \in Q$,

$$L = K^{(r)} := J^{-1} K K_q^r J \qquad (r \in Q^*). \tag{7.30}$$

Note that this together with (7.25) defines $K^{(r)}$ for all $r \in Q$. We then have, by (7.24) and Lemma 7.4,

$$\sum{}^* = \frac{1}{q} \sum_{r \in Q^*} \sum_{n \in Q} \bar{\epsilon}_q(1 - nrgh\tau_1/q) G(K^{(r)}, \chi).$$

By (3.8), $\delta \leq \alpha - 1$ and this then reduces to

$$\sum{}^* = \sum_{r \in Q^*} G(K^{(r)}, \chi) \tag{7.31}$$

and so

$$F_q = \chi_2(q) G(K^{(0)}, \chi) + \sum_{r \in Q^*} G(K^{(r)}, \chi) \qquad (q \nmid \tau_2, q^2 | \tau_1). \tag{7.32}$$

Observe that $\operatorname{div} K^{(0)} = \tau_1$ and $\operatorname{div} K^{(r)} = \tau_1/q$ $(r \in Q^*)$. We have assumed here that τ_1/q is unbranched. If that is not the case, then $\delta = \alpha$ and the analysis shows that $\sum^* = 0$, as was to be expected.

Finally, if $\alpha = 1$ then $h' = h$ and there is only one solution of (7.28) modulo h, namely $u = v + gh$, since we must have $(u, q) = 1$. Then, by (7.30),$L = K^{(1)}$. This holds for all $n \in Q$ with the exception of $n = n_v \in Q$, where

$$n_v gh\tau_1/q = 1 \pmod{q}.$$

Then we have

$$\sum{}^* = \frac{1}{q} \sum_{n \neq n_v} \bar{\epsilon}_q(v + gh) \bar{\epsilon}_q(1 - ngh\tau_1/q) G(K^{(1)}, \chi). \tag{7.33}$$

By (3.8) $\delta \leq \alpha$ so that we obtain

$$F_q = \begin{cases} \chi_2(q) G(K^{(0)}, \chi) & (\delta = 1), \\ q_\alpha G(K^{(1)}, \chi) + \chi_2(q) G(K^{(0)}, \chi) & (\delta = 0). \end{cases} \tag{7.34}$$

Note that $K^{(r)} \sim K^{(1)}$ for all $r \in Q^*$.

(b) Now suppose that $0 < \beta \leq \alpha - 1$, so that $q | h$, $h | \tau_1/q$ and $q \nmid v$.

If $\beta = \alpha - 1$, then $h' = h$ and (7.28) is solved by taking $u = v$ and $L = K^{(0)}$ for each $n \in Q$. We obtain

$$\sum{}^{*} = \frac{1}{q}\bar{\epsilon}_q(v) \sum_{n \in Q} \bar{\epsilon}_q(1 - nv\tau_1/q) G(K^{(0)}, \chi) \tag{7.35}$$

and $\delta \leq \alpha$, by (3.8). Thus, since $\sum_1 = 0$,

$$F_q = \begin{cases} 0 & \text{for } \delta = \alpha = \beta + 1, \\ \bar{\epsilon}_q(v) G(K^{(0)}, \chi) & \text{for } \delta \leq \alpha - 1 = \beta. \end{cases} \tag{7.36}$$

We now assume that $0 < \beta \leq \alpha - 2$, so that $h' = qh$, $q|h$ and $q|\tau_1/q$. Moreover $\delta \leq \alpha - 1$ since we assume that τ_1/q is unbranched. From (7.28)we deduce that we can take

$$u = v + rgh \ (r \in Q),$$

which gives $L = K^{(r)}$ $(r \in Q)$ and so

$$F_q = \sum{}^{*} = \sum_{r \in Q} G(K^{(r)}, \chi) \qquad (0 < \beta \leq \alpha - 2). \tag{7.37}$$

(c) Finally, suppose that $\beta \geq \alpha > 0$, so that $h' = h/q$ and (7.28) gives

$$u \equiv v \pmod{h'}.$$

This determines a unique cusp class of divisor τ_1/q, which we may take to be represented by $L = W^{v\tau_1/q} = K^{(0)}$. But we have to satisfy $L_n \sim K$ and this implies that

$$mv^2\tau_1/q \equiv 0 \pmod{h},$$

and therefore $m = n = 0$. Hence

$$F_q = \sum{}^{*} = \begin{cases} q^{-1}\bar{\epsilon}_q(v) G(K^{(0)}, \chi) & \text{for } \beta = \alpha > 0, \\ q^{-1} G(K^{(0)}, \chi) & \text{for } \beta > \alpha > 0. \end{cases} \tag{7.38}$$

Note that $\delta \leq \beta$ in either case.

We collect these results in

Theorem 7.4. *let q be a prime factor of N and let*

$$K = W^{v\tau_1}, \quad K_q = W^{gh\tau_1},$$

where τ_1 is unbranched for χ, $(v, \tau_2) = 1$ and g is the greatest factor of τ_2 prime to q. When $q \nmid \tau_2$ we may assume that $q|v$. Let

$$K^{(r)} = J_q^{-1} K K_q^r J_q \qquad (r \in Q)$$

and let α, β and δ be as defined in (7.17). Let

$$F_q = G(K,\chi)\,|\,T_q.$$

Then (i)

$$F_q = q^{k-1}\chi_1(q)G(K^q K_q,\chi) \quad \text{for } q\nmid\tau_1. \tag{7.39}$$

If (ii) $q|\tau_1$ then

$$F_q = \sum_{r\in Q}\lambda_r(K)G(K^{(r)},\chi), \tag{7.40}$$

where the value of $\lambda_r(K)$ ($r\in Q$) are given in the following table.

Row	β	δ	$\lambda_0(K)$	$\lambda_1(K)$	$\lambda_r(K)$ $(r>1)$
1	$\beta\geq\alpha+1>1$	$\delta\leq\beta$	$1/q$	0	0
2	$\beta=\alpha>0$	$\delta\leq\beta$	$\bar{\epsilon}_q(v)/q$	0	0
3	$\beta=\alpha-1$	$\delta=\alpha$	$\chi_2(q)$	0	0
4	$\beta=\alpha-1=0$	$\delta\leq\alpha-1$	$\chi_2(q)$	q_α	0
5	$\beta=\alpha-1>0$	$\delta\leq\alpha-1$	$\bar{\epsilon}_q(v)$	0	0
6	$0=\beta\leq\alpha-2$	$\delta\leq\alpha-1$	$\chi_2(q)$	1	1
7	$0<\beta\leq\alpha-2$	$\delta\leq\alpha-1$	1	1	1
8	$0\leq\beta\leq\alpha-2$	$\delta=\alpha$	$\chi_2(q)$	0	0

Proof: Rows 1 and 2 follow from (7.38), rows 3-5 from (7.34,36), row 6 from (7.32) and row 7 from (7.37). The upper bounds for δ in rows 1, 2, 6 and 7 are the maximum values possible when τ_1/q is unbranched. In rows 3-5 α is the maximum value possible when $\beta=\alpha-1$. Note also that, in row 3, $\chi_2(q)=0$ when $\beta>0$. In row 8, τ_1/q is branched and so the only contribution comes from $\sum_1$ in (7.27). Further, we always have

$$F_q \in \mathcal{E}(\tau_1/q) \quad \text{when } \beta>0. \tag{7.41}$$

The theorem confirms the results obtained in Theorem 6.1 of [5] for $m=0$, where $N=q$.

8. The action of $T_q(q|N)$ on the eigenforms $f_\epsilon(\tau_1,q)$

We define α,β and δ as in (7.17). As in §5, ϵ is an arbitrary character modulo $h=(\tau_1,\tau_2)$. We then define γ and ρ by

$$q^\gamma||N(\epsilon), \quad q^\rho||N(\epsilon\bar{\epsilon}_q), \tag{8.1}$$

for any prime q dividing N. Then, necessarily,

$$\gamma\leq\min(\alpha,\beta), \quad \delta\leq\max(\alpha,\beta), \quad \rho\leq\max(\gamma,\delta). \tag{8.2}$$

When $q|\tau_1$ we shall write η for a character modulo h' determined by ϵ; here $h'=(\tau_1/q, q\tau_2)$ as usual. Put

$$\xi_q(\eta)=\begin{cases}0 & \text{if } q^\alpha||N(\eta),\\ 1 & \text{if } q^\alpha\nmid N(\eta).\end{cases} \tag{8.3}$$

Then we have

Theorem 8.1. *Let τ_1 be unbranched for the character χ and let*

$$f_\epsilon(z,\tau_1) = \sum_{v\in\mathcal{K}(\tau_1)} \epsilon(v)G(W^{v\tau_1},\chi;z) \tag{8.4}$$

as in §5, so that f_ϵ is an eigenform for all the Hecke operator T_n with $(n,N)=1$. Let q be any prime number dividing N. Then

$$f_\epsilon(z,\tau_1)\,|\,T_q = \begin{cases} q^{k-1}\chi_1(q)\bar{\epsilon}(q)f_\epsilon(z,\tau_1) & \text{if } q\nmid\tau_1, \\ \chi_2(q)\epsilon(q)f_\epsilon(z,\tau_1)+q_\alpha\xi_q(\eta)f_\eta(z,\tau_1/q) & \text{if } q|\tau_1. \end{cases} \tag{8.5}$$

Here $\eta=\epsilon$ except that $\eta=\epsilon\bar{\epsilon}_q$ when $0<\delta\le\beta=\alpha$ or $0<\delta\le\beta=\alpha-1$.

The exceptional cases correspond essentially to rows 2 and 5 of the table.

Proof: When $q\nmid\tau_1$ we use (7.39) and note that $K^qK_q = W^{u\tau_1}$ where $u=vq+gh$, so that $\epsilon(v)=\epsilon(u)\bar{\epsilon}(q)$ and the required result follows.

When $q|\tau_1$ it will be sufficient to carry out the proof in the following two cases, the other cases being similar or more straightforward.

Let $\beta=\alpha>0$. Then by row 2 of the table we have

$$f_\epsilon(z,\tau_1)\,|\,T_q = \frac{1}{q}\sum_{v\in\mathcal{K}(\tau_1)} \bar{\epsilon}_q(v)\epsilon(v)G(W^{v\tau_1/q},\chi;z) \quad (v,\tau_2)=1. \tag{8.6}$$

Here $h'=h/q$ so that $\phi(h)=qq_\alpha\phi(h')$. Note that $W^{u\tau_1/q}\sim W^{v\tau_1/q}$ if and only if $u\equiv v \pmod{h'}$. Hence in the summation we may replace v by $v+rgh'$, letting r run through Q and v through $\mathcal{K}(\tau_1/q)$, excluding values of r such that q divides $v+rgh'$ when $\alpha=1$.

Write $\eta=\epsilon\bar{\epsilon}_q$, so that $\rho\le\alpha$. We have, since $q^{\alpha-1}||h'$,

$$\sum_{r\in Q}\eta(v+rgh') = qq_\alpha\xi_q(\eta)\eta(v) \tag{8.7}$$

and so

$$f_\epsilon(z,\tau_1)\,|\,T_q = q_\alpha\xi_q(\eta)f_\epsilon(z,\tau_1/q).$$

Note that in this case $\chi_2(q)=0$.

Let $0=\beta\le\alpha-2$ and $\delta\le\alpha-1$. Then, by row 6 of the table,

$$\begin{aligned} f_\epsilon(\tau_1,q)\,|\,T_q &= \chi_2(q)\sum_{v\in\mathcal{K}(\tau_1)}\epsilon(v)G(W^{v\tau_1/q},\chi;z) \\ &\quad+\sum_{r\in Q^*}\sum_{v\in\mathcal{K}(\tau_1)}\epsilon(v)G(W^{(v+rgh)\tau_1/q},\chi;z). \end{aligned} \tag{8.8}$$

Here we have $q|v$, since $\beta = 0$, and $h' = qh$. The first term on the right of (8.8) is $\chi_2(q)\epsilon(q)f_\epsilon(\tau_1, z)$. In the second term $\epsilon(v) = \epsilon(u)$, where $u = v + rgh$, and runs through the $\phi(h') = (q-1)\phi(h)$ residues prime to h' as v runs through $\mathcal{K}(\tau_1)$ and r runs through Q^*. Hence the second term is $f_\eta(z, \tau_1/q)$, where $\eta = \epsilon$ and can be regarded as a character modulo h'. The result follows since, by (8.3), we have $\xi_q(\eta) = 1 = q_\alpha$; for $\alpha > 1$.

We note that, in conclusion, that in rows 3 and 8 of the table, where we have taken $\eta = \epsilon$, we could equally well have taken $\eta = \epsilon\bar{\epsilon}_q$, since, in either case $\xi_q(\eta) = 0$.

When N is squarefree $h = (\tau_1, \tau_2) = 1$ and $\epsilon(v)$ takes on the value 1 for all v. the results are then much simpler as shown in (8.10) below, where we note that $f_\epsilon(z, \tau_1) =: f(z, \tau_1) = G(W^{\tau_1}, \chi; z)$.

$$f(z,\tau_1) \,|\, T_q = \begin{cases} \chi_2(q)f(z,\tau_1) + (1-q^{-1})f(z,\tau_1/q) & (q|\tau_1,\ q\nmid N(\chi)), \\ \chi_2(q)f(z,\tau_1) & (q|\tau_1,\ q|N(\chi)), \\ q^{k-1}\chi_1(q)f(z,\tau_1) & (q\nmid \tau_1). \end{cases} \tag{8.9}$$

From this we deduce

Theorem 8.2. *Let N be squarefree and let χ be a primitive character modulo N. Then, for each $\tau_1|N$. $G(W^{\tau_1}, \chi)$ is an eigenform for all the operators T_n ($n \in \mathbb{N}$), with eigenvalues*

$$\lambda(n; \tau_1, \chi) := \lambda(n; \tau_1, 1, \chi) = \sum_{n=ad} a^{k-1}\chi_1(a)\chi_2(d). \tag{8.10}$$

Proof: Since the operators commute and, in particular,

$$T_{q^r} = (T_q)^r \quad (r \in \mathbb{N}),$$

this follows from (8.9) and (5.16). We also note that, in this case, the Dirichlet series associated with $G(W^{\tau_1}, \chi)$ is

$$L(s-k+1, \chi_1)L(s, \chi_2),$$

the characters χ_1 and χ_2 depending, of course on the divisor τ_1.

REFERENCES

[1] J.-M.Deshouillers, H.Iwaniec, Kloosterman sums and Fourier coefficients of cusp forms, Inventiones Math. **70** (1982), 219–288.

[2] E. Hecke, Ueber Modulfunktionen und die Dirichletschen Reihen mit Eulerscher Produktentwicklung. II, Math. Ann. **114** (1937), 316–351.

[3] H.Petersson, Ueber die lineare Zerlegung der den ganzen Modulformen von höherer Stufe entsprechenden Dirichletreihen in vollständige Eulersche Produkte, Acta Math. **80** (1948), 191–221.
[4] R.A.Rankin, *Modular forms and functions*, Cambridge University Press, 1977.
[5] R.A. Rankin, The adjoint Hecke operator. I, to appear in J. Madras University.
[6] R.A. Rankin, The adjoint Hecke operator. II, in: *Number theory and related topics*, Oxford University Press, 1989.
[7] B. Schoeneberg, *Elliptic modular functions*, Springer-Verlag, 1974.

Robert A. Rankin
Department of Mathematics
University of Glasgow
Glasgow G12 8QW
Scotland

Some Binary Partition Functions

BRUCE REZNICK

Dedicated to Professor Paul T. Bateman on the occasion of his retirement

1. Introduction and Overview

For $d \geq 2$, the *d-th binary partition function*, $b(d;n)$, is the number of representations

$$n = \sum_{i=0}^{\infty} \epsilon_i 2^i,\ \epsilon_i \in \{0,1,\ldots,d-1\}; \tag{1.1}$$

the usual (Euler) binary partition function is $b(\infty;n) = \lim_{d\to\infty} b(d;n)$. This paper explores various arithmetic and analytic properties of the $b(d;n)$'s. For small values of d, $b(d;n)$ is familiar:

$$b(2;n) = 1 \text{ (Euler [E1, p.333])}, \qquad (1.2)(i)$$

$$b(3;n) = s(n+1) \text{ (Thm. 5.2)}, \qquad (1.2)(ii)$$

$$b(4;n) = \lfloor n/2 \rfloor + 1 \text{ (Problem B2, 1983 Putnam [KAH])}. \qquad (1.2)(iii)$$

In (ii), $s(n)$ denotes the Stern sequence; no other $b(d;n)$'s appear in [**S1**].

Euler [**E2**, p.288] defined $b(\infty;n)$ and computed its values for $n \leq 37$. Some recurrences for $b(\infty;n)$ and, in effect, $b(2^r;n)$ were studied by Tanturri [**T1,T2,T3**] in the 1910s. In 1940, Mahler [**M**] established that $\log b(\infty;n) \sim (\log n)^2/(\log 4)$; this asymptotic estimate was refined by de Bruijn [**B**] in 1948. Knuth [**K**] also investigated the growth of $\log b(\infty;n)$ in 1966, and gave some other recurrences for $b(\infty;n)$. In 1969, Churchhouse [**C4**] discussed the behavior of $b(\infty;n) \pmod{2^r}$. Let $\nu_2(m)$ denote the largest power of 2 dividing m. Then 2 divides $b(\infty;n)$ for $n \geq 2$, 4 divides $b(\infty;n)$ if and only if $\nu_2(n)$ or $\nu_2(n-1)$ is positive and even, and 8 never divides $b(\infty;n)$. Churchhouse conjectured that, for all even m,

$$\nu_2(b(\infty;4m) - b(\infty;m)) = \lfloor (3\nu_2(m)+4)/2 \rfloor. \tag{1.3}$$

Author supported in part by the National Science Foundation

This conjecture was proved by Rödseth [**R5**], Gupta (thrice) [**G1,G2, G4**] and generalized by Hirschhorn and Loxton [**HL**] in 1976.

The m-ary partition function is defined by replacing 2 by m in (1.1) and eliminating the restriction on the ϵ_i's. Mahler and de Bruijn actually studied the asymptotics of the m-ary partition function. The proof of (1.3) was generalized to $m > 2$ by Rödseth, Gupta [**G3**], Andrews [**A1**] and Gupta and Pleasants [**GP**]. Restricted m-ary partition functions ($\epsilon_i = 0$ for $i \geq t$) also appeared in Gupta and Pleasants, and Dirdal [**D1,D2**]. Analysis of their generating functions shows that they are equal to the number of m-ary partitions with $\epsilon_i < m^t$ for all i (see Thm. 3.2(i) for $m = 2$.) A nice summary of this work can be found in Ch. 10.2 of [**A2**], and its exercises.

Here is the plan for the rest of the paper.

In section two, we give an infinite product representation for $F_d(x)$, the generating function of $b(d;n)$. We derive some simple relationships among the F_d's and deduce the resulting recurrences on $b(d;n)$, $b(2d;n)$ and $b(\infty;n)$, which often depend on the parity of d and n. Clearly, $b(d;n)$ is non-decreasing in d; the monotonicity in n depends on the parity of d. We show that $b(2k;2n) = b(2k;2n+1) < b(2k;2n+2)$ and that $b(2k+1;2n) \geq b(2k+1;2n+1) < b(2k+1;2n+2)$, with strict inequality in the first case if $n \geq k$. In other words, $b(2k;n)$ is an increasing staircase, and $b(2k+1;n)$ starts that way but eventually zigzags. By reducing $F_d(x)$ in $(\mathbb{Z}/2\mathbb{Z})[[x]]$, we show that $b(d;n)$ is odd if and only if n is congruent to 0 or 1 (mod d). We conclude the section with an alternate interpretation of $b(d;n)$, which was suggested to us by Richard Stanley.

In section three, we discuss the special case $d = 2^r$. We show that $F_{2^r}(x)$ is rational, and that $b(2^r;n)$ is the number of partitions of n into powers of $2 \leq 2^{r-1}$. We give a closed form for $b(2^r;2^{r-1}s+t)$, $0 \leq t \leq 2^{r-1}-1$: it is a polynomial in s of degree $r-1$, in fact, a linear combination of $\binom{s+r-1-j}{r-1}$'s, $0 \leq j \leq r-1$. Each such polynomial has the same leading coefficient, so $b(2^r;n) \sim (2^{r(r-1)/2}(r-1)!)^{-1}n^{r-1}$. We conclude the section by reinterpreting some early work of Tanturri on $b(2^r;n)$.

In section four, we consider the asymptotic growth of $b(2k;n)$. We show that $b(2k;n) = \Theta(n^{\lambda(2k)})$ for $\lambda(2k) = \log_2 k$ (that is, there exist $\alpha > \beta > 0$ and n_0 so that $\alpha n^{\lambda(2k)} > b(2k;n) > \beta n^{\lambda(2k)}$ for $n \geq n_0$.) We also show that $b(2k+1;n)$ is not $\Theta(n^{\lambda(2k+1)})$ for any $\lambda(2k+1)$. (The previous result implies that $\lambda(2k+1) = \log_2(k+\frac{1}{2})$, so $b(2k+1;2^r)$ would be $\Theta((k+\frac{1}{2})^r)$. However, the recurrences imply that $b(2k+1;2^r)$ satisfies a monic linear recurrence in r with integer coefficients, and $b(2k+1;2^r) = \Theta(\tau^r)$ implies that τ is an algebraic integer–see Cor. 1.7.) We also compute $\mu_i(2k+1)$ so that, for suitable $\alpha_i > 0$, $\alpha_1 n^{\mu_1(2k+1)} > b(2k+1;n) > \alpha_2 n^{\mu_2(2k+1)}$. Since $\lambda(2k+2) > \mu_1(2k+1)$ for $k \geq 1$ and $\mu_2(2k+1) > \lambda(2k)$ for $k \geq 2$, it follows that $b(d+1;n)/b(d;n) \to \infty$ for all $d \geq 3$.

In section five, we use known properties of the Stern sequence to give more specific information about the growth of $b(3;n)$ and $b(6;n)$. We show that $\mu_1(3)$ and $\mu_2(3)$ are best possible, and that $n^{-\lambda(6)}b(6;n)$ does not converge, even though $b(6;n) = \Theta(n^{\lambda(6)})$. This is, in effect, a result of Carlitz [**C3**], which was suggested by a question of P. T. Bateman.

We conclude, in section six, with acknowledgments and some open questions, and, in an appendix, give a table of $b(d;n)$ for $0 \leq n \leq 32$ and $2 \leq d \leq 9$ and $d = \infty$.

We shall repeatedly use a familiar result on linear recurrences with constant coefficients, which goes back to Lagrange and Euler.

Linear Recurrence Theorem. *Suppose*

$$p(t) = t^r + c_1 t^{r-1} + \cdots + c_r = t^\kappa \prod_{i=1}^{s} (t - \lambda_i)^{r_i}, \tag{1.4}$$

where $c_i \in \mathbb{C}$, $0 \neq \lambda_i \in \mathbb{C}$, $r_i \geq 1$ and the λ_i's are distinct, and suppose (x_n) is a sequence satisfying the recurrence

$$x_{n+r} + c_1 x_{n+r-1} + \cdots + c_r x_n = 0, n \geq 0. \tag{1.5}$$

Then there exist polynomials h_i, of degree $r_i - 1$, so that

$$x_n = \sum_{i=1}^{s} h_i(n)\lambda_i^n \quad \text{for} \quad n \geq \kappa. \tag{1.6}$$

The simplest proof of the Linear Recurrence Theorem involves generating functions and partial fractions; one version is in [**R2**].

Corollary 1.7. *Keeping the previous notation, suppose (x_n) is a real sequence satisfying (1.5) and for some $\tau > 0$, α, $\beta > 0$ and all $n \geq n_0$,*

$$\alpha\tau_n \geq x_n \geq \beta\tau_n. \tag{1.8}$$

Then $\tau = max|\lambda_i|$ and $p(\tau) = 0$. If $p \in \mathbb{Z}[t]$, then τ is an algebraic integer.

Proof: Let $M = \max\{|\lambda_j|\}$, let $d = \max\{\deg h_j : |\lambda_j| = M\}$ and reindex so that $\lambda_j = M\epsilon_j$, $|\epsilon_j| = 1$, and $\deg h_j = d$ precisely for $1 \leq j \leq k$. Then, $h_j(n) = \alpha_j n^d + o(n^d)$, where $\alpha_j \neq 0$. Finally, let

$$H(n) = \sum_{j=1}^{k} \alpha_j \epsilon_n^j. \tag{1.9}$$

Then by the Linear Recurrence Theorem,

$$x_n = H(n)n^d M^n + o(n^d M^n). \tag{1.10}$$

If $|\omega| = 1$, $\omega \neq 1$, then

$$|\sum_{n=1}^{N} \omega^n| = \frac{|(\omega^{N+1} - \omega)|}{|\omega - 1|} \leq \frac{2}{|\omega - 1|} < \infty. \tag{1.11}$$

Since $|H(n)| \leq \Sigma|\alpha_j| = A$, and

$$\begin{aligned}\sum_{i=1}^{N} |H(n)|^2 &= \left(\sum_{j=1}^{k} |\alpha_j|^2\right) N + \sum_{j \neq \ell} \alpha_j \bar{\alpha}_\ell \left(\sum_{n=1}^{N} \epsilon_j^n \bar{\epsilon}_\ell^n\right) \\ &= B^2 N + \mathcal{O}(1),\end{aligned} \tag{1.12}$$

$\varlimsup_{n \to \infty} |H(n)| \geq B > 0$. Thus for all $\epsilon > 0$, there are infinitely many n with

$$(A + \epsilon)n^d M^n \geq x_n \geq (B - \epsilon)n^d M^n. \tag{1.13}$$

It follows from (1.8) that $d = 0$ and $M = \tau$.

Suppose $p(\tau) \neq 0$, then $\epsilon_j \neq 1$ for all j, and by (1.11),

$$|\sum_{n=1}^{N} H(n)| = |\sum_{j=1}^{N} \alpha_j \sum_{n=0}^{N} \epsilon_j^n| \leq \sum_{j=1}^{k} \frac{2|\alpha_j|}{|1 - \epsilon_j|} < \infty. \tag{1.14}$$

But by (1.8), $\varliminf_{n \to \infty} H(n) \geq \beta > 0$. This is a contradiction, so $p(\tau) = 0$. ∎

2. Basic Properties of $b(d; n)$

The following infinite product formulas for the generating functions of $b(d; n)$ and $b(\infty; n)$ are immediate from (1.1):

$$\begin{aligned}F_d(x) = \sum_{n=0}^{\infty} b(d; n)x^n &= \prod_{j=0}^{\infty} (1 + x^{2^j} + \cdots + x^{(d-1)2^j}) \\ &= \prod_{j=0}^{\infty} \frac{1 - x^{d \cdot 2^j}}{1 - x^{2^j}};\end{aligned} \tag{2.1}$$

$$F_\infty(x) = \prod_{j=0}^{\infty} \frac{1}{1 - x^{2^j}}. \tag{2.2}$$

The following theorem summarizes some elementary manipulations of the generating functions in (2.1) and (2.2).

Theorem 2.3.

(i) $F_d(x)F_\infty(x^d) = F_\infty(x)$,

(ii) $F_{2k}(x) = (1-x)^{-1}F_k(x^2)$,

(iii) $(1-x)F_d(x) = (1-x^d)F_d(x^2)$,

(iv) $F_k(x) = (1-x^k)F_{2k}(x)$.

Theorem 2.3 leads to many recurrences. For convenience, we shall construe $b(d;n)$ to be 0 when n is negative.

Theorem 2.4.

(i) $b(\infty;n) = \sum_{r=0}^{\lfloor n/d\rfloor} b(d;n-dr)b(\infty;r)$,

(ii) $b(2k;n) = \sum_{j=0}^{\lfloor n/2\rfloor} b(k;j)$,

(iii) $b(2k;2n) = b(2k;2n+1) = \sum_{j=0}^{k-1} b(2k;n-j)$,

(iv) $b(2k+1;2n) = \sum_{j=0}^{k} b(2k+1;n-j)$,

(v) $b(2k+1;2n+1) = \sum_{j=0}^{k-1} b(2k+1;n-j)$,

(vi) $b(2k;n) = b(k;n) + b(2k;n-k)$,

(vii) $b(2k;n) - b(2k;n-2) = b(k;\lfloor n/2\rfloor)$,

(viii) $b(2k+1;2n) - b(2k+1,2n-1) = b(2k+1;n)$,

(ix) $b(2k+1;2n) - b(2k+1,2n+1) = b(2k+1;n-k)$,

(x) $b(2k;n) = \sum_{r=0}^{\lfloor n/k\rfloor} b(k;n-rk)$.

Proof: Expanding Thm. 2.3(i), we have

$$\sum_{m=0}^{\infty} b(d;m)x^m \sum_{r=0}^{\infty} b(\infty;r)x^{dr} = \sum_{n=0}^{\infty} b(\infty;n)x^n; \tag{2.5}$$

part (i) follows from comparing the coefficient of x^n on both sides of (2.5). Similarly, Thm. 2.3(ii) expands to

$$F_{2k}(x) = \sum_{n=0}^{\infty} b(2k;n)x^n = (1+x+x^2+\cdots)\sum_{i=0}^{\infty} b(k;i)x^{2i}, \tag{2.6}$$

which implies (ii). Thm. 2.3(iii) is equivalent to:

$$F_d(x) = \sum_{n=0}^{\infty} b(d;n)x^n = (1 + x + \cdots + x^{d-1}) \sum_{i=0}^{\infty} b(d;i)x^{2i}. \qquad (2.7)$$

The term x^n occurs on the right when $n = j + 2i$, where $0 \le j \le d-1$, so $b(d;n)$ is the sum of those $b(d;n-j)$'s in which $n-j$ is an even integer. Parts (iii) through (v) arise by considering the varying parities of d and n. We obtain (vi) by writing out Thm. 2.3(iv). Finally, (vii), (viii), (ix) and (x) result from iterating (ii), (iv), (v) and (vi). ∎

Several comments about these recurrences are in order. Since $b(\infty;0) = 1$, we could use Thm. 2.4(i) to define $b(d;n)$ recursively. Also, when $r = 2$, this becomes (by (1.2)(i)),

$$b(\infty;n) = b(\infty;0) + \cdots + b(\infty;\lfloor n/2 \rfloor), \qquad (2.8)$$

This equation is in Tanturri [**T2**], but also follows easily from

$$b(\infty;n) = b(\infty, n-2) + b(\infty;\lfloor n/2 \rfloor), \qquad (2.9)$$

which is implicit in Euler [**E2**]. Churchhouse iterated (2.8) to express $b(\infty;2^r n)$ in terms of $\{b(\infty;j) : 0 \le j \le n\}$, and generalizations of this idea represent much of the literature on binary (and m-ary) partitions.

There are combinatorial proofs for many of these recurrences. For example, if $\epsilon_j \in \{0,\ldots,2k-1\}$, then $\epsilon_j = 2\nu_j + \eta_j$, where $\nu_j \in \{0,\ldots,k-1\}$ and $\eta_j \in \{0,1\}$. So, $n = \sum \epsilon_j 2^j = 2(\sum \nu_j 2^j) + (\sum \eta_j 2^j)$, and for every partition $n = 2s + t$, there are $b(k;s) \cdot 1$ ways of writing $s = \sum \nu_j 2^j$ and $t = \sum \eta_j 2^j$; this proves (ii).

We turn to the monotonicity properties of $b(d;n)$.

Theorem 2.10.

(i) $b(d;n) \le b(d+1;n)$,

(ii) $1 \le b(d;n)$,

(iii) $b(d;n) = b(\infty;n)$ *if* $d > n$,

(iv) $b(d;n) = b(\infty;n) - b(\infty;n-d)$ *if* $n \ge d > n/2$.

Proof: Any solution of (1.1) satisfies $0 \le \epsilon_i \le d-1 \le d$, whence (i); (ii) follows by induction and (1.2)(i). For (iii) and (iv), we use Thm. 2.4(i):

$$b(\infty;n) = b(d;n)b(\infty;0) + b(d;n-d)b(\infty,1) + b(d;n-2d)b(\infty;2) + \cdots. \qquad (2.11)$$

Recall that $b(\infty;0) = b(\infty;1) = 1$. If $d > n$, then (2.11) implies (iii), if $n \ge d > n/2$, then $b(\infty;n) = b(d;n) + b(d;n-d)$, but $b(d;n-d) = b(\infty;n-d)$ by (iii) since $d > n-d$, thus (iv). These can also be proved combinatorially. ∎

Theorem 2.12.

(i) $b(2k;2n) = b(2k;2n+1)$,

(ii) $b(2k;2n) > b(2k;2n-1)$,

(iii) $b(2k+1;2n) > b(2k+1,2n-1)$,

(iv) $b(2k+1;2n) \geq b(2k+1,2n+1)$, *with ">" if* $n \geq k$.

Proof: Parts (i), (iii) and (iv) follow from Thm. 2.4(iii), (viii) and (ix). For (ii), Thm. 2.4(vii) implies that

$$b(2k;2n) - b(2k;2n-1) = b(2k;2n) - b(2k;2n-2) = b(k;n). \quad \blacksquare \tag{2.13}$$

The generating functions allow us to determine the parity of $b(d;n)$.

Theorem 2.14.

$$b(d;n) \equiv 1 \pmod 2 \iff n \equiv 0,1 \pmod d$$

Proof: We reduce (2.1) (mod 2), viewing $F_d(x)$ as an element of $(\mathbb{Z}/2\mathbb{Z})[[x]]$:

$$F_d(x) = \prod_{j=0}^{\infty} \frac{(1-x^{d\cdot 2^j})}{(1-x^{2^j})} \equiv \prod_{j=0}^{\infty} \frac{(1+x^{d\cdot 2^j})}{(1+x^{2^j})}. \tag{2.15}$$

Since $\prod(1+x^{2^j}) = (1-x)^{-1} \equiv (1+x)^{-1}$ in $(\mathbb{Z}/2\mathbb{Z})[[x]]$, (2.15) becomes

$$\sum_{n=0}^{\infty} b(d;n)x^n = \frac{1+x}{1+x^d} \equiv (1+x)(1+x^d+x^{2d}+\cdots). \quad \blacksquare \tag{2.16}$$

This is consistent with (1.2): $b(2;n) = 1$ is always odd; $b(3;n)$ is even when n is a multiple of 3 (Stern); $b(4;n)$ is odd when $\lfloor n/2 \rfloor$ is even. There is a vaguely similar formula for any prime p, based on the identity $(\varphi(x))^p = \varphi(x^p)$ for $\varphi \in (\mathbb{Z}/p\mathbb{Z})[[x]]$. Theorem 2.3(i) implies that:

$$F_p(x)(F_\infty(x))^{p-1} \equiv 1 \pmod p. \tag{2.17}$$

Let $v(m)$ denote the number of 1's in the usual (unique) binary representation of m. Richard Stanley **[S2]** has made the following interesting observation to the author.

Theorem 2.18. *Suppose ω is any primitive d-th root of unity. Then,*

$$b(d;n) = \sum (-\omega)^{v(m_1)}(-\omega^2)^{v(m_2)} \cdots (-\omega^{d-1})^{v(m_{d-1})}, \tag{2.19}$$

where the sum is taken over all (ordered) sums $n = m_1 + \cdots + m_{d-1}$.

Proof: For all z,

$$\sum_{n=0}^{\infty} z^{v(n)} x^n = \prod_{j=0}^{\infty} (1 + z x^{2^j}). \tag{2.20}$$

Since ω is a primitive d-th root of unity,

$$1 + x^{2^j} + x^{2\cdot 2^j} + \cdots + x^{(d-1)\cdot 2^j} = \prod_{\ell=1}^{d-1} (1 - \omega^\ell x^{2^j}); \tag{2.21}$$

hence by (2.1), (2.20) and (2.21),

$$\begin{aligned} F_d(x) &= \prod_{j=0}^{\infty} (1 + x^{2^j} + x^{2\cdot 2^j} + \cdots + x^{(d-1)\cdot 2^j}) \\ &= \prod_{j=0}^{\infty} \prod_{\ell=1}^{d-1} (1 - \omega^\ell x^{2^j}) = \prod_{\ell=1}^{d-1} \sum_{m=0}^{\infty} (-\omega^\ell)^{v(m)} x^m, \end{aligned} \tag{2.22}$$

from which (2.19) follows. ∎

Let $d = 3$ and $\omega = \exp(4\pi i/3)$. Then $\epsilon = -\omega = \exp(\pi i/3)$ is a primitive sixth root of 1, as is $-\omega^2 = \epsilon^{-1}$; we have $(m_1, m_2) = (j, n - j)$, and

$$b(3; n) = \sum_{j=0}^{n} \epsilon^{v(j) - v(n-j)}. \tag{2.23}$$

It follows that the positivity of $b(3; n)$ reflects upon the distribution of $\{v(j) - v(n - j)\}$ (mod 6).

3. The Case $d = 2^r$

If $d = 2^r$, then the infinite product in (2.1) telescopes:

$$F_{2^r}(x) = \sum_{n=0}^{\infty} b(2^r; n) x^n = \prod_{j=0}^{r-1} \frac{1}{1 - x^{2^j}}. \tag{3.1}$$

Theorem 3.2.

(i) $b(2^r;n)$ *is the number of partitions of* n *into* $1, 2, 2^2, \ldots, 2^{r-1}$.

(ii) $\sum_{j=0}^{2^r-1} (-1)^{v(j)} b(2^r;n-j) = 0$ *for* $n \geq 1$.

(iii) $b(2^{r+s};n) = \sum_{j=0}^{\lfloor n/2^r \rfloor} b(2^r;n-j\cdot 2^r)b(2^s;j)$.

Proof:
(i) This is immediate from (3.1).
(ii) Note that

$$(F_{2^r}(x))^{-1} = \prod_{j=0}^{r-1}(1-x^{2^j}) = \sum_{k=0}^{2^r-1}(-1)^{v(k)}x^k, \tag{3.3}$$

(c.f. (2.20)); (ii) follows upon multiplying both sides of (3.3) by $F_{2^r}(x)$.
(iii) Manipulation of (3.1) (or iteration of Theorem 2.3(ii) with $k = 2^s$) shows that $F_{2^{r+s}}(x) = F_{2^r}(x)F_{2^s}(x^{2^r})$, which leads directly to (iii). ∎

We remark that (i) connects $b(d;n)$ with the literature on restricted binary partitions. We may use (ii) with the Linear Recurrence Theorem to describe a closed form for $b(2^r;n)$. By (3.3), (1.6) holds with the λ_i's equal to the 2^u-th primitive roots of unity, $0 \leq u \leq r-1$; for such a λ_i, h_i has degree $r-1-u$ (see Thm. 3.6.) Finally, if r or s equals 1, then (iii) reduces to Thm. 2.4(ii) or (x). A version of (iii) seems to be in **[T2,T3]**.

We now introduce an important reparameterization for $b(2^r;n)$. For $0 \leq t \leq 2^{r-1}-1$, let

$$f(r,t)(s) = b(2^r;2^{r-1}s+t). \tag{3.4}$$

Using our recurrences, it is easy to compute $f(r;t)$ for small r:

$$f(1,0)(s) = b(2;s) = 1, \tag{3.5)(i}$$

$$f(2,0)(s) = b(4;2s) = f(2,1)(s) = b(4;2s+1) = s+1, \tag{3.5)(ii}$$

$$f(3,0)(s) = b(8;4s) = f(3;1)(s) = b(8;4s+1) = (s+1)^2, \tag{3.5)(iii}$$

$$f(3,2)(s) = b(8;4s+2) = f(3;3)(s) = b(8;4s+3) = (s+1)(s+2). \tag{3.5)(iv}$$

Theorem 3.6. *For all* $r \geq 1$, $0 \leq t \leq 2^{r-1}-1$, *and* $s \geq 0$,

$$f(r,t)(s) = \sum_{j=0}^{r-2} a_j(r,t)\binom{s+r-1-j}{r-1}, \tag{3.7}$$

where the $a_j(r,t)$'s are defined recursively by

$$a_0(1,0) = a_0(2,1) = 1, \tag{3.8)(i}$$

$$a_j(r+1,2k) = a_j(r+1,2k+1) = \sum_{t=0}^{k} a_j(r,t) + \sum_{t=k+1}^{2^{r-1}-1} a_{j-1}(r,t), \tag{3.8)(ii}$$

and we take $a_{-1}(r,t) = a_{r-1}(r,t) = 0$ in (3.8)(ii) when appropriate.

Proof: Clearly, (3.7) holds for $r = 2$; assume it holds for r. Theorem 2.4(ii), rephrased and applied to $b(2^{r+1}; 2^r s + t) = f(r+1,t)(s)$ gives:

$$\begin{aligned} f(r+1,2k)(s) = f(r+1,2k+1)(s) &= \sum_{i=0}^{2^{r-1}s+k} b(2^r;i) \\ &= \sum_{t=0}^{k}\sum_{u=0}^{s} f(r,t)(u) + \sum_{t=k+1}^{2^{r-1}-1}\sum_{u=0}^{s-1} f(r,t)(u), \end{aligned} \tag{3.9}$$

By the induction hypothesis and a familiar binomial identity, we obtain:

$$\begin{aligned} &f(r+1,2k)(s) = f(r+1;2k+1)(s) \\ &= \sum_{t=0}^{k}\sum_{u=0}^{s}\sum_{j=0}^{r-2} a_j(r,t)\binom{u+r-1-j}{r-1} \\ &\quad + \sum_{t=k+1}^{2^{r-1}-1}\sum_{u=0}^{s-1}\sum_{j=0}^{r-2} a_j(r,t)\binom{u+r-1-j}{r-1} \\ &= \sum_{t=0}^{k}\sum_{j=0}^{r-2} a_j(r,t)\binom{s+r-j}{r} \\ &\quad + \sum_{t=k+1}^{2^{r-1}-1}\sum_{j=0}^{r-2} a_j(r,t)\binom{s+r-1-j}{r} \\ &\sum_{j=0}^{r-1}\left(\sum_{t=0}^{k} a_j(r,t) + \sum_{t=k+1}^{2^r-1} a_{j-1}(r,t)\right)\binom{s+r-j}{r}. \end{aligned} \tag{3.10}$$

It follows from the last expression and (3.8)(ii) that (3.7) holds for $r+1$. ∎

Since $a_0(r,t) = b(\infty;t)$, it seems unlikely that there is a closed-form expression for the $a_j(r,t)$'s. We can rephrase (3.5) in these terms:

$$f(1,0)(s) = \binom{s}{0}, \tag{3.11)(i}$$

$$f(2,0)(s) = f(2,1)(s) = \binom{s+1}{1}, \tag{3.11)(ii}$$

$$f(3,0)(s) = f(3;1)(s) = \binom{s+2}{2} + \binom{s+1}{2}, \tag{3.11)(iii}$$

$$f(3,2)(s) = f(3;3)(s) = 2\binom{s+2}{2}. \tag{3.11)(iv}$$

Corollary 3.12.
(i) For $r \geq 1$, $f(r,t)(s)$ is a polynomial in s of degree $r-1$ with leading coefficient $2^{(r-1)(r-2)/2}\{(r-1)!\}^{-1}$.

(ii)
$$\lim_{n\to\infty} \frac{b(2^r;n)}{n^{r-1}} = \frac{1}{2^{r(r-1)/2}(r-1)!}.$$

Proof:
(i) Each $\binom{s+r-1-j}{r-1}$ is a polynomial in s of degree $r-1$ with leading coefficient $\{(r-1)!\}^{-1}$. An easy induction shows that $\sum_j a_j(r,t) = 2^{(r-1)(r-2)/2}$.
(ii) Let $u = r-1$ and fix t. For $n \in \{2^{r-1}s + t : 0 \leq s < \infty\}$, we have

$$n^{-u}b(2^r;n) = (2^u s + t)^{-u}\{2^{u(u-1)/2}(u!)^{-1}s^u + o(s^u)\}. \tag{3.13}$$

Since $u(u-1)/2 - u^2 = -r(r-1)/2$, (3.13) implies that (ii) holds for n in every sequence $\{2^{r-1}s + t\}$, and so for n in general. ∎

One can also prove (ii) by looking at the coefficient of $(1-z)^{-(r-1)}$ in the Laurent series for $F_{2^r}(z)$ at $z=1$.

A. Tanturri wrote a series of papers **[T1,T2,T3]** during World War I on binary partitions. His formulas are written in the now obscure symbolic notation of Peano, and, perhaps, have not become generally known for that reason. In **[T2]**, he defines $D(2^r;n)$ to be the number of partitions of n into powers of 2 such that the largest is 2^r. There is a clear bijection between these partitions and the partitions of $n-2r$ with parts taken from the set $\{1,2,\ldots,2^r\}$. Thus, by Thm. 3.2(i),

$$D(2^r;n) = b(2^{r+1};n-2^r). \tag{3.14}$$

(This also follows from $D(2^r;n) = b(2^{r+1};n) - b(2^r;n)$ and Thm. 2.4(vi).)

Proposition 3.15. *(Tanturri)*

(*i*)
$$\sum_{r=0}^{\infty} D(2^r;n) = \sum_{r=0}^{\infty} b(2^{r+1};n-2^r) = b(\infty;n),$$

(*ii*)
$$\sum_{r=0}^{\infty}(-1)^r D(2^r;n) = \sum_{r=0}^{\infty}(-1)^r b(2^{r+1};n-2^r) = 0 \quad \textit{for} \quad n \geq 1.$$

Proof: Since every binary partition of $n \geq 1$ contains a largest power of 2, (i) is immediate. For (ii), let $E(2^r;n)$ denote the number of partitions of n into powers of 2, in which 2^r is the largest power and occurs exactly once. By replacing the unique 2^r with two 2^{r-1}'s, we obtain a partition of n in which the largest power is 2^{r-1}, which occurs more than once. Conversely, in any such partition, two 2^{r-1}'s may be coalesced into one 2^r. Thus, for $r \geq 1$, $E(2^r;n) = D(2^{r-1};n) - E(2^{r-1};n)$, and for $n \geq 2$,

$$\sum_{r=0}^{\infty}(-1)^r D(2^r;n) = \sum_{r=0}^{\infty}(-1)^r (E(2^r;n) + E(2^{r+1},n)). \tag{3.16}$$

Since $E(2^r;n) = 0$ for $r > \log_2 n$, the sum in (3.16) converges; since $E(1;n) = 0$ for $n \geq 2$, the sum is zero. ∎

4. The Growth of $b(d;n)$

In this section we discuss the growth of $b(d;n)$ as $n \to \infty$; there is a dichotomy depending on the parity of d. We have seen that $b(2^r;n) \sim c \cdot n^{r-1}$. This generalizes partially to $b(2k;n) = \Theta(n^{\log_2 k})$, but no such relation holds for $b(2k+1;n)$. These results were announced in [**R1**].

Here is a sketch of the argument. We define intervals $I_r = I_r(d)$ so that, if $2n$, $2n+1 \in I_{r+1}$, then $n - j \in I_r$, $0 \leq j \leq (d-1)/2$. We then use the recurrences of Thm. 2.4 (iii), (iv), (v) to estimate $b(d;n)$ on I_{r+1} in terms of $b(d;n)$ on I_r. Finally, we turn these estimates into bounds in terms of n^λ. We need two straightforward lemmas.

Lemma 4.1. *For $r \geq r_0 = \lceil \log_2 d \rceil$, let $I_r = I_r(d) = [2^r - (d-1), 2^{r+1}]$. If $2n$ or $2n+1$ belongs to I_{r+1}, then $n-j$ belongs to I_r for $0 \leq j \leq (d-1)/2$.*

Proof: By hypothesis, $n \leq 2^{r+1}$ and

$$n - j \geq 2^r - (d-1)/2 - j \geq 2^r - (d-1). \quad ∎ \tag{4.2}$$

Lemma 4.3. *Suppose there exist γ_1, γ_2, σ, $\tau > 0$ so that for $n \in I_r$, $r \geq r_0$,*

$$\gamma_1 \sigma^r \geq b(d;n) \geq \gamma_2 \tau^r. \tag{4.4}$$

Then there exist new constants $\delta_i > 0$ so that for all sufficiently large n,

$$\delta_1 n^{\log_2 \sigma} \geq b(d;n) \geq \delta_2 n^{\log_2 \tau}. \tag{4.5}$$

Proof: If $r \geq r_0 + 1$, then $2^r - (d-1) \geq 2^{r-1}$, so if $n \in I_r$, then $2^{r+1} \geq n \geq 2^{r-1}$. Since $\rho = (2^r)^{\log_2 \rho}$ for any $\rho > 0$, it follows that

$$\rho n^{\log_2 \rho} = (2n)^{\log_2 \rho} \geq (2^r)^{\log_2 \rho} \geq (n/2)^{\log_2 \rho} = \rho^{-1} n^{\log_2 \rho}. \tag{4.6}$$

Thus, (4.6) with $\rho = \sigma$ and τ, and (4.4) combine to give (4.5). ∎

Theorem 4.7. *For all $k \geq 1$, $b(2k;n) = \Theta(n^{\lambda(2k)})$, where*

$$\lambda(2k) = \log_2 k. \tag{4.8}$$

Proof: We must find α and $\beta > 0$ and n_0 so that

$$\alpha n^{\lambda(2k)} \geq b(2k;n) \geq \beta n^{\lambda(2k)}, n \geq n_0. \tag{4.9}$$

By Lemma 4.3, it suffices to show that for $r \geq r_0$ there exist $\gamma_i > 0$ so that:

$$\gamma_1 k^r \geq b(2k;n) \geq \gamma_2 k^r \quad \text{for} \quad n \in I_r. \tag{4.10}$$

Let

$$M(2k;r) = \max\{b(2k;n) : n \in I_r\}, \tag{4.11)(i}$$

$$m(2k;r) = \min\{b(2k;n) : n \in I_r\}. \tag{4.11)(ii}$$

In Thm. 2.4(iii), if the argument on the left hand side comes from I_{r+1}, then the arguments on the right hand side come from I_r by Lemma 4.1. Thus, for $n \in I_{r+1}$:

$$kM(2k;r) \geq b(2k;n) \geq km(2k;r). \tag{4.12}$$

Taking the maximum and minimum in (4.12) for $n \in I_{r+1}$, we have

$$kM(2k;r) \geq M(2k;r+1), \tag{4.13)(i}$$

$$m(2k;r+1) \geq km(2k;r). \tag{4.13)(ii}$$

It follows from (4.13) that, for $n \in I_r$,

$$\begin{aligned} M(2k;r_0)k^{r-r_0} &\geq M(2k;r) \geq b(2k;n) \geq m(2k;r) \\ &\geq m(2k;r_0)k^{r-r_0}. \end{aligned} \tag{4.14}$$

This is an inequality of shape (4.10), which completes the proof. ∎

One fundamental difference between the even and the odd case is that, in Thm. 2.4(iv) and (v), the number of terms in the recurrences for $b(2k+1;2n)$ and $b(2k+1;2n+1)$ depend on the parity of n and k. The proof of the following theorem is quite oblique, and a more direct proof would be desirable.

Theorem 4.15. *There do not exist ν, α and $\beta > 0$ so that for $n \geq N$,*

$$\alpha n^{\nu} \geq b(2k+1;n) \geq \beta n^{\nu}. \tag{4.16}$$

Proof: Suppose to the contrary, that (4.16) holds, and let

$$R = \sum_{j=0}^{N-1} b(2k+1;j). \tag{4.17}$$

Then by Thm. 2.4(ii), for $n \geq N$, we would have

$$R + \alpha \sum_{j=N}^{n} j^{\nu} \geq \sum_{j=0}^{n} b(2k+1;j) = b(4k+2;2n) \geq R + \beta \sum_{j=N}^{n} j^{\nu}. \tag{4.18}$$

By the usual estimates for $\sum j^{\nu}$, (4.18) implies that, for suitable constants c_i and sufficiently large n,

$$c_1 + c_2 n^{\nu+1} \geq b(4k+2;2n) \geq c_3 + c_4 n^{\nu+1}. \tag{4.19}$$

In view of Thm. 4.7, it follows that

$$\nu + 1 = \lambda(4k+2) = \log_2(2k+1). \tag{4.20}$$

Thus, for t sufficiently large, (4.16) and (4.20) imply that

$$\alpha(k+\frac{1}{2})^t \geq b(2k+1;2^t) \geq \beta(k+\frac{1}{2})^t. \tag{4.21}$$

Let $M = M_{2k+1} = [m_{ij}]$, $0 \leq i, j \leq 2k$ denote the matrix in which

$$m_{ij} = 1 \quad \text{if} \quad \lceil i/2 \rceil \leq j \leq \lfloor i/2 \rfloor + k, \; m_{ij} = 0 \quad \text{otherwise}. \tag{4.22}$$

Thus, the even rows of M contain a block of $(k+1)$ 1's and the odd rows contain a block of k 1's, and these blocks sidestep their way from northwest to southeast. Define the $(2k+1)$-column vector V_t by:

$$V_t = (b(2k+1;2^t), b(2k+1;2^t-1), \ldots, b(2k+1;2^t-2k))^T. \tag{4.23}$$

Then by construction, and Thm. 2.4 (iv), (v),

$$V_{t+1} = MV_t, \quad t \geq 0. \tag{4.24}$$

Hence for $t \geq 0$,

$$V_t = M^t V_0. \tag{4.25}$$

Let

$$p(t) = det[xI - M] = x^{2k+1} + c_1 x^{2k} + \cdots + c_{2k+1} \tag{4.26}$$

denote the characteristic polynomial of M. Then $p \in \mathbb{Z}[x]$ and, by the Cayley-Hamilton theorem, $p(M) = 0$. Thus, for $t \geq 0$,

$$M^{t+2k+1} + c_1 M^{t+2k} + \cdots + c_{2k+1} M^t = 0. \tag{4.27}$$

It follows from (4.25) that for $t \geq 0$,

$$V_{t+2k+1} + c_1 V_{t+2k} + \cdots + c_{2k+1} V_t = (0, 0, \ldots, 0)^T. \tag{4.28}$$

Let $x_t = b(2k+1; 2^t)$; taking the first component of (4.28), we obtain

$$x_{t+2k+1} + c_1 x_{t+2k} + \cdots + c_{2k+1} x_t = 0 \text{ for } t \geq 0. \tag{4.29}$$

That is, (1.5) holds for (x_t). But by Cor. 1.7, (4.21) and (4.29) imply that $k + \frac{1}{2}$ is an algebraic integer. This contradiction completes the proof. ∎

In any event, the monotonicity of $b(d; n)$ in d implies that, for suitable constants and sufficiently large n,

$$\alpha n^{\log_2(k+1)} \geq b(2k+2; n) \geq b(2k+1; n) \geq b(2k; n) \geq \beta n^{\log_2 k}. \tag{4.30}$$

We can improve on (4.30) by using a lemma, which we do not give in its greatest generality.

Lemma 4.31. *Suppose $M = [a_{ij}]$ is a real 2×2 matrix, $a_{ij} > 0$, with eigenvalue $\lambda > 0$, and associated eigenvector (v_1, v_2), $v_i > 0$. Suppose for all $r \geq 0$, the sequences $(f_i(r))$ and $(h_i(r))$ satisfy the inequalities:*

$$f_1(r) \geq f_2(r) > 0, \tag{4.32)(i}$$
$$f_1(r+1) \leq a_{11} f_1(r) + a_{21} f_2(r), \tag{4.32)(ii}$$
$$f_2(r+1) \leq a_{12} f_1(r) + a_{22} f_2(r); \tag{4.32)(iii}$$

$$h_1(r) \geq h_2(r) > 0, \tag{4.33)(i}$$
$$h_1(r+1) \geq a_{11} h_1(r) + a_{21} h_2(r), \tag{4.33)(ii}$$
$$h_2(r+1) \geq a_{12} h_1(r) + a_{22} h_2(r). \tag{4.33)(iii}$$

Then there exist $c > 0$ and $c' > 0$ so that for all $r \geq 0$,

$$c\lambda^r \geq f_1(r) \geq f_2(r), \tag{4.34}$$
$$h_1(r) \geq h_2(r) \geq c'\lambda^r. \tag{4.35}$$

Proof: First, we take the (v_1, v_2) linear combination of (4.32)(ii) and (iii), which preserves the inequality. Since $(v_1, v_2)^T$ is an eigenvector,

$$\begin{aligned} &v_1 f_1(r+1) + v_2 f_2(r+1) \\ &\leq (v_1 a_{11} + v_2 a_{12}) f_1(r) + (v_1 a_{21} + v_2 a_{22}) f_2(r) \\ &= \lambda(v_1 f_1(r) + v_2 f_2(r)) \end{aligned} \tag{4.36}$$

Since $f_2(r) > 0$, (4.36) iterated r times implies that

$$v_1 f_1(r) \leq v_1 f_1(r) + v_2 f_2(r) \leq (v_1 f_1(0) + v_2 f_2(0))\lambda^r. \tag{4.37}$$

Thus, (4.34) holds with $c = f_1(0) + v_1^{-1} v_2 f_2(0) > 0$. Similar reasoning applied to (4.33) leads to

$$v_1 h_1(r) + v_2 h_2(r) \geq (v_1 h_1(0) + v_2 h_2(0))\lambda^r, \tag{4.38}$$

and, since $h_1(r) \geq h_2(r)$, (4.38) implies that

$$(v_1 + v_2) h_1(r) \geq (v_1 + v_2) h_2(0)\lambda^r. \tag{4.39}$$

By (4.33)(iii) and (4.39),

$$h_2(r+1) \geq a_{12} h_1(r) + a_{22} h_2(r) \geq a_{12} h_1(r) \geq a_{12} h_2(0)\lambda^r. \tag{4.40}$$

Thus, (4.35) holds for $c' = \min\{h_2(0), a_{12} h_2(0)\lambda^{-1}\} > 0$. ∎

Theorem 4.41. *There exist $\mu_i(2k+1)$ and α and $\beta > 0$ so that for $n \geq n_0$,*

$$\alpha n^{\mu_1(2k+1)} \geq b(2k+1; n) \geq \beta n^{\mu_2(2k+1)}. \tag{4.42}$$

Moreover,

$$\lambda(2k+2) > \mu_1(2k+1), k \geq 1 \qquad (4.43)(i)$$
$$\mu_2(2k+1) > \lambda(2k), k \geq 2. \qquad (4.43)(ii)$$

Proof: We mimic the proof of Thm. 4.7. Let

$$M^e(2k+1; r) = \max\{b(2k+1; n) : n \in I_r,\ n \text{ even}\}, \qquad (4.44)(i)$$
$$M^o(2k+1; r) = \max\{b(2k+1; n) : n \in I_r,\ n \text{ odd}\}, \qquad (4.44)(ii)$$
$$m^e(2k+1; r) = \min\{b(2k+1; n) : n \in I_r,\ n \text{ even}\}, \qquad (4.44)(iii)$$
$$m^o(2k+1; r) = \min\{b(2k+1; n) : n \in I_r,\ n \text{ odd}\}. \qquad (4.44)(iv)$$

By Thm. 2.12(iii) and (iv), $b(2k+1;2m) \geq b(2k+1;2m\pm 1)$, hence

$$M^e(2k+1;r) \geq M^o(2k+1;r), \tag{4.45)(i}$$
$$m^e(2k+1;r) \geq m^o(2k+1;r). \tag{4.45)(ii}$$

As before, (4.42) follows if we can find $c_i > 0$ and σ, τ so that for $r \geq r_0$,

$$c_1\tau^r \geq M^e(2k+1;r), \tag{4.46)(i}$$
$$m^o(2k+1;r) \geq c_2\sigma r. \tag{4.46)(ii}$$

We divide into two cases, depending on $k \pmod 2$.

First suppose $k = 2s$, $s \geq 1$. Then, Thm. 2.4 (iv), (v) becomes

$$b(4s+1;2n) = b(4s+1;n) + \cdots + b(4s+1;n-2s), \tag{4.47)(i}$$
$$b(4s+1;2n+1) = b(4s+1;n) + \cdots + b(4s+1;n-(2s-1)). \tag{4.47)(ii}$$

There are $2s+1$ $(n-j)$'s on the right hand side of (4.47)(i), either $s+1$ or s of them are even, and the rest odd. Taking the most extreme cases, we obtain the following estimates for $2n \in I_{r+1}$:

$$b(4s+1;2n) \leq (s+1)M^e(4s+1;r) + sM^o(4s+1;r), \tag{4.48)(i}$$
$$b(4s+1;2n) \geq sm^e(4s+1;r) + (s+1)m^o(4s+1;r). \tag{4.48)(ii}$$

Similarly, there are $2s$ $(n-j)$'s on the right-hand side of (4.47)(ii), so s of them are even and s are odd, and

$$b(4s+1;2n+1) \leq sM^e(4s+1;r) + sM^o(4s+1;r), \tag{4.49)(i}$$
$$b(4s+1;2n+1) \geq sm^e(4s+1;r) + sm^o(4s+1;r). \tag{4.49)(ii}$$

In (4.48) and (4.49), we take the maximum over $2n$, $2n+1 \in I_{r+1}$ in (i) and the minimum in (ii) and, in view of (4.45), obtain two systems like (4.32) and (4.33), (with f_1, $f_2 = M^e$, M^o and h_1, $h_2 = m^e, m^o$):

$$M^e(4s+1;r+1) \leq (s+1)M^e(4s+1;r) + sM^o(4s+1;r), \tag{4.50)(i}$$
$$M^o(4s+1;r+1) \leq sM^e(4s+1;r) + sM^o(4s+1;r); \tag{4.50)(ii}$$

$$m^e(4s+1;r+1) \geq sm^e(4s+1;r) + (s+1)m^o(4s+1;r), \tag{4.51)(i}$$
$$m^o(4s+1;r+1) \geq sm^e(4s+1;r) + sm^o(4s+1;r). \tag{4.51)(ii}$$

Observe that the matrices

$$M_1 = \begin{bmatrix} s+1 & s \\ s & s \end{bmatrix}, \quad M_2 = \begin{bmatrix} s & s \\ s+1 & s \end{bmatrix}, \tag{4.52}$$

have characteristic equations

$$p_1(x) = x^2 - (2s+1)x + s,\ p_2(x) = x^2 - 2sx - s, \tag{4.53}$$

respectively. We choose their larger eigenvalues:

$$\lambda_1 = \lambda_1(s) = \frac{1}{2}((2s+1) + (4s^2+1)^{1/2}), \qquad (4.54)(\mathrm{i})$$

$$\lambda_2 = \lambda_2(s) = s + (s^2+s)^{1/2}. \qquad (4.54)(\mathrm{ii})$$

For $s \geq 1$, each row of each $M_i - \lambda_i I$ has one positive and one negative entry, so the associated eigenvector has positive components. Thus the hypotheses of Lemma 4.31 are satisfied, upon identifying (4.50) and (4.51) with (4.32) and (4.33). We conclude from (4.34) and (4.35) that:

$$M^o(4s+1;r) \leq M^e(4s+1;r) \leq c\lambda_1^r, \qquad (4.55)(\mathrm{i})$$

$$m^e(4s+1;r) \geq m^o(4s+1;r) \geq c'\lambda_2^r. \qquad (4.55)(\mathrm{ii})$$

By the previous argument, it follows that (4.42) holds with

$$\mu_1(4s+1) = \log_2 \lambda_1(s),\ \mu_2(4s+1) = \log_2 \lambda_2(s). \tag{4.56}$$

We wish to establish (4.43) in this case. Considering (4.8), (4.54) and (4.56), we exponentiate both sides of (4.43) to base 2, obtaining

$$2s+1 > \frac{1}{2}((2s+1) + (4s^2+1)^{1/2}),\ s \geq 1, \qquad (4.57)(\mathrm{i})$$

$$s + (s^2+s)^{1/2} > 2s,\ s \geq 1. \qquad (4.57)(\mathrm{ii})$$

These inequalities may be routinely verified, completing the proof.

The identical reasoning holds when $k = 2s+1, s \geq 0$, with slight numerical changes. We have

$$b(4s+3;2n) = b(4s+1;n) + \cdots + b(4s+1;n-2s-1), \qquad (4.58)(\mathrm{i})$$

$$b(4s+3;2n+1) = b(4s+1;n) + \cdots + b(4s+1;n-2s). \qquad (4.58)(\mathrm{ii})$$

Arguing as before, but without the details, we find that

$$M^e(4s+3;r+1) \le (s+1)M^e(4s+3;r) + (s+1)M^o(4s+3;r), \tag{4.59)(i}$$

$$M^o(4s+3;r+1) \le (s+1)M^e(4s+3;r) + sM^o(4s+3;r); \tag{4.59)(ii}$$

$$m^e(4s+3;r+1) \ge (s+1)m^e(4s+3;r) + (s+1)m^o(4s+3;r), \tag{4.60)(i}$$

$$m^o(4s+3;r+1) \ge sm^e(4s+3;r) + (s+1)m^o(4s+3;r). \tag{4.60)(ii}$$

The matrices

$$M_3 = \begin{bmatrix} s+1 & s+1 \\ s+1 & s \end{bmatrix}, \quad M_4 = \begin{bmatrix} s+1 & s \\ s+1 & s+1 \end{bmatrix}, \tag{4.61}$$

again satisfy the hypotheses of Lemma 4.31 with eigenvalues

$$\lambda_3 = \lambda_3(s) = \frac{1}{2}((2s+1) + (4s^2+8s+5)^{1/2}), \tag{4.62)(i}$$

$$\lambda_4 = \lambda_4(s) = s+1+(s^2+s)^{1/2}, \tag{4.62)(ii}$$

and we conclude that (4.42) holds, where

$$\mu_1(4s+3) = \log_2 \lambda_3(s),\ \mu_2(4s+3) = \log_2 \lambda_4(s). \tag{4.63}$$

Again, verification of (4.43) reduces to two more easy inequalities:

$$2s+2 > \frac{1}{2}((2s+1) + (4s^2+8s+5)^{1/2}),\ s \ge 0, \tag{4.64)(i}$$

$$s+1+(s^2+s)^{1/2} > 2s+1,\ s \ge 1. \ \blacksquare \tag{4.64)(ii}$$

No claim is made that $\mu_1(2k+1)$ and $\mu_2(2k+1)$ are best possible for all k, although we show that this is true for $k=1$ (see Thm. 5.13.)

Corollary 4.65. *If $d \ge 3$, then*

$$\lim_{n\to\infty} \frac{b(d+1;n)}{b(d;n)} = \infty.$$

The omission of $d=2$ is intentional. It is easy to check that $b(3;2^r-1) = 1$ for all r, hence $b(3;n)/b(2;n) = 1$ infinitely often.

5. Two Special Cases

In this section we discuss in greater detail the growth of $b(3;n)$ and $b(6;n)$. We have seen that $b(2r;n)$ is very well-behaved. On the other hand, $b(3;n)$ is quite irregular, and its growth is described most easily in terms of a closely related sequence.

The Stern sequence was first studied [**S3**] in the 1850s by M. Stern, a student of Eisenstein, and has reappeared sporadically in the literature. (See [**R2**] for a more extensive bibliography.) It is defined recursively:

$$s(0) = 0, s(1) = 1, s(2n) = s(n), s(2n+1) = s(n) + s(n+1), n \geq 1. \quad (5.1)$$

The Stern sequence, *per se.* was apparently first defined in de Rham [**R4**]. The block of terms $\{s(2^r), s(2^r+1), \ldots, s(2^{r+1})\}$ formed the r-th row in the Stern-Brocot array, which was studied by Stern, Lucas [**L3**], Lehmer [**L1**], et al. We construe results about these terms as results about the Stern sequence.

Theorem 5.2.

$$b(3;n) = s(n+1) \quad \textit{for} \quad n \geq 0. \quad (5.3)$$

Proof: By Thm. 2.4(iv), (v), we have the recurrences:

$$b(3;2n) = b(3;n) + b(3;n-1); b(3;2n+1) = b(3;n) \quad \text{for} \quad n \geq 1. \quad (5.4)$$

Together with the initial condition $b(3;0) = 1$, (5.4) determines $b(3;n)$ for all n. A comparison with (5.1) shows that (5.3) holds for $n \leq 1$. An easy induction now shows that it holds for all n:

$$b(3;2n) = b(3;n) + b(3;n-1) = s(n+1) + s(n) = s(2n+1), \quad (5.5)(\text{i})$$

$$b(3;2n+1) = b(3;n) = s(n+1) = s(2n+2). \blacksquare \quad (5.5)(\text{ii})$$

An incorrect version of Thm. 5.2 appeared in [**C1,C2,C3,L2**]. Carlitz defined $\theta_0(n)$ to be the number of odd Stirling numbers $S(n,2r)$ of the second kind, and proved it was also the number of odd binomial coefficients $\binom{t}{u}$ so that $t+u=n$. He showed that $\theta_0(n)$ satisfied the same recurrences as $b(3;n)$ and gave its generating function:

$$G(x) = \sum_{n=0}^{\infty} \theta_0(n) x^n = \prod_{j=0}^{\infty} (1 + x^{2^j} + x^{2^{j+1}}). \quad (5.6)$$

Thus, $G(x) = F_3(x)$ and $\theta_0(n) = b(3;n)$. From this, "it is clear that $\theta_0(n)$ is the number of partitions

$$n = n_0 + n_1 \cdot 2 + n_2 \cdot 2^2 + \cdots \quad (0 \leq n_j \leq 2) \quad (5.7)$$

subject to the following conditions: if $n_0 = 1$, then $n_1 \leq 1$, if $n_1 = 2$, then $n_2 \leq 1$, if $n_2 = 2$, then $n_3 \leq 1$, and so on." [C1, p.62]. Since $\theta_0(n)$ is, in fact, the number of such partitions without the given conditions, there is an error. Apparently, (5.6) was viewed as counting partitions of n in which the j-th part was chosen from $\{0, 2^j, 2^{j+1}\}$. If $n_0 = 1$, then 2^0 was chosen in the 0-th part, and there is only one part left in which to select 2^1, etc. The error first presents itself at $n = 5$, where $(n_0, n_1, n_2) = (1, 2, 0)$ or $(1, 0, 1)$. Note that the first violates the "conditions", but the second occurs twice in this alternate interpretation: $5 = 2^0 + 2^2$ and 2^0 is taken in the 0-th part, but 2^2 may be taken either in the first or second part.

We need the following classical facts about the Stern sequence.

Proposition 5.8. *(Lucas, Lehmer) For $r \geq 0$, let $I_r = [2^r, 2^{r+1}]$, and define $M_r = \max\{s(n) : n \in I_r\}$ and $m_r = \min\{s(n) : n \in I_r\}$ Then*

$$M_r = F_{r+2}, m_r = 1, \tag{5.9}$$

where F_n is the n-th Fibonacci number ($F_0 = 0$, $F_1 = 1$, $F_{n+1} = F_n + F_{n-1}$.)

Proof: Since $s(2^r) = 1$ for all r and $s(n) = b(3; n-1) \geq 1$, $m_r = 1$. By (5.1), $s(2n \pm 1) - s(2n) = s(n \pm 1)$, so $M_r = s(m)$, where m is odd. As $s(4n+1) = s(2n+1) + s(n)$ and $s(4n+3) = s(2n+1) + s(n+1)$,

$$M_r \leq M_{r-1} + M_{r-2}. \tag{5.10}$$

We see from Table 7.1 that $M_0 = 1$, $M_1 = 2$, $M_2 = 3$, and $M_3 = 5$. Define the sequence a^r by $a_{2r} = (2^{2r+2} - 1)/3$ and $a_{2r+1} = (2^{2r+3} + 1)/3$. (That is, a_t is the integer closest to $\frac{4}{3}2^t$.) Then $a_0 = 1$, $a_1 = 3$, $a_2 = 5$, $a_3 = 11$, etc., and $a_t \in I_t$. We wish to show that $M_t = s(a_t)$ for all t. From their definitions, $a_{2r} = 2a_{2r-1} - 1$ and $a_{2r+1} = 2a_{2r} + 1$, hence

$$s(a_{2r}) = s(a_{2r-1}) + s(a_{2r-1} - 1) = s(a_{2r-1}) + s(a_{2r-2}), \quad (5.11)(\text{i})$$
$$s(a_{2r+1}) = s(a_{2r}) + s(a_{2r} + 1) = s(a_{2r}) + s(a_{2r-1}). \quad (5.11)(\text{ii})$$

Since $M_r = s(a_r)$ for $r \leq 3$, by (5.10) and (5.11),

$$M_{r-1} + M_{r-2} \geq M_r \geq s(a_r) = s(a_{r-1}) + s(a_{r-2}) = M_{r-1} + M_{r-2}. \tag{5.12}$$

Since $M_0 = 1 = F_2$ and $M_1 = 2 = F_3$, it follows that $M_t = F_{t+2}$. ∎

It can be shown (see [**R2**]) that, for $0 \leq k \leq 2^r$, $s(2^r + k) = s(2^{r+1} - k)$. Thus, $M_r = F_{r+2}$ also equals $s(b_r)$, where b_r is the integer closest to $\frac{5}{3}2^r$.

Theorem 5.13. *Let $\varphi = \frac{1}{2}(1+\sqrt{5})$ and $\mu = \log_2 \varphi$. Then for $n \geq 1$, and some $c > 0$,*

$$cn^{\mu} \geq b(3;n) \geq 1, \tag{5.14}$$

and $\mu = \mu_1(3)$ is best possible.

Proof: The lower bound is clear since $b(3;2^r-1) = s(2^r) = 1$. By Thm. 5.2 and Prop. 5.8, $b(3;n) \leq F_{r+1}$ for $n \leq 2^r - 1$. The Binet formula for the Fibonacci numbers implies that $F_{r+1} = (\varphi/\sqrt{5})\varphi^r + o(1)$, hence

$$b(3;n) \leq (\varphi/\sqrt{5})\varphi^r + o(1) \leq (\varphi/\sqrt{5})n^{\mu} + o(1). \tag{5.15}$$

On the other hand, in the notation of the last theorem,

$$b(3;a_r-1) = F_{r+2} = (\varphi^2/\sqrt{5})\varphi^r + o(1). \tag{5.16}$$

Since $a_r - 1 \approx \frac{4}{3}2^r$, the constant μ cannot be reduced. ∎

Theorem 5.17. *Suppose $u = 2t$ is even. Then for $r \geq 0$,*

$$b(6;u\cdot 2^r-1) = (b(6;u-1) - \frac{1}{2}b(3;u-1))3^r + \frac{1}{2}b(3;u-1). \tag{5.18}$$

Proof: Let $A_r = b(6;u\cdot 2^r - 1)$. Then by Thm. 2.4(ii), (iv) and (v),

$$\begin{aligned} A_{r+1} &= \sum_{j=0}^{u2^r-1} b(3;j) = \sum_{j=0}^{t2^r-1} \{b(3;2j) + b(3;2j+1)\} \\ &= \sum_{j=0}^{t2^r-1} \{b(3;j) + b(3;j-1) + b(3;j)\} = 3A_r - b(3;t\cdot 2^r-1). \end{aligned} \tag{5.19}$$

Since $b(3;t\cdot 2^r - 1) = s(t\cdot 2^r) = s(2t) = b(3;u-1)$, (5.19) implies that

$$A_{r+1} - \frac{1}{2}b(3;u-1) = 3\{A_r - \frac{1}{2}b(3;u-1)\}, \tag{5.20}$$

from which (5.18) is immediate. ∎

This theorem has interesting consequences for the behavior of

$$\hat{b}(6;n) = b(6;n)n^{-\lambda(6)} = b(6;n)n^{-\log_2 3}. \tag{5.21}$$

Proposition 5.22 (Carlitz). $\lim_{n\to\infty} \hat{b}(6;n)$ *does not exist.*

Proof: We apply Thm. 5.19 with $u = 2$ and $u = 6$, obtaining:

$$b(6;2\cdot 2^r - 1) = \frac{1}{2}(3^r + 1), \tag{5.23)(i}$$

$$b(6;6\cdot 2^r - 1) = 3^{r+1} + 1. \tag{5.23)(ii}$$

Thus,

$$\hat{b}(6;2\cdot 2^r - 1) = \frac{1}{2}(3^r + 1)(2^{r+1} - 1)^{-\log_2 3} \to 1/6 \cong .1667 \tag{5.24)(i}$$

$$\hat{b}(6;6\cdot 2^r - 1) = (3^{r+1} + 1)(3\cdot 2^{r+1} - 1)^{-\log_2 3} \to 3^{-\log_2 3} \cong .1753. \blacksquare \tag{5.24)(ii}$$

Carlitz writes in [**C3**,p.151]: "P. T. Bateman has suggested that it would be of interest to examine the sum function

$$S(n) = \sum_{k=0}^{n} \theta_0(k)." \tag{5.25}$$

We have seen that $S(n) = b(6;2n-2)$. The computations (5.24) appear, in effect, in [**C3**,p.152].

It can be shown that there exists a continuous function ψ on [**1,2**] so that, if $u = m/2^k$ is a dyadic rational in [**1,2**], then $\hat{b}(6;u2^r) \to \psi(u)$. A proof of this result will appear elsewhere [**R3**].

6. Open Questions and Acknowledgements

We believe there is still much to learn about binary partition functions, let alone their analogues for bases other than 2.

What other recurrences are satisfied by binary partition functions? How can the various properties of $b(\infty;n)$ be regarded as the limit of properties of finite $b(d;n)$'s? For example, Knuth remarks that

$$b(\infty;4n)^2 - b(\infty;4n-2)b(\infty;4n+2) = b(\infty;2n)^2 \tag{6.1}$$

is an immediate consequence of (2.9); similarly, by Thm. 2.4(vii),

$$b(4d;4n)^2 - b(4d;4n-2)b(4d;4n+2) = b(2d;2n)^2. \tag{6.2}$$

There do not seem to be easy generalizations of Thm. 2.14 to moduli greater than 2; whenever the answer is known, the set

$$\mathcal{A}(d;m,a)=\{n:b(d;n)\equiv a \pmod{m}\} \tag{6.3}$$

is either finite or has a well-defined positive density. This is clear for $m=2$. Since $f(r,t)(s)$ is an integer valued polynomial of degree $r-1$, it is easy to show that $f(r,t)(s) \pmod{m}$ is periodic for all s and m. It follows that $\mathcal{A}(2^r;m,a)$ is a finite union of disjoint arithmetic progressions. In [**R2**] we compute the density of $\mathcal{A}(3;m,a)$, which is determined by the primes which divide m and a. For example, if p is prime, $\mathcal{A}(3;p,0)$ has density $1/(p+1)$ and, if $1\le a\le p-1$, then $\mathcal{A}(3;p,a)$ has density $p/(p^2-1)$. Churchhouse's results on $b(\infty;n) \pmod{4}$ also imply that $\mathcal{A}(\infty;4,0)$ has density $1/3$ and $\mathcal{A}(\infty;4,2)$ has density $2/3$. We risk a conjecture.

Conjecture 6.4. *For all d, a and m, $\mathcal{A}(d;m,a)$ has well-defined density $\alpha=\alpha(d,m,a)$.*

The asymptotic analysis of $b(d;n)$ in section four begs a number of questions. What are the values of (or estimates for)

$$\alpha(2k)=\varliminf_{n\to\infty} b(k;n)n^{-\log_2 k}, \tag{6.5)(i}$$

$$\beta(2k)=\varlimsup_{n\to\infty} b(k;n)n^{-\log_2 k}; \tag{6.5)(ii}$$

are these ever equal, except when $k=2^r$? What are the actual values of

$$\lambda_1(2k+1)=\varlimsup_{n\to\infty}\log(b(2k+1;n))/(\log n), \tag{6.6)(i}$$

$$\lambda_2(2k+1)=\varliminf_{n\to\infty}\log(b(2k+1;n)/(\log n)? \tag{6.6)(ii}$$

Is $c\cdot n^{\lambda}$ the "correct" bound; that is, is it true that, for all $k\ge 1$,

$$\infty>\varlimsup_{n\to\infty} b(2k+1;n)n^{-\lambda_1(2k+1)}, \tag{6.7)(i}$$

$$\varliminf_{n\to\infty} b(2k+1;n)n^{-\lambda_2(2k+1)}>0? \tag{6.7)(ii}$$

Are there any Churchhouse-like formulas (viz. (1.3)) for $b(d;n)$'s, $d<\infty$? How does $b(d;n)/b(\infty;n)$ behave; for which d_n is $b(d_n;n)\sim b(\infty;n)/2$? Are there more combinatorial interpretations for the recurrences?

We thank Paul Bateman for providing a supportive atmosphere at the University of Illinois for number theory, without which this paper would never have been attempted. We also thank the organizers of the conference for their efforts and the editors of this collection, for their generosity in considering this manuscript.

7. Appendix

Here is a table of $b(d;n)$ for $2 \le d \le 9$ and $d = \infty$, and $0 \le n \le 32$.

$n\backslash d$	2	3	4	5	6	7	8	9	∞
0	1	1	1	1	1	1	1	1	1
1	1	1	1	1	1	1	1	1	1
2	1	2	2	2	2	2	2	2	2
3	1	1	2	2	2	2	2	2	2
4	1	3	3	4	4	4	4	4	4
5	1	2	3	3	4	4	4	4	4
6	1	3	4	5	5	6	6	6	6
7	1	1	4	4	5	5	6	6	6
8	1	4	5	8	8	9	9	10	10
9	1	3	5	6	8	8	9	9	10
10	1	5	6	9	10	12	12	13	14
11	1	2	6	7	10	10	12	12	14
12	1	5	7	12	13	16	16	18	20
13	1	3	7	8	13	14	16	16	20
14	1	4	8	12	14	19	20	22	26
15	1	1	8	9	14	15	20	20	26
16	1	5	9	17	18	24	25	30	36
17	1	4	9	12	18	20	25	26	36
18	1	7	10	18	21	28	30	35	46
19	1	3	10	14	21	22	30	31	46
20	1	8	11	23	26	34	36	44	60
21	1	5	11	15	26	29	36	38	60
22	1	7	12	22	28	39	42	50	74
23	1	2	12	16	28	30	42	44	74
24	1	7	13	28	33	46	49	62	94
25	1	5	13	19	33	38	49	52	94
26	1	8	14	27	36	52	56	68	114
27	1	3	14	20	36	40	56	59	114
28	1	7	15	32	40	59	64	81	140
29	1	4	15	20	40	49	64	68	140
30	1	5	16	29	41	64	72	88	166
31	1	1	16	21	41	48	72	76	166
32	1	6	17	38	46	72	81	106	202

Table 7.1

REFERENCES

[A1] Andrews, G., Congruence properties of the m-ary partition function, J. Number Theory, **3** (1971), 104-110. MR 42#3043.

[A2] Andrews, G., The Theory of Partitions. Encyc. of Math. and Appl., v.2, Addison-Wesley, Reading 1976, MR 58#27738.

[B] de Bruijn, N. G., On Mahler's partition problem, Indag. Math. **10** (1948), 210-220. MR 10, 16d.

[C1] Carlitz, L., A problem in partitions related to the Stirling numbers, Riv. Mat. Univ. Parma **5** (1964), 61-75. MR 34#158.

[C2] Carlitz, L., A problem in partitions related to the Stirling numbers, Bull. Amer. Math. Soc. **70** (1964), 275-278. MR 28#1135.

[C3] Carlitz, L., Generating functions and partition problems. Pp. 144-169, in Proc. Sympos. Pure Math. VIII, Amer. Math. Soc., Providence, 1965, MR 31 #72.

[C4] Churchhouse, R. F., Congruence properties of the binary partition function, Proc. Camb. Phil. Soc. **66** (1969), 371-376. MR 40#1356.

[D1] Dirdal, G., On restricted m-ary partitions, Math. Scand. **37** (1975), 51-60. MR 52#10582.

[D2] Dirdal, G., Congruences for m-ary partitions, Math. Scand. **37** (1975), 76-82. MR 52#10583.

[E1] Euler, L., Introductio in analysin infinitorum. Lausanne, 1748, in Opera Omnia Series Prima Opera Math. vol. 8, B. G. Teubner, Leipzig, 1922.

[E2] Euler, L., De partitione numerorum, Nov. com. acad. sci. Petro. **3** (1750/1751), 125-169. In Opera Omnia Series Prima Opera Math. vol. 2, pp. 254-294, B. G. Teubner, Leipzig, 1915.

[G1] Gupta, H., Proof of the Churchhouse conjecture concerning binary partitions, Proc. Camb. Phil. Soc. **70** (1971), 53-56. MR 45#4986.

[G2] Gupta, H., A simple proof of the Churchhouse conjecture concerning binary partitions, Ind. J. Pure Appl. Math. **3** (1972), 791-794. MR 48#8377

[G3] Gupta, H., On m-ary partitions, Proc. Camb. Phil. Soc **71** (1972), 343-345. MR 45#204.

[G4] Gupta, H., A direct proof of the Churchhouse conjecture concerning binary partitions, Ind. J. Math. **18** (1976), 1-5. MR 82b:10013.

[GP] Gupta, H. and P. A. B. Pleasants, Partitions into powers of m, Ind. J. Pure Appl. Math. **10** (1979), 655-694. MR 80f:10014.

[HL] Hirschhorn, M. D. and J. H. Loxton, Congruence properties of the binary partition function, Proc. Camb. Phil. Soc. **78** (1975), 437-442. MR 52#3045.

[KAH] Klosinski, L. F., G. L. Alexanderson and A. P. Hillman, The William Lowell Putnam Mathematical Competition, Amer. Math. Monthly

91 (1984), 487- 495.
[K] Knuth, D. E., An almost linear recurrence, Fib. Q. **4** (1966), 117-128. MR 33#7317.
[L1] Lehmer, D. H., On Stern's diatomic series, Amer. Math. Monthly **36** (1929), 59-67.
[L2] Lind, D. A., An extension of Stern's diatomic series, Duke Math. J. **36** (1969), 55-60. MR 30 #6810.
[L3] Lucas, É., Sur les suites de Farey, Bull. Soc. Math. France (1877 -1878), 118-119.
[M] Mahler, K., On a special functional equation, J. London Math. Soc. **15** (1940), 115-123. MR 2, 133e.
[R1] Reznick, B., Digital representations using the greatest integer function, Trans. Amer. Math. Soc. **312** (1989), 355-375. MR 89g:11010.
[R2] Reznick, B., Congruence properties of the Stern sequence. in preparation.
[R3] Reznick, B., Stern measures. in preparation.
[R4] de Rham, G., Un peu de mathématiques à propos d'une courbe plane, Elem. Mat. **2** (1947), 73-76, 89-97. MR 9-246.
[R5] Rödseth, O., Some arithmetical properties of m-ary partitions, Proc. Camb. Phil. Soc. **68** (1970), 447-453. MR 41#5319.
[S1] Sloane, N. J. A., A Handbook of Integer Sequences, Academic Press, New York (1973). MR 50#9760.
[S2] Stanley, R. P. Private communication.
[S3] Stern, M. A., Ueber eine zahlentheoretische funktion, J. für Math. **55** (1858), 193-220.
[T1] Tanturri, A, Della partizione dei numeri. Ambi, terni quaterne e cinquine di data somma, Att. delle Sci. di Torino **52** (1916/7), 902-918.
[T2] Tanturri, A, Sul numero delle partizioni d'un numero in potenze di 2, Att. delle Sci. di Torino **54** (1918), 97-110.
[T3] Tanturri, A., Sul numero delle partizioni d'un numero in potenze di 2, Att. Accad. Naz. Lincei **27** (1918), 399-403.

Bruce Reznick
Department of Mathematics
University of Illinois
1409 West Green Street
Urbana, Illinois 61801

On the Minimal Level of Modular Forms

H. M. STARK

To P. T. Bateman on his 70th birthday

0. Introduction

In this paper, we will deal with a very practical problem. The problem is that of finding the lowest level of a given modular form which is known to have some level. For forms on principal (homogeneous) congruence subgroups, the answer is the greatest common divisor of the widths at all cusps [6, p. 82]. This is often not an easy calculation to make unless the transformation formulas are already known. However, frequently another congruence group appears: namely the Γ_0 type group. Here, the calculation of finding the minimal level turns out to be much easier and involves just the inversion formula. Since it is often the case that if one non-trivial transformation is known for the form in question, it is the inversion formula, this will be very convenient. This is the subject of Theorem 1.

As an example of the application of Theorem 1, we will consider general η-products in Theorem 2 and find the best possible Γ_0 levels for these. Since the transformation formulas for the η-function are so complicated, Theorem 2 serves as a good example of the ease with which Theorem 1 may be applied. Theorem 1 has the drawback that it does not prove that a level exists at all. For η-products, we give a new very simple method of showing that high levels exist by using not the η-multiplier but the much simpler theta multiplier. This method combined with the application of Theorem 1 provides an excellent new way to deal with combinations of η-functions and θ-functions.

As it turns out, the main result applies to Hilbert modular forms as well and, surprisingly, the proof has almost no extra complications once

---

Supported in part by the NSF

the theorem is phrased correctly. In particular, the different of the field makes no appearance at all. In fact, the theorem applies to fields that aren't totally real and forms that aren't analytic also, but to save ourselves notational troubles, we will consider only totally real fields here. It should be remarked however that once the notation is set up, Theorem 1 holds in the more general setup, without change. In this paper, we consider forms with integral weight vectors. There is also a half integral weight version of Theorem 1. It is presented in full generality in [8].

First let us introduce the customary notation for Hilbert modular forms. If $R = \begin{pmatrix} r & s \\ t & u \end{pmatrix}$ is a matrix with non-zero determinant and z a complex number, we let $R \circ z = (rz+s)/(tz+u)$. Let K be a totally real field whose conjugate fields are denoted by $K^{(1)}, \ldots, K^{(n)}$ and let $\mathcal{H} = \mathcal{H}_1 \times \cdots \times \mathcal{H}_1$ (n times) where $\mathcal{H}_1$ is the usual upper half plane. Thus $z = (z_1, \ldots, z_n)$ is in $\mathcal{H}$ if and only if $\operatorname{Im}(z_j) > 0$ for all j, $1 \le j \le n$. A matrix $R = \begin{pmatrix} r & s \\ t & u \end{pmatrix}$ in $GL_2(K)^{++}$ (totally positive determinant) acts on $\mathcal{H}$ via

$$R \circ z = \left(R^{(1)} \circ z_1, \ldots, R^{(n)} \circ z_n\right).$$

Hilbert modular forms allow weight vectors. For an integral weight vector $k = (k_1, \ldots, k_n)$, we introduce the "slash operator", $| = |_k$:

$$(f \mid R)(z) = (\det R)^{k/2}(tz+u)^{-k} f(R \circ z)$$

where we have used the vector notation,

$$(\det R)^{k/2}(tz+u)^{-k} = \prod_{j=1}^{n} \left(\det R^{(j)}\right)^{k_j/2} \left(t^{(j)} z_j + u^{(j)}\right)^{-k_j}.$$

As always, if R and S are in $GL_2(K)^{++}$, then $(f \mid R) \mid S = f \mid (RS)$. This is the reason that the slash notation is so useful.

Let $\mathcal{N}$ be an integral ideal of K and let

$$\Gamma_0(\mathcal{N}) = \left\{ \begin{pmatrix} a & b \\ c & d \end{pmatrix} \in SL_2(\mathcal{O}_K) \mid c \equiv 0 \pmod{\mathcal{N}} \right\}.$$

Also, we let χ denote a *numerical* character (mod $\mathcal{N}$). (In other words, χ is a character of numbers rather than of ideals.) In this paper, we are interested in how functions defined on $\mathcal{H}$ transform under various groups. Thus we will let $\mathcal{M}_0(\mathcal{N}, k, \chi)$ be the set of all $f(z)$ such that $(f \mid A)(z) = \chi(d) f(z)$ for all $A = \begin{pmatrix} a & b \\ c & d \end{pmatrix}$ in $\Gamma_0(\mathcal{N})$ and we will not worry about whether the functions are analytic or whether they are nice at cusps. This is the reason why the results of this paper actually apply to non-analytic modular forms and to any number field and functions defined on the appropriate combination of ordinary and quaternionic upper half planes. It must be noted that (at least when $K = \mathbf{Q}$), all the lemmas leading to the proof of Theorem 1 are well known. However, since they are all rather short, and in one case (Lemma 6), many proofs in print ignore a vital detail, we include proofs here.

1. The level determiner theorem

There are three main facts about the groups and characters which we will need. We present them as three lemmas. If H and K are both subgroups of a bigger group G, we denote by $\langle H, K\rangle$ the subgroup of G generated by all the elements of H and K. It is the smallest subgroup of G containing both H and K.

Lemma 1. *If $\mathcal{N} = (\mathcal{N}_1, \mathcal{N}_2)$, then $\Gamma_0(\mathcal{N}) = \langle \Gamma_0(\mathcal{N}_1), \Gamma_0(\mathcal{N}_2)\rangle$.*

Proof: In fact we will show that $A = \begin{pmatrix} a & b \\ c & d \end{pmatrix}$ in $\Gamma_0(\mathcal{N})$ can be written as $A = A_1A_2$ with A_j in $\Gamma_0(\mathcal{N}_j)$. Determine e so that $ce + d = d_1$ is relatively prime to $\mathcal{N}_1$ either via the Chinese remainder theorem, or with overkill, by the theorem on primes in progressions in number fields. This gives

$$\begin{pmatrix} a & b \\ c & d \end{pmatrix}\begin{pmatrix} 1 & e \\ 0 & 1 \end{pmatrix} = \begin{pmatrix} * & * \\ c & d_1 \end{pmatrix}.$$

Next, $(\mathcal{N}_1, d_1\mathcal{N}_2) = (\mathcal{N}_1, \mathcal{N}_2) = \mathcal{N}$, and so there are integers c_1 in $\mathcal{N}_1$ and d_1c_2 in $d_1\mathcal{N}_2$ (i. e. c_2 in $\mathcal{N}_2$) whose sum is c in $\mathcal{N}$: $c_1 + d_1c_2 = c$. Therefore

$$\begin{pmatrix} * & * \\ c & d_1 \end{pmatrix}\begin{pmatrix} 1 & 0 \\ -c_2 & 1 \end{pmatrix} = \begin{pmatrix} * & * \\ c_1 & d_1 \end{pmatrix}$$

is in $\Gamma_0(\mathcal{N}_1)$ and we may take $A_2^{-1} = \begin{pmatrix} 1 & e \\ 0 & 1 \end{pmatrix}\begin{pmatrix} 1 & 0 \\ -c_2 & 1 \end{pmatrix}$ in $\Gamma_0(\mathcal{N}_2)$. Q.E.D.

Lemma 2. *If $\mathcal{N} = (\mathcal{N}_1, \mathcal{N}_2)$, and χ is definable (mod $\mathcal{N}_1$) and χ is definable (mod $\mathcal{N}_2$), then χ is definable (mod $\mathcal{N}$).*

Proof: Let $\mathcal{N}'$ be any ideal divisible by both $\mathcal{N}_1$ and $\mathcal{N}_2$. If for $\mathcal{A}|\mathcal{N}'$, we denote $H(\mathcal{A})$ the subgroup of $(\mathcal{O}_K/\mathcal{N}')^*$ of elements $a\mathcal{N}'$ with $a \equiv 1 \pmod{\mathcal{A}}$, then by the Chinese remainder theorem again $H(\mathcal{N}) = H(\mathcal{N}_1)H(\mathcal{N}_2)$ and thus $\chi = 1$ on $H(\mathcal{N})$. Hence χ is definable (mod $\mathcal{N}$). Q.E.D.

Let

$$\Gamma_0^0(\mathcal{N},\mathcal{B}) = \left\{ \begin{pmatrix} a & b \\ c & d \end{pmatrix} \in \Gamma_0(\mathcal{N}) \mid b \equiv 0 \pmod{\mathcal{B}} \right\}$$

and let the translation subgroup of $\Gamma_0(\mathcal{N})$ be denoted by

$$T = \left\{ \begin{pmatrix} 1 & t \\ 0 & 1 \end{pmatrix} \mid t \in \mathcal{O}_K \right\}.$$

Note that $\left[f \middle| \begin{pmatrix} 1 & t \\ 0 & 1 \end{pmatrix} \right](z) = f(z+t)$. If $f(z+t) = f(z)$ for all t in $\mathcal{O}_K$, we will say that f is invariant under T.

Lemma 3. *For any ideals $\mathcal{A}$ and $\mathcal{B}$, we have $\Gamma_0(\mathcal{A}) = \langle \Gamma_0^0(\mathcal{A},\mathcal{B}), T \rangle$.*

Proof: Indeed, $\Gamma_0(\mathcal{A}) = T\Gamma_0^0(\mathcal{A},\mathcal{B})T$ as may be seen from the two relations,

$$\begin{pmatrix} a' & b' \\ c' & d' \end{pmatrix} = \begin{pmatrix} a & b \\ c & d \end{pmatrix} \begin{pmatrix} 1 & t \\ 0 & 1 \end{pmatrix} = \begin{pmatrix} a & at+b \\ c & ct+d \end{pmatrix}$$

and

$$\begin{pmatrix} 1 & t' \\ 0 & 1 \end{pmatrix} \begin{pmatrix} a' & b' \\ c' & d' \end{pmatrix} = \begin{pmatrix} a'+c't' & b'+d't' \\ c' & d' \end{pmatrix}.$$

We first choose t so that $d' = ct+d$ is relatively prime to $\mathcal{B}$ and then we choose t' so that $b'+d't'$ is divisible by $\mathcal{B}$. Q.E.D.

Lemma 4. *If $f(z)$ is in $\mathcal{M}_0(N,k,\chi)$ where $N \gg 0$ and χ is defined (mod N), then $f \middle| \begin{pmatrix} 0 & -1 \\ N & 0 \end{pmatrix}$ is in $\mathcal{M}_0(N,k,\overline{\chi})$.*

Proof: If $A = \begin{pmatrix} a & b \\ c & d \end{pmatrix}$ is in $\Gamma_0(N)$ then

$$\begin{pmatrix} 0 & -1 \\ N & 0 \end{pmatrix} \begin{pmatrix} a & b \\ c & d \end{pmatrix} = \begin{pmatrix} -c & -d \\ Na & Nb \end{pmatrix} = \begin{pmatrix} d & -c/N \\ -Nb & a \end{pmatrix} \begin{pmatrix} 0 & -1 \\ N & 0 \end{pmatrix}$$

and therefore

$$\left[f \middle| \begin{pmatrix} 0 & -1 \\ N & 0 \end{pmatrix} \right] \middle| A = \chi(a) f \middle| \begin{pmatrix} 0 & -1 \\ N & 0 \end{pmatrix} = \overline{\chi}(d) f \middle| \begin{pmatrix} 0 & -1 \\ N & 0 \end{pmatrix}.$$

Q.E.D.

Corollary 1. *If f is in $\mathcal{M}_0(N,k,\chi)$ where $N \gg 0$ and χ is defined (mod N), then $f \middle| \begin{pmatrix} 0 & -1 \\ N & 0 \end{pmatrix}$ is invariant under T.*

(Note that f is invariant under T also.)
We can now state the level determiner theorem.

Theorem 1. *Suppose that $f(z)$ is in $\mathcal{M}_0(\mathcal{N}', k, \chi)$ for some $\mathcal{N}'$, with character χ defined (mod $\mathcal{N}'$) and that $f(z)$ is not identically zero. Let $\mathcal{N}$ be the greatest common divisor of all totally positive integral N such that $f\Big|\begin{pmatrix} 0 & -1 \\ N & 0 \end{pmatrix}$ is invariant under T. Then $\mathcal{N}|\mathcal{N}'$, χ is definable (mod $\mathcal{N}$) and $f(z)$ is in $\mathcal{M}_0(\mathcal{N}, k, \chi)$.*

The proof will be based on two further lemmas.

Lemma 5. *Suppose that $f(z)$ is in $\mathcal{M}_0(\mathcal{N}, k, \chi)$ and that $M \gg 0$ is in $\mathcal{O}_K$. Then $f(Mz)$ is in $\mathcal{M}_0(M\mathcal{N}, k, \chi)$.*

Proof: Let $g(z) = \left[f\Big|\begin{pmatrix} M & 0 \\ 0 & 1 \end{pmatrix}\right](z)$ which is a constant multiple of $f(Mz)$. For $\begin{pmatrix} a & b \\ c & d \end{pmatrix}$ in $\Gamma_0(M\mathcal{N})$, we have

$$\begin{pmatrix} M & 0 \\ 0 & 1 \end{pmatrix}\begin{pmatrix} a & b \\ c & d \end{pmatrix} = \begin{pmatrix} Ma & Mb \\ c & d \end{pmatrix} = \begin{pmatrix} a & Mb \\ c/M & d \end{pmatrix}\begin{pmatrix} M & 0 \\ 0 & 1 \end{pmatrix}$$

with $\begin{pmatrix} a & Mb \\ c/M & d \end{pmatrix}$ in $\Gamma_0(\mathcal{N})$. Thus,

$$g\Big|\begin{pmatrix} a & b \\ c & d \end{pmatrix} = f\Big|\begin{pmatrix} M & 0 \\ 0 & 1 \end{pmatrix}\begin{pmatrix} a & b \\ c & d \end{pmatrix} = \chi(d)f\Big|\begin{pmatrix} M & 0 \\ 0 & 1 \end{pmatrix} = \chi(d)g.$$

Q.E.D.

Note that not only is the function g constructed in Lemma 5 translation invariant by integers but if b is in $\mathcal{O}_K$, then

$$g(z + \frac{b}{M}) = g(z).$$

The heart of the proof of Theorem 1 lies in the fact that Lemma 5 has a converse.

Lemma 6. *Suppose that $g(z)$ is not identically zero and is in $\mathcal{M}_0(M\mathcal{N}, k, \chi)$ where $M \gg 0$ is in $\mathcal{O}_K$ and χ is defined (mod $M\mathcal{N}$). If for all integers b, $g(z + (b/M)) = g(z)$, then χ is definable (mod $\mathcal{N}$) and $g(z/M)$ is in $\mathcal{M}_0(\mathcal{N}, k, \chi)$.*

Proof: Let $f(z) = \left[g\Big|\begin{pmatrix} 1 & 0 \\ 0 & M \end{pmatrix}\right](z)$ which is a constant multiple of $g(z/M)$. By hypothesis, f is invariant under T. Thus for integral b,

$$f\Big|\begin{pmatrix} 1 & b \\ 0 & 1 \end{pmatrix} = \chi(1)f.$$

For $\begin{pmatrix} a & b \\ c & d \end{pmatrix}$ in $\Gamma_0^0(\mathcal{N}, M)$, we have

$$\begin{pmatrix} 1 & 0 \\ 0 & M \end{pmatrix}\begin{pmatrix} a & b \\ c & d \end{pmatrix} = \begin{pmatrix} a & b \\ Mc & Md \end{pmatrix} = \begin{pmatrix} a & b/M \\ Mc & d \end{pmatrix}\begin{pmatrix} 1 & 0 \\ 0 & M \end{pmatrix}$$

where $\begin{pmatrix} a & b/M \\ Mc & d \end{pmatrix}$ is in $\Gamma_0(M\mathcal{N})$. Hence

$$f\Big|\begin{pmatrix} a & b \\ c & d \end{pmatrix} = g\Big|\begin{pmatrix} 1 & 0 \\ 0 & M \end{pmatrix}\begin{pmatrix} a & b \\ c & d \end{pmatrix} = \chi(d)g\Big|\begin{pmatrix} 1 & 0 \\ 0 & M \end{pmatrix} = \chi(d)f.$$

We have thus established that

$$f\Big|\begin{pmatrix} a & b \\ c & d \end{pmatrix} = \chi(d)f \tag{1}$$

holds for $\begin{pmatrix} a & b \\ c & d \end{pmatrix}$ in $\Gamma_0^0(\mathcal{N}, M)$ and for $\begin{pmatrix} a & b \\ c & d \end{pmatrix}$ in T. Since these two groups generate $\Gamma_0(\mathcal{N})$, it would seem we are done, and indeed many published proofs of this result stop here. However, there are still difficulties with the character χ. We still need to show that χ is definable (mod $\mathcal{N}$) before we are finished. Let $d \equiv 1 \pmod{\mathcal{N}}$ where $(d, M\mathcal{N}) = 1$. We have to show that $\chi(d) = 1$. For this, choose an integer t such that the matrix

$$\begin{pmatrix} 1 & t \\ 0 & 1 \end{pmatrix}\begin{pmatrix} 1 & 0 \\ d-1 & 1 \end{pmatrix}\begin{pmatrix} 1 & 1 \\ 0 & 1 \end{pmatrix} = \begin{pmatrix} 1 & t \\ 0 & 1 \end{pmatrix}\begin{pmatrix} 1 & 1 \\ d-1 & d \end{pmatrix}$$
$$= \begin{pmatrix} * & dt+1 \\ d-1 & d \end{pmatrix}$$

has $dt + 1 \equiv 0 \pmod{M}$. Note that $(d, M) = 1$ so that this can be done. Then by (1), $\chi(d) = 1$ and so χ is definable (mod $\mathcal{N}$). Q.E.D.

Proof of Theorem 1: If $\mathcal{N}'|N'$ where $N' \gg 0$, then $f\Big|\begin{pmatrix} 0 & -1 \\ N' & 0 \end{pmatrix}$ is invariant under T by Corollary 1. Since $\mathcal{N}'$ is the greatest common divisor of all such N', we see that $\mathcal{N}|\mathcal{N}'$. Now let $N \gg 0$ be such that $f\Big|\begin{pmatrix} 0 & -1 \\ N & 0 \end{pmatrix}$ is invariant under T. Choose $M \gg 0$ so that $\mathcal{N}'|MN$ and let $g = f\Big|\begin{pmatrix} 0 & -1 \\ MN & 0 \end{pmatrix}$ which is in $\mathcal{M}_0(MN, k, \overline{\chi})$ by Lemma 4. For integral b, we have

$$\begin{pmatrix} 0 & -1 \\ MN & 0 \end{pmatrix}\begin{pmatrix} 1 & b/M \\ 0 & 1 \end{pmatrix} = \begin{pmatrix} 0 & -1 \\ MN & bN \end{pmatrix} = \begin{pmatrix} 0 & -1 \\ N & 0 \end{pmatrix}\begin{pmatrix} M & b \\ 0 & 1 \end{pmatrix}$$
$$= \begin{pmatrix} 0 & -1 \\ N & 0 \end{pmatrix}\begin{pmatrix} 1 & b \\ 0 & 1 \end{pmatrix}\begin{pmatrix} M & 0 \\ 0 & 1 \end{pmatrix}.$$

Thus

$$\begin{aligned} g\left|\begin{pmatrix}1 & b/M\\ 0 & 1\end{pmatrix}\right. &= f\left|\begin{pmatrix}0 & -1\\ MN & 0\end{pmatrix}\right.\begin{pmatrix}1 & b/M\\ 0 & 1\end{pmatrix} \\ &= \left\{\left[f\left|\begin{pmatrix}0 & -1\\ N & 0\end{pmatrix}\right.\right]\left|\begin{pmatrix}1 & b\\ 0 & 1\end{pmatrix}\right.\right\}\left|\begin{pmatrix}M & 0\\ 0 & 1\end{pmatrix}\right. \\ &= \left\{f\left|\begin{pmatrix}0 & -1\\ N & 0\end{pmatrix}\right.\right\}\left|\begin{pmatrix}M & 0\\ 0 & 1\end{pmatrix}\right. \\ &= g. \end{aligned}$$

Therefore, by Lemma 6, $\overline{\chi}$, and hence χ, is definable (mod N) and

$$g\left|\begin{pmatrix}1 & 0\\ 0 & M\end{pmatrix}\right. = \left\{f\left|\begin{pmatrix}0 & -1\\ MN & 0\end{pmatrix}\right.\right\}\left|\begin{pmatrix}1 & 0\\ 0 & M\end{pmatrix}\right. = f\left|\begin{pmatrix}0 & -1\\ N & 0\end{pmatrix}\right.$$

is in $\mathcal{M}_0(N, k, \overline{\chi})$. By Lemma 4 again, f is in $\mathcal{M}_0(N, k, \chi)$. The Theorem follows from Lemmas 1 and 2. Q.E.D.

2. η-products

As an example of the application of Theorem 1, we will take $K = \mathbf{Q}$ and look at products of η-functions. That is, we will look at products of functions of the form $\eta(az)^b$, where

$$\eta(z) = q^{1/24}\prod_{n=1}^{\infty}(1-q^n), \quad q = e^{2\pi i z}.$$

Let

$$f(z) = \prod_{j=1}^{J}\eta(a_j z)^{b_j} = q^{\frac{1}{24}\sum a_j b_j}\prod_{j=1}^{J}\prod_{n=1}^{\infty}(1-q^{a_j n})^{b_j}.$$

The weight for $f(z)$ is

$$k = \frac{1}{2}\sum b_j$$

which we will assume is in $\mathbf{Z}$ so that we are dealing with an integral weight. In order for $f(z)$ to transform under any $\Gamma_0(N)$, it must be translation invariant by integers. Thus the only $f(z)$ that can occur must have

$$\sum a_j b_j \equiv 0 \pmod{24}.$$

We now show that if this condition is satisfied, $f(z)$ does transform under some $\Gamma_0(N)$. The proof of this fact will be simplicity itself, but the level

N' that we find will frequently not be optimal; the optimal level will then by found from Theorem 1. For non-zero integral c, we denote by χ_c the Kronecker symbol corresponding to $\mathbf{Q}(\sqrt{c})$. In case c is a perfect square, $\chi_c = \chi_1$ is the trivial character. These symbols appear in the theta function transformation formula. We will take as known the transformation formula for $f(z) = \eta(24Mz)$ (see[7]): for $A = \begin{pmatrix} a & b \\ c & d \end{pmatrix}$ in $\Gamma_0(576M)$, we have

$$f(A \circ z) = \chi_{12M}(d) j(A, z) f(z) \tag{2}$$

where $j(A, z)$ is the theta multiplier and all we need to know about it in this paper is that for A in $\Gamma_0(4)$,

$$j(A, z)^2 = \chi_{-1}(d)(cz + d).$$

Lemma 7. *Let*

$$f(z) = \prod_{j=1}^{J} \eta(a_j z)^{b_j} = q^{\frac{1}{24}\sum a_j b_j} \prod_{j=1}^{J} \prod_{n=1}^{\infty} (1 - q^{a_j n})^{b_j}$$

with weight $k = \frac{1}{2}\sum b_j$ *which we assume to be integral. A necessary and sufficient condition for* $f(z)$ *to be in some* $\mathcal{M}_0(N, k, \chi)$ *is that*

$$\sum a_j b_j \equiv 0 \pmod{24}. \tag{3}$$

If this condition is satisfied, then $f(z)$ *is in* $\mathcal{M}_0(N', k, \chi_D)$ *where* N' *is 24 times the least common multiple of all the* a_j *and* $D = (-1)^k \prod a_j^{b_j}$.

Proof: We have just seen that for f to be in any $\mathcal{M}_0(N, k, \chi)$, f must be translation invariant by integers and hence that (3) is necessary. Conversely, suppose that (3) holds. Set

$$g(z) = f(24z) = \prod_{j=1}^{J} \eta(24 a_j z)^{b_j}.$$

Let $N'' = 24N'$ be 576 times the least common multiple of the a_j. Then for $A = \begin{pmatrix} a & b \\ c & d \end{pmatrix}$ in $\Gamma_0(N'')$, we have by (2),

$$\begin{aligned} g(A \circ z) &= \chi_{12}(d)^{2k} \prod \chi_{a_j}(d)^{b_j} j(A, z)^{2k} g(z) \\ &= \chi_{-1}(d)^k \prod \chi_{a_j}(d)^{b_j} (cz + d)^k g(z). \end{aligned}$$

Thus $g(z)$ is in $\mathcal{M}_0(N'', k, \chi_K)$.

The product expansion for $g(z)$ is

$$g(z) = q^{\sum a_j b_j} \prod_{j=1}^{J} \prod_{n=1}^{\infty} \left(1 - q^{24 a_j n}\right)^{b_j}.$$

Since (3) holds, we see from Lemma 6 that $g(z)$ can be shifted downwards by a factor of 24 to $f(z) = g(z/24)$, and that $f(z)$ is in $\mathcal{M}_0(N', k, \chi_D)$ and χ_D is definable (mod N'). Q.E.D.

This sets up Theorem 2 which then gives the best possible level for $f(z)$. For the proof, the transformation formula,

$$(iNz/a)^{-b/2}\eta\left(-\frac{a}{Nz}\right)^b = \eta(Nz/a)^b = q^{\frac{Nb}{24a}} \prod_{n=1}^{\infty} \left(1 - q^{\frac{N}{a}n}\right)^b$$

will be needed. We first give three simple examples of forms which frequently appear due to their low levels and weights.

For our first example, we take

$$f(z) = \eta(z)\eta(23z).$$

As we have just seen, $f(z)$ is in $\mathcal{M}_0(23 \cdot 24, 1, \chi_{-23})$. From the inversion formula, we see that $f \left| \begin{pmatrix} 0 & -1 \\ N & 0 \end{pmatrix} \right.$ is a constant times

$$\eta(Nz)\eta(Nz/23) = q^{\frac{1}{24}(N + \frac{N}{23})} \prod_{n=1}^{\infty} \left(1 - q^{Nn}\right)\left(1 - q^{\frac{N}{23}n}\right).$$

Thus $N = 23$ is minimal and $f(z)$ is in $\mathcal{M}_0(23, 1, \chi_{-23})$. This is the cusp form of weight one which corresponds to either of the two L-functions on $\mathbf{Q}(\sqrt{-23})$ with non-trivial ideal class characters.

For our second example, take

$$f(z) = \eta(z)^2\eta(11z)^2.$$

By Lemma 7, $f(z)$ is in $\mathcal{M}_0(11 \cdot 24, 2, \chi_1)$. Here $f \left| \begin{pmatrix} 0 & -1 \\ N & 0 \end{pmatrix} \right.$ is a constant times

$$\eta(Nz)^2\eta(Nz/11)^2 = q^{\frac{1}{12}(N + \frac{N}{11})} \prod_{n=1}^{\infty} \left(1 - q^{Nn}\right)^2 \left(1 - q^{\frac{N}{11}n}\right)^2.$$

Here $N = 11$ is minimal and $f(z)$ is in $\mathcal{M}_0(11, 2, \chi_1)$. This is the cusp form of weight two which corresponds to the elliptic curves over $\mathbf{Q}$ with minimal conductor.

For our third example, take

$$f(z) = [\eta(37z)/\eta(z)]^2 .$$

Here Lemma 7 says that $f(z)$ is in $\mathcal{M}_0(37 \cdot 24, 0, \chi_1)$. This time $f \Bigg| \begin{pmatrix} 0 & -1 \\ N & 0 \end{pmatrix}$ is a constant times

$$[\eta(Nz/37)/\eta(Nz)]^2 = q^{\frac{1}{12}(\frac{N}{37}-N)} \prod_{n=1}^{\infty} (1 - q^{Nn})^2 \left(1 - q^{\frac{1}{37}n}\right)^2 .$$

Here $N = 37$ is minimal and $f(z)$ is in $\mathcal{M}_0(37, 0, \chi_1)$. This is a modular function. This last example is a special case of a theorem of Rademacher [**4**] which was later generalized by Newman [**3**]. We will discuss Newman's theorem immediately after the proof of Theorem 2 is concluded. We will also discuss a frequently quoted result of Hecke [**1**] at that point.

Theorem 2 (General η-product, integral weight). *Suppose that*

$$f(z) = \prod_{j=1}^{J} \eta(a_j z)^{b_j}$$

where the a_j are positive integers and the b_j are integers. Let

$$k = \frac{1}{2} \sum b_j \in \mathbf{Z}$$

and suppose that

$$\sum a_j b_j \equiv 0 \pmod{24} \tag{4}$$

so that $f(z)$ is in $\mathcal{M}_0(N', k, \chi_D)$ where N' and D are given in Lemma 7. Then the minimal Γ_0 level for $f(z)$ is the smallest common multiple N of all the a_j such that

$$\sum \frac{N}{a_j} b_j \equiv 0 \pmod{24}. \tag{5}$$

Proof: The hypotheses continue to be satisfied if we collect all the terms with the same a_j's. Thus we may assume that the a_j are distinct and that the b_j are non-zero. Here we have $f \Bigg| \begin{pmatrix} 0 & -1 \\ N & 0 \end{pmatrix}$ is a constant multiple of

$$\prod_{j=1}^{J} \eta(Nz/a_j)^{b_j} = q^{\frac{1}{24} \sum \frac{N}{a_j} b_j} \prod_{j=1}^{J} \prod_{n=1}^{\infty} \left(1 - q^{\frac{N}{a_j} n}\right)^{b_j} .$$

For this expression to be translation invariant by integers, every power of q in the collected series expansion must be integral. The lead power is

$$\frac{1}{24}\sum \frac{N}{a_j}b_j,$$

and this gives us the condition (5). Now every term in the collected series expansion of the product

$$\prod_{j=1}^{J}\prod_{n=1}^{\infty}\left(1-q^{\frac{N}{a_j}n}\right)^{b_j} \tag{6}$$

must be integral. If we suppose that the a_j are ordered so that

$$a_1 > a_2 > a_3 > \cdots > a_J,$$

then the first term past q^0 in the expansion is

$$-b_1 q^{\frac{N}{a_1}}.$$

Thus N must be a multiple of a_1. This allows us to remove the $j = 1$ term from the product (6) without disturbing the integrality conditon. In this way, we see inductively that N must be a common multiple of all the a_j and this proves Theorem 2. Q.E.D.

A special case of Theorem 2 is a result of Newman [3]. Newman showed that in the notation of Theorem 2, if $k = 0$ and D is a square, then $f(z)$ is a modular function on $\Gamma_0(N)\backslash\mathcal{H}$ when the two congruences (4) and (5) are satisfied. We see that D is a square is exactly the condition to make $\chi = \chi_1$. Also, Theorem 2 shows that N is the best possible level. It is interesting to compare the methods of proof. Following Rademacher [4], Newman proves that $f(z)$ is invariant under $\Gamma_0(N)$ directly. This requires dealing with the η-multiplier system and this in turn leads to dealing with results on Dedekind sums. The method used here seems much easier to me. The reason is that the η-multiplier does not behave nicely under shift transformations ($z \mapsto az$) whereas the theta multiplier does. Thus the extra shift by a factor of 24 allows an easy proof of a transformation at some level via the theta multiplier and we have seen how to shift back down.

Finally, another special case of Theorem 2 generalizes a result of Hecke [1]. Let n be an odd positive integer and let

$$H_n(z) = \frac{\eta(nz)}{\eta(z)^n}.$$

The condition that n be odd is so that we are dealing with integral weight $k = (1-n)/2$. By Lemma 7, $H_n(z)$ is in $\mathcal{M}_0(24n, k, \chi)$ where $\chi = \chi_D$ and $D = (-1)^{(n-1)/2}n$. By Theorem 2, $H_n(z)$ is then in $\mathcal{M}_0(N, k, \chi)$ where $N = n$ if $3 \nmid n$ and $N = 3n$ if $3|n$. In this case, the Kronecker symbol χ_D is also realizable as a Jacobi symbol, $\chi(d) = (\frac{d}{n})$. Hecke's result is the case where n is a prime greater than 3. Hecke proves his result by comparing his function with two different products of Eisenstein series. I believe the method here to be more straightforward. Hecke's result is usually used to produce the three examples before Theorem 2 since

$$\eta(z)\eta(23z) = H_{23}(z)\Delta(z),$$
$$\eta(z)^2\eta(11z)^2 = H_{11}(z)^2\Delta(z),$$
$$[\eta(37z)/\eta(z)]^2 = H_{37}(z)^2\Delta(z)^3.$$

Concluding Remarks

We must point out that the forms in Theorem 2 are of level N and that this too is best possible. If for some $M|N$, $f(z)$ is of level M, then f would transform nicely under the group $G = \langle\Gamma_0(N), \Gamma(M)\rangle$. According to a beautiful result of Newman [2] that the only groups intermediate to two Γ_0 groups are themselves Γ_0 groups, G must be a Γ_0 group itself and is clearly of lower level if $M \neq N$. This contradicts Theorem 2 and so N is the best level for $f(z)$. This consequence can be produced more easily, but Newman's result is so nice that I can't resist using it. In the generalization to Hilbert modular groups, Reiner and Swift [5] have shown that Newman's result continues to hold, providing that $(N, 6) = 1$. It was at the Bateman conference that I learned of the two papers [2] and [5] from Prof. Newman. I wish to thank him for the references.

There is a half integral weight analogue of Theorem 1. At first sight, this might not seem possible. For example, $\eta(z)$ is itself a form of weight 1/2 on the full modular group in violation of everything in this paper. It is also in violation of [7] when imprecisely referenced. The catch is that the η-multiplier system is rather complicated and in particular is not translation invariant. A seemingly more serious example is $\eta(24z)$ which is translation invariant and being a "shift" of a function on $\Gamma_0(1)$ by a factor of 24, $\eta(24z)$ transforms to a multiple of itself under $\Gamma_0(24)$. This too, violates Theorem 1 which predicts the best level is level $24^2 = 576$ as well as [7] which shows that the level is 576. Again, the catch is that the multiplier system for $\eta(24z)$ on $\Gamma_0(24)$ is still not nice enough to allow Theorem 1 to hold. However, on $\Gamma_0(576)$, the multiplier system for $\eta(24z)$ is the theta multiplier times a character. The theta multiplier system is nice enough, and it is for this system that there is a half integral analogue of Theorem

1 and it is also for this system that the theorems of [7] are derived. This explains the result of 24^2. Even in integral weight, we must be careful. For example, $\eta(12z)^2$ is a translation invariant modular form of weight one which transforms to multiples of itself under $\Gamma_0(12)$. But these multiples still involve more than just a character of d and according to Theorem 2, the best level with just a character for the multiplier is level 144.

REFERENCES

The parenthetical additions after each reference are the review numbers in the Leveque and Guy collected Reviews in Number Theory.

[1] Erich Hecke, Herleitung des Euler-Produktes der Zetafunktionen und einiger L-Reihen aus ihrer Funktionalgleichung, Math. Ann. **119** (1944), 266–287. (F65–3)

[2] Morris Newman, Structure theorems for modular subgroups, Duke Math. J. **22** (1955), 25–32. (F05–8)

[3] Morris Newman, Construction and application of a class of modular functions, Proc. London Math. Soc. (3) **9** (1959), 373–387. (F110–60)

[4] Hans A. Rademacher, The Ramanujan identities under modular substitutions, Trans. Amer. Math. Soc. **51** (1942), 609–636. (P60–1)

[5] Irving Reiner and J. D. Swift, Congruence subgroups of matrix groups, Pacific J. Math. **6** (1956), 529–540. (F15–8)

[6] B. Schoeneberg, *Elliptic Modular Functions*, Springer–Verlag, New York 1974. (F2–214)

[7] J.-P. Serre and H. M. Stark, Modular forms of weight 1/2, in: *Modular Functions of One Variable, VI*, Springer Lecture Notes 627 (1977), pp. 27–67. (F36–207)

[8] H. M. Stark, Examples of modular forms in number fields, to appear.

H. M. Stark
Department of Mathematics
University of California, San Diego
La Jolla, CA 92093

Inequalities for Heights Of Algebraic Subspaces And the Thue-Siegel Principle

THOMAS STRUPPECK AND JEFFREY D. VAALER

Dedicated to Professor Paul Bateman on the occasion of his 70th birthday

1. Introduction

Let k be an algebraic number field and let k^N denote the vector space of $N \times 1$ column vectors over k. In his fundamental paper **[16]** W.M. Schmidt introduced a concept of height on linear subspaces $\mathcal{A}$ of k^N. The idea is to apply the so-called Weil-height (or the absolute Weil-height) to the vector of Grassmann coordinates of any basis for $\mathcal{A}$. In this way Schmidt was able to formulate and prove many of the basic theorems of Diophantine approximation in a very general setting. In the present paper we prove a new inequality for heights on subspaces and apply it to the problem of constructing certain auxiliary polynomials in two variables. Let $H(\mathcal{A})$ denote the height of the subspace $\mathcal{A} \subseteq k^N$, which is precisely defined in section 2. We adopt the convention that $H(\{\vec{0}\}) = 1$. Our inequality has both a local and global version and the global version can be stated as follows.

Theorem 1. *Let $\mathcal{A} \subseteq k^N$ and $\mathcal{B} \subseteq k^N$ be subspaces. If $\langle \mathcal{A}, \mathcal{B} \rangle$ denotes the subspace spanned over k by $\mathcal{A} \cup \mathcal{B}$, then*

$$H(\langle \mathcal{A}, \mathcal{B} \rangle) H(\mathcal{A} \cap \mathcal{B}) \leq H(\mathcal{A}) H(\mathcal{B}) . \tag{1.1}$$

The research of both authors was supported by grants from the National Science Foundation, DMS-8601279 and DMS-8701396, respectively.

The inequality (1.1) has a variety of applications. In Theorem 7 of section 2 we use it to derive a slightly less sharp but more convenient form of Siegel's Lemma concerning small solutions of simultaneous linear equations. (While this manuscript was in preparation we learned that the inequality (1.1) has also been obtained recently and independently by W.M. Schmidt.)

If $\mathcal{L}$ denotes the lattice of all subspaces of k^N then (1.1) shows that the map

$$\mathcal{A} \to \log H(\mathcal{A}) , \qquad \mathcal{A} \in \mathcal{L} ,$$

is a submodular function on $\mathcal{L}$. In section 3 we prove a new and rather specialized inequality for submodular functions. This inequality together with the local form of (1.1) is then applied to the problem of constructing auxiliary polynomials in two variables.

Let α_1 and α_2 be two nonzero algebraic numbers and let $K = k(\alpha_1, \alpha_2)$ have degree $r \geq 1$ over k. We will be interested in the problem of finding nontrivial polynomials $P(x_1, x_2)$ in $k[x_1, x_2]$ such that $\deg_{x_1}(P) \leq N_1 - 1$, $\deg_{x_2}(P) \leq N_2 - 1$, the polynomial P together with all of its low order partial derivatives vanish at the point (α_1, α_2), and yet the height of P (that is, the height of the vector of coefficients of P) is not too large. The construction of such polynomials is an important step in the method developed by Bombieri [**3**], [**5**] and by Bombieri and Mueller [**4**] for obtaining effective measures of irrationality for certain algebraic numbers, (see especially the discussion in [**5**, p.35]). As is well known, polynomials with the desired properties can be determined by using some form of Siegel's Lemma. We now describe this procedure and our results in more detail. In doing so we use notation identical to that developed already in [**18**] and described in section 2 below. In particular, if P is a polynomial in $k[x_1, x_2]$ we write $h(P)$ for the height of its vector of coefficients. If α is an algebraic number we write

$$h_1(\alpha) = \prod_v \max\{1, |\alpha|_v\} ,$$

(as in [**7**]), for the inhomogeneous height of α.

Let $0 < \theta_1 < 1$, $0 < \theta_2 < 1$, and let N_1 and N_2 be positive integers. We define $\Gamma = \Gamma(\theta_1, \theta_2, N_1, N_2)$ to be the set

$$\Gamma = \left\{(m_1, m_2) \in \mathbf{Z}^2 : 0 \leq m_1 \ , \ 0 \leq m_2 \ \text{ and } \ \frac{m_1}{\theta_1 N_1} + \frac{m_2}{\theta_2 N_2} < 1\right\} . \tag{1.2}$$

We write $|\Gamma|$ for the cardinality of Γ which is obviously positive. Then define $Z = Z(\theta_1, \theta_2, N_1, N_2, \alpha_1, \alpha_2)$ to be the $|\Gamma| \times N_1 N_2$ matrix

$$Z = \left(\binom{n_1}{m_1}\binom{n_2}{m_2}(\alpha_1)^{n_1 - m_1}(\alpha_2)^{n_2 - m_2}\right) \tag{1.3}$$

where $(m_1, m_2) \in \Gamma$ indexes rows and columns are indexed by all ordered pairs (n_1, n_2) with $n_1 = 0, 1, 2, \ldots, N_1 - 1$, and $n_2 = 0, 1, 2, \ldots, N_2 - 1$. If (m_1, m_2) is a point in $\mathbf{Z}^2$ with nonnegative coordinates we write

$$D^{(m_1, m_2)} = (m_1!)^{-1}(m_2!)^{-1}\frac{\partial^{m_1+m_2}}{\partial x_1^{m_1}\partial x_2^{m_2}}$$

for the corresponding partial differential operator. Now a more precise statement of our problem is as follows. We wish to determine a polynomial $P(x_1, x_2)$ in $k[x_1, x_2]$ such that $\deg_{x_1}(P) \leq N_1 - 1$, $\deg_{x_2}(P) \leq N_2 - 1$, P is not identically zero, and

$$(D^{(m_1,m_2)}P)(\alpha_1, \alpha_2) = 0 \tag{1.4}$$

for each point $(m_1, m_2) \in \Gamma$. Moreover, we would like to establish the existence of such a polynomial P with $h(P)$ bounded from above by a suitable function of θ_1, θ_2, N_1, N_2, α_1 and α_2. Of course we may identify the polynomial P with its vector $\vec{\xi}$ of coefficients and then $\vec{\xi} \neq \vec{0}$ is to be determined in $k^{N_1 N_2}$. It is easy to verify that P satisfies (1.4) if and only if its vector $\vec{\xi}$ of coefficients satisfies $Z\vec{\xi} = \vec{0}$. Thus our problem is to prove the existence of $\vec{\xi} \neq \vec{0}$ in $k^{N_1 N_2}$ such that $Z\vec{\xi} = \vec{0}$ and $h(\vec{\xi})$ is not too large.

In order to insure that the system of linear equations $Z\vec{x} = \vec{0}$ has a nontrivial solution in $k^{N_1 N_2}$ it is sufficient to require that $N_1 N_2 - r|\Gamma| \geq 1$, where $r = [K : k]$. The reason for this is that when we express the entries of Z in terms of a basis for K over k, each equation in the linear system $Z\vec{x} = \vec{0}$ gets replaced by r equations having coefficients in k. Therefore we assume for the remainder of this section that $N_1 N_2 - r|\Gamma| \geq 1$.

Let F be a number field which is a Galois extension of k and also a Galois extension of K. Let $\sigma_1, \sigma_2, \ldots, \sigma_r$ be the distinct embeddings of K in F which fix k. We assemble the matrices $\sigma_1(Z), \sigma_2(Z), \ldots, \sigma_r(Z)$ into an $r|\Gamma| \times N_1 N_2$ matrix

$$Y = \begin{pmatrix} \sigma_1(Z) \\ \sigma_2(Z) \\ \cdots \\ \sigma_r(Z) \end{pmatrix}.$$

If rank $(Y) = r|\Gamma|$ we may apply the invariant form of Siegel's Lemma given in [6] as Theorem 12. By that result there exists a basis $\{\vec{\xi}_1, \vec{\xi}_2, \ldots, \vec{\xi}_L\}$ for the null space

$$\mathcal{X} = \left\{\vec{x} \in k^{N_1 N_2} : Z\vec{x} = \vec{0}\right\} \tag{1.5}$$

such that

$$\sum_{\ell=1}^{L} \log h(\vec{\xi}_\ell) \leq L \log c_k + \log H(Y) . \tag{1.6}$$

Here c_k is a field constant, $H(Y)$ denotes the height of the matrix Y (which is defined in section 2), and $L = \dim(\mathcal{X}) = N_1N_2 - r|\Gamma|$. Now in general the matrix Y will not have maximum rank and so this approach involves some loss of generality. On the other hand it is easy to show that $\operatorname{rank}(Z) = |\Gamma|$. Therefore, without making any assumption on the rank of Y, we may apply Theorem 7 of section 2. By that more general result we deduce the existence of a basis $\{\vec{\xi}_1, \vec{\xi}_2, \ldots, \vec{\xi}_L\}$ for $\mathcal{X}$ such that

$$\sum_{\ell=1}^{L} \log h(\vec{\xi}_\ell) \leq L \log c_k + r \log H(Z)\,. \tag{1.7}$$

In (1.7) we have $N_1N_2 - r|\Gamma| \leq L = \dim(\mathcal{X}) = N_1N_2 - \operatorname{rank}(Y) \leq N_1N_2 - |\Gamma|$.

It remains to estimate log $H(Z)$. For this purpose we use the special inequalities on submodular functions obtained in section 3 and section 4. In this way we establish the following technical result. Let $u(x)$ be defined by $u(x) = 0$ if $x \leq 0$ or $1 \leq x$ and by

$$u(x) = \frac{1}{4}x^2 \log\left(\frac{1-x^2}{16x^2}\right) + \frac{1}{2}x \log\left(\frac{1+x}{1-x}\right) + \frac{1}{4}\log(1-x^2) \tag{1.8}$$

if $0 < x < 1$. If α is an algebraic number let $\Phi_\alpha(x)$ be defined by $\Phi_\alpha(x) = 0$ if $x \leq 0$ or $1 \leq x$ and by

$$\Phi_\alpha(x) = x(1-x)\log h_1(\alpha) + u(x) \tag{1.9}$$

if $0 < x < 1$. Both u and Φ_α are absolutely continuous.

Theorem 2. *Let* Γ *and* Z *be defined by* (1.2) *and* (1.3) *respectively. Then* $\operatorname{rank}(Z) = |\Gamma|$ *and*

$$\begin{aligned} \log H(Z) \leq{} & (N_1)^2\theta_2 N_2 \int_0^1 \Phi_{\alpha_1}(\theta_1 x)\,dx \\ & + \theta_1 N_1 (N_2)^2 \int_0^1 \Phi_{\alpha_2}(\theta_2 x)\,dx \\ & + \frac{1}{2}(N_1+N_2)^2 \log\{4h_1(\alpha_1)h_1(\alpha_2)\}\,. \end{aligned} \tag{1.10}$$

The integrals on the right of (1.10) can be evaluated explicitly. Using [7, equation (2.4)] they can also be estimated by

$$\int_0^1 \Phi_{\alpha_i}(\theta_i x)\,dx \leq \frac{1}{2}\theta_i\left(1 - \frac{2}{3}\theta_i\right)\log h_1(\alpha_i) + \frac{1}{6}(\theta_i)^2 \log\left(\frac{1}{4\theta_i}\right) + \frac{11}{36}(\theta_i)^2\,. \tag{1.11}$$

In fact a simpler estimate follows from [**7**, equation (2.3)], we have

$$\int_0^1 \Phi_{\alpha_i}(\theta_i x)\,dx \le \frac{1}{2}\theta_i\Big(1-\frac{2}{3}\theta_i\Big)\log\{2h_1(\alpha_i)\}\ . \tag{1.12}$$

Of course (1.11) is useful when α_i is a root of unity and therefore $\log h_1(\alpha_i) = 0$. However, for many applications the bound (1.12) is entirely adequate. In section 6 we indicate how our polynomial construction can be used to derive a simple form of the Thue-Siegel principle. For this purpose we use the estimate

$$\begin{aligned}\log H(Z) \le \Big(\frac{1}{2}\theta_1\theta_2 N_1 N_2\Big)\Big(N_1\log\{2h_1(\alpha_1)\} + N_2\log\{2h_1(\alpha_2)\}\Big)\\ + \frac{1}{2}(N_1+N_2)^2\log\{4h_1(\alpha_1)h_1(\alpha_2)\}\ ,\end{aligned} \tag{1.13}$$

which follows immediately from (1.10) and (1.12).

2. Heights of Subspaces

We suppose that the algebraic number field k has degree d over $\mathbf{Q}$. Our notation for places of k, completions, normalized absolute values and heights will be identical to that which was developed in [**18**], section 2. Here we briefly review the most important facts.

If v is a place of k we write k_v for the completion of k at v. Then $d_v = [k_v, \mathbf{Q}_v]$ will denote the local degree. At each place v of k we introduce two absolute values $|\ \ |_v$ and $\|\ \ \|_v$ as follows:

(i) if $v \mid p$, where p is a finite rational prime, then $\|p\|_v = p^{-1}$,
(ii) if $v \mid \infty$ then $\|\ \ \|_v$ is the usual Euclidean absolute value on $k_v = \mathbf{R}$ or $k_v = \mathbf{C}$.

Obviously $\|\ \ \|_v$ for $v \mid p$ extends the usual p-adic absolute value. The absolute values $\|\ \ \|_v$ and $|\ \ |_v$ are related by $\|\ \ \|_v^{d_v/d} = |\ \ |_v$. If $\alpha \in k$, $\alpha \ne 0$, then the product formula asserts that $\prod_v |\alpha|_v = 1$.

Let

$$\vec{x} = \begin{pmatrix} x_1 \\ x_2 \\ \cdots \\ x_N \end{pmatrix}$$

be a column vector in $(k_v)^N$. We extend our absolute value $\|\ \ \|_v$ to vectors by

(i) if $v \nmid \infty$ then

$$\|\vec{x}\|_v = \max_{1\le n\le N} \|x_n\|_v\ ,$$

(ii) if $v \mid \infty$ then

$$\|\vec{x}\|_v = \left(\sum_{n=1}^{N} \|x_n\|_v^2\right)^{1/2}.$$

We also extend $|\ \ |_v$ to vectors by setting $|\vec{x}|_v = \|\vec{x}\|_v^{d_v/d}$.

If $\vec{x}$ is a (column) vector in k^N we define a height $h(\vec{x})$ for such vectors by

$$h(\vec{x}) = \prod_v \left\{\max_{1\le n\le N} |x_n|_v\right\}.$$

If $\alpha \neq 0$ and $\alpha \in k$, then by the product formula we have $h(\alpha\vec{x}) = h(\vec{x})$. Thus h is an absolute height on the projective space $\mathbf{P}_k^{N-1}$.

Let $X = (x_{nm})$ be an $N \times M$ matrix over k_v. If $I \subseteq \{1, 2, \ldots, N\}$ is a subset of cardinality $|I| = L$, we write

$${}_IX = (x_{nm}), \qquad n \in I,\ m = 1, 2, \ldots, M,$$

for the corresponding $L \times M$ submatrix. Similarly, if $J \subseteq \{1, 2, \ldots, M\}$ with $|J| = L$, we write

$$X_J = (x_{nm}), \qquad n = 1, 2, \ldots, N,\ m \in J,$$

for the corresponding $N \times L$ submatrix. Now suppose that $\operatorname{rank}(X) = M \le N$. Then the $\binom{N}{M}$ numbers

$$\left\{\det({}_IX) : I \subseteq \{1, 2, \ldots, N\},\ |I| = M\right\}$$

are the Grassmann coordinates of X. We use these to define the local height $H_v(X)$ as follows:

(i) if $v \nmid \infty$ then

$$H_v(X) - \max_{|I|=M} |\det {}_IX|_v,$$

(ii) if $v \mid \infty$ then

$$H_v(X) = \left(\sum_{|I|=M} \|\det {}_IX\|_v^2\right)^{d_v/2d}.$$

If $v \mid \infty$ we also have

$$H_v(X) = |\det X^*X|_v^{1/2}$$

by the Cauchy-Binet formula, where X^* denotes the complex conjugate transpose of X. If $\vec{x}$ is a (column) vector in $(k_v)^N$ we write $H_v(\vec{x})$ for the local height of $\vec{x}$ where we regard $\vec{x}$ as an $N \times 1$ matrix. We note the identity $H_v(\vec{x}) = \|\vec{x}\|_v^{d_v/d} = |\vec{x}|_v$.

Let $X = (x_{nm})$ be an $N \times M$ matrix with entries in k and $\text{rank}(X) = M \leq N$. In this case we may apply H_v at each place v of k. We define the *global height* $H(X)$ by

$$H(X) = \prod_v H_v(X) .$$

If W is an $M \times M$ nonsingular matrix over k then

$$H(XW) = \prod_v \{|\det W|_v H_v(X)\} = H(X) \tag{2.1}$$

by the product formula. This identity allows us to view H as a height on subspaces. More precisely, let $\mathcal{X} \subseteq k^N$ be a vector subspace of dimension M, $1 \leq M \leq N$. Suppose that the columns of the $N \times M$ matrix X form a k-basis for $\mathcal{X}$. Then let Y be another $N \times M$ matrix with columns which form a k-basis for $\mathcal{X}$. Obviously $XW = Y$ for some $M \times M$ nonsingular matrix W and by (2.1) we have $H(X) = H(Y)$. Thus we define *the height $H(\mathcal{X})$ of the subspace $\mathcal{X}$* by $H(\mathcal{X}) = H(X)$. Our previous remarks show that this is well defined. Heights on vectors or one-dimensional subspaces occured already in a paper of Siegel [**17**] and were used later by Northcott [**14**] and Weil [**19**]. Heights on subspaces of k^N were first introduced by Schmidt [**16**] who used them to formulate and prove very general theorems in Diophantine approximation. In fact Schmidt's height on $\mathcal{X} \subseteq k^N$ would be equal to $H(\mathcal{X})^d$, $d = [k : \mathbb{Q}]$, in our notation. The height we have defined is often described as an *absolute height* because it does not depend on the field k which contains the entries of X.

Concerning matrices which are not of full rank we adopt the following convention: if $X = (x_{nm})$ is an $N \times M$ matrix with entries in k_v and $\text{rank}(X) < M \leq N$ we set $H_v(X) = 0$. However, if $\mathcal{X} \subseteq k^N$ is the zero dimensional subspace $\mathcal{X} = \{\vec{0}\}$ we set $H(\mathcal{X}) = 1$. Let $X = (x_{nm})$ be an $N \times M$ matrix over k_v or k with $\text{rank}(X) = N \leq M$. We extend our heights H_v and H to X by applying them to the transpose of X.

The subspace $\mathcal{X} \subseteq k^N$ may occur as the null space of a system of linear forms. Let A be an $(N-M) \times N$ matrix over k with $\text{rank}(A) = N - M < N$ and such that

$$\mathcal{X} = \left\{\vec{x} \in k^N : A\vec{x} = \vec{0}\right\} .$$

By the duality theorem of Brill-Gordan [**9**] (see also [**10**, Theorem I, p.294]) there exists a constant $\gamma \in k$, $\gamma \neq 0$, such that for very subset $I \subseteq \{1, 2, \ldots, N\}$, $|I| = M$,

$$\det{}_I X = \gamma(-1)^{\varepsilon(J)} \det A_J ,$$

where J is the complement of I and $\varepsilon(J) = \sum_{j \in J} j$. Using the product formula we obtain the following *duality principle*:

$$H(\mathcal{X}) = \prod_v H_v(X) = \prod_v \{|\gamma|_v H_v(A)\} = H(A) . \tag{2.2}$$

Next we suppose that v is a fixed place of $\boldsymbol{k}$ and $\mathcal{A} \subseteq (\boldsymbol{k}_v)^N$ is a subspace of dimension L, $1 \le L \le N$. Let $\{\vec{a}_1, \vec{a}_2, \ldots, \vec{a}_L\}$ be a basis for $\mathcal{A}$ as a vectorspace over $\boldsymbol{k}_v$. We write $A = (\vec{a}_1\, \vec{a}_2\, \cdots\, \vec{a}_L)$ for the $N \times L$ matrix having $\vec{a}_\ell$ as its ℓth column. Then we define an $N \times N$ matrix $P_v = P_v(A)$ as follows:

(i) if $v \mid \infty$ then

$$P_v = A(A^*A)^{-1}A^* ,$$

where A^* is the complex conjugate transpose if $\boldsymbol{k}_v = \mathbf{C}$ (and, of course, the transpose of A if $\boldsymbol{k}_v = \mathbf{R}$).

(ii) If $v \nmid \infty$ we select $J = J_v \subseteq \{1, 2, \ldots, N\}$ so that $|J| = L$ and

$$\max_{|I|=L} |\det{}_I A|_v = |\det{}_J A|_v . \tag{2.3}$$

Then we define

$$P_v = A({}_J A)^{-1}{}_J(\mathbf{1}_N) ,$$

where $\mathbf{1}_N$ denotes the $N \times N$ identity matrix.

We note that in our definition of P_v at finite places v the maximum on the left of (2.3) may occur at several distinct subsets. For definiteness we may assume that the subsets $I \subseteq \{1, 2, \ldots, N\}$ with $|I| = L$ are ordered in some way and then select J among all possibilities to be the first with respect to this ordering.

If $\{\vec{a}_1{}', \vec{a}_2{}', \ldots, \vec{a}_L{}'\}$ is another basis for $\mathcal{A}$, if A' is the corresponding $N \times L$ matrix, then $A = A'W$ for some nonsingular $L \times L$ matrix W with entries in $\boldsymbol{k}_v$. A simple calculation shows that $P_v(A) = P_v(A')$. Therefore we may write $P_v = P_v(\mathcal{A})$, since P_v depends on the subspace $\mathcal{A}$ and not on our choice of a basis for $\mathcal{A}$.

The matrix P_v acts as a projection operator onto the subspace $\mathcal{A}$. In particular we have

$$P_v\vec{x} \in \mathcal{A} \quad \text{for all} \quad \vec{x} \in (\boldsymbol{k}_v)^N , \tag{2.4}$$

and

$$P_v\vec{x} = \vec{x} \quad \text{for all} \quad \vec{x} \in \mathcal{A} . \tag{2.5}$$

The trivial verification of (2.4) and (2.5) is given in [**18**, section 4]. Using (2.4) and (2.5), or by direct calculation, we also find that

$$P_v^2 = P_v . \tag{2.6}$$

Let M be a integer with $L < M \leq N$ and let B be an $N \times (M - L)$ matrix over k_v. The matrices A and B may be assembled into blocks of an $N \times M$ matrix C by setting

$$C = (A\ B)\ . \tag{2.7}$$

If $P_v = P_v(\mathcal{A})$ is projection onto the L-dimensional subspace $\mathcal{A}$ spanned by the columns of A, then we have

$$H_v(C) = H_v(A)H_v\big((\mathbf{1}_N - P_v)B\big)\ . \tag{2.8}$$

The identity (2.8) is proved in [**18**, Lemma 4] where it is assumed that $\operatorname{rank}(C) = M$. In fact (2.8) holds generally since the local height of a matrix which is not of full rank is zero.

Lemma 3. *Let $\mathcal{A} \subseteq \mathcal{B} \subseteq (k_v)^N$ be subspaces and let X be an $N \times M$ matrix over k_v, $1 \leq M \leq N$. Then we have*

$$H_v\Big((\mathbf{1}_N - P_v(\mathcal{B}))X\Big) \leq H_v\Big((\mathbf{1}_N - P_v(\mathcal{A}))X\Big) \leq H_v(X)\ . \tag{2.9}$$

Proof. We may assume that $\operatorname{rank}(X) = M$, for otherwise each of the heights in (2.9) is zero. We will prove the second inequality in (2.9) first, and in doing so we may also assume that $\mathcal{A}$ has positive dimension. Let

$$1 \leq L = \dim(\mathcal{A}) \leq N\ ,$$

and let $\mathcal{X}$ be the M dimensional subspace spanned by the columns of X. If $\mathcal{A} \cap \mathcal{X}$ contains a nonzero vector then $\operatorname{rank}((\mathbf{1}_N - P_v(\mathcal{A}))X) < M$ and so the second inequality in (2.9) is trivial. Hence we may assume that $\mathcal{A} \cap \mathcal{X} = \{\vec{0}\}$. Let A be an $N \times L$ matrix having columns which span $\mathcal{A}$. Then $L + M \leq N$ and the matrix $C = (A\ X)$ has full rank, that is, $\operatorname{rank}(C) = L + M$. Using (2.8) we have

$$H_v(C) = H_v(A)H_v\Big((\mathbf{1}_N - P_v(\mathcal{A}))X\Big) \tag{2.10}$$

and by a basic inequality for heights, (see [**6**, equation (2.6)]),

$$H_v(C) \leq H_v(A)H_v(X)\ . \tag{2.11}$$

Of course (2.10) and (2.11) show that

$$H_v\Big((\mathbf{1}_N - P_v(\mathcal{A}))X\Big) \leq H_v(X)\ . \tag{2.12}$$

Since $\mathcal{A} \subset \mathcal{B}$ we have

$$P_v(\mathcal{B})P_v(\mathcal{A}) = P_v(\mathcal{A})$$

by (2.4) and therefore

$$(\mathbf{1}_N - P_v(\mathcal{B}))X = (\mathbf{1}_N - P_v(\mathcal{B}))(\mathbf{1}_N - P_v(\mathcal{A}))X . \tag{2.13}$$

We apply (2.12) with $\mathcal{A}$ replaced by $\mathcal{B}$ and X replaced by $(\mathbf{1}_N - P_v(\mathcal{A}))X$. We also use (2.13). It follows that

$$\begin{aligned} H_v\Big((\mathbf{1}_N - P_v(\mathcal{B}))X\Big) &= H_v\Big((\mathbf{1}_N - P_v(\mathcal{B}))(\mathbf{1}_N - P_v(\mathcal{A}))X\Big) \\ &\le H_v\Big((\mathbf{1}_N - P_v(\mathcal{A}))X\Big) , \end{aligned}$$

and this establishes the first inequality in (2.9).

If $C = (A\ B)$ is a matrix partitioned as in (2.7) then by the inequality referred to in our proof of Lemma 3,

$$H_v((A\ B)) \le H_v(A)H_v(B) . \tag{2.14}$$

We now prove a more elaborate inequality for matrices partitioned into three blocks. This inequality is the local form of (1.1) and will be fundamental for our later applications.

Theorem 4. *Let A, B, and C be three matrices with entries in k_v. We assume that A is $N \times L_1$, B is $N \times L_2$, C is $N \times L_3$, and $L_1 + L_2 + L_3 \le N$. Then we have*

$$H_v((A\ B\ C))H_v(C) \le H_v((A\ C))H_v((B\ C)) . \tag{2.15}$$

Proof. Clearly we may assume that

$$\operatorname{rank}\{(A\ B\ C)\} = L_1 + L_2 + L_3 .$$

Let $\mathcal{C}$ be the subspace of $(k_v)^N$ spanned by the columns of C, let $\mathcal{B}$ be the subspace of $(k_v)^N$ spanned by the columns of $(B\ C)$. Then $\mathcal{C} \subseteq \mathcal{B} \subseteq (k_v)^N$ and by (2.8) we have

$$\begin{aligned} &H_v((A\ B\ C))H_v(C) \\ &\quad \le H_v\Big((\mathbf{1}_N - P_v(\mathcal{B}))A\Big)H_v((B\ C))H_v(C) . \end{aligned} \tag{2.16}$$

Applying Lemma 3 and (2.8) again to the right hand side of (2.16) we find that

$$\begin{aligned} &H_v\Big((\mathbf{1}_N - P_v(\mathcal{B}))A\Big)H_v((B\ C))H_v(C) \\ &\leq H_v\Big((\mathbf{1}_N - P_v(\mathcal{C}))A\Big)H_v((B\ C))H_v(C) \\ &= H_v((A\ C))H_v((B\ C))\ . \end{aligned} \tag{2.17}$$

The Theorem plainly follows from (2.16) and (2.17).

If the three matrices A, B and C in Theorem 4 have entries in k then (2.15) holds at each place v and therefore (2.15) holds also for the global height. When we formulate this remark in terms of subspaces we obtain the statement of Theorem 1, which we repeat here as a corollary.

Corollary 5. *Let $\mathcal{A} \subseteq k^N$ and $\mathcal{B} \subseteq k^N$ be subspaces and let $\langle \mathcal{A}, \mathcal{B}\rangle$ denote the subspace of k^N which is spanned over k by $\mathcal{A} \cup \mathcal{B}$. Then we have*

$$H(\langle \mathcal{A}, \mathcal{B}\rangle)H(\mathcal{A}\cap\mathcal{B}) \leq H(\mathcal{A})H(\mathcal{B})\ . \tag{2.18}$$

Proof. Obviously we may assume that $\mathcal{A}$ and $\mathcal{B}$ have positive dimension. Let $\mathcal{C} = \mathcal{A}\cap\mathcal{B}$ and assume that $\mathcal{C}$ has positive dimension L_3. Then we may select an $N \times L_3$ matrix C having columns which span $\mathcal{C}$ over k. If $\mathcal{A} \subseteq \mathcal{B}$ or $\mathcal{B} \subseteq \mathcal{A}$ the inequality (2.18) is trivial so we may assume that $\mathcal{A}$ has dimension $L_1 + L_3$ and $\mathcal{B}$ has dimension $L_2 + L_3$, with $L_1 \geq 1$, $L_2 \geq 1$. Thus we can determine $N \times L_1$ and $N \times L_2$ matrices A and B such that $\mathcal{A}$ is spanned over k by the columns of $(A\ C)$ and $\mathcal{B}$ is spanned over k by the columns of $(B\ C)$. It follows that $\langle\mathcal{A},\mathcal{B}\rangle$ is spanned over k by the columns of $(A\ B\ C)$. The desired inequality now follows immediately from (2.15) and our definition of heights on subspaces.

If $\mathcal{C} = \{\vec{0}\}$ we may select A and B as above so that their columns span $\mathcal{A}$ and $\mathcal{B}$ respectively over k. Now the inequality (2.18) follows immediately from (2.14).

Corollary 6. *For $i = 1, 2, \ldots, r$ let A_i be an $m_i \times N$ matrix over k with $\operatorname{rank}(A) = m_i \leq N$. Let B be a $(\sum_{i=1}^r m_i) \times N$ matrix over k formed by assembling the matrices $A_1, A_2, \ldots, A_r$ as blocks:*

$$B = \begin{pmatrix} A_1 \\ A_2 \\ \cdots \\ A_r \end{pmatrix}\ .$$

If $\mathcal{B} = \{\vec{x} \in k^N : B\vec{x} = \vec{0}\}$ then

$$H(\mathcal{B}) \leq \prod_{i=1}^r H(A_i)\ . \tag{2.19}$$

Remark. If the matrix B has $\operatorname{rank}(B) = \sum_{i=1}^{r} m_i \leq N$, then by the duality principle $H(\mathcal{B}) = H(B)$. The inequality (2.19) follows immediately from the analog of (2.14) for matrices which are partitioned vertically. The useful feature of the corollary is that we do *not* require the matrix B to have maximum rank.

Proof. Let

$$\mathcal{A}_i = \left\{ \vec{x} \in k^N : A_i \vec{x} = \vec{0} \right\}$$

so that

$$H(\mathcal{A}_i) = H(A_i) \tag{2.20}$$

and

$$\mathcal{B} = \bigcap_{i=1}^{r} \mathcal{A}_i \; . \tag{2.21}$$

Since the height of any subspace of k^N is greater than or equal to one, the inequality (2.18) implies that

$$H(\mathcal{B}) \leq \prod_{i=1}^{r} H(\mathcal{A}_i) \; . \tag{2.22}$$

We may now appeal to (2.20) to establish the corollary.

The inequality obtained in Corollary 6 can be used to prove a slightly more general form of the Siegel's Lemma which was given as Theorem 12 of [6].

Theorem 7. *Let K be a finite extension of the number field k with $[K : k] = r$. Suppose that A is an $M \times N$ matrix over K with* $\operatorname{rank}(A) = M$ *and $Mr < N$. Let $\mathcal{A} \subseteq k^N$ be the subspace*

$$\mathcal{A} = \left\{ \vec{x} \in k^N : A\vec{x} = \vec{0} \right\}$$

and assume that $\mathcal{A}$ has dimension L over k. Then $N - Mr \leq L \leq N - M$ and there exist L linearly independent vectors $\vec{\xi}_1, \vec{\xi}_2, \ldots, \vec{\xi}_L$ in $\mathcal{A} \cap (O_k)^N$ such that

$$\prod_{\ell=1}^{L} h(\vec{\xi}_\ell) \leq (c_k)^L H(A)^r \; . \tag{2.23}$$

Here $c_k = (\frac{2}{\pi})^{s/d} |\Delta_k|^{1/2d}$ where s is the number of complex places of k and Δ_k is the discriminant of k.

Proof. Since $\mathcal{A}$ is a subspace of k^N with dimension L it follows immediately from [6, Theorem 8] that there are linearly independent vectors $\vec{\xi}_1, \vec{\xi}_2, \ldots, \vec{\xi}_L$ in $\mathcal{A} \cap (O_k)^N$ such that

$$\prod_{\ell=1}^{L} h(\vec{\xi}_\ell) \leq (c_k)^L H(\mathcal{A}) \; . \tag{2.24}$$

It remains to estimate $H(\mathcal{A})$.

Let F be an algebraic number field such that $k \subseteq K \subseteq F$, F is a Galois extension of k with Galois group $G(F/k)$, and F is also a Galois extension of K with Galois group $G(F/K)$. Then $G(F/K)$ is a subgroup of $G(F/k)$ having index r. Let $\sigma_1, \sigma_2, \ldots, \sigma_r$ be elements of $G(F/k)$ which form a set of distinct coset representatives of $G(F/K)$ in $G(F/k)$. For each σ_i we write $\sigma_i(A) = (\sigma_i(a_{mn}))$ for the corresponding $M \times N$ matrix. We assemble these matrices as blocks of the $Mr \times N$ matrix

$$B = \begin{pmatrix} \sigma_1(A) \\ \sigma_2(A) \\ \cdots \\ \sigma_r(A) \end{pmatrix} . \tag{2.25}$$

Of course B has entries in F and each block $\sigma_i(A)$ has rank equal to M. Therefore we may apply Corollary 6 to the subspace

$$\mathcal{B} = \{\vec{y} \in F^N : B\vec{y} = \vec{0}\}$$

and conclude that

$$H(\mathcal{B}) \leq \prod_{i=1}^{r} H(\sigma_i(A)) = H(A)^r . \tag{2.26}$$

Let $\omega_1, \omega_2, \ldots, \omega_r$ be a basis for K over k. Then each entry a_{mn} in the matrix A can be written in the form

$$a_{mn} = \sum_{j=1}^{r} \omega_j c_{mn}(j) \tag{2.27}$$

with each $c_{mn}(j)$ in k. Since

$$\sum_{n=1}^{N} a_{mn} x_n = \sum_{j=1}^{r} \omega_j \left\{ \sum_{n=1}^{N} c_{mn}(j) x_n \right\} \tag{2.28}$$

it follows that a vector $\vec{x}$ in k^N satisfies $A\vec{x} = \vec{0}$ if and only if $\vec{x}$ is a solution to the system

$$\sum_{n=1}^{N} c_{mn}(j) x_n = 0 , \qquad m = 1, 2, \ldots M , \ j = 1, 2, \ldots, r .$$

Therefore we define $M \times N$ matrices $C(j)$ for each j, $1 \leq j \leq r$, by

$$C(j) = \left(c_{mn}(j)\right) ,$$

where $m = 1, 2, \ldots M$ indexes rows and $n = 1, 2, \ldots, N$ indexes columns. We assemble the matrices $C(j)$ as blocks of the $Mr \times N$ matrix

$$D = \begin{pmatrix} C(1) \\ C(2) \\ \cdots \\ C(r) \end{pmatrix} .$$

If we apply the automorphisms σ_i to each side of (2.28) we obtain a system of equations which can be written as

$$\Omega D = B , \tag{2.29}$$

where Ω is an $Mr \times Mr$ nonsingular matrix. In fact (2.29) is equivalent to [6, equation (5.18)] and $\Omega = (\sigma_i(\omega_j)) \otimes \mathbf{1}_M$. (We define the tensor product of matrices below as (2.32).) Of course (2.29) shows that

$$\mathcal{B} = \left\{\vec{y} \in F^N : D\vec{y} = \vec{0}\right\} \tag{2.30}$$

and (2.28) implies that

$$\mathcal{A} = \left\{\vec{x} \in k^N : D\vec{x} = \vec{0}\right\} .$$

Let

$$X = (\vec{\xi}_1 \ \vec{\xi}_2 \ \cdots \ \vec{\xi}_L)$$

be an $N \times L$ matrix with entries in k having columns which form a basis over k for $\mathcal{A}$. In view of (2.30) the columns of X also form a basis over F for $\mathcal{B}$. It follows that

$$H(\mathcal{A}) = H(X) = H(\mathcal{B}) \tag{2.31}$$

since our heights do not depend on the number field containing the entries of X. The bound (2.23) now follows from (2.24), (2.26) and (2.31). Of course

$$\begin{aligned} \dim(\mathcal{A}) = L &= N - \operatorname{rank}(D) \\ &= N - \operatorname{rank}(B) , \end{aligned}$$

and plainly $M \leq \operatorname{rank}(B) \leq Mr$.

If the matrix B occuring in (2.25) has $\operatorname{rank}(B) = Mr$ then $L = N - Mr$ and (2.23) can be replaced by the sharper bound

$$\prod_{i=1}^{N-Mr} h(\vec{\xi}_i) \leq (c_k)^{N-Mr} H(B) .$$

This is the form of Siegel's Lemma given as Theorem 12 of **[6]**. The advantage of our Theorem 7 is that we make no assmption concerning the rank of B. This is especially convenient in our application to the problem of constructing polynomials in two variables with prescribed vanishing.

Next we consider heights on tensor products. Let A and B be $N_1 \times M_1$ and $N_2 \times M_2$ matrices respectively with entries in k_v. We denote by $A \otimes B$ the tensor (or Kronecker) product of A and B. That is, $A \otimes B$ is the $N_1N_2 \times M_1M_2$ matrix

$$A \otimes B = \begin{pmatrix} a_{11}B & a_{12}B & \cdots & a_{1M_1}B \\ a_{21}B & a_{22}B & \cdots & a_{2M_1}B \\ & \cdots & & \\ a_{N_11}B & a_{N_12}B & \cdots & a_{N_1M_1}B \end{pmatrix}, \tag{2.32}$$

where $A = (a_{nm})$. Alternatively,

$$A \otimes B = (a_{n_1m_1}b_{n_2m_2})$$

where rows are indexed by ordered pairs (n_1, n_2), $1 \le n_1 \le N_1$, $1 \le n_2 \le N_2$, and columns are indexed by ordered pairs (m_1, m_2), $1 \le m_1 \le M_1$, $1 \le m_2 \le M_2$. The arrangement of rows and columns is determined by (2.32). Basic properties of the tensor product of matrices are given in **[11**, section 8.2].

Theorem 8. *Let A and B be $N_1 \times M_1$ and $N_2 \times M_2$ matrices respectively with entries in k_v. If $1 \le M_1 \le N_1$ and $1 \le M_2 \le N_2$, then*

$$H_v(A \otimes B) = H_v(A)^{M_2} H_v(B)^{M_1} . \tag{2.33}$$

Proof. Since $\mathrm{rank}(A \otimes B) = \mathrm{rank}(A)\,\mathrm{rank}(B)$ the identity (2.33) holds with both sides equal to zero if either $\mathrm{rank}(A) < M_1$ or $\mathrm{rank}(B) < M_2$. Therefore we may assume that both A and B have full rank.

If $v \mid \infty$ we have

$$\begin{aligned} H_v(A \otimes B) &= \left|\det\{(A \otimes B)^*(A \otimes B)\}\right|_v^{1/2} \\ &= \left|\det\{(A^* \otimes B^*)(A \otimes B)\}\right|_v^{1/2} \\ &= \left|\det\{A^*A \otimes B^*B\}\right|_v^{1/2} \\ &= \left|\det\{A^*A\}^{M_2} \det\{B^*B\}^{M_1}\right|_v^{1/2} \\ &= H_v(A)^{M_2} H_v(B)^{M_1} . \end{aligned}$$

If $v \nmid \infty$ we select $I \subseteq \{1, 2, \ldots, N_1\}$, $J \subseteq \{1, 2, \ldots, N_2\}$ so that $|I| = M_1$, $|J| = M_2$, $|\det{}_I A|_v = H_v(A)$ and $|\det{}_J B|_v = H_v(B)$. It follows that

$$\begin{aligned} H_v(A \otimes B) &\geq \Big|\det\{\ {}_{I\times J}(A \otimes B)\}\Big|_v \\ &= \Big|\det\{{}_I A \otimes {}_J B\}\Big|_v \\ &= \Big|\det({}_I A)^{M_2} \det({}_J B)^{M_1}\Big|_v \\ &= H_v(A)^{M_2} H_v(B)^{M_1} . \end{aligned} \tag{2.34}$$

To obtain an upper bound we note that

$$\begin{aligned} H_v(A \otimes B) &= H_v\big(A({}_I A)^{-1}({}_I A) \otimes B({}_J B)^{-1}({}_J B)\big) \\ &= H_v\Big(\big(A({}_I A)^{-1} \otimes B({}_J B)^{-1}\big)({}_I A \otimes {}_J B)\Big) \\ &= H_v\big(A({}_I A)^{-1} \otimes B({}_J B)^{-1}\big) \big|\det\{{}_I A \otimes {}_J B\}\big|_v \\ &= H_v\big(A({}_I A)^{-1} \otimes B({}_J B)^{-1}\big) H_v(A)^{M_2} H_v(B)^{M_1} . \end{aligned} \tag{2.35}$$

Finally we claim that the matrices $A({}_I A)^{-1}$ and $B({}_J B)^{-1}$ have entries in the ring

$$O_v = \big\{\alpha \in k_v : |\alpha|_v \leq 1\big\} .$$

In fact this was shown already in the proof of [**18**, Lemma 7]. From the definition of the tensor product we find that the entries of $A({}_I A)^{-1} \otimes B({}_J B)^{-1}$ and hence the Grassmann coordinates of this matrix are in O_v. Thus we have

$$H_v\big(A({}_I A)^{-1} \otimes B({}_J B)^{-1}\big) \leq 1 \tag{2.36}$$

and therefore (2.33) follows from (2.34), (2.35) and (2.36).

3. Submodular Functions

Let $(\mathcal{L}, \wedge, \vee)$ be a lattice. That is, $\mathcal{L}$ is a set and $\wedge$ and $\vee$ are binary, associative, commutative, idempotent operations defined on $\mathcal{L} \times \mathcal{L}$. A function $f : \mathcal{L} \to \mathbf{R}$ is *submodular* if the inequality

$$f(x \vee y) + f(x \wedge y) \leq f(x) + f(y) \tag{3.1}$$

holds for all points x and y in $\mathcal{L}$. We say that f is a *modular* function if (3.1) holds with equality for all points x and y in $\mathcal{L}$. A wide variety of examples, results and applications of submodular functions can be found in [**2**], [**12**], and [**13**].

Suppose, for example, that $\mathcal{L}$ is the set of all subspaces of the k-vectorspace k^N. If $\mathcal{A}$ and $\mathcal{B}$ are subspaces let $\mathcal{A} \vee \mathcal{B} = \langle \mathcal{A}, \mathcal{B} \rangle$ (that is, $\mathcal{A} \vee \mathcal{B}$ is the subspace spanned over k by $\mathcal{A} \cup \mathcal{B}$) and $\mathcal{A} \wedge \mathcal{B} = \mathcal{A} \cap \mathcal{B}$. It is well known that $(\mathcal{L}, \wedge, \vee)$ forms a lattice and Corollary 5 shows that the function

$$f(\mathcal{A}) = \log H(\mathcal{A})$$

is submodular on $\mathcal{L}$.

Throughout the remainder of this section we will restrict our attention to the finite distributive lattices $(\mathcal{P}(S), \cup, \cap)$. Here S is a finite set, $\mathcal{P}(S)$ is the collection of all subsets of S and the binary operations are union and intersection. It is known (see [**1**, p. 59]) that every finite distributive lattice is isomorphic to a sublattice of $(\mathcal{P}(S), \cup, \cap)$ for an S of suitable finite cardinality. Thus a function $f : \mathcal{P}(S) \to \mathbf{R}$ is submodular if

$$f(I \cup J) + f(I \cap J) \leq f(I) + f(J) \tag{3.2}$$

for all subsets $I \subseteq S$ and $J \subseteq S$. Also, throughout this section we will assume that submodular functions $f : \mathcal{P}(S) \to \mathbf{R}$ satisfy the additional condition that

$$f(\emptyset) = 0 \,. \tag{3.3}$$

If $g : \mathcal{P}(S) \to \mathbf{R}$ satisfies (3.2) then $f(I) = g(I) - g(\emptyset)$, $I \subseteq S$, plainly satisfies (3.2) and (3.3). Thus there will be no significant loss of generality in assuming that submodular functions on $\mathcal{P}(S)$ satisfy both (3.2) and (3.3).

If S is a finite set, a *measure* μ on $\mathcal{P}(S)$ is a function $\mu : \mathcal{P}(S) \to [0, \infty)$ which is modular and satisfies $\mu(\emptyset) = 0$. Obviously a measure μ is determined by its values on singletons. If $\mu : \mathcal{P}(S) \to [0, \infty)$ is a measure and $I \subseteq S$ then

$$\mu(I) = \sum_{i \in I} \mu(\{i\}) \,.$$

Let S and T be two finite sets. We assume that

$$\mu : \mathcal{P}(S) \to [0, \infty) \quad \text{and} \quad \nu : \mathcal{P}(T) \to [0, \infty)$$

are measures and

$$d : \mathcal{P}(S) \to \mathbf{R} \quad \text{and} \quad e : \mathcal{P}(T) \to \mathbf{R}$$

are submodular functions. We wish to use the two triples $(S; \mu, d)$ and $(T; \nu, e)$, which consist of a finite set, a measure, and a submodular function, and construct from them a new triple $(S \times T; \eta, f)$ where η is a measure on $\mathcal{P}(S \times T)$ and f is a submodular function on $\mathcal{P}(S \times T)$. Moreover, we desire that for each $I \subseteq S$ and $J \subseteq T$ we have

$$\eta(I \times J) = \mu(I)\nu(J)$$

and

$$f(I \times J) = d(I)\nu(J) + \mu(I)e(J)\ . \tag{3.4}$$

The existence and uniqueness of η is trivial. Indeed, η is simply the usual product measure of analysis. In this very elementary situation it is given by

$$\eta(K) = \sum_{(i,j)\in K} \mu(\{i\})\nu(\{j\}) \tag{3.5}$$

for all $K \subseteq S\times T$. The existence of a submodular function $f : \mathcal{P}(S\times T) \to \mathbf{R}$ satisfying (3.4) is more complicated and, it turns out, not unique. As we will see, the submodular function f which we construct here satisfies a certain extremal condition and this makes it useful in our later applications.

We begin by constructing a submodular function $f_1 : \mathcal{P}(S \times T) \to \mathbf{R}$ which satisfies

$$f_1(I \times J) = d(I)\nu(J) \tag{3.6}$$

for all rectangles $I \times J \subseteq S \times T$. If $K \subseteq S \times T$ and $K = \emptyset$ we set $f_1(K) = 0$. If $K \neq \emptyset$ then K can be represented as a union of *nonempty* rectangles:

$$K = \bigcup_{m=1}^{M} (I_m \times J_m)\ . \tag{3.7}$$

Of course K may have many representations in this form. However, we now require that the sets $I_1, I_2, \ldots, I_M$ be distinct and that the sets $J_1, J_2, \ldots, J_M$ be disjoint. It is easy to prove that every nonempty set $K \subseteq S \times T$ has such a decomposition and that it is unique up to a permutation of the rectangles $I_m \times J_m$. Having determined such a unique representation we set

$$f_1(K) = \sum_{m=1}^{M} d(I_m)\nu(J_m)\ . \tag{3.8}$$

It is clear that f_1 is well defined and it is also clear that f_1 satisfies (3.6) when K is a rectangle.

Lemma 9. *Let $I_1, \ldots, I_N$ be subsets of S and let $J_1, \ldots, J_N$ be a disjoint collection of subsets of T (but possibly some of the sets $I_1, \ldots, I_N$, $J_1, \ldots, J_N$ are empty). If*

$$K = \bigcup_{n=1}^{N} (I_n \times J_n) \tag{3.9}$$

then

$$f_1(K) = \sum_{n=1}^{N} d(I_n)\nu(J_n)\ . \tag{3.10}$$

Proof. In fact neither (3.9) nor (3.10) are changed if empty rectangles are removed from the collection $I_1 \times J_1, \ldots, I_N \times J_N$. Thus without loss of generality we may assume that each set I_n and J_n is not empty. Then by reordering the rectangles if necessary we may suppose that $I_1, I_2, \ldots, I_M$ are distinct and that among the subsets $I_1, \ldots, I_N$ there are exactly M distinct subsets of S. Now set

$$J'_m = \bigcup_{\substack{n=1 \\ I_n = I_m}}^{N} J_n \ ,$$

where $m = 1, 2, \ldots, M$. Obviously $J'_1, J'_2, \ldots, J'_M$ are disjoint and no J'_m is empty. As

$$K = \bigcup_{m=1}^{M} (I_m \times J'_m)$$

is the unique representation of K used in the definition of f_1 we have

$$\begin{aligned} f_1(K) &= \sum_{m=1}^{M} d(I_m)\nu(J'_m) \\ &= \sum_{m=1}^{M} d(I_m) \sum_{\substack{n=1 \\ I_m = I_n}}^{N} \nu(J_n) \\ &= \sum_{n=1}^{N} \Big\{ \sum_{\substack{m=1 \\ I_m = I_n}}^{M} d(I_m) \Big\} \nu(J_n) \\ &= \sum_{n=1}^{N} d(I_n)\nu(J_n) \ . \end{aligned}$$

This proves the lemma.

Theorem 10. *The function $f_1 : \mathcal{P}(S \times T) \to \mathbf{R}$ is submodular.*

Proof. Let

$$K = \bigcup_{m=1}^{M} (I_m \times J_m) \quad \text{and} \quad L = \bigcup_{n=1}^{N} (I'_n \times J'_n) \tag{3.11}$$

be two nonempty subsets of $S \times T$. In (3.11) we assume that $I_1, I_2, \ldots, I_M$ are distinct and nonempty and that $J_1, J_2, \ldots, J_M$ are disjoint and nonempty. We make a similar assumption on the representation of L. It follows that

$$K \cap L = \bigcup_{m=1}^{M} \bigcup_{n=1}^{N} (I_m \cap I'_n \times J_m \cap J'_n)$$

and that

$$J_m \cap J'_n\ , \qquad m = 1, 2, \ldots, M\ ,\ n = 1, 2, \ldots,, N$$

is a disjoint collection of sets. Also, it is easy to check that

$$K \cup L = \left\{ \bigcup_{m=1}^{M} \left(I_m \times \left\{ J_m \setminus \bigcup_{n=1}^{N} J'_n \right\} \right) \right\}$$

$$\bigcup \left\{ \bigcup_{m=1}^{M} \bigcup_{n=1}^{N} (I_m \cup I'_n \times J_m \cap J'_n) \right\}$$

$$\bigcup \left\{ \bigcup_{n=1}^{N} \left(I'_n \times \left\{ J'_n \setminus \bigcup_{m=1}^{M} J_m \right\} \right) \right\} .$$

Since $J_1, J_2, \ldots, J_M$ and $J'_1, J'_2, \ldots, J'_N$ are both disjoint collections we find that the collection consisting of

$$J_m \setminus \bigcup_{n=1}^{N} J'_n\ , \qquad m = 1, 2, \ldots, M\ , \text{ and}$$

$$J_m \cap J'_n\ , \qquad m = 1, 2, \ldots, M\ ,\ n = 1, 2, \ldots, N\ , \text{ and}$$

$$J'_n \setminus \bigcup_{m=1}^{M} J_m\ , \qquad n = 1, 2, \ldots, N\ ,$$

is also a disjoint collection of subsets of L. Obviously some of the sets in these disjoint collections may be empty. However, by Lemma 9 we have

$$f_1(K \cap L) = \sum_{m=1}^{M} \sum_{n=1}^{N} d(I_m \cap I'_n)\nu(J_m \cap J'_n) \tag{3.12}$$

and

$$\begin{aligned} f_1(K \cup L) = & \sum_{m=1}^{M} d(I_m)\nu\left(J_m \setminus \bigcup_{n=1}^{N} J'_n \right) \\ & + \sum_{m=1}^{M} \sum_{n=1}^{N} d(I_m \cup I'_n)\nu(J_m \cap J'_n) \\ & + \sum_{n=1}^{N} d(I'_n)\nu\left(J'_n \setminus \bigcup_{m=1}^{M} J_m \right) . \end{aligned} \tag{3.13}$$

As d is submodular it satisfies

$$d(I_m \cup I'_n) + d(I_m \cap I'_n) \leq d(I_m) + d(I'_n) \ . \tag{3.14}$$

Combining (3.12), (3.13), (3.14) and using the fact that ν is nonnegative valued, we deduce the inequality

$$\begin{aligned}
&f_1(K \cup L) + f_1(K \cap L) \\
&\leq \sum_{m=1}^{M} d(I_m)\nu\Big(J_m \setminus \bigcup_{n=1}^{N} J'_n\Big) + \sum_{m=1}^{M}\sum_{n=1}^{N} d(I_m)\nu(J_m \cap J'_n) \\
&\quad + \sum_{m=1}^{M}\sum_{n=1}^{N} d(I'_n)\nu(J_m \cap J'_n) + \sum_{n=1}^{N} d(I'_n)\nu\Big(J'_n \setminus \bigcup_{m=1}^{M} J_m\Big) \\
&= \sum_{m=1}^{M} d(I_m)\nu\Big(J_m \setminus \bigcup_{n=1}^{N} J'_n\Big) + \sum_{m=1}^{M} d(I_m)\nu\Big(J_m \cap \Big\{\bigcup_{n=1}^{N} J'_n\Big\}\Big) \\
&\quad + \sum_{n=1}^{N} d(I'_n)\nu\Big(\Big\{\bigcup_{m=1}^{M} J_m\Big\} \cap J'_n\Big) + \sum_{n=1}^{N} d(I'_n)\nu\Big(J'_n \setminus \bigcup_{m=1}^{M} J_m\Big) \\
&= f_1(K) + f_1(L) \ .
\end{aligned}$$

Of course this is exactly the condition (3.2) that f_1 be submodular (the condition (3.3) being included in the definition of f_1).

We define a second function $f_2 : \mathcal{P}(S \times T) \to \mathbf{R}$ in a similar manner but with the roles of S and T reversed. Thus if $K \subseteq S \times T$ and $K = \emptyset$ we set $f_2(K) = 0$. If $K \neq \emptyset$ we represent K as

$$K = \bigcup_{n=1}^{N} (I_n \times J_n)$$

where $I_1, I_2, \ldots, I_N$ are disjoint nonempty subsets of S and $J_1, J_2, \ldots, J_N$ are distinct nonempty subsets of T. As before this representation is unique apart from a permutation of the rectangles $I_1 \times J_1, \ldots, I_N \times J_N$. Then we set

$$f_2(K) = \sum_{n=1}^{N} \mu(I_n)e(J_n) \tag{3.15}$$

so that by our remarks f_2 is well defined. If $K = I \times J$ is a nonempty rectangle then

$$f_2(I \times J) = \mu(I)e(J) \ . \tag{3.16}$$

Applying Theorem 10 with S and T interchanged we find that f_2 is also submodular. Finally we define $f : \mathcal{P}(S \times T) \to \mathbf{R}$ by

$$f(K) = f_1(K) + f_2(K) . \tag{3.17}$$

Then f is plainly submodular and, in view of (3.6) and (3.16), the function f satisfies (3.4).

We define a *submodular triple* to be a finite set S together with a measure $\mu : \mathcal{P}(S) \to [0, \infty)$ and a submodular function $d : \mathcal{P}(S) \to \mathbf{R}$. If $(S; \mu, d)$ and $(T; \nu, e)$ are two submodular triples we write $\mu \times \nu = \eta$ for the product measure given by (3.5), $d \times \nu = f_1$ for the product submodular function given by (3.8), and $\mu \times e = f_2$ for the product submodular function given by (3.15). Then $(S \times T, \eta, f)$ is also a submodular triple where $f = f_1 + f_2 = d \times \nu + \mu \times e$. Next we will show that f satisfies an extremal property. For this we need a further definition. A subset $K \subseteq S \times T$ is said to be *coherent* if it is either empty or it can be represented as a union of nonempty rectangles,

$$K = \bigcup_{m=1}^{M} (I_m \times J_m) , \tag{3.18}$$

such that the sequence $\{I_m\}_{m=1}^{M}$ is either monotone increasing or monotone decreasing, and the sequence $\{J_m\}_{m=1}^{M}$ is either monotone increasing or monotone decreasing. Of course if $\{I_m\}_{m=1}^{M}$ and $\{J_m\}_{m=1}^{M}$ are both increasing or if they are both decreasing then K is in fact a rectangle.

Theorem 11. *Let $(S; \mu, d)$ and $(T; \nu, e)$ be two submodular triples and let $g : \mathcal{P}(S \times T) \to \mathbf{R}$ be a submodular function. If*

$$g(I \times J) = f(I \times J) \tag{3.19}$$

for all rectangles $I \times J \subseteq S \times T$, where $f = d \times \nu + \mu \times e$, then

$$g(K) \le f(K) \tag{3.20}$$

for all coherent sets $K \subseteq S \times T$.

Proof. Without loss of generality we may assume that K is not empty and that K is not a rectangle. Then K is given by (3.18) and we may assume that $M \ge 2$,

$$I_1 \subseteq I_2 \subseteq \cdots \subseteq I_M ,$$
$$J_1 \supseteq J_2 \supseteq \cdots \supseteq J_M .$$

If $R_m = I_m \times J_m$ then for each integer L, $1 \leq L \leq M-1$, we have

$$\begin{aligned}\left(\bigcup_{\ell=1}^{L} R_\ell\right) \cap R_{L+1} &= \bigcup_{\ell=1}^{L}(R_\ell \cap R_{L+1}) \\ &= \bigcup_{\ell=1}^{L}((I_\ell \cap I_{L+1}) \times (J_\ell \cap J_{L+1})) \\ &= \bigcup_{\ell=1}^{L}(I_\ell \times J_{L+1}) \\ &= I_L \times J_{L+1} \\ &= R_L \cap R_{L+1}\ .\end{aligned}$$

As g is submodular we deduce that

$$g\left(\bigcup_{\ell=1}^{L+1} R_\ell\right) + g(R_L \cap R_{L+1}) \leq g\left(\bigcup_{\ell=1}^{L} R_\ell\right) + g(R_{L+1})\ . \tag{3.21}$$

Then we write (3.21) as

$$g\left(\bigcup_{\ell=1}^{L+1} R_\ell\right) \leq g\left(\bigcup_{\ell=1}^{L} R_\ell\right) + g(R_{L+1}) - g(R_L \cap R_{L+1}) \tag{3.22}$$

and iterate the inequality (3.22) using $L = M-1, M-2, \ldots, 1$. In this way we find that

$$g(K) = g\left(\bigcup_{m=1}^{M} R_m\right) \leq \sum_{m=1}^{M} g(R_m) - \sum_{m=1}^{M-1} g(R_m \cap R_{m+1})\ . \tag{3.23}$$

Next we will show that there is equality in the inequality (3.23) when g is replaced by the submodular function $f_1 = d \times \nu$ or by $f_2 = \mu \times e$. We give the argument for this only in the case of f_1 as the argument for f_2 is very similar. We have

$$K = \bigcup_{m=1}^{M} R_m = \bigcup_{m=1}^{M} (I_m \times (J_m \setminus J_{m+1}))$$

where $J_{M+1} = \emptyset$. Since the sequence of sets $\{J_m \setminus J_{m+1}\}_{m=1}^M$ is disjoint we may apply Lemma 9 to deduce that

$$\begin{aligned}
f_1(K) &= f_1\left(\bigcup_{m=1}^{M} R_m\right) \\
&= \sum_{m=1}^{M} d(I_m)\nu(J_m \setminus J_{m+1}) \\
&= \sum_{m=1}^{M} d(I_m)\nu(J_m) - \sum_{m=1}^{M-1} d(I_m)\nu(J_{m+1}) \qquad (3.24) \\
&= \sum_{m=1}^{M} f_1(R_m) - \sum_{m=1}^{M-1} f_1(I_m \times J_{m+1}) \\
&= \sum_{m=1}^{M} f_1(R_m) - \sum_{m=1}^{M-1} f_1(R_m \cap R_{m+1}) .
\end{aligned}$$

Finally, we recall that $f = f_1 + f_2$ and therefore (3.24) implies that

$$\begin{aligned}
f(K) &= f\left(\bigcup_{m=1}^{M} R_m\right) \\
&= \sum_{m=1}^{M} f(R_m) - \sum_{m=1}^{M-1} f(R_m \cap R_{m+1}) .
\end{aligned} \qquad (3.25)$$

Since $f(R_m) = g(R_m)$ and $f(R_m \cap R_{m+1}) = g(R_m \cap R_{m+1})$ the desired inequality (3.20) follows immediately from (3.23) and (3.25).

Before considering applications of Theorem 11 we briefly remark on the results of this section. Obviously it would be of interest to prove Theorem 11 for a wider class of subsets K. In fact this can be done for certain special subsets which are not in general coherent. Suppose for example that $\{I_m\}_{m=1}^M$ is a sequence of disjoint, nonempty subsets of S and $\{J_m\}_{m=1}^M$ is a sequence of disjoint, nonempty subsets of T. Let f and g be as in the statement of Theorem 11 and set $K_0 = \bigcup_{m=1}^{M}(I_m \times J_m)$. Then

$$\begin{aligned} g(K_0) &\leq \sum_{m=1}^{M} g(I_m \times J_m) \\ &= \sum_{m=1}^{M} \{f_1(I_m \times J_m) + f_2(I_m \times J_m)\} \\ &= f_1(K_0) + f_2(K_0) \\ &= f(K_0) , \end{aligned}$$

and therefore the conclusion of Theorem 11 holds. It is easy to see, however, that K_0 is not necessarily coherent. At present it is an open problem to determine the precise collection of subsets $K \subseteq S \times T$ for which Theorem 11 holds.

Another interesting problem is to identify special submodular triples for which the conclusion of Theorem 11 holds for *all* subsets $K \subseteq S \times T$. In the remainder of our paper we consider submodular functions which occur as the logarithm of the height of certain matrices. Such special submodular functions may satisfy additional conditions which can be used to establish a stronger from of Theorem 11.

4. An Inequality for Heights

In this section we give an example of two submodular triples and a submodular function g which satisfy the hypotheses of Theorem 11. Let S and T be finite, nonempty sets. Let A be an $|S| \times |S|$ nonsingular matrix with entries in an algebraic number field. We assume that the rows and columns of A are indexed by the elements of S. Similarly, we suppose that B is a $|T| \times |T|$ nonsingular matrix with entries in the same number field. Obviously the function $I \to |I|$ which maps a subset $I \subseteq S$ to its cardinality is a measure on $\mathcal{P}(S)$. Next we define $\varphi : \mathcal{P}(S) \to \mathbf{R}$ by

$$\varphi(I) = \log H({}_I A) .$$

By Corollary 5 the function φ is submodular on $\mathcal{P}(S)$ and so $(S; |\ \ |, \varphi)$ is a submodular triple. Similarly, we define $\psi : \mathcal{P}(T) \to \mathbf{R}$ by $\psi(J) = \log H({}_J B)$ and then $(T; |\ \ |, \psi)$ is a submodular triple. By Theorem 10 the functions $f_1 = \varphi \times |\ \ |$, $f_2 = |\ \ | \times \psi$, and $f = \varphi \times |\ \ | + |\ \ | \times \psi$ are all submodular on $\mathcal{P}(S \times T)$. Of course the product measure $|\ \ | \times |\ \ |$ on $\mathcal{P}(S \times T)$ is again the cardinality of a subset and we continue to denote it by $|\ \ |$. If $I \times J \subseteq S \times T$ is a rectangle then by (3.4) we have

$$f(I \times J) = |J| \log H({}_I A) + |I| \log H({}_J B) . \tag{4.1}$$

If $K \subseteq S \times T$ is an arbitrary subset then $f(K)$ is defined by (3.8) and (3.15).

From (2.32) we recall that $(A \otimes B)$ is an $|S|\ |T| \times |S|\ |T|$ matrix having rows and columns indexed by $S \times T$. If $K \subseteq S \times T$ we define $g : \mathcal{P}(S \times T) \to \mathbf{R}$ by

$$g(K) = \log H\left({}_K(A \otimes B)\right) . \tag{4.2}$$

Again Corollary 5 shows that g is submodular on $\mathcal{P}(S \times T)$. If $I \times J \subseteq S \times T$ is a rectangle then by Theorem 8 we have

$$\begin{aligned} g(I \times J) &= \log H\left({}_{I\times J}(A \otimes B)\right) \\ &= \log H\left(({}_I A) \otimes ({}_J B)\right) \\ &= |J| \log H({}_I A) + |I| \log H({}_J B) \\ &= f(I \times J) . \end{aligned}$$

Thus we have shown that the submodular triples $(S; |\ \ |, \varphi)$ and $(T; |\ \ |, \psi)$ and the submodular function $g : \mathcal{P}(S \times T) \to \mathbf{R}$ defined by (4.2) satisfy the hypotheses of Theorem 11. Obviously this proves the following result.

Theorem 12. *Let f and g be submodular functions on $\mathcal{P}(S \times T)$ as defined in this section. If $K \subseteq S \times T$ is a coherent subset then $g(K) \leq f(K)$.*

5. Proof of Theorem 2

We will need to estimate sums by integrals and for this purpose it will be convenient to use the following lemma.

Lemma 13. *Let $N \geq 0$ be an integer, let β and γ be real numbers, and suppose that $F : \mathbf{R} \to \mathbf{R}$ is an absolutely continuous function. Then*

$$\left|\sum_{n=0}^{N} F([\beta n + \gamma]) - \int_0^N F(\beta x + \gamma)\, dx\right| \leq \|F\|_\infty + N\left(\tfrac{1}{4}|\beta| + 1\right)\|F'\|_\infty ,$$

where $\|\ \ \|_\infty$ is the sup norm.

Proof. Let $\sum_{n=0}^{N}{}^{*}$ denote a sum in which the terms corresponding to $n = 0$ and $n = N$ are each multiplied by the factor 1/2. Since $F(\beta x + \gamma)$ is absolutely continuous we have

$$\begin{aligned} &\tfrac{1}{2}F(\beta n + \gamma) + \tfrac{1}{2}F(\beta(n+1) + \gamma) \\ &\quad = \int_n^{n+1} \left(\frac{d}{dx}\{F(\beta x + \gamma)(x - [x] - \tfrac{1}{2})\}\right) dx \\ &\quad = \int_n^{n+1} F(\beta x + \gamma)\, dx + \beta \int_n^{n+1} F'(\beta x + \gamma)(x - [x] - \tfrac{1}{2})\, dx . \end{aligned}$$

Summing this identity over $n = 0, 1, 2, \ldots, N-1$ we find that

$$\sum_{n=0}^{N}{}^{*} F(\beta n + \gamma) - \int_0^N F(\beta x + \gamma)\, dx = \beta \int_0^N F'(\beta x + \gamma)(x - [x] - \tfrac{1}{2})\, dx \; . \tag{5.1}$$

It follows using (5.1) that

$$\begin{aligned} \Big| \sum_{n=0}^{N} F([\beta n + \gamma]) - \int_0^N F(\beta x + \gamma)\, dx \Big| &\leq \tfrac{1}{2}|F([\gamma])| + \tfrac{1}{2}|F([\beta N + \gamma])| \\ &\quad + \Big| \sum_{n=0}^{N}{}' \{ F([\beta n + \gamma]) - F(\beta n + \gamma) \} \Big| \\ &\quad + \Big| \sum_{n=0}^{N}{}' F(\beta n + \gamma) - \int_0^N F(\beta x + \gamma)\, dx \Big| \\ &\leq \|F\|_\infty + \sum_{n=0}^{N}{}' \int_{[\beta n + \gamma]}^{\beta n + \gamma} |F'(x)|\, dx \\ &\quad + |\beta| \int_0^N |F'(\beta x + \gamma)(x - [x] - \tfrac{1}{2})|\, dx \\ &\leq \|F\|_\infty + N\|F'\|_\infty + \tfrac{1}{4}|\beta| N \|F'\|_\infty \; , \end{aligned}$$

and this proves the lemma.

We now proceed with the proof of Theorem 2. If $N \geq 1$ is an integer and α is an algebraic number we set

$$S_N = \{0, 1, 2, \ldots, N-1\} \; .$$

We also define an $N \times N$ matrix $A(\alpha, N)$ by

$$A(\alpha, N) = \left(\binom{n}{m} \alpha^{n-m} \right) \; ,$$

where $m \in S_N$ indexes rows and $n \in S_N$ indexes columns. It is clear that $A(\alpha, N)$ is upper triangular and satisfies $\det\{A(\alpha, N)\} = 1$. If $I \subseteq \mathcal{P}(S_N)$ we set

$$\varphi_\alpha(I) = \log H\big({}_I A(\alpha, N) \big)$$

so that $(S_N; |\ \ |, \varphi_\alpha)$ is a submodular triple.

If α_1 and α_2 are algebraic numbers, if $N_1 \geq 1$ and $N_2 \geq 1$ are integers we may apply Theorem 11 to the pair of submodular triples $(S_{N_1}; |\ \ |, \varphi_{\alpha_1})$ and $(S_{N_2}; |\ \ |, \varphi_{\alpha_2})$. As in section 4 let $g : \mathcal{P}(S_{N_1} \times S_{N_2}) \to \mathbf{R}$ be defined by

$$g(\Lambda) = g_{\alpha_1,\alpha_2}(\Lambda) = \log H\big({}_{\Lambda}(A(\alpha_1, N_1) \otimes A(\alpha_2, N_2))\big)$$

for each $\Lambda \subseteq S_{N_1} \times S_{N_2}$. If

$$f = f_{\alpha_1,\alpha_2} = \varphi_{\alpha_1} \times |\ \ | + |\ \ | \times \varphi_{\alpha_2}$$

then $g(\Lambda) \leq f(\Lambda)$ whenever $\Lambda \subseteq S_{N_1} \times S_{N_2}$ is a coherent subset. In particular, if $\Gamma = \Gamma(\theta_1, \theta_2, N_1, N_2)$ is defined by (1.2) then it follows easily that Γ is coherent and

$$_{\Gamma}\big(A(\alpha_1, N_1) \otimes A(\alpha_2, N_2)\big) = Z\ ,$$

where Z is defined by (1.3). Thus we have

$$\log H(Z) = g(\Gamma) \leq f(\Gamma) = f_1(\Gamma) + f_2(\Gamma)\ , \tag{5.2}$$

where $f_1 = \varphi_{\alpha_1} \times |\ \ |$ and $f_2 = |\ \ | \times \varphi_{\alpha_2}$.

To complete the proof we must estimate $f_1(\Gamma)$. Of course a similar estimate holds for $f_2(\Gamma)$ and when these estimates are combined with (5.2) we obtain the desired inequality (1.10). We write $[x]$ for the greatest integer less than or equal to x and $\langle x \rangle$ for the greatest integer strictly less than x. In general f_1 is defined by (3.8) and therefore we write Γ as

$$\Gamma = \bigcup_{n_2=0}^{\langle \theta_2 N_2 \rangle} \left(\left\{0, 1, 2, \ldots, \left\langle \theta_1 N_1 \left(1 - \frac{n_2}{\theta_2 N_2}\right)\right\rangle\right\} \times \{n_2\} \right)\ .$$

It follows that

$$f_1(\Gamma) = \sum_{n_2=0}^{\langle \theta_2 N_2 \rangle} \varphi_{\alpha_1} \left(\left\{0, 1, 2, \ldots, \left\langle \theta_1 N_1 \left(1 - \frac{n_2}{\theta_2 N_2}\right)\right\rangle\right\} \right)\ . \tag{5.3}$$

If M is an integer, $1 \leq M \leq N_1$, then by an inequality of Bombieri and Vaaler [7, Theorem 4] we have

$$\varphi_{\alpha_1}(\{0, 1, 2, \ldots, M-1\}) \leq (N_1)^2 \Phi_{\alpha_1}\left(\frac{M}{N_1}\right) \tag{5.4}$$

where Φ_{α_1} is defined by (1.9). When (5.3) and (5.4) are combined with the trivial identity $\langle x\rangle = -[-x]-1$ we obtain the bound

$$f_1(\Gamma) \le (N_1)^2 \sum_{n_2=0}^{\langle\theta_2 N_2\rangle} \Phi_{\alpha_1}\left(-N_1^{-1}\left[\theta_1 N_1\left(\frac{n_2}{\theta_2 N_2}-1\right)\right]\right) . \qquad (5.5)$$

In order to estimate the right hand side of (5.5) we apply Lemma 13 with $F(x) = (N_1)^2\Phi_{\alpha_1}(-N_1^{-1}x)$, $\beta = \theta_1 N_1(\theta_2 N_2)^{-1}$ and $\gamma = -\theta_1 N_1$. This leads to the inequality

$$\begin{aligned} f_1(\Gamma) \le (N_1)^2\theta_2 N_2 \int_0^1 \Phi_{\alpha_1}(\theta_1 x)\,dx \\ + (N_1)^2\|\Phi_{\alpha_1}\|_\infty + N_1(\tfrac{1}{4}\theta_1 N_1 + \theta_2 N_2)\|\Phi'_{\alpha_1}\|_\infty . \end{aligned} \qquad (5.6)$$

A brief calculation shows that

$$\|\Phi_{\alpha_1}\|_\infty \le \tfrac{1}{4}\log\{2h_1(\alpha_1)\} ,$$

and

$$\|\Phi'_{\alpha_1}\|_\infty \le \log\{2h_1(\alpha_1)\} .$$

Now (5.6) implies that

$$f_1(\Gamma) \le (N_1)^2\theta_2 N_2 \int_0^1 \Phi_{\alpha_1}(\theta_1 x)\,dx + \{\tfrac{1}{2}(N_1)^2 + \theta_2 N_1 N_2\}\log\{2h_1(\alpha_1)\} .$$

Finally, an identical bound holds for $f_2(\Gamma)$ but with the indices 1 and 2 reversed. Combining estimates for $f_1(\Gamma)$ and $f_2(\Gamma)$ and (5.2) proves the theorem.

6. The Thue-Siegel Principle

In this section we briefly indicate how our polynomial construction can be used to obtain a basic result — the Thue-Siegel principle — on rational approximation of algebraic numbers. The method we follow here is due to Bombieri [**3**]. Our purpose is to show how (1.13) can be applied and so some arguments are only sketched. Further details can be found in [**3**] and [**5**].

Let K be a finite extension of the number field k, $[K:k]=r$, $r\ge 3$, and let $\bar{v}$ be a fixed place of k. We assume that K has an embedding in the completion $k_{\bar{v}}$. Then we may identify K with its embedding in $k_{\bar{v}}$ and if α generates K over k we may ask how well α can be approximated by elements $\beta \in k$ with respect to the normalized absolute value $|\ \ |_{\tilde{v}}$ on $k_{\tilde{v}}$. To begin with we have the Liouville bound

$$\big(2h_1(\alpha)h_1(\beta)\big)^{-r} \le |\alpha - \beta|_{\tilde{v}} \,. \tag{6.1}$$

In fact, the inequality (6.1) can be sharpened in various ways. By the well known result of K.F. Roth [15], generalized to the present situation, for every $\varepsilon > 0$ there exists a constant $C_0 > 0$ so that

$$C_0\big(h_1(\beta)\big)^{-2-\varepsilon} \le |\alpha - \beta|_{\tilde{v}}$$

for all $\beta \in k$. However, the constant C_0 in this inequality cannot be effectively computed from knowledge of α and ε. Thus an important problem is to sharpen the Liouville bound (6.1) in such a way that all constants can be effectively computed.

Now suppose that α_1 and α_2 are both generators of K over k (and we do not exclude the possibility that $\alpha_1 = \alpha_2$). Since there is no significant difference in approximating α_i or $(\alpha_i)^{-1}$ by elements of k, we will assume that $|\alpha_i|_{\tilde{v}} \le 1$ for $i = 1, 2$. Next we suppose that β_1 and β_2 are elements of k with

$$|\alpha_i - \beta_i|_{\tilde{v}} < 1\,, \qquad i = 1, 2\,. \tag{6.2}$$

For $R > 0$ we define

$$\eta_i = \eta_i(R) = \frac{\log\{|\alpha_i - \beta_i|_{\tilde{v}}^{-1}\}}{\log\{4h_1(\beta_i)\} + R\log\{2h_1(\alpha_i)\}} \tag{6.3}$$

so that

$$|\alpha_i - \beta_i|_{\tilde{v}} = \Big\{\big(2h_1(\alpha_i)\big)^R\big(4h_1(\beta_i)\big)\Big\}^{-\eta_i(R)} \tag{6.4}$$

for $i = 1, 2$. Obviously $\eta_i(R)$ is a positive decreasing function of R and the Liouville bound shows that $\eta_i(r) < r$. The Thue-Siegel principle asserts that, under suitable conditions, the product $\eta_1(r)\eta_2(r)$ is not much larger than $2r$. Therefore, if a pair (α_1, β_1) can be found such that $\eta_1(r)$ is approximately r then for all other pairs (α_2, β_2), subject to suitable conditions, the value of $\eta_2(r)$ will be not much larger than 2. In practice we actually bound the product $\eta_1(R)\eta_2(R)$ and then choose the parameter R so as to optimize the result. Also, for technical reasons it is often necessary to select R somewhat larger than r and this inevitably weakens the final estimates. We now give a precise formulation of these remarks.

Theorem 14. *Let $K, k, r, \tilde{v}, \alpha_1, \alpha_2, \beta_1$ and β_2 be as above. Let δ and R be positive real numbers satisfying the inequality $r + \frac{1}{2}r^2R\delta < R$. If*

$$\frac{\log\{4h_1(\beta_1)\} + R\log\{2h_1(\alpha_1)\}}{\log\{4h_1(\beta_2)\} + R\log\{2h_1(\alpha_2)\}} < \delta \tag{6.5}$$

then

$$\eta_1(R)\eta_2(R) \le \frac{2r(R+1)}{\{R^{1/2} - (r + \frac{1}{2}r^2 R\delta)^{1/2}\}^2}, \tag{6.6}$$

where $\eta_1(R)$ and $\eta_2(R)$ are defined by (6.3).

Before proving Theorem 14 we consider one of its implications. With α_1 and β_1 as above we set

$$\mu_1 = \inf \left\{ \frac{2r(R+1)}{\eta_1(R)(R^{1/2} - r^{1/2})^2} \right\}, \tag{6.7}$$

where the infimum is taken over all R such that $r < R$. If $\mu_1 < r$ we say that (α_1, β_1) is an *anchor pair* for the data K, k and $\tilde{v}$. In certain special cases anchor pairs have been constructed by Bombieri [3], [5] and by Bombieri and Mueller [4]. For arbitrary K, k and $\tilde{v}$, subject to the condition that K have an embedding in $k_{\tilde{v}}$, there is no general method known for constructing anchor pairs and in fact it is not known that they always exist. Their importance arises from the following corollary.

Corollary 15. *If (α_1, β_1) is an anchor pair for the data K, k and $\tilde{v}$, then μ_1 is an effective measure of irrationality for all generators α_2 of K over k with respect to the absolute value $|\ \ |_{\tilde{v}}$.*

Proof of the Corollary. Let α_2 generate K over k and let $\beta_2 \in k$. Without loss of generality we may assume that $|\alpha_2|_{\tilde{v}} \le 1$ and $|\alpha_2 - \beta_2|_{\tilde{v}} < 1$. If $\varepsilon > 0$ we must determine constants

$$C_i = C_i(\alpha_1, \alpha_2, \beta_1, \varepsilon) > 0, \qquad i = 1, 2,$$

so that if $h_1(\beta_2) > C_1$ then

$$C_2\big(h_1(\beta_2)\big)^{-\mu_1-\varepsilon} \le |\alpha_2 - \beta_2|_{\tilde{v}}. \tag{6.8}$$

It is trivial that the infimum on the right of (6.7) is achieved at a number $R_1 > r$ and R_1 can plainly be calcuated from the anchor pair (α_1, β_1). With $\varepsilon > 0$ we may determine $\delta > 0$ so that $r + \frac{1}{2}r^2 R_1\delta < R_1$ and also

$$\frac{2r(R_1+1)}{\eta_1(R_1)\{R_1^{1/2} - (r + \frac{1}{2}r^2 R_1\delta)^{1/2}\}^2} < \mu_1 + \varepsilon. \tag{6.9}$$

Obviously δ can be computed from α_1, β_1 and ε. We set

$$C_1 = \frac{1}{4}\big(2h_1(\alpha_2)\big)^{-R_1}\Big\{\big(2h_1(\alpha_1)\big)^{R_1}\big(4h_1(\beta_1)\big)\Big\}^{1/\delta} \tag{6.10}$$

and

$$C_2 = \left\{4\bigl(2h_1(\alpha_2)\bigr)^{R_1}\right\}^{-\mu_1-\varepsilon} . \tag{6.11}$$

If $h_1(\beta_2) > C_1$ then the inequality (6.5) holds. It follows that (6.6) also holds with $R = R_1$. In view of (6.9) we find that

$$\eta_2(R_1) < \mu_1 + \varepsilon$$

and therefore

$$\left\{\bigl(2h_1(\alpha_2)\bigr)^{R_1}\bigl(4h_1(\beta_2)\bigr)\right\}^{-\mu_1-\varepsilon} \le |\alpha_2 - \beta_2|_{\bar{v}} .$$

That is, the inequality (6.8) has been established with C_1 and C_2 given by (6.10) and (6.11) respectively.

We now outline the proof of Theorem 14 following the method of Bombieri [5] but incorporating our general inequality (1.13). To begin with we may assume that $2 \le \eta_i(R)$ for $i = 1, 2$, since $\eta_i(R) \le \eta_i(r) < r$ for each $j = 1, 2$, by the Liouville inequality and the right hand side of (6.6) is already larger than $2r$. Next we select positive parameters θ_i, $i = 1, 2$, by setting

$$\theta_1 = \left(\frac{2R\eta_2(R)}{r(R+1)\eta_1(R)}\right)^{1/2} \quad \text{and} \quad \theta_2 = \left(\frac{2R\eta_1(R)}{r(R+1)\eta_2(R)}\right)^{1/2} . \tag{6.12}$$

Since $2 \le \eta_i(R) \le \eta_i(r) < r$ it follows immediately that $\theta_i < 1$ for $i = 1, 2$. Finally, we select positive integers N_i, $i = 1, 2$, by

$$N_i = \left[\frac{N}{\log\{4h_1(\beta_i)\} + R\log\{2h_1(\alpha_i)\}}\right] \tag{6.13}$$

where N is a large positive integer. Since

$$\lim_{N\to\infty} \frac{N_2}{N_1} = \frac{\log\{4h_1(\beta_1)\} + R\log\{2h_1(\alpha_1)\}}{\log\{4h_1(\beta_2)\} + R\log\{2h_1(\alpha_2)\}} < \delta$$

we may suppose that N is so large that $N_2 \le \delta N_1$. Later we will let $N \to \infty$.

Let Γ be defined by (1.2), Z be defined by (1.3), and let

$$\mathcal{X} = \mathcal{X}(\theta_1, \theta_2, N_1, N_2, \alpha_1, \alpha_2) \subseteq k^{N_1 N_2}$$

be the subspace defined by (1.5). As before we identify the vectors in $\mathcal{X}$ with the vector of coefficients of polynomials $P(x_1, x_2)$ in $k[x_1, x_2]$ which satisfy $\deg_{x_1}(P) \leq N_1 - 1$, $\deg_{x_2}(P) \leq N_2 - 1$ and the vanishing conditions (1.4). If P is in $\mathcal{X}$ *and if* P *satisfies the condition* $P(\beta_1, \beta_2) \neq 0$ then

$$\min\left(\theta_1 N_1 \log\{|\beta_1 - \alpha_1|_{\bar{v}}^{-1}\}\ ,\ \theta_2 N_2 \log\{|\beta_2 - \alpha_2|_{\bar{v}}^{-1}\}\right)$$
$$\leq \log(N_1 N_2) + \log h(P) + N_1 \log\{2h_1(\beta_1)\} + N_2 \log\{2h_1(\beta_2)\}\ . \tag{6.14}$$

This follows from the product formula as in [**5**, p.39]. In order to apply the inequality (6.14) we must verify the condition $P(\beta_1, \beta_2) \neq 0$ and this is a decidedly nontrivial step. To accomplish it we apply Dyson's Lemma [**5**, p.41] with $t = (\theta_1\theta_2)^{1/2}$ and $\theta = (\theta_1/\theta_2)^{1/2}$. It follows that for each polynomial P in $\mathcal{X}$ which is not identically zero there exists a lattice point $\vec{n} = \vec{n}(P)$ in Γ such that

$$(D^{(n_1,n_2)}P)(\beta_1, \beta_2) \neq 0\ . \tag{6.15}$$

Moreover, the lattice point $\vec{n}$ is such that the quantity

$$\kappa = \kappa(P) = \frac{n_1}{\theta_1 N_1} + \frac{n_2}{\theta_2 N_2}$$

satisfies

$$\kappa < \left(\frac{rN_1 + \frac{1}{2}(r-1)r(R+1)N_2}{RN_1}\right)^{1/2}\ .$$

Using the inequality $N_2 \leq \delta N_1$ and the hypothesis of the theorem we find that

$$\kappa < \left(\frac{r + \frac{1}{2}r^2 R\delta}{R}\right)^{1/2} < 1\ . \tag{6.16}$$

In view of (6.16) we may apply the basic inequality (6.14) to the polynomial $(D^{(n_1,n_2)}P)(x_1, x_2)$. But now the vanishing condition

$$\left(D^{(m_1,m_2)}(D^{(n_1,n_2)}P)\right)(\alpha_1, \alpha_2) = 0$$

holds only for

$$\frac{m_1 + n_1}{\theta_1 N_1} + \frac{m_2 + n_2}{\theta_2 N_2} < 1\ ,$$

or equivalently for

$$\vec{m} \in \Gamma\left(\theta_1(1-\kappa), \theta_2(1-\kappa), N_1, N_2\right)\ . \tag{6.17}$$

Also, the height of $(D^{(n_1,n_2)}P)(x_1, x_2)$ may be somewhat larger than the height of P. A simple calculation shows that

$$\begin{aligned} h(D^{(n_1,n_2)}P) &\leq \binom{N_1}{n_1}\binom{N_2}{n_2} h(P) \\ &\leq 2^{N_1+N_2} h(P)\ . \end{aligned} \tag{6.18}$$

Thus when we apply (6.14) to the polynomial $(D^{(n_1,n_2)}P)(x_1,x_2)$ and take (6.16), (6.17) and (6.18) into account we find that

$$\left(1-\left\{\frac{r+\frac{1}{2}r^2R\delta}{R}\right\}^{1/2}\right)$$
$$\min\left(\theta_1N_1\log\{|\beta_1-\alpha_1|_{\bar v}^{-1}\},\,\theta_2N_2\log\{|\beta_2-\alpha_2|_{\bar v}^{-1}\}\right)$$
$$\le\log(N_1N_2)+\log h(P)+N_1\log\{4h_1(\beta_1)\}+N_2\{4h_1(\beta_2)\}\,. \tag{6.19}$$

The inequality (6.19) holds for any nontrivial polynomial P in $\mathcal{X}$. The bound (1.7), which follows from Theorem 7, implies that there exists a nontrivial polynomial P_1 in $\mathcal{X}$ such that

$$\begin{aligned}\log h(P_1)&\le\log c_k+\frac{r\log H(Z)}{L}\\&\le\log c_k+\frac{r\log H(Z)}{N_1N_2-r|\Gamma|}\,.\end{aligned}$$

From the estimate (1.13) we conclude that

$$\begin{aligned}\log h(P_1)\le\log c_k&\\&+\left\{\frac{\frac{1}{2}r\theta_1\theta_2N_1N_2}{N_1N_2-r|\Gamma|}\right\}\left(N_1\log\{2h_1(\alpha_1)\}+N_2\log\{2h_1(\alpha_2)\}\right)\\&+\left\{\frac{\frac{1}{2}r(N_1+N_2)^2}{N_1N_2-r|\Gamma|}\right\}\log\{4h_1(\alpha_1)h_1(\alpha_2)\}\,.\end{aligned}\tag{6.20}$$

To complete the proof we use the polynomial P_1 in (6.19), apply the estimate (6.20), and let $N\to\infty$. We have

$$\lim_{N\to\infty}\left\{\frac{\frac{1}{2}r\theta_1\theta_2N_1N_2}{N_1N_2-r|\Gamma|}\right\}=R$$

and therefore

$$\left(1-\left\{\frac{r+\frac{1}{2}r^2R\delta}{R}\right\}^{1/2}\right)\min\left(\theta_1\eta_1(R),\,\theta_2\eta_2(R)\right)\le 2\,. \tag{6.21}$$

When we combine (6.12) and (6.21) we obtain the desired inequality (6.6). This proves the theorem.

REFERENCES

1. G. Birkhoff, "Lattice Theory", 3rd ed. American Math. Soc. Colloquium Pub., vol. XXV, American Math. Soc., Providence, R.I., 1967.
2. B. Bollobás, "Combinatorics", Cambridge University Press, 1988.
3. E. Bombieri, *On the Thue-Siegel-Dyson Theorem*, Acta Math. **148** (1982), 255–296.
4. E. Bombieri and J. Mueller, *On effective measures of irrationality for $\sqrt[r]{a/b}$ and related numbers*, J. Reine Angew. Math. **342** (1983), 173–196.
5. E. Bombieri, *Lectures on the Thue Principle*, Analytic Number Theory and Diophantine Problems, Progress in Mathematics Volume 70, A.C. Adolphson, J.B. Conrey, A. Ghosh, R.I. Yager, ed., Birkhäuser, 1987.
6. E. Bombieri and J. Vaaler, *Siegel's lemma*, Invent. Math. **73** (1983), 11–32.
7. E. Bombieri and J. Vaaler, *Polynomials with low height and prescribed vanishing*, Analytic Number Theory and Diophantine Problems, Progress in Mathematics Volume 70, A.C. Adolphson, J.B. Conrey, A. Ghosh, R.I. Yager, ed., Birkhäuser, 1987.
8. G.V. Chudnovsky, *Number Theoretic Applications of Polynomials with Rational Coefficients Defined by Extremality Conditions*, Arithmetic and Geometry, vol. 1, M. Artin and J. Tate, ed., Birkhäuser, 1983, 61–106.
9. P. Gordan, *Über den grössten gemeinsamen factor*, Math. Ann. **7** (1873), 443–448.
10. W.V.D. Hodge and D. Pedoe, "Methods of Algebraic Geometry", vol. 1,

18. J.D. Vaaler, *Small zeros of quadratic forms over number fields*, Trans. AMS. **302** (1987), 281–296.
19. A. Weil, *Arithmetic on algebraic varieties*, Ann. of Math. **53** (1951), 412–444.

Thomas Struppeck
Department of Mathematics
Rutgers University
New Brunswick, NJ 08903

Jeffrey D. Vaaler
Department of Mathematics
The University of Texas at Austin
Austin, TX 78712

The Abstract Prime Number Theorem For Algebraic Function Fields

WEN-BIN ZHANG

Dedicated to Professor Paul Bateman on his seventieth birthday

0. Introduction

The main purpose of this paper is to establish the "abstract prime number theorem" for algebraic function fields. This theorem has its original motivation in enumeration theorems on algebraic function fields and has been investigated by several authors [**4,7,8,9**].

Let C be an algebraic curve of genus g defined over the Galois field $GF[q]$ of q elements. Let N_m be the number of points of C (including those at infinity) with coordinates in $GF[q^m]$. A well-known theorem of A. Weil [**13**] states that

$$N_m = q^m - \sum_{i=1}^{2g} \alpha_i^m + 1$$

where the α_i are algebraic integers with $|\alpha_i| = q^{\frac{1}{2}}$. In [**3**], Bombieri obtained an analogue for the curve C of Selberg's formula in classical prime number theory. Using this formula and a lemma of Wirsing, he gave a more elementary proof of the purely asymptotic conclusion $N_m \sim q^m$. Andrews [**1**] followed his method.

In [**9**], Knopfmacher, motivated by the works of Bombieri, Fogels [**7**], Reichardt [**12**], and himself [**10**], developed the concept of an additive arithmetic semigroup satisfying Axiom $A^{\#}$. We recall that an additive arithmetic semigroup G is, by definition, a free commutative semigroup with identity element 1 such that G has a countable free generating set P

of "primes" p and such that G admits an integer-valued degree mapping $\partial : G \to N \cup \{0\}$ satisfying:

(1) $\partial(1) = 0$ and $\partial(p) > 0$ for all $p \in P$,
(2) $\partial(ab) = \partial(a) + \partial(b)$ for all $a, b \in G$, and
(3) the total number $\bar{G}(n)$ of elements of degree n in G is finite for each $n \geq 0$.

According to Knopfmacher, G satisfies Axiom $A^{\#}$ if there exist constants $A > 0$, $q > 1$, and ν with $0 \leq \nu < 1$ such that

$$\bar{G}(n) = Aq^n + O(q^{\nu n}) \quad \text{as } n \to \infty$$

Knopfmacher [9] proved an abstract prime number theorem for algebraic function fields ;which seeks to cover the main asymptotic consequences of the works mentioned above. The theorem states that, for any $\alpha > 1$,

$$\bar{P}(n) = \frac{q^n}{n} + O\left(\frac{q^n}{n^\alpha}\right) \quad \text{as } n \to \infty,$$

where $\bar{P}(n)$ is the total number of primes of degree n in G. In his paper, he gave two proofs, one by complex analysis and one by elementary methods. His elementary proof also followed Bombieri's method.

However, Knopfmacher's proof by complex analysis does not show that the generating function

$$Z^{\#}(y) := \sum_{n=0}^{\infty} \bar{G}(n) y^n = \prod_{m=1}^{\infty} (1 - y^m)^{-\bar{P}(m)} \tag{0.1}$$

has no zeros on the circle $|y| = q^{-1}$. Also, Bombieri's use of Wirsing's lemma in his argument is not valid[1]. We shall give two examples in Section 1 to establish these claims.

Therefore, the abstract prime number theorem (henceforth, P.N.T.) is in question. A condition which guarantees that $Z^{\#}(y)$ has no zeros on $|y| = q^{-1}$ is needed and we shall give one in Theorem 3.1. This theorem is sharp in the general case. Moreover, we shall give three versions of the abstract prime number theorem, i.e., Theorems 2.3 and 2.5, which assume the nonvanishing of $Z^{\#}(y)$ on $|y| = q^{-1}$, and Theorem 3.6, without the assumption. The last one gives rise to the problem of determining the quantity θ in the remainder term more precisely in terms of ν (see (3.7)).

[1]Professor Bombieri has communicated to us another elementary proof for $N_m \sim q^m$ that avoids the objection cited here. This result will appear in Rend. Sc. fis. mat. e nat.-Lincei.

We shall give an example in Section 5 to show that, in the general case, there is not too much we can say about this.

In the following discussion, we need the function $\Lambda(a)$ defined on G by

$$\Lambda(a) = \begin{cases} \partial(p), & \text{if } a \text{ is a prime-power } p^r \neq 1, \\ 0, & \text{otherwise,} \end{cases}$$

which is the analogue for G of the von Mongoldt function in classical prime number theory. We put

$$\bar{\Lambda}(m) = \sum_{\partial(a)=m} \Lambda(a) = \sum_{\substack{p\in P, r\geq 1 \\ \partial(p^r)=m}} \partial(p),$$

the counterpart of the number N_m for an algebraic curve C. Therefore, we have

$$\bar{\Lambda}(n) = \sum_{r|n} \frac{n}{r}\bar{P}(\frac{n}{r})$$

and by the Möbius inversion formula,

$$n\bar{P}(n) = \sum_{r|n} \bar{\Lambda}(r)\mu(\frac{n}{r}),$$

where μ is the Möbius function on $\mathbf{N}$. This shows that instead of investigating $\bar{P}(n)$ we may study $\bar{\Lambda}(n)$. In particular, we shall discuss the upper estimate for $\bar{\Lambda}(n)$ and show briefly how to deduce Chebyshev type and Mertens type estimates as well as Bombieri's analogue of Selberg's formula.

The author thanks Professor K.–H. Indlekofer for drawing attention to this subject through his lecture at the Bateman Conference and a preprint. The author also extends thanks to Professor H. G. Diamond for his comments.

1. The upper estimate for $\bar{\Lambda}(n)$

Here we shall study the upper estimate for $\bar{\Lambda}(n)$ from which we can deduce Chebyshev type estimates for algebraic function fields. Chebyshev showed that there exist numbers $\alpha > 0$ and $\beta < \infty$ such that the weighted prime counting function ψ satisfies

$$\alpha \leq \liminf_{x\to\infty} \frac{\psi(x)}{x}, \quad \limsup_{x\to\infty} \frac{\psi(x)}{x} \leq \beta.$$

The prime number theorem asserts that $\alpha = \beta = 1$.

In the following discussion, for brevity, we shall use additive convolution of arithmetic functions. Let $f(n)$ and $g(n)$ be two arithmetic functions defined for all non-negative integers. The function $h(n)$ defined by setting

$$h(n) = \sum_{k=0}^{n} f(k)g(n-k), \quad n = 0, 1, 2, \dots .$$

is called the additive convolution of f and g and denoted by $f \star g$. It is easy to see that the additive convolution is associative and commutative. We also define an operator L on all arithmetic functions f by setting

$$(Lf)(n) = nf(n), \quad n = 0, 1, 2, \dots .$$

We shall prove the following theorem. Our method of proof follows the general idea of Diamond [5] for deducing the Chebyshev type estimates for Beurling generalized prime numbers.

Theorem 1.1. *Suppose there exist constants $A > 0$, $q > 1$, $\gamma > 1$, and $c > 0$ such that*

$$|\bar{G}(n) - Aq^n| \le cq^n/n^\gamma, \quad n = 1, 2, \dots . \tag{1.1}$$

Then $\bar{\Lambda}(n) \ll q^n$.

To prove Theorem 1.1, we need the following lemma.

Lemma 1.2. *Assume (1.1). Set $c_1 = \sum_{n=1}^{\infty} n^{-\gamma}$. We fix positive integers k, ℓ, and n_0 such that*

$$c\ell^{-\gamma} \le \frac{1}{3}A, \tag{1.2}$$

$$k \ge \ell + \frac{2}{A}(2^{\gamma+2}cc_1 + |1 - A|) + 1, \tag{1.3}$$

and

$$\frac{k - \ell}{n_0 + k - \ell} \le \frac{1}{4\gamma}. \tag{1.4}$$

Define arithmetic functions U and M by setting

$$U(n) = \begin{cases} 1, & \text{if } n = 0, \\ -q^k, & \text{if } n = k, \\ 0, & \text{otherwise}, \end{cases}$$

and

$$M(n) = \begin{cases} 0, & \text{if } n < n_0, \\ q^n n^{-\gamma}, & \text{if } n \ge n_0. \end{cases}$$

Then the function $V = \bar{G} \star U \star M$ *is non- negative and* $V(n) \to \infty$ *as* $n \to \infty$.

Proof: We first note that $\bar{G} \star M(n) = 0$ and hence $V(n) = 0$ if $n < n_0$. Then we have

$$V(n) = \begin{cases} \bar{G} \star M(n), & if\ n < k, \\ \bar{G} \star M(n) - q^k \bar{G} \star M(n-k), & if\ n \geq k. \end{cases}$$

For $n_0 \leq n < k + n_0$, it is easy to see that

$$V(n) = \bar{G} \star M(n) \geq 0$$

Therefore we may assume that $n \geq n_0 + k$. Then we have

$$\begin{aligned} V(n) = & \sum_{s=n-k+1}^{n} \bar{G}(n-s) M(s) \\ & + \sum_{s=n_0}^{n-k} (\bar{G}(n-s) - q^k \bar{G}(n-k-s)) M(s) \\ = & \sum_1 + \sum_2, \end{aligned} \tag{1.5}$$

say. We shall show that $\sum_1$ is dominant and that $\sum_2$ is negligible.

We denote

$$\bar{G}(n) = q^n (A + a_n n^{-\gamma}),\ n = 1, 2, \ldots\ ,$$

where $|a_n| \leq c$, $n = 1, 2, \ldots$ by (1.1). Therefore we have

$$\begin{aligned} \sum_1 & \geq q^n \left\{ \left(A + \frac{a_{k-1}}{(k-1)^\gamma} \right) \frac{1}{(n-k+1)^\gamma} + \cdots + \left(A + \frac{a_\ell}{\ell^\gamma} \right) \frac{1}{(n-\ell)^\gamma} \right\} \\ & \geq \frac{2Aq^n}{3} \frac{1}{(n-k)^\gamma} \left\{ \left(1 - \frac{1}{n-k+1} \right)^\gamma \right. \\ & \quad \left. + \left(1 - \frac{2}{n-k+2} \right)^\gamma + \cdots + \left(1 - \frac{k-\ell}{n-\ell} \right)^\gamma \right\} \\ & \geq \frac{2Aq^n}{3} \frac{k-\ell}{(n-k)^\gamma} \left(1 - \frac{k-\ell}{n-\ell} \right)^\gamma \\ & > \frac{2Aq^n}{3} \frac{k-\ell}{(n-k)^\gamma} \left(1 - \gamma \frac{k-\ell}{n-\ell} \right) > \frac{Aq^n}{2} \frac{k-\ell}{(n-k)^\gamma} \end{aligned}$$

since, by (1.2),

$$A + \frac{a_n}{n^\gamma} \geq \frac{2A}{3} \quad \text{for} \quad n \geq \ell,$$

and, by (1.4),

$$1-\gamma\frac{k-\ell}{n-\ell}\geq 1-\gamma\frac{k-\ell}{n_0+k-\ell}\geq\frac{3}{4}.$$

We now rewrite $\sum_2$ in the form

$$\begin{aligned}
&q^n\left\{\sum_{s=n_0}^{n-k}\left(\frac{a_{n-s}}{(n-s)^\gamma}-\frac{a_{n-k-s}}{(n-k-s)^\gamma}\right)\frac{1}{s^\gamma}+\left(A+\frac{a_k}{k^\gamma}-1\right)\frac{1}{(n-k)^\gamma}\right\}\\
&=q^n\left\{\sum_{s=n_0}^{n-k}\frac{a_{n-s}}{(n-s)^\gamma}\frac{1}{s^\gamma}-\left(\sum_{s=n_0}^{n-k-1}\frac{a_{n-k-s}}{(n-k-s)^\gamma}\frac{1}{s^\gamma}+\frac{1-A}{(n-k)^\gamma}\right)\right\}\\
&=q^n\left(\sum_{21}-\sum_{22}\right),
\end{aligned}$$

say. If $\frac{1}{2}(n-k)<n_0$, then we have

$$|\sum_{21}|\leq\frac{2^\gamma}{(n-k)^\gamma}\sum_{s=n_0}^{n-k}\frac{c}{(n-s)^\gamma}\leq\frac{2^\gamma cc_1}{(n-k)^\gamma}$$

and, in the same way,

$$|\sum_{22}|\leq\frac{2^\gamma cc_1+|1-A|}{(n-k)^\gamma}.$$

If $\frac{1}{2}(n-k)\geq n_0$, then

$$\begin{aligned}
|\sum_{22}|&\leq\frac{2^\gamma c}{(n-k)^\gamma}\sum_{s=n_0}^{[\frac{1}{2}(n-k)]}s^{-\gamma}\\
&\quad+\frac{2^\gamma c}{(n-k)^\gamma}\sum_{s=[\frac{1}{2}(n-k)]+1}^{n-k-1}\frac{1}{(n-k-s)^\gamma}+\frac{|1-A|}{(n-k)^\gamma}\\
&\leq\frac{2^{\gamma+1}cc_1+|1-A|}{(n-k)^\gamma},
\end{aligned}$$

and, in the same way,

$$|\sum_{21}|\leq\frac{2^{\gamma+1}cc_1}{(n-k)^\gamma}.$$

Therefore we have the estimate

$$|\sum_2|\leq q^n\frac{2^{\gamma+2}cc_1+|1-A|}{(n-k)^\gamma}. \tag{1.7}$$

From (1.5), (1.6), and (1.7), we arrive at

$$V(n) \geq \frac{q^n}{(n-k)^\gamma}\left(\frac{A(k-\ell)}{2} - 2^{\gamma+2}cc_1 - |1-A|\right) > \frac{A}{2}\frac{q^n}{(n-k)^\gamma}$$

for $n \geq n_0 + k$. This completes the proof of the lemma. ∎

Proof of Theorem 1.1: We first consider the associated power series

$$\Lambda^{\#}(y) = \sum_{n=1}^{\infty} \bar{\Lambda}(n) y^n \tag{1.8}$$

of the arithmetic function $\bar{\Lambda}(n)$. From (0.1), it is easy to see that

$$\Lambda^{\#}(y) = y\frac{d}{dy}Z^{\#}(y)/Z^{\#}(y). \tag{1.9}$$

This can be written, in an additive convolution version, as $\bar{\Lambda} \star \bar{G} = L\bar{G}$, an analogue of Chebyshev's identity in classical prime number theory. We convolve each side of $\bar{\Lambda} \star \bar{G} = L\bar{G}$ by $U \star M$, where the functions U and M are defined in Lemma 1.2, and obtain that

$$\bar{\Lambda} \star \bar{G} \star U \star M(n) = L\bar{G} \star U \star M(n). \tag{1.10}$$

We then show that the magnitude of the right-hand side is $O(q^n)$. Actually, for $n \geq k+1$, we have

$$\begin{aligned} L\bar{G} \star U(n) &= n\bar{G}(n) - q^k(n-k)\bar{G}(n-k) \\ &= Akq^n + O\left(\frac{q^n}{(n-k)^{\gamma-1}}\right), \end{aligned}$$

by (1.1). Therefore, for $n \geq n_0$,

$$L\bar{G} \star U \star M(n) = \sum_{n_0 \leq s \leq n} O(q^{n-s})\frac{q^s}{s^\gamma} = O(q^n). \tag{1.11}$$

Finally, from (1.10) and (1.11), for n sufficiently large,

$$\bar{\Lambda}(n-n_1)V(n_1) \leq \bar{\Lambda} \star V(n) \leq Kq^n,$$

where $V = \bar{G} \star U \star M$. This implies that $\bar{\Lambda}(n-n_1) \leq Kq^n$, i.e., $\bar{\Lambda}(n) \leq Kq^{n+n_1}$, since for fixed n_1 sufficiently large $V(n_1) \geq 1$ by Lemma 1.2 ∎

Therefore we deduce the upper estimate $\bar{\Lambda}(n) \ll q^n$ in Theorem 1.1 under rather weak condition (1.1). It is interesting that, unlike the situation in a multiplicative arithmetic semigroup (see [5,14]), in the general case, the lower estimate $\bar{\Lambda}(n) \gg q^n$ requires almost as strong hypotheses as those under which an abstract prime number theorem can be deduced. We shall see this in Example 4.1 and Theorem 3.2.

We conclude this section by showing how to deduce the Chebyshev type estimates from Theorem 1.1. Let $\bar{\psi}(n) = \sum_{s=1}^{n} \bar{\Lambda}(s)$. Then, from Theorem 1.1, we have $\bar{\psi}(n) \ll q^n$. Moreover, we have

$$\bar{\Lambda}(n) + \sum_{1 \leq s \leq n-1} \bar{\Lambda}(s)\bar{G}(n-s) = n\bar{G}(n). \tag{1.12}$$

By (1.1), $\bar{G}(n) = Aq^n + a_n q^n n^{-\gamma}$ with $a_n \ll 1$ and $\gamma > 1$. Dividing both sides of (1.12) by q^n, we get

$$\frac{\bar{\Lambda}(n)}{q^n} + A \sum_{1 \leq s \leq n-1} \frac{\bar{\Lambda}(s)}{q^s} + \sum_{1 \leq s \leq n-1} \frac{a_{n-s}}{(n-s)^\gamma} \frac{\bar{\Lambda}(s)}{q^s} = nA + \frac{a_n}{n^{\gamma-1}}$$

and hence

$$\sum_{1 \leq s \leq n} \frac{\bar{\Lambda}(s)}{q^s} = n + O(1) \tag{1.13}$$

by Theorem 1.1. This is a Mertens type estimate. From (1.13), we can easily deduce that

$$\sum_{1 \leq s \leq n} \bar{\psi}(s) q^{-s} = \frac{q}{q-1} n + O(1)$$

and then $\bar{\psi}(n) \gg q^n$ by a simple tauberian argument. Therefore, Chebyshev type estimates hold. Finally, assume that

$$\bar{G}(n) = Aq^n + O\left(\frac{q^n}{n^\gamma}\right)$$

holds with $\gamma > 3$. Using the additive convolution techniques and the general idea of Diamond [6] for proving Selberg's formula in classical prime number theory, from the convolution identity

$$(L\bar{\Lambda} + \bar{\Lambda} \star \bar{\Lambda}) \star \bar{G} = L^2 \bar{G},$$

we can easily deduce

$$n\bar{\Lambda}(n) + \sum_{1 \leq s \leq n-1} \bar{\Lambda}(s)\bar{\Lambda}(n-s) = 2nq^n + O(q^n).$$

This can be rewritten as

$$n\frac{\bar{\Lambda}(n)}{q^n} + \sum_{1 \leq s \leq n-1} \frac{\bar{\Lambda}(s)}{q^s} \frac{\bar{\Lambda}(n-s)}{q^{n-s}} = 2n + O(1), \tag{1.14}$$

an analogue [3] of the well-known Selberg's formula.

However, the argument in [3] deducing $\bar{\Lambda}(n) \sim q^n$ (the abstract prime number theorem) from (1.13) and (1.14) is not correct as Example 4.6 will show.

2. The abstract prime number theorem and the zeros of the generating function

From (0.1), we can see that $\prod_{m=1}^{\infty}(1-y^m)^{-\bar{P}(m)}$ converges absolutely and hence the generating function $Z^{\#}(y)$ has no zeros in the disk $\{|y| < q^{-1}\}$ if $\bar{G}(n) \ll q^n$. However, whether the zero-free region of $Z^{\#}(y)$ may be extended to include the circle $|y| = q^{-1}$ is, in the general case, a complicated problem which requires strong hypotheses as Example 4.1 will show. The following Theorems 2.1 and 2.3 give the connection of the abstract prime number theorem with the zeros of $Z^{\#}(y)$ on $|y| = q^{-1}$.

Theorem 2.1. *Suppose that there exist constants $q > 1$ and $\gamma > 1$ such that*

$$\bar{\Lambda}(n) \sim q^n \quad (P.N.T) \tag{2.1}$$

and

$$\bar{G}(n) - q\bar{G}(n-1) = O\left(\frac{q^n}{n^\gamma}\right) \tag{2.2}$$

holds. Then $Z^{\#}(y)$ has no zeros on the circle $|y| = q^{-1}$.

Remark: Condition (1.1) implies (2.2) and, conversely, condition (2.2) implies that $\bar{G}(n) = Aq^n + O(q^n n^{-\gamma+1})$.

To prove Theorem 2.1, we need the following lemma.

Lemma 2.2. *Let γ be a constant satisfying $1 < \gamma < 2$. Then we have*

$$\sum_{n=1}^{\infty} \frac{x_1^n - x_2^n}{n^\gamma} \ll (x_1 - x_2)^{\gamma-1}, \quad 0 \le x_2 < x_1 < 1.$$

Proof: Actually, we have

$$\sum_{n=1}^{\infty} \frac{x_1^n - x_2^n}{n^\gamma} = \sum_{n \le (x_1-x_2)^{-1}} \frac{x_1^n - x_2^n}{n^\gamma} + \sum_{n > (x_1-x_2)^{-1}} \frac{x_1^n - x_2^n}{n^\gamma} = \sum_1 + \sum_2,$$

say. It is easy to see that

$$\sum_1 \le (x_1 - x_2) \sum_{n \le (x_1-x_2)^{-1}} \frac{1}{n^{\gamma-1}} \le (x_1 - x_2)\left(1 + \int_1^{(x_1-x_2)^{-1}} x^{-\gamma+1}dx\right)$$
$$\ll (x_1 - x_2)^{\gamma-1}$$

and that

$$\sum_2 \le \sum_{n > (x_1-x_2)^{-1}} n^{-\gamma} \le (x_1 - x_2)^\gamma + \int_{(x_1-x_2)^{-1}}^{\infty} x^{-\gamma}dx \ll (x_1 - x_2)^{\gamma-1} \quad \blacksquare$$

Proof of Theorem 2.1: We consider the function

$$Z(y) := (1-qy)Z^{\#}(y) = 1 + \sum_{n=1}^{\infty}(\bar{G}(n) - q\bar{G}(n-1))y^n$$

which has a continuous continuation to the circle $|y| = q^{-1}$ by (2.2). It suffices to show that $Z(y)$ has no zeros on $|y| = q^{-1}$.

On the one hand, by (0.1), we have

$$\Lambda^{\#}(y) - \frac{qy}{1-qy} = \sum_{n=1}^{\infty}(\bar{\Lambda}(n) - q^n)y^n = y\frac{Z'(y)}{Z(y)}, \quad |y| < q^{-1}.$$

We set $a_n = \bar{\Lambda}(n) - q^n$. Then, by (2.1), $a_n = o(q^n)$. It turns out that

$$\log(Z(re^{i\theta})) = \sum_{n=1}^{\infty} \frac{a_n}{n} r^n e^{in\theta},$$

since $Z(0) = 1$. Therefore, for any given $\epsilon > 0$, we have

$$\begin{aligned} |Z(re^{i\theta})| &= \exp\left\{\operatorname{Re}\sum_{n=1}^{\infty} \frac{a_n q^{-n}}{n}(rq)^n e^{in\theta}\right\} \\ &\geq e^{-c}\exp\left\{-\epsilon\sum_{n=1}^{\infty}\frac{(rq)^n}{n}\right\} = e^{-c}(1-rq)^{\epsilon}, \end{aligned} \tag{2.3}$$

where $c = c(\epsilon)$ is a constant, since $\operatorname{Re} a_n q^{-n} e^{in\theta} > -\epsilon$ for $n \geq n_0$.

On the other hand, by (2.2), we have

$$\begin{aligned} |Z(r_1e^{i\theta}) - Z(re^{i\theta})| &\leq \sum_{n=1}^{\infty} |\bar{G}(n) - q\bar{G}(n-1)|(r_1^n - r^n) \\ &\ll \sum_{n=1}^{\infty} \frac{1}{n^{\gamma}}((qr_1)^n - (qr)^n) \end{aligned}$$

for $0 \leq r < r_1 < q^{-1}$. It follows that

$$|Z(r_1e^{i\theta}) - Z(re^{i\theta})| \ll (qr_1 - qr)^{\gamma-1} \tag{2.4}$$

from Lemma 2.2.

Now suppose Theorem 2.1 is false and $Z(q^{-1}e^{i\theta}) = 0$. Then, upon letting $r_1 \to q^{-1}$ in (2.4), we would obtain

$$|Z(re^{i\theta})| \ll (1-qr)^{\gamma-1}.$$

Taking $\epsilon = (\gamma - 1)/2$ in (2.3), we would have

$$e^{-c}(1 - qr)^{(\gamma-1)/2} \leq K(1 - qr)^{\gamma-1}$$

or

$$e^{-c}/K \leq (1 - qr)^{(\gamma-1)/2};$$

this is certainly absurd for r sufficiently close to q^{-1}. ∎

Conversely, we have the following result which is a "conditional" abstract prime number theorem and an inverse of Theorem 2.1 in some sense. Moreover, it is a generalization of a conjecture of Bateman and Diamond [**2**] for Beurling generalized prime numbers.

Theorem 2.3. *Suppose that there exists a constant $q > 1$ such that*

$$\sum_{n=1}^{\infty} q^{-2n} n^2 |\bar{G}(n) - q\bar{G}(n-1)|^2 < \infty. \tag{2.5}$$

If $Z(y) = (1 - qy)Z^{\#}(y)$ is continuous on the closed disk $\{|y| \leq q^{-1}\}$ and has no zeros on the circle $|y| = q^{-1}$ then

$$\bar{\Lambda}(n) \sim q^n \quad (P.N.T.).$$

Proof: We first note that the condition (2.5) implies the absolute convergence of $Z^{\#}(y) = \prod_{m=1}^{\infty}(1 - y^m)^{-\bar{P}(m)}$ for $|y| < q^{-1}$ and hence that $Z^{\#}(y)$ has no zeros in this disk. We then have

$$\Lambda^{\#}(y) = \frac{qy}{1 - qy} + y\frac{Z'(y)}{Z(y)} = \sum_{n=1}^{\infty} \bar{\Lambda}(n) y^n, \quad |y| < q^{-1}.$$

Therefore,

$$\bar{\Lambda}(n) = q^n + \frac{1}{2\pi i}\int_{|y|=r} \frac{Z'(y)}{Z(y)} y^{-n} dy$$

with $0 < r < q^{-1}$. We note that

$$\begin{aligned}
&\int_{-\pi}^{\pi} |Z'(re^{i\theta}) - Z'(r_1 e^{i\theta})|^2 d\theta \\
&= 2\pi \sum_{n=1}^{\infty} n^2 (\bar{G}(n) - q\bar{G}(n-1))^2 q^{-2n+2}((qr)^{n-1} - (qr_1)^{n-1})^2 \\
&\longrightarrow 0
\end{aligned}$$

as $r, r_1 \to q^{-1}-$ by (2.5), since $((qr)^{n-1} - (qr_1)^{n-1})^2 < 1$. Therefore, there exists a function $F(\theta) \in L_2[-\pi, \pi]$ such that $Z'(re^{i\theta}) \to F(\theta)$ in $L_2[-\pi, \pi]$ as $r \to q^{-1}-$. It follows that

$$\lim_{r \to q^{-1}-} \frac{1}{2\pi i} \int_{|y|=r} \frac{Z'(y)}{Z(y)} y^{-n} dy = \frac{q^{n-1}}{2\pi} \int_{-\pi}^{\pi} e^{-i(n-1)\theta} \frac{F(\theta)}{Z(q^{-1}e^{i\theta})} d\theta$$

since $Z(y)$ is continuous in $\{|y| \le q^{-1}\}$ and has no zeros on it. Therefore, we have

$$\bar{\Lambda}(n) = q^n + \frac{q^{n-1}}{2\pi} \int_{-\pi}^{\pi} e^{-i(n-1)\theta} \frac{F(\theta)}{Z(q^{-1}e^{i\theta})} d\theta = q^n + o(q^n),$$

for the last integral tends to zero as $n \to \infty$ by the Riemann–Lebesgue lemma. ∎

Corollary 2.4. *Suppose there exist constants $q > 1$, $A > 0$, and $\gamma > \frac{3}{2}$ such that*

$$\bar{G}(n) = Aq^n + O(q^n/n^\gamma).$$

If $Z(y)$ has no zeros on the circle $|y| = q^{-1}$ then $\bar{\Lambda}(n) \sim q^n$ holds.

The next result is also a "conditional" abstract prime number theorem but with a remainder term.

Theorem 2.5. *Suppose that there exist constants $A > 0$, $q > 1$, and ν with $0 \le \nu < 1$ such that*

$$\bar{G}(n) = Aq^n + O(q^{\nu n}) \tag{2.6}$$

holds. If the function $Z^{\#}(y)$ has no zeros on the circle $|y| = q^{-1}$, then

$$\bar{\Lambda}(n) = q^n + O(q^{\theta n}) \tag{2.7}$$

holds for some θ with $\nu < \theta < 1$.

Remark: This theorem is essentially a result in [4]. However, we should note that the condition (2.6) with $\nu \ge \frac{1}{2}$ does not guarantee that $Z^{\#}(y)$ has no zeros on the circle $|y| = q^{-1}$ as Example 4.1 will show. Therefore, the main statement (2) in [4] is wrong for $\nu \ge \frac{1}{2}$. Also, (2.7) can be improved (see [8]).

Proof: We note that the function $Z(y) = (1 - qy)Z^{\#}(y)$ is holomorphic in the disk $\{|y| < q^{-\nu}\}$ by (2.6) and has no zeros in the disk $\{|y| \le q^{-1}\}$. Therefore, there exist some constant θ_1 with $\nu < \theta_1 < 1$ such that $Z(y)$ has no zeros in $\{|y| < q^{-\theta_1}\}$. If we shift the integration path in the formula

$$\bar{\Lambda}(n) = q^n + \frac{1}{2\pi i} \int_{|y|=r} \frac{Z'(y)}{Z(y)} y^{-n} dy$$

with $r < q^{-1}$ to a circle with $r = q^{-\theta}$ where $\theta_1 < \theta < 1$ and without zeros of $Z^{\#}(y)$ on it, then we arrive at the conclusion. ∎

Although Theorem 2.5 is "conditional", it applies to some natural examples given in [**11**] in which the fact $Z^{\#}(y)$ has no zeros on $|y| = q^{-1}$ is known.

3. The abstract prime number theorem

The key to establish the abstract prime number theorem is to show that the generating function $Z^{\#}(y)$ has no zeros on the circle $|y| = q^{-1}$. For this we have the following theorem; which is sharp.

Theorem 3.1. *If there exist constants $q > 1$ and $A > 0$ such that*

$$\sum_{n=1}^{\infty} (\bar{G}(n) - Aq^n)^2 q^{-n} < \infty \tag{3.1}$$

then $Z^{\#}(y)$ has no zeros on the circle $|y| = q^{-1}$.

To prove Theorem 3.1, we first note that, by (3.1),

$$\bar{G}(n) = Aq^n + o(q^{n/2}).$$

Therefore, $Z^{\#}(y)$ has an analytic continuation in the disk $\{|y| < q^{-\frac{1}{2}}\}$ as a meromorphic function with the only singularity a pole of order one at $y = q^{-1}$. We divide the proof of Theorem 3.1 into several lemmas.

Lemma 3.2. *$Z^{\#}(y)$ has no zeros on the circle $|y| = q^{-1}$ except at the point $y = -q^{-1}$ where it has a zero of order at most one.*

Proof: Consider the associated "zeta function" $\zeta(s) := Z^{\#}(q^{-s})$ with $\mathrm{Re}\, s = \sigma > \frac{1}{2}$. Then

$$\zeta(\sigma + it) = \prod_{m=1}^{\infty} (1 - q^{-m(\sigma+it)})^{-\bar{P}(m)}$$

for $\sigma > 1$ and we have

$$\zeta(\sigma) = Z^{\#}(q^{-\sigma}) = \sum_{n=0}^{\infty} \bar{G}(n) q^{-n\sigma} = \frac{A/\log q}{\sigma - 1}(1 + O(\sigma - 1)).$$

Since

$$(\zeta(\sigma))^3 |\zeta(\sigma + it)|^4 |\zeta(\sigma + 2it)|$$

$$= \exp\left\{\sum_{m=1}^{\infty} \bar{P}(m) \sum_{k=1}^{\infty} \frac{1}{k} q^{-mk\sigma}(3 + 4\cos(tkm\log q) + \cos(2tkm\log q))\right\}$$
$$\geq 1$$

for $\sigma > 1$, $\zeta(\sigma + it)$ has no zeros on the line $\sigma = 1$ except possibly at those points with $t = m\pi/\log q$, $m = \pm 1, \pm 3, \ldots$ where it has a zero of order at most one. Therefore $Z^{\#}(y)$ has no zeros on the circle $|y| = q^{-1}$ except possibly at the point $y = -q^{-1}$ where it has at most a simple zero. ∎

Lemma 3.3. *Suppose that $Z^{\#}(y)$ has a zero at $y = -q^{-1}$. Let*

$$Z^{\#}(y)Z^{\#}(-y) = 1 + \sum_{n=1}^{\infty} H(n)y^n.$$

Then

$$H(n) = o(q^{n/2}). \tag{3.2}$$

Proof: We first show that

$$\lim_{r, r_1 \to q^{-\frac{1}{2}}-} \int_{-\pi}^{\pi} |Z^{\#}(re^{i\theta}) - Z^{\#}(r_1 e^{i\theta})|^2 d\theta = 0. \tag{3.3}$$

Actually, we have

$$Z^{\#}(y) = g(y) + f(y)$$

where

$$g(y) = \frac{Aqy}{1 - qy}, \quad f(y) = 1 + \sum_{n=1}^{\infty} (\bar{G}(n) - Aq^n)y^n.$$

The function $f(y)$ is holomorphic in the disk $\{|y| < q^{-\frac{1}{2}}\}$ by (3.1) and we have

$$\int_{-\pi}^{\pi} |f(re^{i\theta}) - f(r_1 e^{i\theta})|^2 d\theta$$
$$= 2\pi \sum_{n=1}^{\infty} (\bar{G}(n) - Aq^n)^2 q^{-n}((q^{\frac{1}{2}}r)^n - (q^{\frac{1}{2}}r_1)^n)^2$$

for $r < q^{-\frac{1}{2}}$, $r_1 < q^{-\frac{1}{2}}$. We note that $((q^{\frac{1}{2}}r)^n - (q^{\frac{1}{2}}r_1)^n)^2 < 1$. Therefore, by (3.1),

$$\lim_{r, r_1 \to q^{-\frac{1}{2}}-} \int_{-\pi}^{\pi} |f(re^{i\theta}) - f(r_1 e^{i\theta})|^2 d\theta = 0.$$

Also note that $g(y)$ is uniformly continuous on the annulus $\{q^{-\frac{2}{3}} \le |y| \le q^{-\frac{1}{2}}\}$. Therefore, (3.3) follows from the inequality

$$\begin{aligned}&|Z^{\#}(re^{i\theta}) - Z^{\#}(r_1 e^{i\theta})|^2 \\ &\le 2(|g(re^{i\theta}) - g(r_1 e^{i\theta})|^2 + |f(re^{i\theta}) - f(r_1 e^{i\theta})|^2).\end{aligned}$$

We now consider $Z^{\#}(y)Z^{\#}(-y)$. By the hypothesis that $Z^{\#}(y)$ has a zero at $y = -q^{-1}$, $Z^{\#}(y)Z^{\#}(-y)$ has no poles at $y = q^{-1}$ and $y = -q^{-1}$. Therefore, it is holomorphic in the disk $\{|y| < q^{-\frac{1}{2}}\}$. We have

$$\begin{aligned}H(n) &= \frac{1}{2\pi i}\int_{|y|=r} \frac{Z^{\#}(y)Z^{\#}(-y)}{y^{n+1}}dy \\ &= \frac{1}{2\pi r^n}\int_{-\pi}^{\pi} e^{-in\theta} Z^{\#}(re^{i\theta})Z^{\#}(-re^{i\theta})d\theta.\end{aligned} \tag{3.4}$$

By (3.3), there exists a function $F(\theta) \in L_1[-\pi, \pi]$ such that

$$\lim_{r\to q^{-\frac{1}{2}}-} \int_{-\pi}^{\pi} |Z^{\#}(re^{i\theta})Z^{\#}(-re^{i\theta}) - F(\theta)|d\theta = 0.$$

Therefore, if we take the limit as $r \to q^{-\frac{1}{2}}-$ on the right–hand side of (3.4), then we obtain that

$$H(n) = \frac{q^{\frac{n}{2}}}{2\pi}\int_{-\pi}^{\pi} e^{-in\theta}F(\theta)d\theta.$$

It follows that $H(n) = o(q^{\frac{n}{2}})$ since the last integral tends to zero as $n \to \infty$ by the Riemann–Lebesgue lemma. ∎

Lemma 3.4. *Suppose that $Z^{\#}(y)$ has a zero at $y = -q^{-1}$. Let*

$$Z_1(y) := \prod_{\substack{m=1 \\ m \text{ odd}}}^{\infty} (1 - y^m)^{-\bar{P}(m)} = 1 + \sum_{n=1}^{\infty} \bar{G}_1(n)y^n.$$

Then

$$\limsup_{n\to\infty} \bar{G}_1(n)q^{-n} > 0. \tag{3.5}$$

Proof: Consider

$$Z^{\#}(y)Z^{\#}(-y) = \prod_{\substack{m=1 \\ m \text{ even}}}^{\infty} (1 - y^m)^{-2\bar{P}(m)} \prod_{\substack{m=1 \\ m \text{ odd}}}^{\infty} (1 - y^{2m})^{-\bar{P}(m)}, \quad |y| < q^{-1}.$$

We have

$$\lim_{y \to q^{-1}} Z^{\#}(y) Z^{\#}(-y) = B > 0,$$

since, at $y = q^{-1}$, $Z^{\#}(y)$ has a pole of order one, whereas $Z^{\#}(-y)$ has a zero of order one and the infinite product is positive for $0 < y < q^{-1}$. Note that $\prod_{\substack{m=1 \\ m \, odd}}^{\infty} (1 - y^{2m})^{-\bar{P}(m)}$ converges for $y = q^{-1}$ since $\prod_{\substack{m=1 \\ m \, odd}}^{\infty} (1 - y^{m})^{-\bar{P}(m)}$ converges for $y = q^{-2}$. This implies that

$$\lim_{y \to q^{-1}-} \prod_{\substack{m=1 \\ m \, even}}^{\infty} (1 - y^m)^{-2\bar{P}(m)} = B_1^2 > 0$$

and hence

$$\lim_{y \to q^{-1}-} \prod_{\substack{m=1 \\ m \, even}}^{\infty} (1 - y^m)^{-\bar{P}(m)} = B_1 > 0.$$

We now write

$$Z^{\#}(y) = Z_1(y) Z_2(y)$$

with

$$Z_2(y) = \prod_{\substack{m=1 \\ m \, even}}^{\infty} (1 - y^m)^{-\bar{P}(m)}, \quad |y| < q^{-1}.$$

Then we have

$$\lim_{y \to q^{-1}-} (1 - qy) Z_1(y) = \lim_{y \to q^{-1}-} \frac{(1 - qy) Z^{\#}(y)}{Z_2(y)} = \frac{A}{B_1} = B_2 > 0. \quad (3.6)$$

We claim that

$$\limsup_{n \to \infty} \bar{G}_1(n) q^{-n} \geq B_2.$$

Otherwise, $\limsup_{n \to \infty} \bar{G}_1(n) q^{-n} < B_2$. Then, for $n \geq n_0$,

$$\bar{G}_1(n) \leq B_3 q^n$$

with a constant B_3 satisfying $\limsup_{n \to \infty} \bar{G}_1(n) q^{-n} < B_3 < B_2$. Hence we would have

$$\begin{aligned} &\lim_{y \to q^{-1}-} (1 - qy) Z_1(y) \\ &\leq \lim_{y \to q^{-1}-} (1 - qy) \left(1 + \sum_{n < n_0} \bar{G}_1(n) y^n + \sum_{n \geq n_0} B_3 q^n y^n \right) \\ &= B_3 < B_2; \end{aligned}$$

this contradicts (3.6). ∎

Proof of Theorem 3.1: We write

$$\prod_{\substack{m=1\\ m \text{ even}}}^{\infty} (1-y^m)^{-2\bar{P}(m)} = 1 + \sum_{n=1}^{\infty} \bar{G}_3(n) y^n, \quad |y| < q^{-1}$$

and

$$Z_1(y^2) = 1 + \sum_{n=1}^{\infty} \bar{G}_4(n) y^n, \quad |y| < q^{-1}.$$

Then we have

$$\bar{G}_4(n) = \begin{cases} \bar{G}_1(\frac{n}{2}), & \text{if } n \text{ is even,} \\ 0, & \text{if } n \text{ is odd.} \end{cases}$$

Now we have

$$Z^{\#}(y) Z^{\#}(-y) = \prod_{\substack{m=1\\ m \text{ even}}}^{\infty} (1-y^m)^{-2\bar{P}(m)} Z_1(y^2) = 1 + \sum_{n=1}^{\infty} H(n) y^n.$$

Therefore,

$$\begin{aligned} H(n) &= \bar{G}_4(n) + \bar{G}_3(1)\bar{G}_4(n-1) + \cdots + \bar{G}_3(n-1)\bar{G}_4(1) + \bar{G}_3(n) \\ &\geq \bar{G}_4(n). \end{aligned}$$

Suppose that Theorem 3.1 is not true and that $Z^{\#}(y)$ has a zero at $y = -q^{-1}$. Then we would have

$$\limsup_{n\to\infty} q^{-n} H(2n) \geq \limsup_{n\to\infty} q^{-n} \bar{G}_1(n) > 0$$

by (3.5); this contradicts (3.2). ∎

Corollary 3.5. *If there exist constants $q > 1$, $A > 0$, and $\gamma > \frac{1}{2}$ such that*

$$\bar{G}(n) = Aq^n + O(q^{\frac{n}{2}} n^{-\gamma})$$

holds, then $Z^{\#}(y)$ has no zeros on the circle $|y| = q^{-1}$.

This Corollary combined with Example 4.1 shows that Theorem 3.1 is sharp.

Furthermore, by combining Corollary 3.5 and Theorem 2.5, we obtain the following theorem.

Theorem 3.6. *Suppose that there exist constants $q > 1$, $A > 0$, and ν with $0 \leq \nu < \frac{1}{2}$ such that*

$$\bar{G}(n) = Aq^n + O(q^{\nu n}),\ n = 1, 2, \ldots$$

or, with $\nu = \frac{1}{2}$, there exists $\gamma > \frac{1}{2}$ such that

$$\bar{G}(n) = Aq^n + O(q^{\frac{n}{2}} n^{-\gamma}),\ n = 1, 2, \ldots.$$

Then

$$\bar{\Lambda}(n) = q^n + O(q^{\theta n}) \tag{3.7}$$

holds for some θ with $\nu < \theta < 1$.

Remark: (3.7) can be improved (see [8]).

4. Two examples

In this section, we will give two examples. The first one shows that Theorem 3.1 is sharp, that Knopfmacher's theorem [8] cannot hold without the hypothesis that $Z^{\#}(y)$ has no zeros on the circle $|y| = q^{-1}$, and that a positive lower bound for $\bar{\Lambda}(n)q^{-n}$ does not exist even with $\bar{G}(n)$ subject to rather strong hypotheses. The second one shows that the argument given in [3] is not valid.

Example 4.1: Let q be a positive integer and $q \geq 2$. Let

$$2q^k \equiv r_k \mod k,\ 0 \leq r_k < k$$

for $k = 1, 2, \ldots$. We set

$$\bar{P}(k) = \begin{cases} \frac{1}{k}(2q^k - r_k) + 1, & \text{if } k \text{ is odd,} \\ 1, & \text{if } k \text{ is even,} \end{cases}$$

(we can even put $\bar{P}(k) = 0$ for k even). Then $\bar{P}(k)$, $k = 1, 2, \ldots$, are all positive integers and $k\bar{P}(k) \ll q^k$. We note that $k\bar{P}(k) > 2q^k$ if k is odd. Therefore, we have

$$\bar{\Lambda}(n) = \sum_{k|n} k\bar{P}(k) = \begin{cases} 2q^n + c_n, & \text{if } n \text{ is odd,} \\ 2q^{\frac{n}{2}} + c_n, & \text{if } n = 2k \text{ with } k \text{ odd,} \\ c_n, & \text{if } 4|n, \end{cases} \tag{4.1}$$

where $c_n > 0$ and $c_n \ll q^{\frac{n}{3}} \log n$. Thus, the P.N.T. does not hold.

It is easy to see that

$$Z^{\#}(y) = \prod_{m=1}^{\infty} (1 - y^m)^{-\bar{P}(m)}$$

converges absolutely in the disk $\{|y| < q^{-1}\}$. We shall prove the following proposition.

Proposition 4.2. *If we write*

$$Z^{\#}(y) = 1 + \sum_{n=1}^{\infty} \bar{G}(n) y^n$$

then there exists a positive constant A such that

$$q^{\frac{n}{2}} n^{-\frac{1}{2}} \ll |\bar{G}(n) - Aq^n| \ll q^{\frac{n}{2}} n^{-\frac{1}{2}} \tag{4.2}$$

holds for n sufficiently large. Moreover, $Z^{\#}(y)$ has a zero of order one at $y = -q^{-1}$.

We devide the prof of Proposition 4.2 into several lemmas.

Let $\mathcal{D}$ be the domain formed by cutting the complex plane along the real axis from $-\infty$ to $-q^{-\frac{1}{2}}$ and from $q^{\frac{1}{2}}$ to $+\infty$ and along the imaginary axis from $-i\infty$ to $-iq^{-\frac{1}{2}}$ and from $iq^{\frac{1}{2}}$ to $i\infty$.

Lemma 4.3. *The function $Z^{\#}(y)$ has an analytic continuation in $\mathcal{D} \cap \{|y| < q^{-\frac{1}{3}}\}$ as a single–valued meromorphic functin with the only singularity a pole of order one at $y = q^{-1}$ and the only zero of order one at $y = -q^{-1}$.*

Proof: By (4.1), we have

$$y \frac{\frac{d}{dy} Z^{\#}(y)}{Z^{\#}(y)} = \Lambda^{\#}(y) = \frac{2qy}{1-(qy)^2} + \frac{2qy^2}{1-(qy^2)^2} + yf(y), \quad |y| < q^{-1},$$

where the function $f(y) = \sum_{n=1}^{\infty} c_n y^{n-1}$ is holomorphic in the disk $\{|y| < q^{-\frac{1}{3}}\}$. It turns out that

$$Z^{\#}(y) = \frac{1+qy}{1-qy} \left(\frac{1+qy^2}{1-qy^2} \right)^{\frac{1}{2}} e^{F(y)}, \quad |y| < q^{-1}, \tag{4.3}$$

where the function $F(y) = \sum_{n=1}^{\infty} c_n n^{-1} y^n$ is, like $f(y)$, holomorphic in the disk $\{|y| < q^{-\frac{1}{3}}\}$. Moreover, in (4.3), the function

$$M(y) := \left(\frac{1+qy^2}{1-qy^2} \right)^{\frac{1}{2}}$$

is the single–valued branch with $M(0) = 1$ of the associated multiple–valued function. The domain where $M(y)$ is holomorphic is $\mathcal{D}$. ∎

We have

$$\bar{G}(n) = \frac{1}{2\pi i} \int_{|y|=r_1} \frac{Z^{\#}(y)}{y^{n+1}} dy$$

where $0 < r_1 < q^{-1}$. From Lemma 4.3, if we shift the integraticn contour to the circle $|y| = q^{-\frac{1}{3}-\epsilon}$ then we will obtain that

$$\bar{G}(n) = -Res_{y=q^{-1}} \frac{Z^{\#}(y)}{y^{n+1}} + \frac{1}{2\pi i} \int_{|y|=q^{-\frac{1}{3}-\epsilon}} Z^{\#}(y) y^{-n-1} dy$$
$$= 2\left(\frac{q+1}{q-1}\right)^{\frac{1}{2}} e^{F(q^{-1})} q^n + O_\epsilon(q^{(\frac{1}{3}+\epsilon)n}).$$

However, it is possible to get the more accurate estimate (4.2) by introducing a complicated integration path $\mathcal{C}$ (Fig. 1)

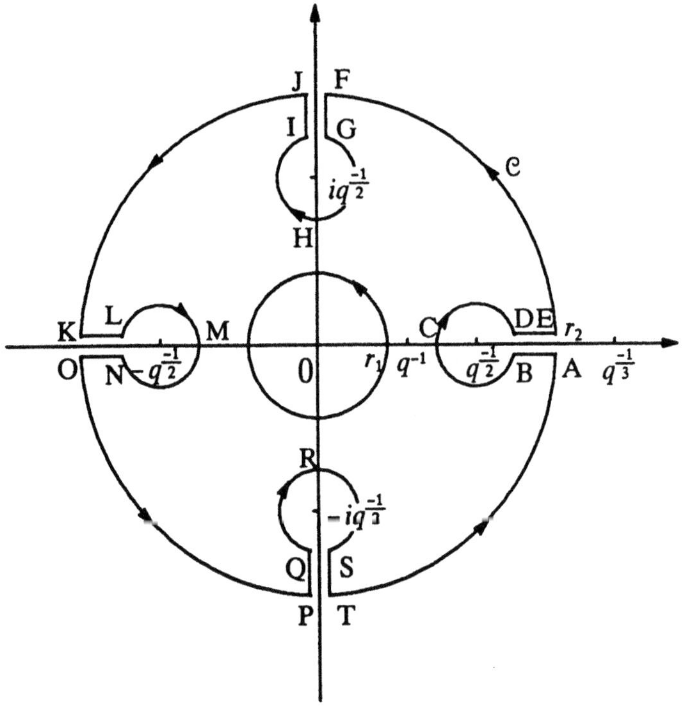

Figure 1

Lemma 4.4. *We have*

$$\bar{G}(n) = Aq^n + \frac{1}{\pi}(I_n^{(1)} + I_n^{(2)} + I_n^{(3)} + I_n^{(4)}) + O_\epsilon(q^{(\frac{1}{3}+\epsilon)n}), \tag{4.4}$$

where

$$I_n^{(1)} = -q^{\frac{n}{2}} \int_1^{q^{\frac{1}{6}-\epsilon}} \alpha^{-n-1} \left(\frac{\alpha^2+1}{\alpha^2-1}\right)^{\frac{1}{2}} \frac{q^{\frac{1}{2}}\alpha+1}{q^{\frac{1}{2}}\alpha-1} \exp\{F(q^{-\frac{1}{2}}\alpha)\} d\alpha,$$

$$I_n^{(2)} = (-1)^{n+1} q^{\frac{n}{2}} \int_1^{q^{\frac{1}{6}-\epsilon}} \alpha^{-n-1} \left(\frac{\alpha^2+1}{\alpha^2-1}\right)^{\frac{1}{2}} \frac{q^{\frac{1}{2}}\alpha+1}{q^{\frac{1}{2}}\alpha-1} \exp\{F(-q^{-\frac{1}{2}}\alpha)\} d\alpha,$$

$$I_n^{(3)} = -\frac{q^{\frac{n}{2}}}{i^n} \int_1^{q^{\frac{1}{6}-\epsilon}} \alpha^{-n-1} \left(\frac{\alpha^2-1}{\alpha^2+1}\right)^{\frac{1}{2}} \frac{1+iq^{\frac{1}{2}}\alpha}{1-iq^{\frac{1}{2}}\alpha} \exp\{F(iq^{-\frac{1}{2}}\alpha)\} d\alpha,$$

$$I_n^{(4)} = (-1)^{n+1} \frac{q^{\frac{n}{2}}}{i^n} \int_1^{q^{\frac{1}{6}-\epsilon}} \alpha^{-n-1} \left(\frac{\alpha^2-1}{\alpha^2+1}\right)^{\frac{1}{2}} \frac{1-iq^{\frac{1}{2}}\alpha}{1+iq^{\frac{1}{2}}\alpha} \exp\{F(-iq^{-\frac{1}{2}}\alpha)\} d\alpha.$$

Proof: We define an integration contour C which consists of the circle $|y| = r_2$ with $r_2 = q^{-\frac{1}{3}-\epsilon}$ cut at points $\pm q^{-\frac{1}{3}-\epsilon}$, $\pm iq^{-\frac{1}{3}-\epsilon}$, the line segments AB and NO on the lower edge of the cut of $\mathcal{D}$ along the real axis, the line segments DE and KL on the upper edge, the segments FG and ST on the right edge of the cut along the imaginary axis, the segments IJ and PQ on the left edge, and the small circles BCD, GHI, LMN, and QRS centered at $q^{-\frac{1}{2}}$, $iq^{-\frac{1}{2}}$, $-q^{-\frac{1}{2}}$, and $-iq^{-\frac{1}{2}}$ respectively with the same radius η sufficiently small. Thus, we have

$$\bar{G}(n) = Aq^n + \frac{1}{2\pi i} \int_C \frac{Z^{\#}(y)}{y^{n+1}} dy \tag{4.5}$$

with $A = 2\left(\frac{q+1}{q-1}\right)^{\frac{1}{2}} \exp\{F(q^{-1})\} > 0$. We shall estimate the last integral on each part of C separately.

It is easy to see that the integrals on the arcs EF, JK, OP, and TA are all $O_\epsilon(q^{(\frac{1}{3}+\epsilon)n})$. To evaluate the integrals on the line segments, we now consider the function $M(y) = \left(\frac{1+qy^2}{1-qy^2}\right)^{\frac{1}{2}}$. We note that $M(y)$ gets a factor -1 when y jumps from AB to DE and so does the integrand $Z^{\#}(y)y^{-n-1}$. We also note that the argument of $1 - q^{\frac{1}{2}}y$ increases by $-\pi$ and hence the argument of $M(y)$ increases by $\pi/2$ when y tours from C to D along the circle BCD. Therefore we have

$$\begin{aligned} \left(\int_{AB} + \int_{DE}\right) \frac{Z^{\#}(y)}{y^{n+1}} dy &= 2\int_{DE} \frac{Z^{\#}(y)}{y^{n+1}} dy \\ &= 2\int_{q^{-\frac{1}{2}}+\eta}^{q^{-\frac{1}{3}-\epsilon}} i \frac{1}{y^{n+1}} \frac{1+qy}{1-qy} \left(\frac{qy^2+1}{qy^2-1}\right)^{\frac{1}{2}} e^{F(y)} dy. \end{aligned} \tag{4.6}$$

Similarly, we have

$$\begin{aligned}\left(\int_{KL}+\int_{NO}\right)\frac{Z^{\#}(y)}{y^{n+1}}dy &= 2\int_{NO}\frac{Z^{\#}(y)}{y^{n+1}}dy\\ &= 2\int_{-q^{-\frac{1}{2}}-\eta}^{-q^{-\frac{1}{3}-\epsilon}} i\frac{1}{y^{n+1}}\frac{1+qy}{1-qy}\left(\frac{qy^2+1}{qy^2-1}\right)^{\frac{1}{2}}e^{F(y)}dy,\end{aligned}\tag{4.7}$$

$$\begin{aligned}\left(\int_{FG}+\int_{IJ}\right)\frac{Z^{\#}(y)}{y^{n+1}}dy &= 2\int_{IJ}\frac{Z^{\#}(y)}{y^{n+1}}dy\\ &= 2\int_{q^{-\frac{1}{2}}+\eta}^{q^{-\frac{1}{3}-\epsilon}} \frac{1}{(it)^{n+1}}\frac{1+itq}{1-itq}\left(\frac{qt^2-1}{qt^2+1}\right)^{\frac{1}{2}}e^{F(it)}dt,\end{aligned}\tag{4.8}$$

and

$$\begin{aligned}\left(\int_{PQ}+\int_{ST}\right)\frac{Z^{\#}(y)}{y^{n+1}}dy &= 2\int_{ST}\frac{Z^{\#}(y)}{y^{n+1}}dy\\ &= 2\int_{-q^{-\frac{1}{2}}-\eta}^{-q^{-\frac{1}{3}-\epsilon}} \frac{1}{(it)^{n+1}}\frac{1+itq}{1-itq}\left(\frac{qt^2-1}{qt^2+1}\right)^{\frac{1}{2}}e^{F(it)}dt.\end{aligned}\tag{4.9}$$

Moreover, on the circle BCD, if we set $y-q^{-\frac{1}{2}}=\eta e^{i\theta}$, $0\le\theta\le 2\pi$, then we have

$$(1-q^{\frac{1}{2}}y)^{-\frac{1}{2}} = iq^{-\frac{1}{4}}\eta^{-\frac{1}{2}}e^{-i\theta/2}$$

and hence

$$\int_{BCD}\frac{Z^{\#}(y)}{y^{n+1}}dy \longrightarrow 0 \quad \text{as } \eta\longrightarrow 0$$

since the circumference of BCD is $2\pi\eta$. Similarly, the integrals on the small circles GHI, LMN, and QRS tend to zero as $\eta\to 0$ too.

From (4.6), (4.7), (4.8), and (4.9), if we let $\eta\to 0$ in (4.5) and take the limit on the right–hand side then we obtain that

$$\bar{G}(n) = Aq^n + \frac{1}{\pi}(I_n^{(1)}+I_n^{(2)}+I_n^{(3)}+I_n^{(4)}) + O_\epsilon(q^{(\frac{1}{3}+\epsilon)n}),$$

where

$$I_n^{(1)} = \int_{q^{-\frac{1}{2}}}^{q^{-\frac{1}{3}-\epsilon}} \frac{1}{y^{n+1}}\frac{1+qy}{1-qy}\left(\frac{qy^2+1}{qy^2-1}\right)^{\frac{1}{2}}e^{F(y)}dy,$$

$$I_n^{(2)} = \int_{-q^{-\frac{1}{2}}}^{-q^{-\frac{1}{3}-\epsilon}} \frac{1}{y^{n+1}}\frac{1+qy}{1-qy}\left(\frac{qy^2+1}{qy^2-1}\right)^{\frac{1}{2}}e^{F(y)}dy,$$

$$I_n^{(3)} = -\int_{q^{-\frac{1}{2}}}^{q^{-\frac{1}{3}-\epsilon}} \frac{1}{i^n t^{n+1}}\frac{1+itq}{1-itq}\left(\frac{qt^2-1}{qt^2+1}\right)^{\frac{1}{2}}e^{F(it)}dt,$$

$$I_n^{(4)} = -\int_{-q^{-\frac{1}{2}}}^{-q^{-\frac{1}{3}-\epsilon}} \frac{1}{i^n t^{n+1}} \frac{1+itq}{1-itq} \left(\frac{qt^2-1}{qt^2+1}\right)^{\frac{1}{2}} e^{F(it)} dt.$$

Now if we make the substitution $y = q^{-\frac{1}{2}}\alpha$ in $I_n^{(1)}$, $t = q^{-\frac{1}{2}}\alpha$ in $I_n^{(3)}$, $y = -q^{-\frac{1}{2}}\alpha$ in $I_n^{(2)}$, and $t = -q^{-\frac{1}{2}}\alpha$ in $I_n^{(4)}$ then the required expressions of $I_n^{(1)}$, $I_n^{(2)}$, $I_n^{(3)}$, and $I_n^{(4)}$ follow. ∎

Lemma 4.5. *We have*

$$n^{-\frac{1}{2}} \ll \int_1^a \alpha^{-n-1}(\alpha-1)^{-\frac{1}{2}} d\alpha \ll n^{-\frac{1}{2}},$$

where a is an arbitrary constant with $a > 1$.

Proof: Actually we have

$$\int_1^{1+\frac{1}{n}} \alpha^{-n-1}(\alpha-1)^{-\frac{1}{2}} d\alpha \geq \left(\frac{1}{n}\right)^{-\frac{1}{2}} \int_1^{1+\frac{1}{n}} \alpha^{-n-1} d\alpha > n^{-\frac{1}{2}}(1-2e^{-1})$$

and, by integration by parts,

$$\begin{aligned}
&\int_1^{1+\frac{1}{n}} \alpha^{-n-1}(\alpha-1)^{-\frac{1}{2}} d\alpha \\
&= 2\left(\frac{1}{n}\right)^{-\frac{1}{2}} \left(1+\frac{1}{n}\right)^{-n-1} + 2(n+1)\int_1^{1+\frac{1}{n}} (\alpha-1)^{-\frac{1}{2}} \alpha^{-n-2} d\alpha \\
&\leq 2n^{-\frac{1}{2}} + 2n^{-\frac{1}{2}}(n+1)\int_1^{1+\frac{1}{n}} \alpha^{-n-2} d\alpha \\
&\ll n^{-\frac{1}{2}}.
\end{aligned}$$

Also, we have

$$\int_{1+\frac{1}{n}}^a \alpha^{-n-1}(\alpha-1)^{-\frac{1}{2}} d\alpha \leq \left(\frac{1}{n}\right)^{-\frac{1}{2}} \int_{1+\frac{1}{n}}^a \alpha^{-n-1} d\alpha \leq n^{-\frac{1}{2}}. \quad ∎$$

Proof of Proposition 4.2: By Lemma 4.4, it remains to show that

$$q^{\frac{n}{2}} n^{-\frac{1}{2}} \ll |I_n^{(1)} + I_n^{(2)} + I_n^{(3)} + I_n^{(4)}| \ll q^{\frac{n}{2}} n^{-\frac{1}{2}} \tag{4.10}$$

holds for n sufficiently large. We rewrite

$$\begin{aligned}
I_n^{(1)} + I_n^{(2)} = q^{\frac{n}{2}} \int_1^{q^{\frac{1}{6}-\epsilon}} \alpha^{-n-1} \left(\frac{\alpha^2+1}{\alpha^2-1}\right)^{\frac{1}{2}} \Bigg(&-\frac{q^{\frac{1}{2}}\alpha+1}{q^{\frac{1}{2}}\alpha-1} \exp\{F(q^{-\frac{1}{2}}\alpha)\} \\
&+(-1)^{n+1} \frac{q^{\frac{1}{2}}\alpha-1}{q^{\frac{1}{2}}\alpha+1} \exp\{F(-q^{-\frac{1}{2}}\alpha)\}\Bigg) d\alpha.
\end{aligned}$$

Hence, by Lemma 4.5,

$$|I_n^{(1)} + I_n^{(2)}| \ll q^{\frac{n}{2}} \int_1^{q^{\frac{1}{6}-\epsilon}} \alpha^{-n-1}(\alpha-1)^{-\frac{1}{2}} d\alpha \ll q^{\frac{n}{2}} n^{-\frac{1}{2}}.$$

Moreover, we have

$$\begin{aligned} &I_n^{(1)} + I_n^{(2)} \\ &\le q^{\frac{n}{2}} \int_1^{q^{\frac{1}{6}-\epsilon}} \alpha^{-n-1} \left(\frac{\alpha^2+1}{\alpha^2-1}\right)^{\frac{1}{2}} \left(-\frac{q^{\frac{1}{2}}\alpha+1}{q^{\frac{1}{2}}\alpha-1} + \frac{q^{\frac{1}{2}}\alpha-1}{q^{\frac{1}{2}}\alpha+1}\right) \exp\{F(q^{-\frac{1}{2}}\alpha)\} d\alpha \\ &= -q^{\frac{n}{2}} \int_1^{q^{\frac{1}{6}-\epsilon}} \alpha^{-n-1} \left(\frac{\alpha^2+1}{\alpha^2-1}\right)^{\frac{1}{2}} \frac{4q^{\frac{1}{2}}\alpha}{q\alpha^2-1} \exp\{F(q^{-\frac{1}{2}}\alpha)\} d\alpha, \end{aligned}$$

since the coefficients of $F(y) = \sum_{n=1}^{\infty} c_n n^{-1} y^n$ are all positive and hence $F(q^{-\frac{1}{2}}\alpha) \ge F(-q^{-\frac{1}{2}}\alpha)$. It follows that

$$I_n^{(1)} + I_n^{(2)} \le -cq^{\frac{n}{2}} n^{-\frac{1}{2}}$$

for some constant $c > 0$ by Lemma 4.5. Therefore

$$q^{\frac{n}{2}} n^{-\frac{1}{2}} \ll |I_n^{(1)} + I_n^{(2)}| \ll q^{\frac{n}{2}} n^{-\frac{1}{2}}.$$

Finally, we have

$$|I_n^{(3)} + I_n^{(4)}| \ll q^{\frac{n}{2}} \int_1^{q^{\frac{1}{6}-\epsilon}} \alpha^{-n-1} d\alpha \ll q^{\frac{n}{2}} n^{-1}.$$

This proves (4.10) and completes the proof of Proposition 4.2. ∎

We now give the second example. In [3], Bombieri first showed that

$$ma_m + \sum_{i=1}^{m-1} a_i a_{m-i} = 2m + O(1)$$

and

$$\sum_{i=1}^{m} a_i = m + O(1),$$

where $a_m = N_m/q^m \ge 0$ (or, in our notation, $\bar{\Lambda}(m)/q^m$, see (1.13) and (1.14)). Introducing $a_k = 1 + r_k$, $R_m = m^{-1} \sum_{k \le m} r_k r_{m-k}$ and

$$\limsup_{m\to\infty} \frac{1}{m} \sum_{k \le m} r_k^2 = A,$$

he then arrived at

$$\frac{1}{m}\sum_{k\le m} R_k^2 \le \frac{A^2}{2} + o(1) \tag{4.11}$$

by Wirsing's lemma. This made it possible to conclude that $r_m \to 0$, i.e., $N_m/q^m \to 1$ as $m \to \infty$. However, the proof of (4.11) is not correct as the following simple example shows.

Example 4.6: Consider the sequence $\{a_n\}$ define by

$$a_m = \begin{cases} 2, & \text{if } m \text{ is odd,} \\ 0, & \text{if } m \text{ is even.} \end{cases}$$

It is easy to see that

$$ma_m + \sum_{i=1}^{m-1} a_i a_{m-i} = 2m,$$

$$\sum_{i=1}^{m} a_i = m + \frac{1-(-1)^m}{2},$$

$$r_m = a_m - 1 = (-1)^{m-1}, \quad R_m = (-1)^m,$$

$$\frac{1}{m}\sum_{1\le k\le m} r_k^2 = 1 = A,$$

$$\sum_{i=1}^{m} R_k = \begin{cases} -1, & \text{if } m \text{ is odd,} \\ 0, & \text{if } m \text{ is even,} \end{cases}$$

and

$$\frac{1}{m}\sum_{i=1}^{m} R_k^2 = 1.$$

This means (4.11) is not true for $\{a_n\}$.

5. The Riemann hypothesis

Weil's theorem [**13**] gave the first proof of the so-called Riemann hypothesis for algebraic curves over Galois fields. We may also consider an analogue of the Riemann hypothesis for additive arithmetic semigroups.

Suppose G is an additive arithmetic semigroup for which

$$\bar{G}(n) = Aq^n + O(q^{\nu n}), \quad n = 1, 2, \ldots \tag{5.1}$$

for some constants $q > 1$, $A > 0$, and ν with $0 \le \nu < 1/2$. Then the associated generating function $Z^{\#}(y)$ has an analytic continuation in the disk $\{|y| < q^{-\nu}\}$ as a meromorphic function with the only singularity a simple pole at $y = q^{-1}$. The Riemann hypothesis for G is the assertion that $Z^{\#}(y)$ has no zeros in the disk $\{|y| < q^{-1/2}\}$. A problem [4] relevant to this hypothesis is to describe more precisely in terms of ν the quantity θ in the remainder term of the abstract prime number theorem (see (3.7)). The following example shows that there is not too much we can say about θ in terms of ν, and that, in the general case, the Riemann hypothesis is not true for additive arithmetic semigroups.

Example 5.1. Let q and η be real numbers with $q > 1$ and $1 > \eta > 0$. Let k be a positive integer and $k \geq 2$. We set

$$S_l(m) = \begin{cases} \sum_{r|\frac{m}{l}} q^r \mu(\frac{m}{lr}), & \text{if } l|m, \\ 0, & \text{if } l\nmid m, \end{cases}$$

for $m = 1, 2, \ldots$, where μ is the Möbius function in $\mathbf{N}$. Let $q_1 = q^{1-\eta}$. We set

$$S_m = \begin{cases} [q^m], & \text{if } m < m_0, \\ \left[q^m - \sum_{r|m} q_1^r \mu(\frac{m}{r}) - \sum_{l=2}^{k} S_l(m)\right], & \text{if } m \geq m_0, \end{cases}$$

for $m = 1, 2, \ldots$, where $[a]$ denotes the largest integer not exceeding a and m_0 is sufficiently large. Plainly,

$$\left|\sum_{r|m} q_1^r \mu(\frac{m}{r})\right| \leq \sum_{s=1}^{m} q_1^s < \frac{q_1^{m+1}}{q_1 - 1},$$

and, if $l|m$,

$$|S_l(m)| = \left|\sum_{r|\frac{m}{l}} q^r \mu(\frac{m}{lr})\right| \leq \sum_{s=1}^{m/l} q^s < \frac{q^{m/l+1}}{q-1}.$$

Therefore, we have

$$q^m - \sum_{r|m} q_1^r \mu(\frac{m}{r}) - \sum_{l=2}^{k} S_l(m)$$
$$> q^m \left(1 - \frac{1}{(q_1 - 1)q^{\eta m - 1}} - \frac{k}{(q-1)q^{m/2-1}}\right) > 1$$

if $m \geq m_0(\eta, k)$, and hence the S_m are positive integers. Moreover, $S_m \leq 2q^m$.

Let

$$S_m \equiv r_m \bmod m, \quad 0 \le r_m < m$$

for $m = 1, 2, \dots$. We then set

$$\bar{P}(m) = \frac{1}{m}(S_m - r_m + m).$$

Then $\bar{P}(m)$, $m = 1, 2, \dots$ are all positive integers and we have

$$m\bar{P}(m) = q^m + m - r_m + \theta_m,$$

with $|\theta_m| < 1$, if $m < m_0$, and

$$m\bar{P}(m) = q^m - \sum_{r|m} q_1^r \mu(\frac{m}{r}) - \sum_{l=2}^{k} S_l(m) + m - r_m + \theta_m,$$

with $|\theta_m| < 1$, if $m \ge m_0$.

The associated generating function is, for $|y| < q^{-1}$,

$$Z^{\#}(y) = \prod_{m=1}^{\infty} (1 - y^m)^{-\bar{P}(m)} = \exp\left\{ \sum_{n=1}^{\infty} \frac{1}{n} \left(\sum_{m|n} m\bar{P}(m) \right) y^n \right\} \tag{5.2}$$

$$= \exp\left\{ \sum_{n=1}^{\infty} \frac{1}{n} \left(\sum_{m|n} \left(q^m - \sum_{r|m} q_1^r \mu(\frac{m}{r}) - \sum_{l=2}^{k} S_l(m) \right) \right) y^n + F(y) \right\},$$

where

$$F(y) = \sum_{n=1}^{\infty} \frac{1}{n} \Bigg(\sum_{m|n} (m - r_m + \theta_m)$$

$$+ \sum_{\substack{m|n \\ m<m_0}} \left(\sum_{r|m} q_1^r \mu(\frac{m}{r}) + \sum_{l=2}^{k} S_l(m) \right) \Bigg) y^n$$

$$= \sum_{n=1}^{\infty} \frac{1}{n} (O(n^{1+\epsilon}) + O(q^{m_0})) y^n$$

is holomorphic in the disk $\{|y| < 1\}$. We note that

$$\sum_{n=1}^{\infty} \left(\sum_{m|n} q^m \right) y^n = \sum_{s=1}^{\infty} \frac{1}{s} \log(1 - qy^s)^{-1}. \tag{5.3}$$

Moreover, since

$$\sum_{m|n}\sum_{r|m} q_1^r \mu\left(\frac{m}{r}\right) = q_1^n,$$

we have

$$\sum_{n=1}^{\infty} \frac{1}{n}\left(\sum_{m|n}\left(-\sum_{r|m} q_1^r \mu\left(\frac{m}{r}\right)\right)\right) y^n$$
$$= -\sum_{n=1}^{\infty} \frac{1}{n}(q_1 y)^n = \log(1 - q_1 y).$$

Similarly, if $l|n$,

$$\sum_{m|n} S_l(m) = \sum_{lm'|n}\sum_{r|m'} q^r \mu\left(\frac{m'}{r}\right) = q^{n/l},$$

and if $l\not| n$,

$$\sum_{m|n} S_l(m) = 0.$$

Therefore,

$$\sum_{n=1}^{\infty} \frac{1}{n}\left(\sum_{m|n}\left(-\sum_{l=2}^{k} S_l(m)\right)\right) y^n$$
$$= -\sum_{l=2}^{\infty}\sum_{\substack{n=1\\ l|n}}^{\infty} \frac{1}{n} q^{n/l} y^n = \sum_{l=2}^{k} \frac{1}{l}\log(1 - q y^l). \tag{5.5}$$

From (5.2), (5.3), (5.4), and (5.5) we obtain

$$Z^{\#}(y) = \frac{1}{1-qy}(1 - q_1 y)\prod_{s=k+1}^{\infty}\frac{1}{(1-qy^s)^{1/s}} e^{F(y)}.$$

This shows that $Z^{\#}(y)$ is a meromorphic function in the disk $\{|y| < q^{-1/(k+1)}\}$ with a simple pole at $y = q^{-1}$ and a simple zero at $y = q_1^{-1} = q^{-1+\eta}$. It is easy to see that

$$\bar{G}(n) = Aq^n + O(q^{\frac{n}{k+1}+\epsilon n})$$

with

$$A = (1 - q^{-\eta})\prod_{s=k+1}^{\infty}\frac{1}{(1-q^{-s+1})^{1/s}} e^{F(q^{-1})} > 0.$$

We note that η and k can be chosen arbitrarily small and arbitrarily large independently. Therefore, this example shows that, no matter how small ν (> 0) in (5.1) is, $Z^{\#}(y)$ may have a zero very close to $|y| = q^{-1}$.

This example also shows that the Riemann hypothesis can hold only for very special algebraic function fields.

REFERENCES

[1] G. E. Andrews, A note on the Bombieri–Selberg formula for algebraic curves, Portugaliae Math. **27** (1968), 75–81.

[2] P. T. Bateman and H. G. Diamond, Asymptotic distribution of Beurling's generalized prime numbers, in "Studies in Numbers Theory," Vol. 6, pp. 152–210, Math. Assoc. Amer., Prentice–Hall, Englewood Cliffs, N.J., 1969, MR39, #4105.

[3] E. Bombieri, Sull'analogo della formula di Selberg nei corpi di funzioni, Atti Accad. Naz. Lincei Rend. Cl. Sci. Fis. Mat. Natur. (8) **35** (1963), 252–257.

[4] S. D. Cohen, The function field abstract prime number theorem, Math. Proc. Camb. Phil. Soc. (7) **106** (1989), 7–12.

[5] H. G. Diamond, Chebyshev estimates for Beurling generalized prime numbers, Proc. Amer. Math. Soc. **39** (1973), 503–508.

[6] H. G. Diamond, Elementary methods in the study of the distibution of prime numbers, Bull. Amer. Math. Soc. (N.S.) **I** (1982), 553–589.

[7] E. Fogels, On the abstract theory of primes I–III, Acta Arith. **10** (1964), 137-182; II, 333–358; III, **11** (1966), 292–331.

[8] K.–H. Indlekofer, The abstract prime number theorem for function fields (preprint).

[9] J. Knopfmacher, An abstract prime number theorem relating to algebraic function fields, Arch. Math. **29** (1977), 271 279.

[10] J. Knopfmacher, Finite modules and algebras over rings of algebraic functions, Bull. London Math. Soc. **8** (1976), 289–293.

[11] J. Knopfmacher, Analytic arithmetic of algebraic function fields. Lecture Notes in Pure and Applied Mathematics 50, Marcel Dekker, New York, Basel, 1979.

[12] H. Reichardt, Der Primdivisorsatz für algebraische Funktionenkörper über einem endlichen Konstanten Körper, Math. Z. **40** (1936), 713-716.

[13] A. Weil, Sur les courbes algébriques et les variétés qui s'en déduisent, Actualités sci. et ind., **1041** (1948).

[14] W.–B. Zhang, Chebyshev type estimates for Beurling generalized prime numbers, Proc. Amer. Math. Soc. **101** (1987), 205–212.

Wen-Bin Zhang
Department of Mathematics
University of Illinois
1409 West Green Street
Urbana, Illinois 61801

Progress in Mathematics

Progress in Mathematics is a series of books intended for professional mathematicians and scientists, encompassing all areas of pure mathematics. This distinguished series, which began in 1979, includes authored monographs and edited collections of papers on important research developments as well as expositions of particular subject areas.

All books in the series are "camera-ready," that is they are photographically reproduced and printed directly from a final-edited manuscript that has been prepared by the author. Manuscripts should be no less than 100 and preferably no more than 500 pages.

Proposals should be sent directly to the editors or to: Birkhäuser Boston, 675 Massachusetts Avenue, Suite 601, Cambridge, MA 02139, U.S.A.

A complete list of titles in this series is available from the publisher.

35 ARTIN/TATE. Arithmetic and Geometry: Papers Dedicated to I.R. Shafarevich on the Occasion of His Sixtieth Birthday, Vol. 1
36 ARTIN/TATE. Arithmetic and Geometry: Papers Dedicated to I.R. Shafarevich on the Occasion of His Sixtieth Birthday. Vol. II
37 BOUTET DE MONVEL/DOUADY/ VERDIER. Mathématique et Physique
38 BERTIN. Séminaire de Théorie des Nombres, Paris 1981–82
39 UENO. Classification of Algebraic and Analytic Manifolds
40 TROMBI. Representation Theory of Reductive Groups
41 STANELY. Combinatorics and Commutative Algebra
42 JOUANOLOU. Théorèmes de Bertini et Applications
43 MUMFORD. Tata Lectures on Theta II
44 KAC. Infinite Dimensional Lie Algebras
45 BISMUT. Large Deviations and the Malliavin Calculus
46 SATAKE/MORITA. Automorphic Forms of Several Variables, Taniguchi Symposium, Katata, 1983
47 TATE. Les Conjectures de Stark sur les Fonctions L d'Artin en s = 0
48 FRÖHLICH. Classgroups and Hermitian Modules
49 SCHLICHTKRULL. Hyperfunctions and Harmonic Analysis on Symmetric Spaces
50 BOREL, ET AL. Intersection Cohomology
51 BERTIN/GOLDSTEIN. Séminaire de Théorie des Nombres, Paris 1982–83

52 GASQUI/GOLDSCHMIDT. Déformations Infinitésimales des Structures Conformes Plates
53 LAURENT. Théorie de la Deuxième Microlocalisation dans le Domaine Complexe
54 VERDIER/LE POTIER. Module des Fibres Stables sur les Courbes Algébriques: Notes de l'Ecole Normale Supérieure, Printemps, 1983
55 EICHLER/ZAGIER. The Theory of Jacobi Forms
56 SHIFFMAN/SOMMESE. Vanishing Theorems on Complex Manifolds
57 RIESEL. Prime Numbers and Computer Methods for Factorization
58 HELFFER/NOURRIGAT. Hypoellipticité Maximale pour des Opérateurs Polynomes de Champs de Vecteurs
59 GOLDSTEIN. Séminaire de Théorie des Nombres, Paris 1983–84
60 PROCESI. Geometry Today: Giornate Di Geometria, Roma. 1984
61 BALLMANN/GROMOV/SCHROEDER. Manifolds of Nonpositive Curvature
62 GUILLOU/MARIN. A la Recherche de la Topologie Perdue
63 GOLDSTEIN. Séminaire de Théorie des Nombres, Paris 1984–85
64 MYUNG. Malcev-Admissible Algebras
65 GRUBB. Functional Calculus of Pseudo-Differential Boundary Problems
66 CASSOU-NOGUÈS/TAYLOR. Elliptic Functions and Rings and Integers
67 HOWE. Discrete Groups in Geometry and Analysis: Papers in Honor of G.D. Mostow on His Sixtieth Birthday
68 ROBERT. Autour de L'Approximation Semi-Classique
69 FARAUT/HARZALLAH. Deux Cours d'Analyse Harmonique
70 ADOLPHSON/CONREY/GHOSH/YAGER. Number Theory and Diophantine Problems: Proceedings of a Conference at Oklahoma State University
71 GOLDSTEIN. Séminaire de Théorie des Nombres, Paris 1985–1986
72 VAISMAN. Symplectic Geometry and Secondary Characteristic Classes
73 MOLINO. Riemannian Foliations
74 HENKIN/LEITERER. Andreotti–Grauert Theory by Integral Formulas
75 GOLDSTEIN. Séminaire de Théorie des Nombres, Paris 1986–87
76 COSSEC/DOLGACHEV. Enriques Surfaces I
77 REYSSAT. Quelques Aspects des Surfaces de Riemann
78 BORHO/BRYLINSKI/MACPHERSON. Nilpotent Orbits, Primitive Ideals, and Characteristic Classes
79 MCKENZIE/VALERIOTE. The Structure of Decidable Locally Finite Varieties
80 KRAFT/PETRIE/SCHWARZ. Topological Methods in Algebraic Transformation Groups
81 GOLDSTEIN. Séminaire de Théorie des Nombres, Paris 1987–1988
82 DUFLO/PEDERSEN/VERGNE. The Orbit Method in Representation Theory: Proceedings of a Conference Held in Copenhagen, August to September 1988
83 GHYS/DE LA HARPE. Sur les Groupes Hyperboliques d'après Mikhael Gromov
84 ARAKI/KADISON. Mappings of Operator Algebras: Proceedings of the Japan-U.S. Joint Seminar, University of Pennsylvania, Philadelphia, Pennsylvania, 1988
85 BERNDT/DIAMOND/HALBERSTAM/HILDEBRAND. Analytic Number Theory: Proceedings of a Conference in Honor of Paul T. Bateman

GPSR Compliance

The European Union's (EU) General Product Safety Regulation (GPSR) is a set of rules that requires consumer products to be safe and our obligations to ensure this.

If you have any concerns about our products, you can contact us on ProductSafety@springernature.com

In case Publisher is established outside the EU, the EU authorized representative is:

Springer Nature Customer Service Center GmbH
Europaplatz 3
69115 Heidelberg, Germany

Batch number: 10725979

Printed by Printforce, the Netherlands